Criticisms of the Einstein Field Equation

This book is dedicated to

"True Progress of Natural Philosophy"

Criticisms of the Einstein Field Equation

The End of the 20th Century Physics

Myron W. Evans • Stephen J. Crothers
Horst Eckardt • Kerry Pendergast

Viva Books
New Delhi | Mumbai | Chennai | Kolkata | Bengaluru | Hyderabad | Kochi | Guwahati

VIVA BOOKS PRIVATE LIMITED, 2017

- 4737/23 Ansari Road, Daryaganj, New Delhi 110 002
  E-mail: vivadelhi@vivagroupindia.net, Tel. 42242200
- 76, Service Industries, Shirvane, Sector 1, Nerul, Navi Mumbai 400 706
  E-mail: vivamumbai@vivagroupindia.net, Tel. 27721273, 27721274
- Megh Towers, Old No. 307, New No. 165, Poonamallee High Road, Maduravoyal
  Chennai - 600 095 E-mail: vivachennai@vivagroupindia.net, Tel. 23780991, 23780992
- B-103, Jindal Towers, 21/1A/3 Darga Road, Kolkata 700 017
  E-mail: vivakolkata@vivagroupindia.net, Tel. 22836381, 22816713
- 7, Sovereign Park Apartments, 56-58, K. R. Road, Basavanagudi, Bengaluru 560 004
  E-mail: vivabangalore@vivagroupindia.net, Tel. 26607409, 26607410
- 101-102 Mughal Marc Apartments, 3-4-637 to 641, Narayanguda, Hyderabad 500 029
  E-mail: vivahyderabad@vivagroupindia.net, Tel. 27564481, 27564482
- Beevi Towers, First Floor, SRM Road, Kaloor, Kochi 682 018, Kerala
  E-mail: vivakochi@vivagroupindia.net, Tel: 0484-2403055, 2403056
- 232, GNB Road, Beside UCO Bank, Silpukhuri, Guwahati 781 003
  E-mail: vivaguwahati@vivagroupindia.net, Tel: 0361-2666386

www.vivagroupindia.com

Published by arrangement with:
Cambridge International Science Publishing
7, Meadow Walk, Great Abington
Cambridge CB1 6AZ, UK

ISBN : 978-81-309-1990-4

Published by Vinod Vasishtha for Viva Books Private Limited, 4737/23, Ansari Road, Daryaganj, New Delhi - 110 002.

Printed and bound by Saurabh Printers Pvt. Ltd., Greater Noida.

Preface

This monograph evolved out of Einstein-Cartan-Evans (ECE) field theory when an investigation was being made of the inhomogeneous field equation. This involved the calculation of a quantity which in the shorthand notation of ECE theory (see www.aias.us) is denoted $\widetilde{R} \wedge q$. Here $\widetilde{R}$ is the Hodge dual of the curvature form and q is the Cartan tetrad form. The symbol $\wedge$ denotes Cartan's wedge product. The Cartan-Evans identity states that:

$$D \wedge \widetilde{T} := \widetilde{R} \wedge q \tag{1}$$

and in the Einstein field equation the Hodge dual torsion $\widetilde{T}$ is zero. In tensor notation Eq. (1) becomes:

$$D_\mu T^{\kappa\mu\nu} := R^{\kappa}{}_{\mu}{}^{\mu\nu} \tag{2}$$

where $T^{\kappa\mu\nu}$ is the torsion tensor in four dimensional space-time, and where $R^{\kappa}{}_{\mu}{}^{\mu\nu}$ is the curvature tensor. The latter can be worked out from the Einstein field equation, so Eq. (2) gives a new test for this equation. The curvature tensor $R^{\kappa}{}_{\mu}{}^{\mu\nu}$ should be zero because the Einstein field equation uses a zero torsion $T^{\kappa\mu\nu}$ by construction. I proceeded to work out $R^{\kappa}{}_{\mu}{}^{\mu\nu}$ by hand, but soon found that this was far too laborious. Thereafter, in mid 2007, Horst Eckardt and his group in Munich wrote code for the evaluation of $R^{\kappa}{}_{\mu}{}^{\mu\nu}$ by computer algebra. This resulted in paper 93 on www.aias.us and came up with the important finding that the Einstein field equation gives a non zero $R^{\kappa}{}_{\mu}{}^{\mu\nu}$ in general, and is therefore wholly incorrect.

This finding was thereafter reinforced by later papers of the ECE series, which showed that every known metric of the Einstein field equation is incorrect. In paper 122 of the ECE series the root cause of this disaster for the now obsolete physics of the twentieth century was discovered, the connection in Riemannian geometry must be antisymmetric in its lower two indices μ and ν. For some obscure reason Einstein and his contemporaries used a symmetric connection, an elementary error. The antisymmetry of the connection is shown easily as follows. Define the commutator of covariant derivatives in any four dimensional space-time:

$$D_{\mu\nu} = -D_{\nu\mu} := [D_\mu, D_\nu]. \tag{3}$$

then well known textbook calculations show that:

$$D_{\mu\nu} V^\rho = R^{\rho}{}_{\sigma\mu\nu} V^\sigma - T^{\lambda}{}_{\mu\nu} D_\lambda V^\rho \tag{4}$$

where $R^{\rho}{}_{\sigma\mu\nu}$ is the curvature tensor, $T^{\lambda}{}_{\mu\nu}$ is the torsion tensor, V^ρ is a vector and $D_\lambda V^\rho$ is its covariant derivative. The torsion tensor is defined by:

$$T^{\lambda}{}_{\mu\nu} = \Gamma^{\lambda}_{\mu\nu} - \Gamma^{\lambda}_{\nu\mu} \tag{5}$$

where $\Gamma^{\lambda}_{\mu\nu}$ is the connection. The latter therefore has the same symmetry as the commutator, i.e. it is antisymmetric in μ and ν:

$$\Gamma^{\lambda}_{\mu\nu} = -\Gamma^{\lambda}_{\nu\mu}. \tag{6}$$

Both the commutator and the connection vanish if μ is the same as ν. This means that the torsion and curvature vanish if μ is the same as ν. The error made by Einstein, and repeated until the evolution of ECE, was:

$$\Gamma^{\lambda}_{\mu\nu} =? \, \Gamma^{\lambda}_{\nu\mu} \neq? \, 0. \tag{7}$$

This error, and its repetition for ninety years, is inexplicable in logic, it is perhaps due to belief retention. When we submitted paper 93 to Physica B in 2007/2008 it was met by crude personal abuse, so this seems indeed to be hostility due to belief retention, a commonplace human failing. This means that academic physics itself comes under the microscope, being a clearly unreliable subject. Schrödinger, and independently, Bauer, were the first to criticise the Einstein field equation in 1918, and these criticisms have echoed down the years. Academic physics seems to have ignored them illogically.

The use of the antisymmetric connection led to important new antisymmetry laws which have been developed in later papers of the ECE series, papers which attract a huge readership of high quality on www.aias.us, but which are anathema to academic physics. The latter has therefore been isolated as a non-scientific remnant by other scientific professions, by industry, government, military staffs and individual scholars in their hundreds of thousands.

Once it is accepted that the connection is antisymmetric in its lower two indices, its Hodge dual may be defined as in the textbooks:

$$\widetilde{\Gamma}^{\lambda}_{\mu\nu} = \frac{1}{2}\|g\|^{1/2}\epsilon_{\mu\nu}{}^{\alpha\beta}\Gamma^{\lambda}_{\alpha\beta} \tag{8}$$

where $\|g\|^{1/2}$ is the square root of the positive value of the determinant of the metric, a weighting factor which cancels out in later calculations. In Eq. (8) the totally antisymmetric unit tensor $\epsilon_{\mu\nu}{}^{\alpha\beta}$ is defined as the flat space tensor. The Hodge dual torsion is the tensor:

$$\widetilde{T}^{\lambda}{}_{\mu\nu} = \widetilde{\Gamma}^{\lambda}_{\mu\nu} - \widetilde{\Gamma}^{\lambda}_{\nu\mu}. \tag{9}$$

Neither the connection nor the Hodge dual connection are tensors, because under the general coordinate transformation they do not transform as tensors. They are nevertheless antisymmetric in their lower two indices. The Hodge dual transformation applied to Eq. (4) produces:

$$\widetilde{D}_{\mu\nu}V^{\rho} = \widetilde{R}^{\rho}{}_{\sigma\mu\nu}V^{\sigma} - \widetilde{T}^{\lambda}{}_{\mu\nu}\,D_{\lambda}V^{\rho} \tag{10}$$

in which the structure of the torsion tensor must be defined as in Eq. (9). It follows that the covariant derivative used in Eq. (10) must be:

$$\widetilde{D}_{\mu}V^{\rho} = \partial_{\mu}V^{\rho} + \widetilde{\Gamma}^{\rho}_{\mu\lambda}V^{\lambda} \tag{11}$$

and that the Hodge dual curvature tensor must be:

$$\widetilde{R}^{\rho}{}_{\sigma\mu\nu} = \partial_{\mu}\widetilde{\Gamma}^{\rho}_{\nu\sigma} - \partial_{\nu}\widetilde{\Gamma}^{\rho}_{\mu\sigma} + \widetilde{\Gamma}^{\rho}_{\mu\lambda}\widetilde{\Gamma}^{\lambda}_{\nu\sigma} - \widetilde{\Gamma}^{\rho}_{\nu\lambda}\widetilde{\Gamma}^{\lambda}_{\mu\sigma}. \tag{12}$$

The two tensors (9) and (12) must therefore give a new identity of differential geometry:

$$D\wedge\widetilde{T}^{a} := d\wedge\widetilde{T}^{a} + \omega^{a}{}_{b}\wedge\widetilde{T}^{b} := \widetilde{R}^{a}{}_{b}\wedge q^{b} \tag{13}$$

which has become known as the Cartan-Evans identity. The original Cartan identity is well known from the textbooks to be:

$$D\wedge T^{a} := d\wedge T^{a} + \omega^{a}{}_{b}\wedge T^{b} := R^{a}{}_{b}\wedge q^{b}. \tag{14}$$

The Cartan-Evans identity (13) shows that the Einstein field equation is incorrect, as explained already in this preface, and this book is dedicated to this demonstration, with chapters by other colleagues who have independently refuted the Einstein field equation and standard cosmology with its pseudoscientific contrivances.

The Cartan-Evans identity is Hodge dual invariant with the Cartan identity. The former identity in tensor notation is:

$$D_{\mu}\widetilde{T}^{a}{}_{\nu\rho} + D_{\rho}\widetilde{T}^{a}{}_{\mu\nu} + D_{\nu}\widetilde{T}^{a}{}_{\rho\mu} := \widetilde{R}^{a}{}_{\mu\nu\rho} + \widetilde{R}^{a}{}_{\rho\mu\nu} + \widetilde{R}^{a}{}_{\nu\rho\mu} \tag{15}$$

which is the same as:

$$D_{\mu}T^{a\mu\nu} := R^{a}{}_{\mu}{}^{\mu\nu} \tag{16}$$

in which the covariant derivative is defined by:

$$D_\mu V^\rho := \partial_\mu V^\rho + \widetilde{\Gamma}^\rho_{\mu\lambda} V^\lambda. \tag{17}$$

A special case of Eq. (16) is:

$$D_\mu T^{\kappa\mu\nu} := R^\kappa{}_\mu{}^{\mu\nu}. \tag{18}$$

The Cartan identity in tensor notation is:

$$D_\mu T^a{}_{\nu\rho} + D_\rho T^a{}_{\mu\nu} + D_\nu T^a{}_{\rho\mu} := R^a{}_{\mu\nu\rho} + R^a{}_{\rho\mu\nu} + R^a{}_{\nu\rho\mu} \tag{19}$$

which is the same as:

$$D_\mu \widetilde{T}^{a\mu\nu} := \widetilde{R}^a{}_\mu{}^{\mu\nu} \tag{20}$$

in which the covariant derivative is

$$D_\mu V^\rho := \partial_\mu V^\rho + \Gamma^\rho_{\mu\lambda} V^\lambda. \tag{21}$$

A special case of Eq. (20) is

$$D_\mu \widetilde{T}^{\kappa\mu\nu} := \widetilde{R}^\kappa{}_\mu{}^{\mu\nu}. \tag{22}$$

Craigcefnparc, Wales
October 2009

Myron W. Evans
The British and Commonwealth Civil List Scientist

Contents

Chapter 1

Introduction

by

Myron W. Evans[1],

Alpha Institute for Advance Study (AIAS)
(www.aias.us, www.atomicprecision.com)

This book is a review of recent criticisms of the well known Einstein field equation of 1915, which is still the basis for ideas such as Big Bang, black holes and dark matter, and of the precision tests of general relativity in the standard physics. The book is divided into chapters by contemporary critics of the equation. Chapter two is a review of the well known Einstein-Cartan-Evans (ECE) variation on relativity theory, which correctly considers the role of space-time torsion and reinstates torsion in its central role. This chapter is paper 100 on the www.aias.us site, the most read site of contemporary physics, and gives a rigorous proof of the Bianchi identity of Cartan geometry and its dual identity. It is shown that the Einstein field equation does not obey these fundamental identities of geometry because of its neglect of torsion. Chapter three is by Stephen Crothers, a leading scholar on solutions of the equation. In his chapter Crothers give several clear arguments as to the fundamental incorrectness of big bang, and exposes glaring errors in the mathematics of such claims. Similarly, Crothers shows with clarity and rigor that claims to the existence of black holes and dark matter cannot be mathematically correct, and so have no significance in physics.

Chapter four is by Horst Eckardt, and uses recently developed computer algebra to show that all mathematical solutions of the Einstein field equation in the presence of finite energy momentum density violate the dual identity proven in chapter two and in several papers on www.aias.us. For each line element solution of the Einstein field equation, computer algebra is used to give all the Christoffel symbols and all elements of the Riemann, Ricci and Einstein tensors. Each line element is checked for metric compatibility, and each line element is checked to see if it obeys the fundamental Ricci cyclic equation, known in the standard literature as the first Bianchi identity. This appellation of the standard physics is a misnomer, because it is neither a true identity nor was it given by Bianchi. It was first given by Ricci and Levi-Civita. The rigorously correct Bianchi identity in geometry was first given by Cartan as is well known, and as is equally well known, must include the torsion ineluctably. Finally in this chapter by Eckardt, the line element solutions are tested against the dual identity proven in chapter two. The results are given in tables and graphs. They are exceedingly intricate, so the computer is used to build up the tables without manual transcription error. The results show clearly that the Einstein field equation is incorrect geometrically - meaning that the standard theory of relativity must be revised fundamentally to place torsion as a central feature of the natural and engineering sciences. This is what ECE theory does.

In the fifth chapter, Kerry Pendergast uses his skills as an educator and writer to put these results in historical and scientific context. In so doing he uses some of the material from his "virtual best seller" on www.aias.us, called "Crystal Spheres" to give a historically accurate description of the evolution of the theory of relativity.

[1] e-mail: emyrone@aol.com

In the remainder of this introductory chapter a brief account is given of the evolution of space-time torsion in geometry and ECE theory. The latter is reviewed in detail in chapter two of this book, (paper 100 of www.aias.us). The fundamental idea of the theory of relativity is that physics must be an objective subject independent of anthropomorphic bias.

This philosophy was given for example by High Renaissance thinkers such as Bernardino Telesio and Francis Bacon, but goes back to classical times. Its most well known manifestation appears in the Idol of the Cave philosophy of Bacon. By this he means the fantasies of the human mind, the word "idol" being used in its original classical Greek meaning of "dream". Bacon asserts that the workings of nature must be manifest through empirical testing of human ideas, in our times "experimental data". Contemporary science is based on this philosophy combined with the earlier philosophy of William of Ockham, that the simpler of two theories is preferred in natural philosophy or physics. The invisible college of Francis Bacon later developed into the Royal Society, which espouses the Baconian philosophy.

In Newtonian times the idea of absolute space and of absolute time was predominant, and the whole of Newtonian mechanics is based on the separation of space and time. The anthropomorphic bias of the human mind makes this separation seem entirely natural from everyday experience. Time seems to be moving in one way, in space we can move forwards an backwards, and the two concepts appeared to be described by Newtonian mechanics with the later addition of the Euler, Laplace and Hamilton equations and so forth in classical dynamics. In the nineteenth century these well known concepts of dynamics were challenged by the then new classical electrodynamics, notably the vector equations of Heaviside. These are the well known equations of classical electrodynamics, developed from the quaternion equations of Maxwell and Faraday's concept of field of force which he named the electro-tonic state. The Heaviside field equations are misnamed "the Maxwell equations" in the standard books in physics, and importantly, are not Newtonian in nature. They are covariant under the Lorentz transformation, not the Galilean transformation of space separated from time.

Heaviside developed his equations a few years before Michelson and Morley proved the absence of the aether, causing a crisis of thought in natural philosophy as is well known. The results of the Michelson Morley experiment were discussed by Heaviside and Fitzgerald, who can be said to be the founders of the theory of relativity. This discussion culminated in the proposal by Fitzgerald of length contraction to explain the results of the Michelson Morley experiment. This was a qualitative proposal made to the journal "Science" and it took a further eight years or so for Lorentz to produce the equations of the Lorentz transform, in which space and time both transform. About this time, 1900, Poincare, Bianchi, Ricci and Levi-Civita began to develop the theory of tensors, a term coined by Hamilton. The Heaviside equations were put into tensor form, the electromagnetic field was shown to be part of a four by four matrix so that electricity and magnetism became part of the same entity. In the years 1900 to 1905 several scientists contributed to special relativity, and in 1905 Einstein proposed the constancy of the speed of light as a cornerstone and showed that the equations of dynamics are Lorentz covariant in special relativity, as well as the Heaviside equations of electro-dynamics. Around the same time Minkowski developed the contra-variant covariant notation and the concept of Minkowski space-time, or flat space-time described by the metric diag (-1, 1, 1, 1). Therefore the discussions between Heaviside and Fitzgerald culminated circa 1906 in the theory of special relativity as still used today. In this theory a frame translates with constant velocity with respect to a second, and the equations of physics retain their tensor form under the Lorentz transform. It is now known that special relativity is the most precise theory in physics, having been rigorously tested in many ways.

The development of general relativity, in which a frame moves in any way, i.e. arbitrarily, with respect to another, is due to Einstein from about 1906 onwards. The basic tensorial idea of general relativity is that the equations of physics retain their form under any type of transformation, and that this transformation is a coordinate transformation in four dimensions (ct, X, Y, Z). General relativity is a philosophical departure from special relativity because in the former subject the metric is no longer the static diag (-1, 1, 1, 1). Also, physics is thought of in terms of a geometry that is not Euclidean. Perhaps this is the greatest achievement of Einstein, the application of Riemann geometry to physics. The fundamental task of development of the Einstein field equation is how to make physics proportional to geometry. This is the easiest way to think of general relativity because it is not an intuitive concept, neither is length contraction nor time dilation, nor space-time nor the Lorentz transform. The well known Einstein field equation of 1915 was finally arrived at independently by Einstein and Hilbert after many discussion with experts in tensor theory and geometry such as Levi-Civita and Grossman. The field equation states that a quantity in Riemann geometry is proportional to the Noether Theorem of physics, the latter being essentially the conservation laws of physics in tensor form, or generally covariant form. The quantity in geometry is what the standard physics calls the second Bianchi identity. The proportionality constant is k, the Einstein constant. The Einstein field equation then follows by assuming that the covariant derivative on both sides can be removed, so that the Einstein tensor itself is proportional through k to a quantity known as the canonical energy momentum density of matter.

This is the equation that is still used in the standard physics to predict such things as big bang, black holes and dark matter. They are all consequences of the Einstein field equation, which has become dogmatic in nature. As early as 1918 Bauer and Schrödinger independently showed major shortcomings of the Einstein field equation and the Eddington experiment is known now to be essentially an exercise in anthropomorphic bias, lacking entirely the precision to prove the field equation as is so often claimed in the standard physics. In the early twenties, Cartan showed that the Riemann geometry itself is incomplete because of its lack of space-time torsion. The torsion was unfortunately eliminated by Ricci and Levi-Civita through their use of the symmetric connection, often attributed to Christoffel.

The Einstein field equation is therefore based on a geometry in which torsion is eliminated arbitrarily. There is no logical justification for this elimination of torsion. In this book, it is shown what happens when the torsion is neglected - essentially a disaster for twentieth century standard physics. The only way in which standard physics can justify its claims is to assert without logic that torsion is a mathematical abstraction. In logic, torsion is no more of an abstraction than curvature, on which the whole of the illogical paraphenalia of big bang, black holes and dark matter is based. Chapter two shows essentially how the torsion is central to a generally covariant unified field theory, the ECE theory. Chapter three by Crothers show shows the vacuum solutions of the Einstein field equation are meaningless, and reveals the basic errors repeated down the twentieth century by the standard physics. Chapter four by Eckardt uses newly developed computer algebra to show precisely how the lack of torsion leads to a basic contradiction with the Bianchi identity of Cartan in its Hodge dual form, and chapter five by Pendergast summarizes the historical context.

Chapter 2

A Review of Einstein-Cartan-Evans (ECE) Field Theory

by

Myron W. Evans[1],

Alpha Institute for Advance Study (AIAS)
(www.aias.us, www.atomicprecision.com)

2.1 Introduction

The well accepted Einstein-Cartan-Evans (ECE) field theory [1, 12] is reviewed in major themes of development from Spring 2003 to present in approximately 103 papers and volumes summarized on www.aias.us and www.atomicprecision.com. Recently a third website, www.telesio-galilei.com, has been associated with these two main websites of the theory. Additionally, these websites contain educational articles by members of the Alpha Institute for Advanced Study (AIAS) and the Telesio-Galilei Association, and also contain an Omnia Opera listing most of the collected works of the present author, including precursor theories to ECE theory from 1992 to present. Most original papers are available by hyperlink for scholarly study. It is seen in detail from the feedback activity sites of the three main sites that ECE theory is fully accepted. All the 103 papers to date are read by someone, somewhere every month, and detailed summaries of the feedback are available on www.aias.us. Additionally ECE theory has been published in the traditional manner: in four journals with anonymous reviewers, (three of them standard model journals), and is constantly internally refereed by AIAS staff. The latter are like minded professionals who have worked voluntarily on ECE theory and in the development of AIAS. Computer algebra (Maxima program) has been developed to check hand calculations of ECE theory and to perform calculations that are too complicated to carry out by hand. Therefore a review of the main themes of development and main discoveries of ECE theory is timely.

The ECE theory is a suggestion for the development of a generally covariant unified field theory based on the principles of general relativity, essentially that natural philosophy is geometry. This principle has been proposed since ancient times in many ways, but its most well known manifestation is probably the work of Albert Einstein from about 1906 to 1915, culminating in the proposal of the well known Einstein-Hilbert (EH) field equation of gravitation. This work by Einstein and contemporaries is very well known, but a brief summary is given here. After several false starts Einstein proposed in 1915 that the so called "second Bianchi identity" of Riemann geometry be proportional to a form of the Noether Theorem in which the covariant derivative vanishes of the

[1] e-mail: emyrone@aol.com

canonical energy-momentum tensor. It is much less well known that in so doing, Einstein used the only type of geometry then available to him: Riemann geometry without torsion. The EH field equation follows from this proposal by Einstein as a special case:

$$G_{\mu\nu} = kT_{\mu\nu} \tag{2.1}$$

where $G_{\mu\nu}$ is the Einstein tensor, k is the Einstein constant, and $T_{\mu\nu}$ is the canonical energy - momentum tensor. Eq. (2.1) is a special case of the Einstein proposal of 1915:

$$D^{\mu}G_{\mu\nu} = kD^{\mu}T_{\mu\nu} = 0 \tag{2.2}$$

where on the left hand side appears geometry, and on the right hand side appears natural philosophy. David Hilbert proposed the same equation at about the same time using Lagrangian principles, but Hilbert's work was motivated by Einstein's ideas, so the EH equation is usually attributed to Einstein. The EH equation applies however only to gravitation, whereas ECE has unified general relativity with the other fields of nature besides gravitation. The other fundamental fields are thought to be the electromagnetic, weak and strong fields. ECE has also unified general relativity with quantum mechanics by discarding the acausality and subjectivity of the Copenhagen School, and by deriving objective and causal wave equations from geometrical first principles. The two major and well accepted achievements of ECE theory are therefore the unification of fields using geometry, and the unification of relativity and quantum mechanics. This review is organized in sections outlining the main themes and discoveries of ECE theory, and into detailed technical appendices dealing with basics. These appendices include flow charts of the inter-relation of the main equations.

In Section 2.2 the geometrical first principles of ECE theory are summarized briefly, the theory is based on a form of geometry developed [13] by Cartan and first published in 1922. This geometry is fully self-consistent and well known - it can be regarded as the standard differential geometry taught in good universities. The dialogue between Einstein and Cartan on this geometry is perhaps not as well known as the dialogue between Einstein and Bohr, but is the basis for the development of ECE theory. It is named "Einstein-Cartan-Evans" field theory because the present author set out to suggest a completion of the Einstein Cartan dialogue. This dialogue was part of the attempt by Einstein and many others to complete general relativity by developing a generally covariant unified field theory on the principles of a given geometry. For many reasons this unification did not come about until Spring of 2003, when ECE theory was proposed. The main obstacles to unification were adherence in the standard model to a U(1) sector for electromagnetism, the neglect of the ECE spin field B(3), inferred in 1992, and adherence to the philosophy of the Copenhagen School. Standard model proponents adhere to these principles at the time of writing, but ECE proponents now adopt a different natural philosophy, since it may be claimed objectively from feedback data that ECE is a new school of thought.

In Section 2.3 the main field and wave equations of ECE are discussed in summary. They are derived from the well known principles of Cartan's geometry. The gravitational, electromagnetic, weak and strong fields are unified by Cartan's geometry, each is an aspect of the same geometry. The field equations are based on the one true Bianchi identity given by Cartan, using different representation spaces. The wave equations are derived from the tetrad postulate, the very fundamental requirement in natural philosophy and relativity theory that the complete vector field be invariant under the general transformation of coordinates. To translate Cartan to Riemann geometry requires use of the tetrad postulate. Therefore both the Bianchi identity and tetrad postulate are fundamentals of standard differential geometry and their use in ECE theory is entirely standard mathematics [13].

In Section 2.4 the unification of phase theory made possible by ECE is summarized in terms of the main discoveries and points of development. The main point of development in this context is the unification of apparently disparate phases such as the electromagnetic phase, the Dirac and Wu Yang phases, and the topological phases. ECE theory presents a unified geometrical approach to each phase, and this approach also gives a straightforward geometrical explanation of the Aharonov-Bohm effects and "non-locality". The electromagnetic phase for example is developed in terms of the B(3) spin field [14] and some glaring shortcomings of the standard model are corrected. Thus, apparently simple and well known effects such as reflection are developed self-consistently with ECE, while in the standard model they are at odds with fundamental symmetry [1, 12]. The standard model development of the Aharonov-Bohm effects is also incorrect mathematically, obscure, controversial and convoluted, while in ECE theory it is straightforward.

In Section 2.5 the ECE laws of classical dynamics and electrodynamics are summarized in the language of vectors, the language used in electrical engineering. The equations of electrodynamics in ECE theory reduce to the four laws: Gauss law of magnetism, Faraday law of induction, Coulomb law and Ampère Maxwell law. In ECE theory they are the same in vector notation as in the familiar Maxwell-Heaviside (MH) field theory, but in ECE are written in a different space-time. In ECE the electromagnetic field is the spinning of space-time, represented by the Cartan torsion, while in MH the field is a nineteenth century concept still used uncritically in the contemporary standard model of natural philosophy. The space-time of MH is the flat and static Minkowski space-time, while in ECE the space-time is dynamic with non-zero curvature and torsion. This difference manifests itself in the relation between the fields and potentials in ECE, a relation which includes the spin connection.

In Section 2.6, spin connection resonance (SCR) is discussed, concentrating as usual on the main discoveries and points of development of the ECE theory. In theory, SCR is of great practical utility because the equations of classical electrodynamics become resonance equations of the type first inferred by the Bernoulli's and Euler. Therefore a new source of electric power has been discovered in ECE theory - this source is the Cartan torsion of space-time. Amplification occurs in principle through SCR, the spin connection itself being the property of the four-dimensional space-time with curvature and torsion which is the base manifold of ECE theory. It is well known [15] that these resonance equations are equivalent to circuits that can be used to amplify electric power. In all probability these circuits were the ones designed by Tesla empirically.

In Section 2.7 the utility of ECE as a unified field theory is illustrated through the effects of gravitation in optics and spectroscopy. These are exemplified by the effect of gravitation on the ring laser gyro (Sagnac effect) and on radiatively induced fermion resonance (RFR). RFR itself is of great potential utility because it is a form of electron and proton spin resonance induced not by a permanent magnet, but by a circularly polarized electromagnetic field. This is known as the inverse Faraday effect (IFE) [16] from which the ECE spin field B(3) was inferred in 1992 [17]. The spin field signals the fact that in a self consistent philosophy, classical electrodynamics must be part of a generally covariant field theory. This is incompatible with the U(1) sector of special relativity still used to describe electrodynamics in the standard model. Any proposal for a unified field theory based on U(1) cannot be generally covariant in all sectors, leaving ECE as the only satisfactory unified field theory at the time of writing.

In Section 2.8 the well known radiative corrections [18] are developed with ECE theory, and a summary of the main points of progress illustrated with the anomalous g factor of the electron and the Lamb shift. It is shown that claims to accuracy of standard model quantum electrodynamics (QED) are greatly exaggerated. The accuracy is limited by that of the Planck constant, the least accurately known fundamental constant appearing in the fine structure constant. There are glaring internal inconsistencies in standards laboratories tables of data on the fundamental constants, and QED is based on a number of what are effectively adjustable parameters introduced by ad hoc procedures such as dimensional renormalization. The concepts used in QED are vastly complicated and are not used in the ECE theory of the experimentally known radiative corrections. The Feynman perturbation method is not used in ECE: it cannot be proven to converge as is well known, i.e. needs many terms of increasing complexity which must be evaluated by computer. So ECE is a fundamental theory of quantized electrodynamics from the first principles of general relativity, while QED is a theory of special relativity needing adjustable parameters, acausal and subjective concepts, and therefore of dubious validity.

In Section 2.9, finally, it is shown that EH theory has several fundamental shortcomings. As described on www.telesio-galilei.com EH has been quite severely criticized down the years by several leading physicists. Notably, Crothers [19] has criticized the methods of solution of EH, and has shown that uncritically accepted concepts are in fact incompatible with general relativity. These include Big Bang, dark hole and dark matter theory and the concept of a Ricci flat space-time. He has also shown that the use of the familiar but mis-named "Schwarzschild metric" is due to lack of scholarship and understanding of Schwarzschild's original papers of 1916. ECE has revealed that the use of the familiar Christoffel symbol is incompatible with the one true Bianchi identity of Cartan. This section suggests a development of the EH equation into one which is self consistent.

Several technical appendices give basic details which are not usually given in standard textbooks, but which are nevertheless important to the student. These appendices also contain flow charts inter-relating the main concepts and equations of ECE.

2.2 Geometrical principles

The ECE theory is based on the two structure equations of Cartan, and the Bianchi identity of Cartan geometry. During the course of development of the theory a useful short-hand notation has been used in which the indices are removed in order to reveal the basic structure of the equations. In this notation the two Cartan structure equations are:

$$T = D \wedge q = d \wedge q + \omega \wedge q \tag{2.3}$$

and

$$R = D \wedge \omega = d \wedge \omega + \omega \wedge \omega \tag{2.4}$$

and the Bianchi identity is:

$$D \wedge T = d \wedge T + \omega \wedge T := R \wedge q. \tag{2.5}$$

In this notation T is the Cartan torsion form, ω is the spin connection symbol, q is the Cartan tetrad form, and R is the Cartan curvature form. The meaning of this symbolism is defined in all detail in the ECE literature [1,12], and the differential form is defined in the standard literature [13]. The purpose of this section is to summarize the main advances in basic geometry made during the development of ECE theory.

The Bianchi identity (2.5) is basic to the field equations of ECE, and its structure has been developed considerably [1,12]. It has been shown to be equivalent to the tensor equation:

$$\begin{aligned} & R^{\lambda}_{\rho\mu\nu} + R^{\lambda}_{\mu\nu\rho} + R^{\lambda}_{\nu\rho\mu} \\ & \quad := \partial_\nu \Gamma^{\lambda}_{\rho\mu} - \partial_\rho \Gamma^{\lambda}_{\nu\mu} + \Gamma^{\lambda}_{\nu\sigma}\Gamma^{\sigma}_{\rho\mu} - \Gamma^{\lambda}_{\rho\sigma}\Gamma^{\sigma}_{\nu\mu} \\ & \quad + \partial_\rho \Gamma^{\lambda}_{\mu\nu} - \partial_\mu \Gamma^{\lambda}_{\rho\nu} + \Gamma^{\lambda}_{\rho\sigma}\Gamma^{\sigma}_{\mu\nu} - \Gamma^{\lambda}_{\mu\sigma}\Gamma^{\sigma}_{\rho\nu} \\ & \quad + \partial_\mu \Gamma^{\lambda}_{\nu\rho} - \partial_\nu \Gamma^{\lambda}_{\mu\rho} + \Gamma^{\lambda}_{\mu\sigma}\Gamma^{\sigma}_{\nu\rho} - \Gamma^{\lambda}_{\nu\sigma}\Gamma^{\sigma}_{\mu\rho} \end{aligned} \tag{2.6}$$

in which a cyclic sum of three Riemann tensors is identically equal to the sum of three fundamental definitions of the same Riemann tensors. These fundamental definitions originate in the commutator of covariant derivatives acting on a four-vector in the base manifold. The latter is four dimensional space-time with BOTH curvature and torsion [1,13]. This operation produces:

$$[D_\mu, D_\nu]V^\rho = R^{\rho}{}_{\sigma\mu\nu}V^\sigma - T^{\lambda}_{\mu\nu}D_\lambda V^\rho \tag{2.7}$$

where the torsion tensor is:

$$T^{\lambda}_{\mu\nu} = \Gamma^{\lambda}_{\mu\nu} - \Gamma^{\lambda}_{\nu\mu}. \tag{2.8}$$

The curvature or Riemann tensor cannot exist without the torsion tensor, and the definition (2.7) has been shown to be equivalent to the Bianchi identity (2.6).

The second advance in basic geometry is the inference [1,12] of the Hodge dual of the Bianchi identity. In short-hand notation this is:

$$D \wedge \widetilde{T} := \widetilde{R} \wedge q \tag{2.9}$$

and is equivalent to:

$$[D_\mu, D_\nu]_{HD}V^\rho = \widetilde{R}^{\rho}{}_{\sigma\mu\nu}V^\sigma - \widetilde{T}^{\lambda}_{\mu\nu}D_\lambda V^\rho \tag{2.10}$$

where the subscript *HD* denotes Hodge dual. From these considerations it may be inferred that the Bianchi identity and its Hodge dual are the tensor equations:

$$D_\mu \widetilde{T}^{\kappa\mu\nu} = \widetilde{R}^{\kappa}{}_{\mu}{}^{\mu\nu} \tag{2.11}$$

and

$$D_\mu T^{\kappa\mu\nu} = R^{\kappa}{}_{\mu}{}^{\mu\nu} \tag{2.12}$$

in which the connection is NOT the Christoffel connection. Computer algebra [1,12] has shown that the tensor $R^{\kappa}{}_{\mu}{}^{\mu\nu}$ is not zero in general for line elements that use the Christoffel symbol, while $T^{\kappa\mu\nu}$ is always zero for the Christoffel symbol. So the use of the latter is inconsistent with the tensor equation (2.12). Therefore the neglect of torsion makes EH theory internally inconsistent, so standard model general relativity and cosmology are also internally inconsistent at a basic level. In short-hand notation the geometry used in EH is:

$$R \wedge q = 0 \tag{2.13}$$

which in tensor notation is known as "the first Bianchi identity":

$$R^{\kappa}{}_{\mu\nu\rho} + R^{\kappa}{}_{\rho\mu\nu} + R^{\kappa}{}_{\nu\rho\mu} = 0 \tag{2.14}$$

in the standard model literature. However, this is not an identity, because it conflicts with equation (2.5), and is true if and only if the Christoffel symbol and symmetric metric are used [1,13]. Eq. (2.14) was actually inferred by Ricci and Levi-Civita, not by Bianchi. So it is referred to in the ECE literature as the Ricci cyclic equation.

In the course of development of ECE theory a similar problem was found with what is referred to in the standard model literature as "the second Bianchi identity". In shorthand notation this is given [13] as:

$$D \wedge R = 0 \tag{2.15}$$

but again this neglects torsion. In tensor notation Eq. (2.15) is:

$$D_\rho R^{\kappa}{}_{\sigma\mu\rho} + D_\mu R^{\kappa}{}_{\sigma\nu\rho} + D_\nu R^{\kappa}{}_{\sigma\rho\mu} = 0. \tag{2.16}$$

It has been shown [1,12] that Eq. (2.15) should be:

$$D \wedge (D \wedge T) := D \wedge (R \wedge q) \tag{2.17}$$

which is found by taking $D\wedge$ on both sides of Eq. (2.15). Eq. (2.17) has been given in tensor notation [1,12], and reduces to Eq. (2.16) when:

$$T^{\lambda}_{\mu\nu} = 0. \tag{2.18}$$

However, Eq. (2.18) is inconsistent with the fundamental operation of the commutator of covariant derivatives on the four vector, Eq. (2.7). So in the ECE literature the torsion is always considered self-consistently. From the fundamentals [13] of Eq. (2.7) there is no a priori reason for neglecting torsion, and in fact the torsion tensor is always non-zero if the curvature tensor is non-zero. This fact precludes the use of the Christoffel symbol, making EH theory self-inconsistent.

These are the main geometrical advances made during the course of the development of ECE theory, which is the only self-consistent theory of general relativity. It has also been pointed out by Crothers [19] that methods of solution of the EH equation are geometrically incorrect, and must be discarded. It is thought that these errors have been repeated uncritically for ninety years because few have the necessary technical ability to understand the geometry of general relativity in sufficient depth, and that the prestige of Einstein has precluded or inhibited due criticism.

2.3 The Field and wave equations of ECE theory

The wave equation of ECE was the first to be developed historically [1,12], and methods of derivation of the wave equation were subsequently simplified and clarified. The field equations were subsequently developed from the Bianchi identity discussed in Section 2.2. This section summarizes the main equations and methods of derivation. More detail of the equations is given in technical appendices. The field equations are relevant to classical gravitation and electrodynamics, and the wave equation to causal and objective quantum mechanics. Full details of derivations are available in the literature [1, 12], the aim of this section is to summarize the main inferences of ECE theory to date.

The Bianchi identity (2.5) and its Hodge dual (2.9) become the homogeneous and inhomogeneous field equations of ECE respectively. These field equations apply to the four fundamental fields of force: gravitational, electromagnetic, weak and strong and can be used to describe the interaction of the fundamental fields on the classical level. For example the electromagnetic field is described by making the fundamental hypothesis:

$$A = A^{(0)}q \tag{2.19}$$

where the shorthand (index-less) notation has been used. Here A represents the electromagnetic potential form and $cA^{(0)}$ is a primordial quantity with the units of volts, a quantity which is proportional to the charge, −e, on the electron. The hypothesis (2.19) implies that:

$$F = A^{(0)}T \tag{2.20}$$

where F is shorthand notation for the electromagnetic field form. The homogeneous ECE field equation of electrodynamics follows from the Bianchi identity (2.5):

$$d \wedge F + \omega \wedge F = A^{(0)}R \wedge q \tag{2.21}$$

and the inhomogeneous ECE field equation follows from the Hodge dual (2.9) of the Bianchi identity:

$$d \wedge \widetilde{F} + \omega \wedge \widetilde{F} = A^{(0)}\widetilde{R} \wedge q. \tag{2.22}$$

Therefore the ECE field equations are duality invariant, a basic symmetry which means that they transform into each other by means of the Hodge dual [1, 12]. The Maxwell-Heaviside (MH) field equations of the standard model do not have this fundamental symmetry and in differential form notation the MH equations are:

$$d \wedge F = 0 \tag{2.23}$$

and

$$d \wedge \widetilde{F} = \widetilde{J}/\epsilon_0 \tag{2.24}$$

where $\widetilde{J}$ denotes the inhomogeneous charge/current density and ϵ_0 is the S. I. vacuum permittivity. Duality symmetry is broken by the fact that there is no homogeneous charge current density (J) in MH theory (the right hand side of Eq. (2.23) is zero). The absence of J in the standard model is made the basis for gauge theory as is well known, and also made the basis for the absence of a magnetic monopole.

The ECE field equations (2.21) and (2.22) are re-arranged as follows in order to define the homogeneous (J) and inhomogeneous ($\widetilde{J}$) charge current densities of ECE theory:

$$d \wedge F = J/\epsilon_0 = A^{(0)}(R \wedge q - \omega \wedge T) \tag{2.25}$$

and

$$d \wedge \widetilde{F} = \widetilde{J}/\epsilon_0 = A^{(0)}(\widetilde{R} \wedge q - \omega \wedge \widetilde{T}). \tag{2.26}$$

Both equations are generally covariant because they originate in the Bianchi identity. The interaction of electromagnetism with gravitation occurs whenever J is non-zero. In MH theory such an interaction cannot be described, because MH theory is developed in Minkowski space-time. The latter has no curvature and in general relativity cannot describe gravitation at all. For all practical purposes in the laboratory there is no interaction of electromagnetism and gravitation, so Eq. (2.25) reduces to:

$$d \wedge F = 0. \tag{2.27}$$

Therefore ECE theory explains in this way why there is no magnetic monopole observable in the laboratory. The standard model has no physical explanation for this, and indeed asserts that gauge theory is mathematical in nature. ECE theory does not use gauge theory, and adopts Faraday's original point of view that the potential A is a physically effective entity. There are therefore important philosophical differences between ECE and the standard model of classical electrodynamics, in which the potential is mathematical in nature.

Therefore the structure of the ECE field equations is a simple one based directly on the Bianchi identity. The structure is seen the most clearly using the shorthand notation of Eqs. (2.25) and (2.26) where all indices are omitted. The notation of classical electrodynamics varies from subject to subject. In advanced field theory the elegant but concise differential form notation is used, and also the tensor notation. In electrical engineering the vector notation is used. In ECE theory all three notations have been developed [1, 12] in all detail, and the ECE field equations developed into a vector form that is identical to the MH equations. The main differences between ECE and MH is firstly that the former is written in a four dimensional space-time with curvature and torsion both present. This is a dynamic space-time whose connection must be more general than the Christoffel connection. The MH equations, although having the same vector form as ECE, are written in the Minkowski space-time of special relativity. This is often referred to as "flat space-time", whose metric is time and space independent. Secondly the relation between the field and potential in ECE includes the connection, whereas in MH the connection is not present. The inclusion of the connection has the all important effect of making the equations of classical electrodynamics resonance equations of the Bernoulli/Euler type. This property means that it is possible to describe well known phenomena such as those first observed by Tesla, and to produce circuits that take electric power from a new source, the Cartan torsion.

The concise tensorial expression of the equations (2.25) and (2.26) is in general [1, 12]

$$D_\mu \widetilde{F}^{a\mu\nu} = A^{(0)} \widetilde{R}^{a}{}_{\mu}{}^{\mu\nu} \tag{2.28}$$

and

$$D_\mu F^{a\mu\nu} = A^{(0)} R^{a}{}_{\mu}{}^{\mu\nu} \tag{2.29}$$

where the covariant derivative appears on one side and a Ricci type curvature tensor on the other. It has been shown [1, 12] that these reduce in the laboratory, and for all practical purposes, to:

$$\partial_\mu \widetilde{F}^{a\mu\nu} = 0 \tag{2.30}$$

and

$$\partial_\mu F^{a\mu\nu} = A^{(0)} R^{a}{}_{\mu}{}^{\mu\nu}. \tag{2.31}$$

The index a in these equations comes from the well known [13] tangent space-time of Cartan geometry. However, it has been shown [1, 12] that Eqs. (2.30) and (2.31) can be written in the base manifold as a special case of Eqs. (2.28) and (2.29), whereupon we arrive at:

$$\partial_\mu \widetilde{F}^{\kappa\mu\nu} = 0 \tag{2.32}$$

and

$$\partial_\mu F^{\kappa\mu\nu} = A^{(0)} R^{\kappa}{}_{\mu}{}^{\mu\nu}. \tag{2.33}$$

Therefore the electromagnetic field tensor in general relativity (ECE theory) develops into a three index tensor. In special relativity (MH theory) it is a two-index tensor as is well known. The equivalents of (2.32) and (2.33) in MH theory are the tensor equations:

$$\partial_\mu \widetilde{F}^{\mu\nu} = 0 \tag{2.34}$$

and

$$\partial_\mu F^{\mu\nu} = J^\nu/\epsilon_0. \tag{2.35}$$

The meaning of the three-index field tensor has been developed [1, 12] in detail. It originates in the well known [18] three index angular energy/ momentum tensor density, $J^{\kappa\mu\nu}$ which is proportional to the three index Cartan torsion tensor. It is well known that the electromagnetic field carries angular momentum which in the Beth effect [20] is experimentally observable Therefore the Cartan torsion tensor is the expression of this well known angular energy/momentum density tensor of Minkowski space-time [18] in a more general manifold with curvature and torsion. The meaning of the vector form of the ECE field equations is further developed in Section 2.5.

The classical field equations of gravitation in ECE are also based directly on the Bianchi identity and its Hodge dual. The EH equation, as argued already, is incompatible with the Bianchi identity in its rigorously correct form, Eq. (2.5), so during the course of development of ECE theory the well known EH equation has been developed with the proportionalities:

$$T^{\kappa\mu\nu} = kJ^{\kappa\mu\nu} \tag{2.36}$$

and

$$R^{\kappa}{}_{\mu}{}^{\mu\nu} = kT^{\kappa}{}_{\mu}{}^{\mu\nu} \tag{2.37}$$

which give:

$$D_\mu J^{\kappa\mu\nu} = T^{\kappa}{}_{\mu}{}^{\mu\nu}. \tag{2.38}$$

This novel field equation of classical gravitation is based directly on the tensorial formulations (2.11) and (2.12) of the Bianchi identity. The Newton inverse square law for example has been derived straightforwardly from Eq. (2.38) in the limit where the connection goes to zero:

$$\partial_\mu J^{\kappa\mu\nu} \doteq T^{\kappa}{}_{\mu}{}^{\mu\nu} \tag{2.39}$$

whereupon we obtain:

$$\boldsymbol{\nabla} \cdot \boldsymbol{g} = kc^2 \rho_m \tag{2.40}$$

an equation which is equivalent to the Newton inverse square law. Here $\boldsymbol{g}$ is the acceleration due to gravity, k is Einstein's constant, ρ_m and is the mass density in kilograms per cubic meter. Similarly the Coulomb inverse square law can be obtained straightforwardly [1, 12] by considering the same type of limit of the inhomogeneous ECE field equation:

$$D_\mu F^{\kappa\mu\nu} = A^{(0)} R^{\kappa}{}_{\mu}{}^{\mu\nu}. \tag{2.41}$$

The appropriate limit in this case is:

$$\partial_\mu F^{\kappa\mu\nu} \doteq A^{(0)} R^{\kappa}{}_{\mu}{}^{\mu\nu} \tag{2.42}$$

and leads to the Coulomb inverse square law:

$$\boldsymbol{\nabla} \cdot \boldsymbol{E} = \rho_e/\epsilon_0 \tag{2.43}$$

where ρ_e is the charge density in coulombs per cubic meter. These procedures illustrate one aspect of the unified nature of ECE, because both laws are obtained from the Bianchi identity. Many other examples of the unification properties of ECE have been discussed [1, 12].

In order to unify the electromagnetic and weak fields in a field equation, the representation space is chosen to be SU(2) instead of O(3) and the parity violating nature of the weak field carefully considered. Similarly the electromagnetic and strong fields are unified with an SU(3) representation space, and we have already discussed the unification of the electromagnetic and gravitational fields. Any permutation or combination of fields may be unified, and several examples have been given [1, 12] in various contexts. These are discussed further in Section 2.7.

The ECE wave equation was developed [1, 12] from the tetrad postulate [13]:

$$D_\mu q^a_\nu = 0 \tag{2.44}$$

via the identity:

$$D^\mu (D_\mu q^a_\nu) := 0. \tag{2.45}$$

This was re-expressed as the ECE Lemma:

$$\square q^a_\lambda = R q^a_\lambda \tag{2.46}$$

in which appears the scalar curvature:

$$R = q^\lambda_a \partial^\mu (\Gamma^\nu_{\mu\lambda} q^a_\nu - \omega^a_{\mu b} q^b_\lambda). \tag{2.47}$$

Here tensor notation is used, $\omega^a_{\mu b}$ being the spin connection and $\Gamma^\nu_{\mu\lambda}$ the general connection. The Lemma becomes the ECE wave equation using a generalization to all fields of the Einstein gravitational equation [1, 13]:

$$R = -kT. \tag{2.48}$$

Here T is an index contracted energy momentum tensor. The main wave equations of physics were all obtained [1, 12] as limits of Eq. (2.46), notably the Proca and Dirac wave equations. In so doing however the causal realist philosophy of Einstein and de Broglie was adhered to. This is the original philosophy of wave mechanics. The Schrödinger and Heisenberg equations were also obtained as non-relativistic quantum limits of the ECE wave equation, but the Heisenberg indeterminacy principle was not used in accord with the basic philosophy of relativity and with recent experimental data [21] which refute the uncertainty principle by as much as nine orders of magnitude.

2.4 Aharonov-Bohm and Phase effects in ECE theory

The well known Aharonov-Bohm (AB) effects have been observed using magnetic, electric and gravitational fields [1, 12] and as shown by ECE theory are ubiquitous for ALL electromagnetic and optical effects, including phase effects: the subject of this section. These must all be explained by general relativity, and not by the obsolete special relativistic methods of the standard model. Therefore it is important to define the various AB conditions in ECE theory. In so doing a unified description of phase effects such as the electromagnetic, Dirac, Wu Yang and Berry phases may also be developed.

In general, the AB condition is defined in ECE theory by the first Cartan structure equation (adopting the index-less short-hand notation [1, 12]):

$$T = D \wedge q := d \wedge q + \omega \wedge q. \tag{2.49}$$

Using the ECE hypothesis:

$$A = A^{(0)} q \tag{2.50}$$

this becomes:

$$F = D \wedge A := d \wedge A + \omega \wedge A \tag{2.51}$$

where F is short-hand for the electromagnetic field form and where A is short-hand for the electromagnetic potential form. The AB effects in ECE theory [1, 12] were developed with the spin connection term $\omega \wedge A$ in

Eq. (2.51). The accepted notation [13] of Cartan geometry uses the tangent space-time indices without the base manifold indices, because the latter are always the same on both sides of an equation of Cartan geometry. So in the standard notation Eq. (2.51) is:

$$F^a = d \wedge A^a + \omega^a{}_b \wedge A^b \tag{2.52}$$

This denotes that the electromagnetic field is a vector-valued two-form and the potential is a vector-valued one-form. In the standard model the spin connection is zero and the standard relation between field and potential is:

$$F = d \wedge A. \tag{2.53}$$

In Eq. (2.53), F is a scalar-valued two-form, and A is a scalar valued one-form [13] The spin connection is zero in Eq. (2.53) because the latter is written in a Minkowski space-time. In the standard model, classical electrodynamics is still represented by the MH equations, which are Lorentz covariant, but not generally covariant. In other words the MH equations are those of special relativity and not general relativity as required by the philosophy of relativity and objectivity. The latter demands that every equation of physics should be an equation of a generally covariant unified field theory. It is well known that the standard model complies with this only in its gravitational sector: the electro-weak and strong fields of the standard model are sectors of special relativity only. The standard model does not comply with general relativity, notably standard model quantum mechanics is philosophically different from relativity (Einstein-Bohr dialogue). ECE complies rigorously with the philosophy of general relativity in all its sectors, and unifies all sectors with geometry as required. In ECE the spin connection is ALWAYS non-zero because the fundamental space-time being used is not a flat space-time, it always contains both torsion and curvature in all sectors of the generally covariant unified field theory [1, 12]. Torsion and curvature are ineluctably inter-related in the Bianchi identity (Section 2.2), and during the course of development of ECE theory it was shown that there is only one true Bianchi identity, which always links torsion to curvature and vice versa. This is an important mathematical advance of ECE theory, another (Section 2.2) being the development of the Hodge dual of the Bianchi identity.

It has been shown [1,12] that there is a fundamental error in the standard model explanation of the magnetic AB effect [22]. In differential form notation the standard explanation is based on the two equations:

$$F = d \wedge A, d \wedge F = 0 \tag{2.54}$$

whose mathematical structure implies:

$$d \wedge (d \wedge A) = 0. \tag{2.55}$$

It is well known that this structure is invariant under the archetypical gauge transformation:

$$A \rightarrow A + d\chi \tag{2.56}$$

because of the Poincaré Lemma:

$$d \wedge d\chi := 0. \tag{2.57}$$

As explained in paper 56 of the ECE series (www.aias.us), the standard model uses the mathematical result (2.57) to claim that:

$$\oint d\chi = \int_s d \wedge d\chi \neq 0. \tag{2.58}$$

This claim is incorrect because it does not agree with the Stokes Theorem. The latter applies [23] in non simply connected spaces. The Poincaré Lemma (2.57) implies therefore that:

$$\oint d\chi = \int_s d \wedge d\chi := 0 \tag{2.59}$$

in all types of spaces, including non simply connected spaces and there cannot be an Aharonov-Bohm effect due to the contour integral of $d\chi$. The incorrect claim of the standard model [22] is that non simply connected spaces allow $\oint d\chi$ to be non-zero. A counter example to this claim was given in paper 56 of www.aias.us. in full detail.

The explanation of the Aharonov-Bohm (AB) effects in ECE theory is not based on the mathematical abstractions of gauge theory but on Einstein's philosophy of relativity and Faraday's philosophy of the potential as a physically effective entity (the electrotonic state). This philosophy of Faraday was also accepted by Maxwell and his followers. The idea that the potential is a mathematical abstraction is based on the perceived redundancy exemplified by Eq. (2.57), and this idea has been made into the basis of the mathematical gauge theory of the standard model, developed in the late twentieth century. It appears in standard model textbooks such as that of Jackson for example [1, 24]. The idea of a mathematical potential and a physical field in classical electrodynamics is contradicted by the well known minimal prescription of field theory and quantum electrodynamics, where the PHYSICAL momentum eA is added to the momentum p. The idea of an abstract potential ran into trouble following the demonstration by Chambers of the first AB effect to be observed, the magnetic AB effect. It is well known that Chambers placed a magnetic iron whisker between the apertures of a Young interferometer and isolated the magnetic field from interfering electron beams. Therefore, if the potential is mathematical as claimed in gauge theory, it should have no effect on the electronic interference pattern. The experimental result showed a shift in the interference pattern, and so contradicts the standard model, meaning that Faraday was correct: the potential is a physically effective entity. The same results were later obtained experimentally in the electric and gravitational AB effects. As argued in this section, various phase effects also indicate the existence of an electromagnetic AB effect if interpreted by general relativity, of which ECE theory is an example.

The AB effect in ECE theory is summarized as follows:

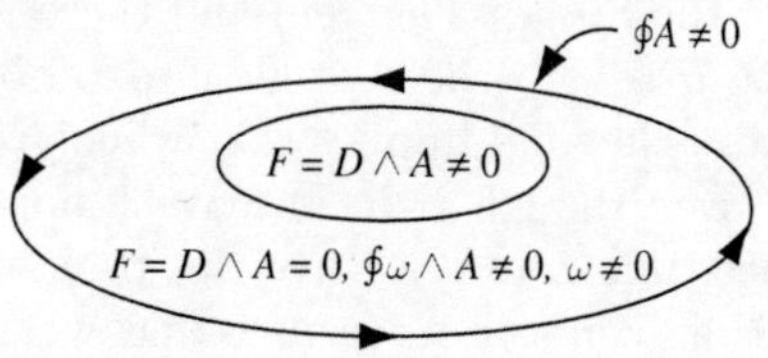

Figure 2.1: ECE Explanation of the Aharonov-Bohm Effect.

It has been shown [1, 12] that the observable phase shift of the Chambers experiment in ECE theory is:

$$\Delta\phi = \frac{e}{\hbar}\Phi \tag{2.60}$$

where

$$\Phi = \oint A := -\int_s \omega \wedge A \tag{2.61}$$

in short-hand or index free notation. In the area between the inner and outer rings in Fig. 2.1:

$$F = D \wedge A = 0, A \neq 0, \omega \neq 0. \tag{2.62}$$

The electromagnetic field (F) is zero by experimental arrangement. However, the potential (A) and the spin connection (ω) are not zero in general in this same region between the inner and outer rings. The phase shift is due therefore to the contour integral around A in Eq. (2.61), as indicated in Fig. 2.1. Therefore ECE theory gives a simple explanation of the AB effects as being due to a physical A and a physical ω. The latter indicates that the ECE space-time is not a Minkowski space-time as in the attempted standard model explanation of the AB effect.

The equivalent of the standard model for Fig. 2.1 is shown in Fig. 2.2. In the standard model the contour

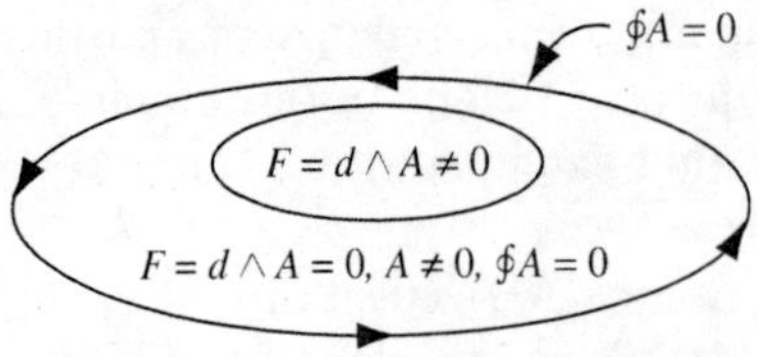

Figure 2.2: Standard Attempt at Explaining the Aharonov-Bohm Effect.

integral of the potential is zero in the area between the inner and outer rings of Fig. 2.2 because:

$$F = d \wedge A = 0, A \neq 0, \tag{2.63}$$

$$\int_s d \wedge A = \oint A = 0. \tag{2.64}$$

So when F is zero in the standard model, so is $d \wedge A$. It is possible therefore for A to be non-zero in the standard model while F is zero, but the incorrect twentieth century idea of a non-physical A means that in the standard model A must have no physical effect. In the end analysis this is pure obscurity and has caused great confusion. Such ideas are bad physics and must be discarded sooner or later. The only clear thing about the attempted standard model explanation of the magnetic AB effect is that in the area between the two rings of Fig. 2.2:

$$\int_s F = \int_s d \wedge A = \oint A = 0. \tag{2.65}$$

So the contour integral of A is zero by the Stokes Theorem and there is no AB effect contrary to experiment. Therefore in the standard model, when F is zero the contour integral of A is always zero even though A itself may be non-zero. In other words Stokes' Theorem implies that when F or $d \wedge A$ is zero in the standard model, the contour integral of A must vanish even though A itself may be non-zero. As we have seen, adding a $d\chi$ in an assumed non simply connected space-time does not solve this problem.

In ECE theory the presence of the spin connection ensures that when F is zero, $d \wedge A$ is not zero in general and the contour integral of A is not zero, meaning a phase shift as observed, Eq. (2.61). The way that such an ECE contour integral must be evaluated has been explained carefully [1, 12]. Therefore the AB effects show that ECE is preferred experimentally over the standard model. This is one out of many experimental advantages of ECE theory over the standard model. A table of about thirty such advantages is available on the www.aias.us website and in the fourth volume of ref. (2.1). As argued already, the standard model has attempted to obfuscate its way out of the AB paradox by adding $d\chi$ to A and claiming that the AB effect is due to a non-zero contour integral of $d\chi$ when the contour integral of A is zero. Paper 56 of ECE (www.aias.us) shows that this claim is incorrect mathematically, and even if it were correct just leads to obscure ideas, notably that [22] space-time itself must be non-simply connected. This is typical of bad physics - the obscurantism of the twentieth century in natural philosophy with its plethora of nigh incomprehensible and unprovable ideas. In contrast, the twenty first century ECE theory explains the AB effect using the older but experimentally provable philosophy of Faraday, Maxwell and Einstein. Therefore one of the key philosophical advances of ECE theory is to discard standard model gauge theory as being obscurantist and meaningless. In so doing, ECE adheres to Baconian philosophy: the theory is fundamentally changed to successfully and simply explain data that clearly refute the old theory (in this case the old theory is gauge theory).

For self-consistency there should be an AB effect whenever there is present a field and its concomitant potential. So an electromagnetic AB effect should be ubiquitous throughout electrodynamics and optics. This is indeed the case, as manifested for example [1, 12] in various well known phase effects interpreted according to general relativity (exemplified in turn by ECE theory). Therefore and in general the electromagnetic AB condition is:

$$\left.\begin{aligned} F = d \wedge A + \omega \wedge A = 0, \\ A \neq 0, \omega \neq 0, \end{aligned}\right\} \tag{2.66}$$

and for the gravitational field the AB condition is:

$$\left.\begin{aligned} T = d \wedge q + \omega \wedge q = 0, \\ q \neq 0, \omega \neq 0. \end{aligned}\right\} \tag{2.67}$$

This short-hand notation has been translated in full detail [1, 12] into three other notations: differential form, tensor and vector because notation is not standardized and different subjects use different notations. In the vector notation of classical electrodynamics [24] and electrical engineering, Eq. (2.66) splits into two equations. The first defines the magnetic field in terms of the vector potential and the spin connection vector. This was developed further in paper 74 of ECE theory (www.aias.us) and published in a standard model journal, Physica B [25]. In paper 74 the context was a balance condition for magnetic motors, but the same equation is also an AB condition. It is:

$$\boldsymbol{B} = \boldsymbol{\nabla} \times \boldsymbol{A} - \boldsymbol{\omega} \times \boldsymbol{A} = \boldsymbol{0}. \tag{2.68}$$

For spin torsion [1, 12] in gravitation the equivalent equation is:

$$\boldsymbol{T} = \boldsymbol{\nabla} \times \boldsymbol{q} - \boldsymbol{\omega} \times \boldsymbol{q} = \boldsymbol{0}. \tag{2.69}$$

In ECE every kind of magnetic field is defined by:

$$\boldsymbol{B} = \boldsymbol{\nabla} \times \boldsymbol{A} - \boldsymbol{\omega} \times \boldsymbol{A} \tag{2.70}$$

for self consistency. The spin connection vector is ubiquitous because it is a property of space-time itself. This is pure relativity of Einstein, but is still missing from the standard model of electrodynamics. The latter is still based on the well known vector development due to Heaviside of the original quaternionic Maxwell equations, and predates the philosophy of relativity.

If an electromagnetic AB effect is being considered the potential in Eq. (2.68) may be modeled by a plane wave as in paper 74 (www.aias.us). In that case the AB condition becomes a Beltrami condition:

$$\boldsymbol{\nabla} \times \boldsymbol{A}^{(1)} = -\kappa \boldsymbol{A}^{(1)} \tag{2.71}$$

$$\boldsymbol{\nabla} \times \boldsymbol{A}^{(2)} = \kappa \boldsymbol{A}^{(2)} \tag{2.72}$$

$$\boldsymbol{\nabla} \times \boldsymbol{A}^{(3)} = 0 \boldsymbol{A}^{(3)} \tag{2.73}$$

which can be developed in turn into a Helmholtz wave equation:

$$(\nabla^2 + \kappa^2) \boldsymbol{A}^{(1)} = \boldsymbol{0}. \tag{2.74}$$

Considering the X component for example:

$$\frac{\partial^2 A_X^{(1)}}{\partial Z^2} + \kappa^2 A_X^{(1)} = 0 \tag{2.75}$$

which is an undamped Bernoulli/Euler resonance equation without a driving force on the right hand side [1, 12]. It is also a free space wave equation without a source. It is however a wave equation in the potential ONLY, there being no magnetic field present by Eq. (2.68). In other words there is no radiated electromagnetic field but there is a radiated potential field. This is an example of an electromagnetic AB effect. In ECE theory the radiated potential without field may have a physical effect, in this case an electrodynamic or optical effect.

These arguments of ECE theory go to the root of what is meant by a photon and what is meant by the electromagnetic field. In the standard model there are two approaches to electromagnetic phenomena. As argued already in this Section, the electromagnetic field F is physical but the electromagnetic potential A is unphysical in the standard model on the classical level, whereas in standard model quantum electrodynamics the minimal prescription is used with a physical potential. Also in the standard model there are other concepts such as virtual photons which occur in Feynman's version of quantum electrodynamics. During the course of ECE development however [1, 12] the claimed accuracy of the Feynman type QED has been shown to be an exaggeration by several orders of magnitude. It is possible to see this through the fact that accuracy of the fine structure constant is limited by the accuracy of the Planck constant (paper 85 on www.aias.us). The standards laboratory data on fundamental constants were shown in this paper to be self-inconsistent. Finally, Feynman's QED method is based on what are essentially adjustable parameters, in other words it is based on obscurantist concepts such as dimensional renormalization, concepts which cannot be proven experimentally and so distill down to parameters that are adjusted to give a good fit of theory to experiment. It is also well known that the series summation used in the Feynman calculus cannot be proven a priori to converge, and thousands of terms have to be evaluated by computer even for the simplest of problems such as one electron interacting with one photon. The situation in quantum chromodynamics is much more complicated and much worse. In QCD it takes Nobel Prizes to prove renormalization, which is just an adjustable parameter. In a subject such as chemistry, such methods are impractical and are never used. They are therefore confined to ultra-specialist physics and even then are of dubious validity. This is typical of bad science, to claim that a theory is fundamental when it is not. It is well known [1, 12] that there are many weaknesses in the standard model of electrodynamics, for example it is still not able to describe the Faraday disk generator of Dec. 26th, 1831 whereas ECE has offered a straightforward explanation.

In ECE the field and potential are both physical [1,12] on both the classical and quantum levels, and in ECE there is no distinction between relativity and wave mechanics. These ideas of natural philosophy all become aspects of the same geometry, and in ECE this is the standard differential geometry of Cartan routinely taught in mathematics. The field, potential and photon are defined by this geometry. In the standard model there is also a distinction between locality and non-locality, a distinction which enters into areas such as quantum entanglement and one photon Young interferometry, in which one photon appears to self-interfere. In ECE [1,12] there is no distinction between locality and non-locality because of the ubiquitous spin connection of general relativity. Thus, in ECE theory, the AB effects are effects of general relativity, and the labels "local" and "non-local" becomes meaningless - all is geometry in four-dimensional space-time.

Having described the essentials of the AB effects, the various phase effects developed during the course of the development of ECE theory [1, 12] have been understood by a similar application of the Stokes theorem:

$$\int_s F = \int_s D \wedge A = \oint A + \int_s \omega \wedge A \tag{2.76}$$

in which the covariant exterior derivative $D\wedge$ appears. The use of this type of Stokes Theorem has been exemplified in volume 1 of ref. (1) by integrating around a helix and by closing the contour in a well defined way. This type of integration was used in the development in ECE theory of the well known Dirac and Wu Yang phases, and in a generalization of the well known Berry phase as for example in the well studied paper 6 of the ECE series (www.aias.us). in which the origin of the Planck constant was discussed. (The extent to which the 103 or so individual ECE papers are studied is measured accurately through the feedback software of www.aias.us, and there can be no doubt that they are all well studied by a high quality of readership.) In the development of the electromagnetic phase with ECE theory [1, 12] it has been demonstrated that the phase is due to the well known B(3) spin field of ECE theory, first inferred in 1992 from the inverse Faraday effect. This generally relativistic development of the electromagnetic phase is closely related to the AB effects and resolves basic problems in the standard model electromagnetic phase [1, 12]. It has therefore been shown that the B(3) field is ubiquitous in optics and electrodynamics, because it derives from the ubiquitous spin connection of space-time itself.

These considerations have also been developed for the topological phases, such as that of Berry, using for self consistency the same methodology as for the electromagnetic, Dirac and Wu Yang phases [1, 12]. These well known phases are again understood in ECE theory in terms of Cartan geometry by use of the Stokes Theorem with $D\wedge$ in place of $d\wedge$. All phase theory in physics becomes part of general relativity, and this methodology has been linked to traditional Lagrangian methods based on the minimization of action.

2.5 Tensor and vector laws of classical dynamics and electrodynamics

The tensor law for the homogeneous field equation has been shown [1,12] to be:

$$\partial_\mu \widetilde{F}^{\kappa\mu\nu} = 0. \tag{2.77}$$

For each κ index the structure of the matrix is:

$$\widetilde{F}^{\mu\nu} = \begin{bmatrix} 0 & cB_X & cB_Y & cB_Z \\ -cB_X & 0 & -E_Z & E_Y \\ -cB_Y & E_Z & 0 & -E_X \\ -cB_Z & -E_Y & E_X & 0 \end{bmatrix} = \begin{bmatrix} 0 & \widetilde{F}^{01} & \widetilde{F}^{02} & \widetilde{F}^{03} \\ \widetilde{F}^{10} & 0 & \widetilde{F}^{12} & \widetilde{F}^{13} \\ \widetilde{F}^{20} & \widetilde{F}^{21} & 0 & \widetilde{F}^{23} \\ \widetilde{F}^{30} & \widetilde{F}^{31} & \widetilde{F}^{32} & 0 \end{bmatrix}. \tag{2.78}$$

The Gauss law of magnetism in ECE theory has been shown to be obtained from:

$$\kappa = \nu = 0 \tag{2.79}$$

and so:

$$\partial_1 \widetilde{F}^{010} + \partial_2 \widetilde{F}^{020} + \partial_3 \widetilde{F}^{030} = 0 \tag{2.80}$$

i.e.:

$$\boldsymbol{\nabla} \cdot \boldsymbol{B} = 0 \tag{2.81}$$

with:

$$\boldsymbol{B} = \boldsymbol{B}_X \boldsymbol{i} + B_Y \boldsymbol{j} + B_Z \boldsymbol{k} \tag{2.82}$$

and:

$$B_X = B^{001}, B_Y = B^{002}, B_Z = B^{003}. \tag{2.83}$$

These are orbital magnetic field components of the Gauss law of magnetism.

In ECE theory the Faraday law of induction is a spin law of electrodynamics defined by:

$$\left.\begin{aligned} \partial_0 \widetilde{F}^{\kappa 01} + \partial_2 \widetilde{F}^{\kappa 21} + \partial_3 \widetilde{F}^{\kappa 31} &= 0 \\ \partial_0 \widetilde{F}^{\kappa 02} + \partial_1 \widetilde{F}^{\kappa 12} + \partial_3 \widetilde{F}^{\kappa 32} &= 0 \\ \partial_0 \widetilde{F}^{\kappa 03} + \partial_1 \widetilde{F}^{\kappa 13} + \partial_2 \widetilde{F}^{\kappa 23} &= 0 \end{aligned}\right\}. \tag{2.84}$$

The ECE Faraday law of induction for all practical purposes is [1,12]:

$$\boldsymbol{\nabla} \times \boldsymbol{E} + \frac{\partial \boldsymbol{B}}{\partial t} = \boldsymbol{0} \tag{2.85}$$

where the spin electric and magnetic components are:

$$\left.\begin{aligned} E_X = E^{332} &= -E^{323}, \; B_X = -B^{110} = B^{101}, \\ E_Y = E^{113} &= -E^{131}, \; B_Y = -B^{220} = B^{202}, \\ E_Z = E^{221} &= -E^{112}, \; B_Z = -B^{330} = B^{303}. \end{aligned}\right\} \tag{2.86}$$

The ECE Ampère-Maxwell law is another spin law [1,12]:

$$\boldsymbol{\nabla} \times \boldsymbol{B} - \frac{1}{c^2}\frac{\partial \boldsymbol{E}}{\partial t} = \mu_0 \boldsymbol{J} \tag{2.87}$$

where the components have been identified as:

$$\left.\begin{aligned} E_X &= E^{101},\ B_X = B^{332},\\ E_Y &= E^{202},\ B_Y = B^{113},\\ E_Z &= E^{303},\ B_Z = B^{221}. \end{aligned}\right\} \tag{2.88}$$

Therefore in these two spin laws different components appear in ECE theory of the electric and magnetic fields. In the MH theory of special relativity these components are not distinguishable.

Finally the Coulomb law has been shown to be [1, 12]:

$$\boldsymbol{\nabla} \cdot \boldsymbol{E} = \rho/\epsilon_0 \tag{2.89}$$

and is an orbital law of electromagnetism as is the Gauss law of magnetism. In ECE theory these individual spin and orbital components are proportional to individual components of the three index Cartan torsion tensor and three index angular energy/momentum density tensor. Therefore ECE theory comes to the important conclusion that there are orbital and spin components of the electric field, and orbital and spin components of the magnetic field. The orbital components occur in the Gauss law of magnetism and Coulomb law and the spin components in the Faraday law of induction and the Ampère-Maxwell law. This information, given by a generally covariant unified field theory, is not available in Maxwell-Heaviside (MH) theory of the un-unified, special relativistic, field.

Therefore each law develops an internal structure which is summarized in Table 2.1. There are two orbital laws (Gauss and Coulomb) and two spin laws (Faraday law of induction and Ampère-Maxwell law). In each law the components of the electric and magnetic fields are proportional to components of the well known [18] angular energy/momentum density tensor. Therefore for example the static electric field is distinguished form the radiated electric field. This is correct experimentally, it is well known that the static electric field exists between two static or unmoving charges, while the radiated electric field requires accelerated charges for its existence. By postulate the components of the electric and magnetic fields are also proportional to components of the Cartan tensor, a rank three tensor in the base manifold (4-D space-time with torsion and curvature).

In tensor notation the inhomogeneous ECE field equation in the base manifold has been shown to be, for all practical purposes [1, 12]:

$$\partial_\mu F^{\kappa\mu\nu} = \frac{1}{\epsilon_0} J^{\kappa\nu} = cA^{(0)} R^{\kappa}{}_{\mu}{}^{\mu\nu}. \tag{2.90}$$

The vacuum is defined by:

$$R^{\kappa}{}_{\mu}{}^{\mu\nu} = 0 \tag{2.91}$$

and is Ricci flat by construction. This result is consistent with the fact that the vacuum solutions of the EH equation are Ricci flat by construction. In a Ricci flat space-time there is no canonical energy momentum density [1, 12] and so there are no electric and magnetic fields because there is no angular energy/momentum density. However, as in the theory of the Aharonov-Bohm effects developed in Section 2.4, there may be non-zero potential and spin connection in a Ricci flat vacuum. Similarly, in the latter type of vacuum the Ricci tensor vanishes but the Christoffel connection and metric of EH theory do not vanish. Crothers has recently criticized the concept of the Ricci flat vacuum [19] as contradicting the Einstein equivalence principle. He has also shown that the mis-named Schwarzschild metric is inconsistent with the concept of a Ricci flat vacuum and with the geometry of the EH equation. Crothers has also argued that ideas such as Big Bang, black holes and dark matter are inconsistent with the EH equation.

The Coulomb law is the case:

$$\nu = 0 \tag{2.92}$$

of Eq. (2.90). During the course of development of ECE theory it has been shown by computer algebra that for all Ricci flat solutions of the EH equation:

$$R^{\kappa}{}_{\mu}{}^{\mu\nu} = 0 \tag{2.93}$$

Table 2.1: Components of the Laws of Classical Electrodynamics

Law	Electric Field	Magnetic Field	Type
Gauss		$B_X = B^{001}$ $B_Y = B^{002}$ $B_Z = B^{003}$	orbital
Coulomb	$E_X = -E^{001}$ $E_Y = -E^{002}$ $E_Z = -E^{003}$		orbital
Faraday	$E_X = E^{332}$ $E_Y = E^{113}$ $E_Z = E^{221}$	$B_X = B^{101}$ $B_Y = B^{202}$ $B_Z = B^{303}$	spin
Ampère-Maxwell	$E_X = -E^{101}$ $E_Y = -E^{202}$ $E_Z = -E^{303}$	$B_X = B^{332}$ $B_Y = B^{113}$ $B_Z = B^{221}$	spin

but for all other solutions of the EH equation the right hand sides of Eq. (2.90) are non zero for the Christoffel connection. This result introduces a basic paradox in the EH equation as discussed already in this review paper.

The Ampère-Maxwell law is the case:

$$\nu = 1, 2, 3 \tag{2.94}$$

in Eq. (2.90) and in tensor notation the ECE Ampère-Maxwell law is:

$$\begin{aligned} \partial_0 F^{\kappa 01} + \partial_2 F^{\kappa 21} + \partial_3 F^{\kappa 31} &= cA^{(0)}(R^{\kappa}{}_{0}{}^{01} + R^{\kappa}{}_{2}{}^{21} + R^{\kappa}{}_{3}{}^{31}) \\ \partial_0 F^{\kappa 02} + \partial_1 F^{\kappa 12} + \partial_3 F^{\kappa 32} &= cA^{(0)}(R^{\kappa}{}_{0}{}^{02} + R^{\kappa}{}_{1}{}^{12} + R^{\kappa}{}_{3}{}^{32}) \\ \partial_0 F^{\kappa 03} + \partial_1 F^{\kappa 13} + \partial_2 F^{\kappa 23} &= cA^{(0)}(R^{\kappa}{}_{0}{}^{03} + R^{\kappa}{}_{1}{}^{13} + R^{\kappa}{}_{2}{}^{23}) \end{aligned} \tag{2.95}$$

Therefore it is inferred that the time-like index is 0 and the space-like indices are 1, 2 and 3. The left hand side of Eq. (2.89) is a scalar and so

$$\kappa = 0 \tag{2.96}$$

is identified with a scalar index. So Eq. (2.89) of the Coulomb law is:

$$\partial_1 F^{010} + \partial_2 F^{020} + \partial_3 F^{030} = cA^{(0)}(R^{0}{}_{1}{}^{10} + R^{0}{}_{2}{}^{20} + R^{0}{}_{3}{}^{30}) \tag{2.97}$$

and is the orbital ECE Coulomb law. In vector notation this law is:

$$\boldsymbol{\nabla} \cdot \boldsymbol{E} = \frac{\rho}{\epsilon_0} \tag{2.98}$$

where:

$$\begin{aligned} &E_X = E^{010}, E_Y = E^{020}, E_Z = E^{030}, \\ &\rho = \epsilon_0 c A^{(0)}(R^{0}{}_{1}{}^{10} + R^{0}{}_{2}{}^{20} + R^{0}{}_{3}{}^{30}). \end{aligned} \tag{2.99}$$

The S.I. units of this law are:

$$A^{(0)} = JsC^{-1}m^{-1}, R = m^{-2}, \epsilon_0 = J^{-1}c^2m^{-1}, \rho = Cm^{-3}. \tag{2.100}$$

In Eq. (2.90):

$$\left.\begin{aligned} cA^{(0)} = JC^{-1} = \text{volts}, \\ \boldsymbol{E} = \text{volt } m^{-1}, \boldsymbol{\nabla}\cdot\boldsymbol{E} = \text{volt } m^{-2}, \\ cA^{(0)}R = \text{volt } m^{-2}, \\ \rho/\epsilon_0 = JC^{-1}m^{-2} = \text{volt } m^{-2}, \end{aligned}\right\} \tag{2.101}$$

thus checking the S. I. units for self consistency. In the Ricci flat vacuum:

$$\boldsymbol{\nabla}\cdot\boldsymbol{E} = 0 \tag{2.102}$$

which is consistent with:

$$R^0{}_1{}^{10} + R^0{}_2{}^{20} + R^0{}_3{}^{30} = 0 \tag{2.103}$$

for vacuum solutions of the EH equation as argued already. However, for complete internal consistency the Christoffel symbol cannot be used, because it is not internally consistent with the Bianchi identity as argued already in this review paper.

It is possible to define a curvature scalar of the Coulomb law as:

$$R_{(0)} := R^0{}_1{}^{10} + R^0{}_2{}^{20} + R^0{}_3{}^{30} \tag{2.104}$$

so that:

$$\boldsymbol{\nabla}\cdot\boldsymbol{E} = \frac{\rho}{\epsilon_0} = cA^{(0)}R_{(0)} \tag{2.105}$$

and that the charge density of the Coulomb law becomes:

$$\rho = cA^{(0)}\epsilon_0 R_{(0)} \tag{2.106}$$

in coulombs per cubic meter. In the Cartesian system of coordinates the electric field components of the Coulomb law are:

$$E_X = E^{010}, E_Y = E^{020}, E_Z = E^{030} \tag{2.107}$$

and are proportional to these same components of the three index angular energy momentum density tensor. They are anti-symmetric in their last two indices:

$$E^{010} = -E^{001}\text{etc.} \tag{2.108}$$

In tensor notation the ECE Ampère-Maxwell law is given by Eq. (2.95), i.e.:

$$\left.\begin{aligned} \partial_\mu F^{\kappa\mu\nu} = cA^{(0)}R^{\kappa}{}_{\mu}{}^{\mu\nu}, \\ \nu = 1, 2, 3 \end{aligned}\right\} \tag{2.109}$$

and in vector notation by:

$$\boldsymbol{\nabla}\times\boldsymbol{B} - \frac{1}{c^2}\frac{\partial\boldsymbol{E}}{\partial t} = \mu_0\boldsymbol{J}. \tag{2.110}$$

In the Cartesian system:

$$\boldsymbol{J} = J_X\boldsymbol{i} + J_Y\boldsymbol{j} + J_Z\boldsymbol{k} \tag{2.111}$$

where:

$$\left.\begin{aligned} J_X &= \frac{A^{(0)}}{\mu_0}\left(R^1{}_0{}^{01} + R^1{}_2{}^{21} + R^1{}_3{}^{31}\right), \\ J_Y &= \frac{A^{(0)}}{\mu_0}\left(R^2{}_0{}^{02} + R^2{}_1{}^{12} + R^3{}_2{}^{32}\right), \\ J_Z &= \frac{A^{(0)}}{\mu_0}\left(R^3{}_0{}^{03} + R^3{}_1{}^{13} + R^3{}_2{}^{23}\right), \end{aligned}\right\} \tag{2.112}$$

and self consistently in the vacuum:

$$J_X = J_Y = J_Z = 0 \tag{2.113}$$

for Ricci flat space-times. As argued this result has been demonstrated by computer algebra [1, 12]. In the Ampère-Maxwell law the electric and magnetic field components are proportional to spin angular energy momentum density tensor components of the electromagnetic field as follows:

$$\left.\begin{aligned} E^{\kappa\mu\nu} &= \frac{c^2}{e\omega}J^{\kappa\mu\nu}, \\ B^{\kappa\mu\nu} &= \frac{c}{e\omega}J^{\kappa\mu\nu}. \end{aligned}\right\} \tag{2.114}$$

The electric field components of the Coulomb law and the magnetic field components of the Gauss law are all orbital angular energy density tensor components of the electromagnetic field. The angular energy momentum density tensor may be defined as [18]:

$$J^{\kappa\mu\nu} = -\frac{1}{2}\left(T^{\kappa\mu}x^\nu - T^{\kappa\nu}x^\mu\right) \tag{2.115}$$

using the symmetric canonical energy momentum density tensor:

$$T^{\kappa\mu} = T^{\mu\kappa} \tag{2.116}$$

and the components of the electric and magnetic fields are components of $J^{\kappa\mu\nu}$ as follows:

$$\begin{aligned} E^{00i} &= \frac{c^2}{e\omega}J^{00i}, i = 1, 2, 3, \text{(orbital)}, \\ E^{ii0} &= \frac{c^2}{e\omega}J^{ii0}, i = 1, 2, 3, \text{(spin)}, \\ B^{112} = \frac{c}{e\omega}J^{112}, B^{221} &= \frac{c}{e\omega}J^{221}, B^{331} = \frac{c}{e\omega}J^{331}. \end{aligned} \tag{2.117}$$

The two index angular energy/momentum tensor of the electromagnetic field is an integral over the three index density tensor. Ryder gives one example of such an integral in Minkowski space-time [18]:

$$M^{\mu\nu} = \int M^{0\mu\nu}d^3x. \tag{2.118}$$

Therefore the four laws of electrodynamics in ECE theory are:

$$\boldsymbol{\nabla}\cdot\boldsymbol{B} = 0, \tag{2.119}$$

$$\boldsymbol{\nabla}\times\boldsymbol{E} + \frac{\partial\boldsymbol{B}}{\partial t} = \boldsymbol{0}, \tag{2.120}$$

$$\boldsymbol{\nabla}\cdot\boldsymbol{E} = \rho/\epsilon_0, \tag{2.121}$$

$$\boldsymbol{\nabla}\times\boldsymbol{B} - \frac{1}{c^2}\frac{\partial\boldsymbol{E}}{\partial t} = \mu_0\boldsymbol{J}, \tag{2.122}$$

and therefore have the same vector structure as the familiar MH equations. However, as argued in this section, the ECE theory gives additional information. In the four laws the components of the magnetic and electric fields are as follows. The Gauss law of magnetism in ECE theory is, for all practical purposes (FAPP):

$$\boldsymbol{\nabla} \cdot \boldsymbol{B} = 0 \tag{2.123}$$

which is an orbital law in which the components of the magnetic field are proportional to orbital components of the angular momentum/energy density tensor and are:

$$\boldsymbol{B} = B^{001}\boldsymbol{i} + B^{002}\boldsymbol{j} + B^{003}\boldsymbol{k}. \tag{2.124}$$

The Faraday law of induction in ECE is a spin law with electric and magnetic field components as follows:

$$\boldsymbol{E} = E^{332}\boldsymbol{i} + E^{113}\boldsymbol{j} + E^{221}\boldsymbol{k}, \tag{2.125}$$

$$\boldsymbol{B} = B^{101}\boldsymbol{i} + B^{202}\boldsymbol{j} + B^{303}\boldsymbol{k}. \tag{2.126}$$

The Coulomb law in ECE is an orbital law with electric field components as follows:

$$\boldsymbol{E} = E^{010}\boldsymbol{i} + E^{020}\boldsymbol{j} + E^{030}\boldsymbol{k}. \tag{2.127}$$

Finally the Ampère-Maxwell law in ECE is a spin law with electric and magnetic field components as follows:

$$\boldsymbol{E} = E^{110}\boldsymbol{i} + E^{220}\boldsymbol{j} + E^{330}\boldsymbol{k}, \tag{2.128}$$

$$\boldsymbol{B} = B^{332}\boldsymbol{i} + B^{113}\boldsymbol{j} + B^{221}\boldsymbol{k}. \tag{2.129}$$

As argued in Section 2.4 of this review paper, the relation between field and potential is different in ECE theory and contains the spin connection [1, 12]. The various notations for the relation between field and potential in ECE theory are collected here for convenience. In the index-less notation:

$$F = d \wedge A + \omega \wedge A \tag{2.130}$$

which is based on the first Cartan structure equation:

$$T = d \wedge q + \omega \wedge q. \tag{2.131}$$

In the standard notation of differential geometry:

$$F^a = d \wedge A^a + \omega^a{}_b \wedge A^b. \tag{2.132}$$

In tensor notation from differential geometry:

$$F^a_{\mu\nu} = (d \wedge A^a)_{\mu\nu} + (\omega^a{}_b \wedge A^b)_{\mu\nu}. \tag{2.133}$$

In the base manifold Eq. (2.133) becomes:

$$F^{\kappa\mu\nu} = \partial^\mu A^{\kappa\nu} - \partial^\nu A^{\kappa\mu} + (\omega^{\kappa\mu}{}_\lambda A^{\lambda\nu} - \omega^{\kappa\nu}{}_\lambda A^{\lambda\mu}) \tag{2.134}$$

In vector notation Eq. (2.134) splits into two equations, one for the electric field and one for the magnetic field:

$$\boldsymbol{E} = -\boldsymbol{\nabla}\phi - \frac{\partial \boldsymbol{A}}{\partial t} + \phi\boldsymbol{\omega} - \omega\boldsymbol{A} \tag{2.135}$$

and

$$\boldsymbol{B} = \boldsymbol{\nabla} \times \boldsymbol{A} - \boldsymbol{\omega} \times \boldsymbol{A} \tag{2.136}$$

For the orbital electric field component of the Coulomb law Eq. (2.135) has the following internal structure:

$$\phi = cA^{00}, \boldsymbol{A} = A^{01}\boldsymbol{i} + A^{02}\boldsymbol{j} + A^{03}\boldsymbol{k}, \tag{2.137}$$

$$\omega = c\omega^{00}{}_0, \boldsymbol{\omega} = (\omega^{01}{}_0\boldsymbol{i} + \omega^{02}{}_0\boldsymbol{j} + \omega^{03}{}_0\boldsymbol{k}). \tag{2.138}$$

This result illustrates that the internal structure of the relation between field and potential is different for each law of electrodynamics in ECE theory. Therefore in a GCUFT such as ECE different types of field and potential exist for each law, and also different types of spin connection.

For the orbital Gauss law of magnetism the internal structure of Eq. (2.136) is:

$$\boldsymbol{A} = A^{01}\boldsymbol{i} + \boldsymbol{A}^{02}\boldsymbol{j} + A^{03}\boldsymbol{k}, \tag{2.139}$$

$$\boldsymbol{\omega} = -(\omega^{01}{}_0\boldsymbol{i} + \omega^{02}{}_0\boldsymbol{j} + \omega^{03}{}_0\boldsymbol{k}). \tag{2.140}$$

For the Ampère-Maxwell law, a spin law, the internal structure of Eqs. (2.135) and (2.136) are again different, and are defined as follows. The structure of Eq. (2.135) is:

$$\begin{aligned}
\phi &= cA^{00} = cA^{01} = cA^{02} = cA^{03},\\
A_X &= A^{01} = A^{11} = A^{21} = A^{31},\\
A_Y &= A^{02} = A^{12} = A^{22} = A^{32},\\
A_Z &= A^{03} = A^{13} = A^{23} = A^{33},\\
\omega_X &= \omega^{11}{}_0 = \omega^{11}{}_1 = \omega^{11}{}_2 = \omega^{11}{}_3,\\
\omega_Y &= \omega^{22}{}_0 = \omega^{22}{}_1 = \omega^{22}{}_2 = \omega^{22}{}_3,\\
\omega_Z &= \omega^{33}{}_0 = \omega^{33}{}_1 = \omega^{33}{}_2 = \omega^{33}{}_3,\\
\omega &= c\omega^{10}{}_0 = c\omega^{10}{}_1 = c\omega^{10}{}_2 = c\omega^{10}{}_3\\
&= c\omega^{20}{}_0 = c\omega^{20}{}_1 = c\omega^{20}{}_2 = c\omega^{20}{}_3\\
&= c\omega^{30}{}_0 = c\omega^{30}{}_1 = c\omega^{30}{}_2 = c\omega^{30}{}_3
\end{aligned} \tag{2.141}$$

and the structure of Eq. (2.136) is:

$$\begin{aligned}
B_X = B^{332} &= \frac{\partial A_Z}{\partial Y} - \frac{\partial A_Y}{\partial Z} + \omega_Z A_Y - \omega_Y A_Z,\\
B_Y = B^{113} &= \frac{\partial A_X}{\partial Z} - \frac{\partial A_Z}{\partial X} + \omega_X A_Z - \omega_Z A_X,\\
B_Z = B^{221} &= \frac{\partial A_Y}{\partial X} - \frac{\partial A_X}{\partial Y} + \omega_Y A_X - \omega_X A_Y.
\end{aligned} \tag{2.142}$$

Finally the internal structures are again different for the Faraday law of induction. In arriving at these conclusions the relation between field and potential in the base manifold is:

$$F^{\kappa\mu\nu} = \partial^\mu A^{\kappa\nu} - \partial^\nu A^{\kappa\mu} + (\omega^{\kappa\mu}{}_\lambda A^{\lambda\nu} - \omega^{\kappa\nu}{}_\lambda A^{\lambda\mu}). \tag{2.143}$$

The Hodge dual of this equation is:

$$\widetilde{F}^{\kappa\mu\nu} = (\partial^\mu A^{\kappa\nu} - \partial^\nu A^{\kappa\mu} + (\omega^{\kappa\mu}{}_\lambda A^{\lambda\nu} - \omega^{\kappa\nu}{}_\lambda A^{\lambda\mu}))_{HD} \tag{2.144}$$

and this is needed to give the results for the homogenous laws. An example of taking the Hodge dual is given below:

$$\begin{aligned}
\widetilde{F}^{001} &= (\partial^0 A^{01} - \partial^1 A^{00} + (\omega^{00}{}_\lambda A^{\lambda 1} - \omega^{01}{}_\lambda A^{\lambda 0}))_{HD}\\
&= \partial^2 A^{03} - \partial^3 A^{02} + (\omega^{02}{}_\lambda A^{\lambda 3} - \omega^{03}{}_\lambda A^{\lambda 2}).
\end{aligned} \tag{2.145}$$

With these rules the overall conclusion is that in a generally covariant unified field theory (GCUFT) such as ECE the four laws of classical electrodynamics can be reduced to the same vector form as the MH laws of un-unified special relativity (nineteenth century), but the four laws are no longer written in a flat, Minkowski space-time. They are written in a four dimensional space-time with torsion and curvature. This procedure reveals the internal structure of the electric and magnetic fields appearing in each law, for example correctly makes the distinction between a static and radiated electric field, and a static and radiated magnetic field. The relation between field and potential also develops an internal structure which is different for each law, but for each law, the vector relation can be reduced to:

$$\boldsymbol{E} = -\boldsymbol{\nabla}\phi - \frac{\partial \boldsymbol{A}}{\partial t} + \phi\boldsymbol{\omega} - \omega\boldsymbol{A} \tag{2.146}$$

and

$$\boldsymbol{B} = \boldsymbol{\nabla} \times \boldsymbol{A} - \boldsymbol{\omega} \times \boldsymbol{A}. \tag{2.147}$$

In a GCUFT, gauge theory is not used because the potential has a physical effect as in the electrotonic state of Faraday. The ECE theory is developed entirely in four dimensions, is entirely self-consistent, and reproduces a range of experimental data [1, 12] which the MH theory cannot explain. The ECE theory is also philosophically consistent with the need to apply the philosophy of relativity to the whole of physics. The latter becomes a unified field theory based on geometry. The first attempts by Einstein to develop general relativity were based on Riemann geometry and restricted to the theory of gravitation. In the philosophy of relativity, however, the basic idea that physics is geometry must be used for every equation of physics, and each equation must be part of the same geometrical framework. This is achieved in a GCUFT such as ECE theory by using Cartan's standard differential geometry. This is a self-consistent geometry that recognizes the existence of space-time torsion in the first Cartan structure equation, and space-time curvature in the second. It is also recognized that there is only one Bianchi identity, and that this must always inter-relate torsion and curvature, both are fundamental to the structure of space-time.

2.6 Spin connection resonance

One of the most important consequences of general relativity applied to electrodynamics is that the spin connection enters into the relation between the field and potential as described in Section 2.5. The equations of electrodynamics as written in terms of the potential can be reduced to the form of Bernoulli Euler resonance equations. These have been incorporated during the course of development of ECE theory into the Coulomb law, which is the basic law used in the development of quantum chemistry in for example density functional code. This process has been illustrated [1, 12] with the hydrogen and helium atoms. The ECE theory has also been used to design or explain circuits which use spin connection resonance to take power from space-time, notably papers 63 and 94 of the ECE series on www.aias.us. In paper 63, the spin connection was incorporated into the Coulomb law and the resulting equation in the scalar potential shown to have resonance solutions using an Euler transform method. In paper 94 this method was extended and applied systematically to the Bedini motor. The method is most simply illustrated by considering the vector form of the Coulomb law deduced in Section 2.5:

$$\boldsymbol{\nabla} \cdot \boldsymbol{E} = \rho/\epsilon_0 \tag{2.148}$$

and assuming the absence of a vector potential (absence of a magnetic field). The electric field is then described by:

$$\boldsymbol{E} = -(\boldsymbol{\nabla} + \boldsymbol{\omega})\phi \tag{2.149}$$

rather than the standard model's:

$$\boldsymbol{E} = -\boldsymbol{\nabla}\phi. \tag{2.150}$$

Therefore Eq. (2.149) in (2.148) produces the equation

$$\nabla^2\phi + \boldsymbol{\omega}\cdot\nabla\phi + (\nabla\cdot\boldsymbol{\omega})\phi = -\frac{\rho}{\epsilon_0} \tag{2.151}$$

which is capable of giving resonant solutions as described in paper 63. The equivalent equation in the standard model is the Poisson equation, which is a limit of Eq. (2.151) when the spin connection is zero. The Poisson equation does not give resonant solutions. It is known from the work of Tesla for example that strong resonances in electric power can be obtained with suitable apparatus, and such resonances cannot be explained using the standard model. A plausible explanation of Tesla's well known results is given by the incorporation of the spin connection into classical electrodynamics. Using spherical polar coordinates and restricting consideration to the radial component:

$$\nabla^2\phi = \frac{\partial^2\phi}{\partial r^2} + \frac{2}{r}\frac{\partial\phi}{\partial r}, \tag{2.152}$$

$$\boldsymbol{\omega}\cdot\nabla\phi = \omega_r\frac{\partial\phi}{\partial r},\ (\nabla\cdot\boldsymbol{\omega})\phi = \frac{\phi}{r^2}\frac{\partial}{\partial r}(r^2\omega_r), \tag{2.153}$$

so that Eq. (2.151) becomes:

$$\frac{\partial^2\phi}{\partial r^2} + \left(\frac{2}{r} + \omega\right)\frac{\partial\phi}{\partial r} + \frac{\phi}{r^2}\left(2r\omega_r + r^2\frac{\partial\omega_r}{\partial r}\right) = \frac{-\rho}{\epsilon_0} \tag{2.154}$$

In paper 63 a spin connection was chosen of the simplest type compatible with its dimensions of inverse meters:

$$\omega_r = -\frac{1}{r} \tag{2.155}$$

and thus giving the differential equation:

$$\frac{\partial^2\phi}{\partial r^2} + \frac{1}{r}\frac{\partial\phi}{\partial r} - \frac{1}{r^2}\phi = \frac{-\rho}{\epsilon_0} \tag{2.156}$$

as a function of r. Eq. (2.156) becomes a resonance equation if the driving term on the right hand side is chosen to be oscillatory, in the simplest instance:

$$\rho = \rho(0)\cos(\kappa_r r). \tag{2.157}$$

To obtain resonance solutions from Eq. (2.156), an Euler transform [1, 12] is needed as follows:

$$\kappa_r r = \exp(i\kappa_r R). \tag{2.158}$$

This is a standard Euler transform extended to a complex variable. This simple change of variable transforms Eq. (2.156) into:

$$\frac{\partial^2\phi}{\partial R^2} + \kappa_r^2\phi = \frac{\rho(0)}{\epsilon_0}\mathrm{Real}(e^{2i\kappa_r R}\cos(e^{i\kappa_r R})) \tag{2.159}$$

which is an undamped oscillator equation as demonstrated in detail in Eq. (2.63), where the domain of validity of the transformed variable was discussed in detail. It is seen from feedback software to www.aias.us that paper 63 has been studied in great detail by a high quality readership, so we may judge that its impact has been extensive. The concept of spin connection resonance has been extended to gravitational theory and magnetic motors and the theory published in standard model journals [25, 27]. In paper 63 the simplest possible form of the spin connection was used, Eq. (2.155) and the resulting Eq. (2.156) was shown to have resonance solutions using a change of variable. There is therefore resonance in the variable R. In paper 90 of www.aias.us this method was made more general by considering the equation

$$\frac{\partial^2 \phi}{\partial r^2} + \left(\frac{2}{r} + \omega_r\right)\frac{\partial \phi}{\partial r} + \frac{\phi}{r^2}\left(2r\omega_r + r^2\frac{\partial \omega_r}{\partial r}\right) = \frac{-\rho}{\epsilon_0} \tag{2.160}$$

which is a more general form of Eq. (2.156). When the spin connection is defined as:

$$\omega_r = \omega_0^2 r - 4\beta \log_e r - \frac{4}{r}. \tag{2.161}$$

Eq. (2.160) becomes a simple resonance equation in r itself:

$$\frac{\partial^2 \phi}{\partial r^2} + 2\beta\frac{\partial \phi}{\partial r} + \omega_0^2 \phi = \frac{-\rho}{\epsilon_0}. \tag{2.162}$$

There is freedom of choice of the spin connection. The latter was unknown in electrodynamics prior to ECE theory and must ultimately be determined experimentally. An example of this procedure is given in paper 94, where spin connection resonance (SCR) theory is applied to a patented device. One of the papers published in the standard model literature [26] applies SCR to magnetic motors that are driven by space-time. It is probable that SCR was also discovered and demonstrated by Tesla [28], but empirically before the emergence of relativity theory. SCR has also been applied to gravitation and published in the standard model literature [27]. So a gradual loosening of the ties to the standard model is being observed at present.

In paper 92 of the ECE series (www.aias.us), Eq. (2.160) was further considered and shown to reduce to an Euler Bernoulli resonance equation of the general type:

$$\frac{d^2 x}{dr^2} + 2\beta\frac{dx}{dr} + \kappa_0^2 x = A\cos(\kappa r) \tag{2.163}$$

in which β plays the role of friction coefficient, κ_0 is a Hooke's law wave-number and in which the right hand side is a cosinal driving term. Eq. (2.160) reduces to Eq. (2.163) when:

$$\omega_r = 2\left(\beta - \frac{1}{r}\right), \kappa_0^2 = \frac{4}{r}\left(\beta - \frac{1}{r}\right) + \frac{\partial \omega_r}{\partial r} \tag{2.164}$$

Therefore the condition udner which the spin connection gives the simple resonance Eq. (2.163) is defined by:

$$\omega_r = \kappa_0^2 - 4\beta \log_e r - \frac{4}{r}. \tag{2.165}$$

Reduction to the standard model Coulomb law occurs when:

$$\beta = \frac{1}{r} \tag{2.166}$$

when

$$\omega_r = 0, \kappa_0^2 = 0. \tag{2.167}$$

In general there is no reason to assume that condition (2.166) always holds. The reason why the standard model Coulomb law is so accurate in the laboratory is that it is tested off resonance. In this off resonant limit the ECE theory has been shown [1, 12] to give the Standard Coulomb law as required by a vast amount of accumulated data of two centuries since Coulomb first inferred the law. In general, ECE theory has been shown to reduce to all the known laws of physics, and in addition ECE gives new information. This is a classic hallmark of a new advance in physics. It is probable that Tesla inferred methods of tuning the Coulomb law (and other laws) to spin connection resonance. Many other reports of such surges in power have been made, and it is now known and accepted by the international community of scientists that they come from general relativity applied to classical electrodynamics.

Paper 94 of the ECE series is a pioneering paper in which the theory of SCR is applied to a patented device in order to explain in detail how the patented device takes energy from space-time. No violation of the laws of conservation of energy and momentum occurs in ECE theory or in SCR theory.

2.7 Effects of gravitation on optics and spectroscopy

In the standard model of electrodynamics the electromagnetic sector is described by the nineteenth century Maxwell-Heaviside (MH) field theory, which in gauge theory is U(1) invariant and Lorentz covariant in a Minkowski space-time. As such MH theory cannot describe the effect of gravitation on optics and spectroscopy because gravitation requires a non-Minkowski space-time. In ECE theory on the other hand all sectors are generally covariant, and during the course of development of ECE theory several effects of gravitation on optics and spectroscopy have been inferred, notably the effect of gravitation on the Sagnac effect, RFR and on the polarization of light grazing a white dwarf. An explanation for the well known Faraday disk generator has also been given in terms of spinning space-time, an explanation which illustrates the fact that the torsion of space-time produces effects not present in the standard model. Gravitation is the curvature of space-time and in ECE theory the interaction of torsion and curvature is determined by Cartan geometry.

The Faraday disk generator has been explained in ECE theory from the basic assumption that the electromagnetic field is the Cartan torsion within a factor:

$$\boldsymbol{F}_{\text{mech}} = A^{(0)} T_{\text{mech}} \tag{2.168}$$

where $cA^{(0)}$ is the primordial voltage. The factor $A^{(0)}$ is considered to originate in the magnet of the Faraday disk generator. The Faraday disk generator consists essentially of a spinning disk placed on a magnet, without the magnet no induction is observed, i.e. no p.d.f. is generated between the center and rim of the disk without a magnet being present. The original experiment by Faraday on 26th Dec. 1831 consisted of spinning a disk on top of a static magnet, but an e.m.f. is also observed if both the disk and the magnet are spun about their common vertical axis. There continues to be no explanation for the Faraday disk generator in the standard model, because in the latter there is no connection between the electromagnetic field and mechanical spin, angular momentum and torsion, while ECE makes this connection in Eq. (2.168). The standard model law of induction of Faraday is:

$$\boldsymbol{\nabla} \times \boldsymbol{E} + \frac{\partial \boldsymbol{B}}{\partial t} = \boldsymbol{0} \tag{2.169}$$

and spinning the magnetic field about its own axis does not produce a non-zero curl of the electric field as required. Clearly, a static magnetic field will not cause induction from Eq. (2.169). So this is a weak point of the standard model, in which induction is caused in the classical textbook description by moving a bar magnet inside a coil, causing a current to appear. In ECE it has been shown [1, 12] that the explanation of the Faraday disk generator is simply:

$$\boldsymbol{F} = \boldsymbol{F}_{e/m} + \boldsymbol{F}_{\text{mech}} \tag{2.170}$$

which in vector notation (section 2.5) produces the law of induction:

$$\boldsymbol{\nabla} \times \boldsymbol{E}_{\text{mech}} + \frac{\partial \boldsymbol{B}_{\text{mech}}}{\partial t} = \boldsymbol{0}. \tag{2.171}$$

Spinning the disk has the following effect in ECE theory.

In the complex circular basis [1, 12] the magnetic flux density in ECE theory is defined by:

$$\boldsymbol{B}^{(1)*} = \boldsymbol{\nabla} \times \boldsymbol{A}^{(1)*} - i\frac{\kappa}{A^{(0)}} \boldsymbol{A}^{(2)} \times \boldsymbol{A}^{(3)} \tag{2.172}$$

$$\boldsymbol{B}^{(2)*} = \boldsymbol{\nabla} \times \boldsymbol{A}^{(2)*} - i\frac{\kappa}{A^{(0)}} \boldsymbol{A}^{(3)} \times \boldsymbol{A}^{(1)} \tag{2.173}$$

$$\boldsymbol{B}^{(3)*} = \boldsymbol{\nabla} \times \boldsymbol{A}^{(3)*} - i\frac{\kappa}{A^{(0)}} \boldsymbol{A}^{(1)} \times \boldsymbol{A}^{(2)} \tag{2.174}$$

where

$$\kappa = \frac{\Omega}{c} \tag{2.175}$$

is a wave-number and Ω is an angular frequency in radians per second. When the disk is stationary the ECE vector potential is [1, 12] proportional by fundamental hypothesis to the tetrad:

$$\boldsymbol{A}^{(1)} = A^{(0)}\boldsymbol{q}^{(1)} \tag{2.176}$$

$$\boldsymbol{A}^{(2)} = A^{(0)}\boldsymbol{q}^{(2)} \tag{2.177}$$

$$\boldsymbol{A}^{(3)} = A^{(0)}\boldsymbol{q}^{(3)}. \tag{2.178}$$

In the complex circular basis the tetrads are:

$$\boldsymbol{q}^{(1)} = \frac{1}{\sqrt{2}}(\boldsymbol{i} - i\boldsymbol{j}), \tag{2.179}$$

$$\boldsymbol{q}^{(2)} = \frac{1}{\sqrt{2}}(\boldsymbol{i} + i\boldsymbol{j}), \tag{2.180}$$

$$\boldsymbol{q}^{(3)} = \boldsymbol{k}, \tag{2.181}$$

and have O(3) symmetry as follows:

$$\boldsymbol{q}^{(1)} \times \boldsymbol{q}^{(2)} = i\boldsymbol{q}^{(3)*}, \tag{2.182}$$

$$\boldsymbol{q}^{(2)} \times \boldsymbol{q}^{(1)} = i\boldsymbol{q}^{(1)*}, \tag{2.183}$$

$$\boldsymbol{q}^{(3)} \times \boldsymbol{q}^{(1)} = i\boldsymbol{q}^{(2)*}. \tag{2.184}$$

In the absence of rotation about Z:

$$\boldsymbol{\nabla} \times \boldsymbol{A}^{(1)} = \boldsymbol{\nabla} \times \boldsymbol{A}^{(2)} = \boldsymbol{0}, \tag{2.185}$$

$$\boldsymbol{A}^{(3)} = A^{(0)}\boldsymbol{k}. \tag{2.186}$$

In the complex circular basis:

$$\boldsymbol{\nabla} \times \boldsymbol{E}^{(1)} + \partial \boldsymbol{B}^{(1)}/\partial t = \boldsymbol{0}, \tag{2.187}$$

$$\boldsymbol{\nabla} \times \boldsymbol{E}^{(2)} + \partial \boldsymbol{B}^{(2)}/\partial t = \boldsymbol{0}, \tag{2.188}$$

$$\boldsymbol{\nabla} \times \boldsymbol{E}^{(3)} + \partial \boldsymbol{B}^{(3)}/\partial t = \boldsymbol{0}. \tag{2.189}$$

Therefore from Eqs. (2.176) to (2.189) the only field present is:

$$\begin{aligned} \boldsymbol{B}^{(3)*} = \boldsymbol{B}^{(3)} &= -iB^{(0)}\boldsymbol{q}^{(1)} \times \boldsymbol{q}^{(2)} \\ &= B_z^{(3)}\boldsymbol{k} = B_z\boldsymbol{k} \end{aligned} \tag{2.190}$$

which is the static magnetic field of the bar magnet.

Now mechanically rotate the disk at an angular frequency Ω to produce:

$$\boldsymbol{A}^{(1)} = \frac{A^{(0)}}{\sqrt{2}}(\boldsymbol{i} - i\boldsymbol{j})\exp(i\Omega t), \tag{2.191}$$

$$\boldsymbol{A}^{(2)} = \frac{A^{(0)}}{\sqrt{2}}(\boldsymbol{i} + i\boldsymbol{j})\exp(-i\Omega t). \tag{2.192}$$

From Eqs. (2.176) to (2.192) electric and magnetic fields are induced in a direction transverse to Z, i.e. in the XY plane of the spinning disk as observed experimentally. However, the Z axis magnetic flux density is unchanged by physical rotation about the same Z axis. This is again as observed experimentally. The (2) component of the transverse electric field spins around the rim of the disk and is defined from Eq. (2.151) as:

$$\boldsymbol{E}^{(2)} = \boldsymbol{E}^{(1)*} = -\left(\frac{\partial}{\partial t} + i\Omega\right)\boldsymbol{A}^{(2)}. \tag{2.193}$$

It can be seen from section 2.5 that $i\Omega$ is a type of spin connection. The latter is caused by mechanical spin, which in ECE is a spinning of space-time itself. The real and physical part of the induced $E^{(1)}$ is:

$$\text{Real}(\boldsymbol{E}^{(1)}) = \frac{2}{\sqrt{2}}A^{(0)}\Omega(\boldsymbol{i}\sin\Omega t - \boldsymbol{j}\cos\Omega t) \tag{2.194}$$

and is proportional to the product of $A^{(0)}$ and Ω, again as observed experimentally. An electromotive force is set up between the center of the disk and the rotating rim, as first observed experimentally by Faraday. This e. m. f. is measured experimentally with a voltmeter at rest with respect to the rotating disk.

The homogeneous law (2.120) of ECE theory is generally covariant [1, 12] by construction, so retains its form in any frame of reference. ECE therefore produces a simple and complete description of the Faraday disk generator in terms of the spinning of space-time, and concomitant spin connection. The latter is therefore demonstrated in classical electrodynamics by the generator. All known experimental features are explained straightforwardly by ECE theory, but cannot be explained by MH theory, in which the spin connection is missing because Minkowski space-time has no connection by construction - it is a "flat" space-time. It is relatively easy to think of electrodynamics as spinning space-time if we think of gravitation as curving space-time. This analysis also gives confidence in the arguments of Section 2.6, where power is obtained from space-time with spin connection resonance.

The same ECE concept just used to explain the Faraday disk generator has been used [1,12] to give a simple explanation of the Sagnac effect (ring laser gyro). Again, the standard model has no satisfactory explanation for the Sagnac effect [1, 12]. Consider the rotation of a beam of light of any polarization around a circle of area πr^2 in the XY plane at an angular frequency ω_1. The rotation is a rotation of space-time itself in ECE theory, described by the rotating tetrad:

$$\boldsymbol{q}^{(1)} = \frac{1}{\sqrt{2}}(\boldsymbol{i} - i\boldsymbol{j})e^{i\omega_1 t}. \tag{2.195}$$

This is rotation around the static platform of the Sagnac interferometer. The fundamental ECE assumption means that this rotation produces the electromagnetic vector potential:

$$\boldsymbol{A}_L^{(1)} = A^{(0)}\boldsymbol{q}^{(1)} \tag{2.196}$$

for left rotation and:

$$\boldsymbol{A}_R^{(1)} = \frac{A^{(0)}}{\sqrt{2}}(\boldsymbol{i} + i\boldsymbol{j})e^{i\omega_1 t} \tag{2.197}$$

for right rotation. When the platform is at rest a beam going around left-wise takes the same time to reach its starting point as a beam going around right-wise. The time delay is zero:

$$\Delta t = 2\pi\left(\frac{1}{\omega_1} - \frac{1}{\omega_1}\right) = 0. \tag{2.198}$$

Eqs. (2.196) and (2.197) do not exist in special relativity because in the MH theory electromagnetism is a nineteenth century entity superimposed on a space-time that is flat and static and never rotates.

Now consider the left - wise rotating beam (2.196) and spin the platform mechanically in the same left-wise direction at an angular frequency Ω. The result is an increase in the angular frequency of the rotating tetrad as follows:

$$\omega_1 \rightarrow \omega_1 + \Omega. \tag{2.199}$$

Similarly consider the left wise rotating beam (2.196) and spin the platform right-wise. The result is a decrease in the angular frequency of the rotating tetrad:

$$\omega_1 \rightarrow \omega_1 - \Omega. \tag{2.200}$$

The time delay between a beam circling left-wise with the platform, and one circling left-wise against the platform is therefore:

$$\Delta t = 2\pi \left(\frac{1}{\omega_1 - \Omega} - \frac{1}{\omega_1 + \Omega} \right) \tag{2.201}$$

which is the Sagnac effect. The angular frequency ω_1 can be calculated from the experimental result [1, 12]:

$$\Delta t = \frac{4\Omega}{c^2} Ar = \frac{4\pi\Omega}{\omega_1^2 - \Omega^2} \tag{2.202}$$

If

$$\Omega \ll \omega_1 \tag{2.203}$$

it is found that

$$\omega_1 = \frac{c}{r} = c\kappa \tag{2.204}$$

Q.E.D. Therefore the Sagnac effect is another result of a spin connection, which in this case can be thought of as the potential (2.196) itself.

Similarly, phase effects such as the Tomita-Chao effect were also described straightforwardly with the same basic concept during the development of ECE theory.

In order to describe the effects of gravitation on optics and spectroscopy a dielectric version of the ECE theory was developed and implemented to find that the polarization of light is changed by light grazing a very massive object such as a white dwarf, and the dielectric theory was also used to demonstrate the effect of gravitation on the Sagnac effect [1, 12]. The standard model is not capable of such descriptions without the use of adjustable parameters in such transient twentieth century artifacts as superstring theory, now being essentially discarded as being untestable experimentally. ECE is far simpler and is also capable of describing data such as the Faraday disk generator and the Sagnac effect straightforwardly. During the course of its development the ECE theory has also been applied to the interaction of three fields [27] and the effect of gravitation on the inverse Faraday effect and its resonance counterpart, known as radiatively induced fermion resonance (RFR).

The interaction of fields in ECE theory is controlled by Cartan geometry, in the particular case of the interaction of gravitation and electromagnetism, there is a very small homogeneous charge current density in the Gauss law and in the Faraday law of induction. For all practical purpose in the laboratory this is not observable. However, it has been shown in ECE theory to result in changes of polarization and other optical properties of light grazing a white dwarf, which is an object many times heavier than the sun. Such changes of polarization are not described by the Einstein-Hilbert equation.

2.8 Radiative corrections in ECE theory

During the course of development of ECE theory the anomalous g factor of the electron and Lamb shifts in hydrogen and helium have been explained satisfactorily in a far simpler manner than the standard model and using the causal and objective principles of Einsteinian relativity. The usual approach to the radiative corrections in quantum electrodynamics (QED) has been criticized [1, 12], especially its claim to accuracy. The QED method of the standard model relies on assumptions that are not present in Einsteinian relativity, and also on adjustable parameters. The Feynman method consists of assuming the existence of virtual particles and on a perturbation method of quantum mechanics which sums thousands of terms of increasing complexity. There is no proof that this sum converges. It is also claimed in standard model QED that the accuracy of the fine structure constant is reproduced theoretically to high precision. However the fine structure constant in S.I. units is:

$$\alpha = \frac{e^2}{4\pi\epsilon\hbar c} \tag{2.205}$$

and its accuracy is limited by the experimental accuracy of the Planck constant. There is no way that a theory can produce a higher accuracy than experiment, and the theoretical value of the g factor of the electron is based on the value of the fine structure constant. Thus g cannot be known with greater accuracy than that of the fine structure constant. These surprising inconsistencies in the standard model data were discussed in detail [1, 12] and a brief summary is given here.

The fundamental constants of physics are agreed upon by treaty and are given on sites such as that of the National Institute for Standards and Technology (www.nist.gov). This site gives:

$$g(\text{exptl.}) = 2.0023193043718 \pm 0.0000000000075 \tag{2.206}$$

$$\hbar(\text{exptl}) = (6.6260693 \pm 0.0000011) \times 10^{-34} Js \tag{2.207}$$

$$e(\text{exptl.}) = (1.60217653 \pm 0.00000014) \times 10^{-19} C \tag{2.208}$$

$$c(\text{exact}) = 2.99792458 \times 10^{8} ms^{-1} \tag{2.209}$$

$$\epsilon_0(\text{exact}) = 8.854187817 \times 10^{-12} J^{-1}C^2m^{-1} \tag{2.210}$$

$$\mu_0(\text{exact}) = 4\pi \times 10^{-7} Js^2C^{-2}m^{-1} \tag{2.211}$$

with relative standard uncertainties. With a sufficiently precise value of:

$$\pi = 3.141592653590 \tag{2.212}$$

gives, from these data:

$$\alpha = 0.007297(34) \tag{2.213}$$

where the result has been rounded off to the relative standard uncertainty of the Planck constant h. This is an experimentally determined uncertainty. The theoretical value of g from ECE theory was found by using Eq. (2.213) in

$$g = 2\left(1 + \frac{\alpha}{4\pi}\right)^2 \tag{2.214}$$

and gives:

$$g(\text{ECE}) = 2.002323(49). \tag{2.215}$$

The experimental value of g is known to a much higher precision than the experimental value of h, and is:

$$g(\text{exptl.}) = 2.0023193043718 \pm 0.0000000000075. \tag{2.216}$$

It is seen that:

$$g(\text{ECE}) - g(\text{exptl.}) = 0.000004 \tag{2.217}$$

which is about the same order of magnitude as the experimental uncertainty of h. Therefore it was shown that ECE theory gives g as precisely as the experimental uncertainty in h will allow. The standard model literature was found to be severely self-inconsistent. For example a much used text by Atkins [29] gives h as:

$$h\,(\text{Atkins}) = 6.62818 \times 10^{-34} Js \tag{2.218}$$

without uncertainty estimates. This is different in the fourth decimal place from the NIST value given above, a discrepancy of four orders of magnitude. Despite this, Atkins gives:

$$\alpha(\text{Atkins}) = 0.00729351 \tag{2.219}$$

which claims to be different from Eq. (2.213) only in the sixth decimal place. Atkins gives the g factor of the electron as:

$$g(\text{Atkins}) = 2.002319314 \tag{2.220}$$

which is different from the NIST value in the eighth decimal place, while it is claimed at NIST that $g(\text{exp})$ from Eq. (2.216) is accurate to the twelfth decimal place. So there is another discrepancy of four orders of magnitude. Ryder on the other hand [18] gives:

$$g(\text{Ryder}) = 2.0023193048 \tag{2.221}$$

which is different from the NIST value in the tenth decimal place, a discrepancy of two orders of magnitude. One could try to explain these discrepancies by increasing accuracy of experimental method over the years, but there is no way in which QED can reproduce g to the tenth decimal place as claimed by Ryder. This is easily seen from the fact that g is calculated theoretically in QED from the fine structure constant, whose accuracy is limited by h as we have argued. There is also no way in which QED can be a fundamental theory as is often claimed in the standard model literature. This is again easily seen from the fact that QED has several assumptions extraneous to the theory of relativity [1, 12]. Examples are virtual particles, acausality (the electron can do what it likes, g backwards in time and so on), dimensional regularization, re-normalization and the hugely elaborate perturbation method known as the Feynman calculus. It is not known whether the series expansion used in the Feynman calculus converges. Its thousands of terms are just worked out by computer in the hope that it converges. In summary:

$$g(\text{Schwinger}) = 2 + \alpha/\pi = 2.002322(8) \tag{2.222}$$

$$g(\text{ECE}) = 2 + \alpha/\pi + \frac{\alpha^2}{8\pi^2} = 2.002323(49) \tag{2.223}$$

$$g(\text{exptl.}) = 2.0023193043718 \pm 0.0000000000075 \tag{2.224}$$

$$g(\text{Atkins}) = 2.002319314 \pm (?) \tag{2.225}$$

$$g(\text{Ryder}) = 2.0023193048 \pm (?) \tag{2.226}$$

and there is little doubt that other textbooks and sources give further different values of g to add to the confusion in the standard model literature. So where does this analysis leave the claims of QED? The Wolfram site claims that QED gives g using the series

$$g = 2\left(1 + \frac{\alpha}{2\pi} - 0.328\left(\frac{\alpha}{2\pi}\right)^2 + 1.181\left(\frac{\alpha}{\pi}\right)^3 - 1.510\left(\frac{\alpha}{\pi}\right)^4 + \ldots + 4.393 \times 10^{-12}\right) \tag{2.227}$$

which is derived from thousands of Feynman diagrams (sic). However, the numbers in Eq. (2.227) all come from the various assumptions of QED, none of which are present in Einsteinian relativity. The latter is causal and objective by construction. An even worse internal inconsistency emerges within the NIST site itself, because the fine structure constant is claimed to be:

$$\alpha(\text{NIST}) = (7.297352560 \pm 0.000000024) \times 10^{-3} \tag{2.228}$$

both experimentally and theoretically. This cannot be true because Eq. (2.228) is different in the eighth decimal place from Eq. (2.213), which is calculated with NIST's OWN data, Eqs. (2.206) to (2.211). So the NIST site is internally inconsistent to several orders of magnitude, because it is at the same time claimed that Eq. (2.228) is accurate to the tenth decimal place. From Eq. (2.207) however it is seen that h at NIST is accurate only to the sixth decimal place, which limits the accuracy of α to this, i.e. four orders of magnitude less precise than claimed.

The theoretical claim for the fine structure constant at NIST comes from QED, which his described as a theory in which an electron emits a virtual photon, which in turn emits virtual electron positron pairs. The virtual positron is attracted and the virtual electron is repelled from the real electron. This process results in a screened charge, a mathematical concept with a limiting value defined as the limit of zero momentum transfer or infinite distance. At high energies the fine structure constant drops to 1/128, and so is not a constant at all. It cannot therefore be claimed to be precise to the relative standard uncertainty of Eq. (2.228), taken directly from the NIST website itself. There is therefore no direct way of proving experimentally the existence of virtual electron positron pairs, or of virtual photons. The experimental claim for the fine structure constant at NIST comes from the quantum Hall effect combined with a calculable cross capacitor to measure standard resistance. The von Klitzing constant:

$$R_\kappa = \frac{\hbar}{e^2} = \frac{\mu_0 c}{2}(\text{sic}) \tag{2.229}$$

is used in this experimental determination.. However, this method is again limited by the experimental accuracy of h. The accuracy of e is only ten times better than h from NIST's own data, and R_κ cannot be more accurate than h. If α were really as accurate as claimed in Eq. (2.228), both h and e would have to be this accurate experimentally, and this is obviously not true.

In view of these severe inconsistencies in the standard model and in view of the many ad hoc and indeed unprovable assumptions of QED, it is considered that the so called "precision tests" of QED are of no utility and no meaning. These include the g factor of the electron, the Lamb shift, the Casimir effect, positronium, and so forth.

The ECE theory of these radiative corrections therefore set out to reproduce what is really known experimentally in the simplest way. These methods are of course those of William of Ockham and Francis Bacon. In the non-relativistic quantum approximation to ECE theory the Schrödinger equation was modified as follows [1,12]:

$$-\frac{\hbar^2}{2m}\nabla^2\left(\frac{\alpha}{2\pi}+\frac{\alpha^2}{16\pi^2}\right)\psi = \frac{e^2}{4\pi\epsilon_0}\left(\frac{1}{r}-\frac{1}{r+r(\text{vac})}\right)\psi \tag{2.230}$$

in which the effect of the vacuum potential is considered to be a shift in the electron to proton distance for each orbital of an atom or molecule, in the simplest case atomic hydrogen (H). Computer algebra was used to show that:

$$\frac{r(\text{vac})(2s)}{r+(r+r(\text{vac}))} - \frac{r(\text{vac})(2p_z, \cos\theta = 1)}{r(r+r(\text{vac}))} = \frac{1}{4\pi}\frac{\hbar}{mc}\frac{1}{r^2} \tag{2.231}$$

so that the simple ECE method of Eq. (2.230) gives the correct qualitative result observed first by Lamb in atomic H. This is known as the Lamb shift. Computer algebra was used to show that the ECE Lamb shift is:

$$\Delta E = \left(\frac{1}{16\pi^{3/2}}\frac{\alpha}{a}\frac{\hbar}{mc}\right)\frac{1}{r} = 0.0353\ cm^{-1} \tag{2.232}$$

in the approximation in which the angular dependence if the Lamb shift is not considered.

The potential energy of the unperturbed H atom in wave-numbers is:

$$V_0 = -\frac{\alpha}{r} \tag{2.233}$$

and the vacuum perturbs this as follows:

$$V = -\frac{\alpha}{r+r(\text{vac})}. \tag{2.234}$$

So the change in potential energy due to the vacuum (i.e. the radiative correction) is positive valued as follows:

$$\Delta V = |V - V_0| = \alpha\left(\frac{1}{r} - \frac{1}{r + r(\text{vac})}\right). \tag{2.235}$$

This equation was obtained by assuming that the Schrödinger equation of H in the presence of the radiative correction due to the vacuum is, to first order in α:

$$-\frac{\hbar^2}{2m}\left(1 + \frac{\alpha}{2\pi}\right)\nabla^2\psi - \frac{e^2}{4\pi\epsilon_0 r}\psi = E\psi \tag{2.236}$$

and that this is equivalent to:

$$-\frac{\hbar^2}{2m}\nabla^2\psi - \frac{e^2}{4\pi\epsilon_0(r + r(\text{vac}))}\psi = E\psi. \tag{2.237}$$

It was assumed that $r(\text{vac})$ is small enough to justify using the analytically known unperturbed wave-functions of H (ψ_0) to a good approximation. So:

$$\psi \sim \psi_0 \tag{2.238}$$

and:

$$\nabla^2\psi_0 = -\frac{4\pi mc}{\hbar}\left(\frac{1}{r} - \frac{1}{r + r(\text{vac})}\right)\psi_0. \tag{2.239}$$

Using computer algebra this approximation gives [1, 12]:

$$\frac{1}{r + r_{2p}(\text{vac})} - \frac{1}{r + r_{2s}(\text{vac})} = \frac{1}{2\pi^{3/2}}\frac{\hbar}{mc}\frac{1}{r^2}. \tag{2.240}$$

The change in potential energy due to the radiative correction of the vacuum is thus:

$$\Delta V = \frac{\alpha}{2\pi^{3/2}}\frac{\hbar}{mc}\frac{1}{r^2} \tag{2.241}$$

and the change in total energy is:

$$\Delta E = \frac{r}{2n^2 a}\Delta V = \left(\frac{1}{16\pi^{3/2}}\frac{\alpha}{a}\frac{\hbar}{mc}\right)\frac{1}{r} = 0.0353\ cm^{-1} \tag{2.242}$$

which is the Lamb shift of atomic H. Here:

$$r = 1.69 \times 10^{-7} m \tag{2.243}$$

From Eq. (240):

$$\frac{r_{2s}(\text{vac}) - r_{2p}(\text{vac})}{(r + r_{2p}(\text{vac}))(r + r_{2s}(\text{vac}))} = \frac{1}{2\pi^{3/2}}\frac{\hbar}{mc}\frac{1}{r^2}. \tag{2.244}$$

Eq. (238) implies:

$$r \gg r_{2s}(\text{vac}) \sim r_{2p}(\text{vac}) \tag{2.245}$$

so in this approximation Eq. (2.244) becomes:

$$r_{2s}(\text{vac}) - r_{2p}(\text{vac}) = \frac{1}{2\pi^{3/2}}\frac{\hbar}{mc} \tag{2.246}$$

i.e.

$$r_{2s}(\text{vac}) - r_{2p}(\text{vac}) = \frac{1}{4\pi^{5/2}} \frac{\hbar}{mc} \tag{2.247}$$

where the standard Compton wavelength is:

$$\frac{h}{mc} = 2.426 \times 10^{-12} m. \tag{2.248}$$

Thus we arrive at:

$$r_{2s}(\text{vac}) - r_{2p}(\text{vac}) = 3.48 \times 10^{-13} m. \tag{2.249}$$

This is a plausible result because the classical electron radius is:

$$r(\text{classical}) = \frac{1}{4\pi\epsilon_0} \frac{e^2}{mc^2} = 2.818 \times 10^{-15} m \tag{2.250}$$

and the Bohr radius is:

$$a = 5.292 \times 10^{-11} m. \tag{2.251}$$

So the radiative correction perturbs the electron orbitals by about ten times the classical radius of the electron and by orders less than the Bohr radius. The ECE theory also shows why the Lamb shift is constant as observed because for a given orientation:

$$\cos\theta = 1 \tag{2.252}$$

the shift is determined completely by 1/r within a constant of proportionality defined by:

$$\zeta = \frac{1}{32\pi^{3/2}} \frac{\alpha}{a} \frac{\hbar}{mc}. \tag{2.253}$$

The angular dependence of the Lamb shift in H was also considered [1, 12] and the method extended to the helium atom. Finally, consideration was given to how radiative corrections may be amplified by spin connection resonance.

Therefore in summary, the accuracy of the fine structure constant is determined experimentally by that of the Planck constant h. The LEAST accurately known constant determines the accuracy of the fine structure constant, as should be well known. There is no way that any theory can determine the fine structure constant more accurately than it is known experimentally. Therefore ECE theory sets out to use the experimental accuracy in α. The latter is determined by the accuracy in h as argued. This was done as simply as possible in accordance with Ockham's Razor. QED on the other hand is hugely elaborate, and its claims to be an accurate fundamental theory are unjustifiable. There can be no experimental justification for the existence of virtual particle pairs because of the gross internal inconsistencies in data reviewed in this section. Additionally, there are several ad hoc assumptions in the theory of QED itself.

2.9 Summary of advances made by ECE theory, and criticisms of the standard model

In this section a summary is given of the main advances of ECE theory over the past five years since inception in Spring 2003, and also a summary of implied criticisms of the current model of physics known as the standard model.

The major advantage of ECE theory is that it relies on the original principles of the theory and philosophy of relativity, without any extraneous input. This approach adheres therefore to the Ockham Razor of philosophy, the simpler the better. It also adheres to the principles of Francis Bacon, that every theory is tested experimentally, and not against another theory.

1. The inverse Faraday effect. This is described by the spinning of space-time and the B(3) field (see www.aias.us Omnia Opera) from first principles. In the standard model the effect cannot be described self consistently and cannot be described without an ad hoc conjugate product of non-linear optics. The latter is introduced phenomenologically in the standard model of non-linear optics, a theory of special relativity. In ECE theory the B(3) spin field indicates that optics and spectroscopy are parts of a generally covariant unified field theory (GCUFT).

2. The Aharonov-Bohm effects. These are described self consistently in ECE through the spin connection using the principles of general relativity. As shown in this review paper, the standard model description of the Aharonov-Bohm (AB) effects is at best controversial and at worst erroneous. A satisfactory description of the AB effects in ECE leads to a new understanding of quantum entanglement and one photon interferometry.

3. The polarization change in light deflected by gravitation. This is not described in the Einstein-Hilbert (EH) equation of the standard model because it is a purely kinematic equation relying on the gravitational attraction between a photon and a mass M, for example the solar mass. In ECE all the optical effects of gravitation are developed self consistently from the Bianchi identity of Cartan geometry.

4. The Faraday disk generator. This is described in ECE through the Cartan torsion of space-time introduced by mechanical spin, this concept is missing entirely from the standard model, which still cannot describe the 1831 Faraday disk generator.

5. The Sagnac effect and ring laser gyro. These are described again by the Cartan torsion of space-time introduced by spinning the platform of the Sagnac interferometer. The Sagnac effect is very difficult to understand using Maxwell-Heaviside theory, but is easily described in ECE theory. The latter offers a far simpler description than other available attempts at explaining the Sagnac effect of 1913.

6. The velocity curve of a spiral galaxy. This is described straightforwardly and simply in ECE theory by introducing again the concept of constant space-time torsion. The spiral galaxies main features cannot be described at all in the standard model. This is because the latter relies on an ad hoc "dark matter" that originates in the EH equation. The latter is self inconsistent as argued in this review paper.

7. The topological phases such as the Berry phase. These are derived in ECE from first principles, and are rigorously inter-related. In the standard model their description is incomplete, and in the case of the electromagnetic phase, erroneous.

8. The electromagnetic phase. This is described self consistently in terms of the B(3) spin field of ECE theory using general relativity. In the standard model the phase is incompletely determined mathematically, and violates parity in simple effects such as reflection.

9. Snell's law, reflection, refraction, diffraction, interferometry and related optical effects. These can be described correctly only in a GCUFT such as ECE. In the standard model the theory of reflection for example, does not fit with parity inversion symmetry due to the neglect of the B(3) spin field.

10. Improvements to the Heisenberg Uncertainty Principle. Various experiments have shown that the principle is incorrect by orders of magnitude, in ECE theory it is developed with causal and objective general relativity and the concept of quantum of action density.

11. The unification of wave mechanics and general relativity. This has been achieved straightforwardly in ECE theory through the use of Cartan geometry. In the standard model it is still not possible to make this basic unification. The Dirac, Proca and other wave equations are limits of the ECE wave equation, which is derived easily from the tetrad postulate of Cartan. So ECE allows the description of the effect of gravitation on such equations, and on such phenomena as the Sagnac effect. This is again not possible in the standard model.

12. Description of particle interaction. This description is achieved with simultaneous ECE equations without assuming the existence either of virtual particles or of the Higgs mechanism. The Higgs boson still has not been verified experimentally, and its energy is not defined theoretically.

13. The photon mass. The Proca equation is derived easily from Cartan geometry using the simple hypothesis that the potential is proportional to the Cartan tetrad. In the standard model the Proca equation is directly incompatible with gauge invariance, a fundamental self-inconsistency of the standard model, one of many self-inconsistencies.

14. Replacement of the gauge principle. The gauge principle is not tenable in a GCUFT such as ECE because the potential in ECE is physically meaningful as in Faraday's original electrotonic state. Abandonment of the gauge principle allows a return to the earlier concepts of relativity without introducing an ad hoc and abstract internal space as in Yang Mills theory. In ECE theory the tetrad postulate is invariant under the general coordinate transform, and this is the principle that governs the potential field in ECE.

15. Description of the electro-weak field without the Higgs mechanism. This becomes possible in a relatively straightforward manner by using simultaneous ECE equations. The Higgs mechanism is ad hoc, and to date unproven experimentally, indeed it is unprovable because an energy cannot be assigned to the Higgs boson. The Higgs boson, having no well defined energy, cannot be proven experimentally by particle collision methods, however powerful the accelerator. No sign of a Higgs boson was found at LEP, and to date no sign at the CERN heavy hadron collider.

16. Description of neutrino oscillations. This is a relatively simple exercise in ECE theory but in the standard model neutrino oscillations remained deeply controversial for years because of adherence to the assumption that the neutrino had no mass. In ECE all particles have mass - a fundamental requirement of relativity.

17. The generally covariant description of the laws of classical electrodynamics. These laws become laws of general relativity and a unified field theory, they are no longer laws of a Minkowski space-time as in the standard model. The concept of spin connection and spin connection resonance make important advances and potentially open up new sources of energy.

18. Derivation of the quark model from general relativity. This has been achieved in ECE theory by using an SU(n) representation space in the wave and field equations. In the standard model the quark theory is one of special relativity. QCD relies on ad hoc concepts such as re-normalization, which as argued in section 2.8, are not internally consistent with data. The situation in QCD is worse than that in QED.

19. Derivation of the quantum theory of electrodynamics. This is achieved using the wave equation and the ECE hypothesis, resulting in a generally covariant version of the Proca equation with non-zero photon mass. In so doing a minimum particle volume is always present, so there are no point particles and no need for re-normalization. Feynman's QED is abandoned as described in Section 2.8.

20. The origin of particle spin. This is traced to geometry and particle spins of all kinds are successfully incorporated into general relativity. This is not possible with the EH equation, which has been shown to be fundamentally flawed.

21. Development of cosmology. The major advantage of considering the Cartan torsion becomes abundantly clear in cosmology, in particular the explanation of the spiral galaxies. Cosmology based on the EH equation has been shown to be meaningless in several different ways.

22. No singularities. This is a flawed concept introduced by incorrect solutions of the EH equation. The latter is itself inconsistent with the Bianchi identity. In ECE theory the concept of Big Bang is replaced with the steady state universe with local oscillations. Similarly there are no black holes and no dark matter. Applications of experimentally untestable string theory to these concepts multiplies the heavily criticized obscurantism of modern physics.

23. Explanation of the red shift. This is a simple optical effect in ECE theory, there can be different red shifts in equidistant objects. ECE also offers a new explanation of the background radiation if indeed it is not an artifact of the Earth's atmosphere as some scholars now think.

24. Spin connection resonance. This concept is made possible in ECE and has been offered as an explanation of Tesla's well known giant resonances and similar reports of over a century of work. The latter cannot be explained in the standard model yet is potentially a source of major new energy.

25. Spinning space-time. This is a key new concept of electrodynamics, akin to curving space-time in gravitation. ECE has made the major discovery that the two concepts are linked ineluctably in relativity, and this has led to the abandonment of the EH equation. A suggested replacement of the equation has been made in recent work.

26. Counter gravitation. It has been shown that this is feasible only by using resonance methods based again on the spin connection and the interaction of gravitation and electromagnetism. It needs a GCUFT such as ECE to begin to describe this interaction of the fundamental fields of force.

27. Gravitational dynamics. These are developed in ECE in the same way as electrodynamics. For example it is relatively easy to show that there is a gravitational equivalent of the Faraday law of induction, as indeed observed recently. A new approach to the derivation of the acceleration due to gravity has also been made possible, an approach based on the rigorous Bianchi identity given by Cartan.

28. Quantum entanglement. These well known quantum effects can be understood using the spin connection of ECE in a similar way to the AB effects. Similarly the argument can be extended to such phenomena as one photon Young interferometry. In the standard model they are very difficult to understand because of the use of a Minkowski space-time with no connection. In the standard model these are mysterious effects with many offered explanations, none convincing.

29. Superconductivity and related fields. The equations governing the behavior of classes of materials are all derived in ECE from geometry, so there is an overall self-consistency which is often missing in the standard model. For example plasma, semiconductors, superconductors, and so forth.

30. Quantum field theory. This is developed in ECE entirely without he use of string theory or super-symmetry. String theory in particular has been heavily criticized because it cannot be tested experimentally and makes no new predictions at all. Such matters as photon mass theory, canonical quantization, and creation annihilation operator theory are all improved by ECE theory.

31. Radiative Corrections. These are understood in a far simpler way in ECE theory as discussed in Section 2.8. The claims of QED theory have been shown to be false by several orders of magnitude, and the complacency of the standard physics community heavily criticized thereby.

32. Fermion resonance. New methods of detecting and developing fermion resonance have been developed and it is shown that such resonance can be induced without the use of magnets. This method is known as radiatively induced fermion resonance (RFR). It has been clearly understood to be due to the B(3) field.

33. Ubiquitous B(3) field. It has been shown that the B(3) field is the one responsible for the general relativistic description of the electromagnetic phase, so it occurs throughout optics and spectroscopy, in everyday phenomena such as reflection.

34. Fundamental advances in geometry. In the course of developing ECE theory it has been shown that there is only one Bianchi identity, not two unrelated identities used in the standard model. It has also been shown rigorously in many ways that the Bianchi identity has a Hodge dual. These properties lead to field equations with duality symmetry. Such a symmetry is not present in the standard model.

35. Self consistency of Cartan's geometry. This has been tested in many ways, and it has been shown that the tetrad postulate is rigorously self consistent and fundamental to physics. Numerous tests of self consistency have been made.

36. Development of gravitational relativity. It has been shown that the correct description of gravitation requires the Bianchi identity of Cartan, which links torsion to curvature. The Bianchi identity used by Einstein has been shown to be incomplete, and using computer algebra, it has been shown that the EH equation is inconsistent with the use of a Christoffel connection and symmetric metric. It has also been shown that claimed solutions of the EH equation are often incorrect mathematically. Finally it has been shown that the Ricci flat space-time is incompatible with the Einsteinian equivalence principle. Therefore the standard model literature has to be read with considerable caution. Many claims of the standard model have not stood up to scrutiny, whereas ECE has developed strongly.

Acknowledgments

The British Government is thanked for the award of a Civil List Pension, and the staff of AIAS for five years of voluntary work on ECE theory and websites. The staff of the Telesio-Galilei Association are thanked for appointments and logistical support.

2.10 Appendix 1: Homogeneous Maxwell-Heaviside equations

In the first of several technical appendices it is shown how to translate the homogeneous Maxwell-Heaviside (MH) from tensor to vector notation, giving details that are rarely found in textbooks. In tensor notation the equation is:

$$\partial_\mu \widetilde{F}^{\mu\nu} = 0 \tag{2.1}$$

and involves the Hodge dual of the 4 x 4 field tensor, defined as follows:

$$\widetilde{F}_{\mu\nu} = \frac{1}{2}\epsilon_{\mu\nu\rho\sigma}F^{\rho\sigma}. \tag{2.2}$$

Indices are raised using the Minkowski metric:

$$\widetilde{F}^{\mu\nu} = g^{\mu\kappa}g^{\nu\rho}\widetilde{F}_{\kappa\rho} \tag{2.3}$$

where:

$$g_{\mu\nu} = g^{\mu\nu} = \begin{bmatrix} 1 & 0 & 0 & 0 \\ 0 & -1 & 0 & 0 \\ 0 & 0 & -1 & 0 \\ 0 & 0 & 0 & -1 \end{bmatrix}. \tag{2.4}$$

Therefore the Hodge dual is:

$$\widetilde{F}^{\mu\nu} = \begin{bmatrix} 0 & cB^1 & cB^2 & cB^3 \\ -cB^1 & 0 & -E^3 & E^2 \\ -cB^2 & E^3 & 0 & -E^1 \\ -cB^3 & -E^2 & E^1 & 0 \end{bmatrix} \tag{2.5}$$

For example:

$$\widetilde{F}_{01} = \frac{1}{2}(\epsilon_{0123}F^{23} + \epsilon_{0132}F^{32}) = F^{23} \tag{2.6}$$

and

$$\widetilde{F}^{01} = g^{00}g^{11}\widetilde{F}_{01} = -\widetilde{F}_{10}. \tag{2.7}$$

The homogeneous laws of classical electrodynamics are the Gauss law and Faraday law of induction. They are obtained as follows by choice of indices. The Gauss law is obtained by choosing:

$$\nu = 0 \tag{2.8}$$

and so

$$\partial_1\widetilde{F}^{10} + \partial_2\widetilde{F}^{20} + \partial_3\widetilde{F}^{30} = 0. \tag{2.9}$$

In vector notation this is

$$\boldsymbol{\nabla}\cdot\boldsymbol{B} = 0. \tag{2.10}$$

The Faraday law of induction is obtained by choosing:

$$\nu = 1, 2, 3 \tag{2.11}$$

and is three component equations:

$$\partial_0 \widetilde{F}^{01} + \partial_2 \widetilde{F}^{21} + \partial_3 \widetilde{F}^{31} = 0 \tag{2.12}$$

$$\partial_0 \widetilde{F}^{02} + \partial_1 \widetilde{F}^{12} + \partial_3 \widetilde{F}^{32} = 0 \tag{2.13}$$

$$\partial_0 \widetilde{F}^{03} + \partial_1 \widetilde{F}^{13} + \partial_2 \widetilde{F}^{23} = 0. \tag{2.14}$$

These can be condensed into one vector equation, which is

$$\boldsymbol{\nabla} \times \boldsymbol{E} + \frac{\partial \boldsymbol{B}}{\partial t} = \boldsymbol{0}. \tag{2.15}$$

The differential form, tensor and vector notations are summarized as follows:

$$\begin{aligned} d \wedge F = 0 \rightarrow \partial_\mu \widetilde{F}^{\mu\nu} = 0 \rightarrow \boldsymbol{\nabla} \cdot \boldsymbol{B} = 0 \\ \boldsymbol{\nabla} \times \boldsymbol{E} + \frac{\partial \boldsymbol{B}}{\partial t} = \boldsymbol{0} \end{aligned} \tag{2.16}$$

The homogeneous laws of classical electrodynamics are most elegantly represented by the differential form notation, but most usefully represented by the vector notation.

2.11 Appendix 2: The inhomogeneous equations

The inhomogeneous laws are the Coulomb law and the Ampère-Maxwell law. In tensor notation they are condensed into one equation:

$$\partial_\mu F^{\mu\nu} = \frac{1}{\epsilon_0} J^\nu \tag{2.1}$$

where the charge current density is:

$$J^\nu = \left(\rho, \frac{\boldsymbol{J}}{c}\right) \tag{2.2}$$

and where the partial derivative is:

$$\partial_\mu = \left(\frac{1}{c}\frac{\partial}{\partial t}, \frac{\partial}{\partial X}, \frac{\partial}{\partial Y}, \frac{\partial}{\partial Z}\right) \tag{2.3}$$

The field tensor is:

$$F^{\mu\nu} = \begin{bmatrix} 0 & -E^1 & -E^2 & -E^3 \\ E^1 & 0 & -cB^3 & cB^2 \\ E^2 & cB^3 & 0 & -cB^1 \\ E^3 & -cB^2 & cB^1 & 0 \end{bmatrix} = \begin{bmatrix} 0 & F^{01} & F^{02} & F^{03} \\ F^{10} & 0 & F^{12} & F^{13} \\ F^{20} & F^{21} & 0 & F^{23} \\ F^{30} & F^{31} & F^{32} & 0 \end{bmatrix} \tag{2.4}$$

and in S.I. units:

$$\epsilon_0 \mu_0 = \frac{1}{c^2}. \tag{2.5}$$

In this notation:

$$\left.\begin{aligned} E_X &= E^1 = F^{10}, \\ E_Y &= E^2 = F^{20}, \\ E_Z &= E^3 = F^{30}, \end{aligned}\right\} \tag{2.6}$$

and so on. The Coulomb law is obtained from choosing:

$$\nu = 0 \tag{2.7}$$

so that:

$$\partial_1 F^{10} + \partial_2 F^{20} + \partial_3 F^{30} = \frac{1}{\epsilon_0} J^0. \tag{2.8}$$

In vector component notation this is:

$$\frac{\partial E_X}{\partial X} + \frac{\partial E_Y}{\partial Y} + \frac{\partial E_Z}{\partial Z} = \frac{1}{\epsilon_0}\rho \tag{2.9}$$

which in vector notation is:

$$\boldsymbol{\nabla} \cdot \boldsymbol{E} = \frac{\rho}{\epsilon_0}. \tag{2.10}$$

The Ampère-Maxwell law is obtained from choosing

$$\nu = 1, 2, 3 \tag{2.11}$$

which gives three equations:

$$\partial_0 F^{01} + \partial_2 F^{21} + \partial_3 F^{31} = \frac{1}{\epsilon_0} J^1 \tag{2.12}$$

$$\partial_0 F^{02} + \partial_1 F^{12} + \partial_3 F^{32} = \frac{1}{\epsilon_0} J^2 \tag{2.13}$$

$$\partial_0 F^{03} + \partial_1 F^{13} + \partial_2 F^{23} = \frac{1}{\epsilon_0} J^3. \tag{2.14}$$

In vector component notation these are:

$$-\frac{1}{c}\frac{\partial E_X}{\partial t} + c\left(\frac{\partial B_Z}{\partial Y} - \frac{\partial B_Y}{\partial Z}\right) = \frac{1}{\epsilon_0} J_X \tag{2.15}$$

$$-\frac{1}{c}\frac{\partial E_Y}{\partial t} + c\left(\frac{\partial B_X}{\partial Z} - \frac{\partial B_Z}{\partial X}\right) = \frac{1}{\epsilon_0} J_Y \tag{2.16}$$

$$-\frac{1}{c}\frac{\partial E_Z}{\partial t} + c\left(\frac{\partial B_Y}{\partial X} - \frac{\partial B_X}{\partial Y}\right) = \frac{1}{\epsilon_0} J_Z. \tag{2.17}$$

The definition of the vector curl is

$$\boldsymbol{\nabla} \times \boldsymbol{B} = \begin{vmatrix} \boldsymbol{i} & \boldsymbol{j} & \boldsymbol{k} \\ \partial/\partial Z & \partial/\partial Y & \partial/\partial Z \\ B_X & B_Y & B_Z \end{vmatrix} \tag{2.18}$$

$$= \left(\frac{\partial B_Z}{\partial Y} - \frac{\partial B_Y}{\partial Z}\right)\boldsymbol{i} - \left(\frac{\partial B_Z}{\partial X} - \frac{\partial B_X}{\partial Z}\right)\boldsymbol{j} + \left(\frac{\partial B_Y}{\partial X} - \frac{\partial B_X}{\partial Y}\right)\boldsymbol{k},$$

so it is seen that the three equations (2.15) to (2.17) can be condensed into one vector equation:

$$\boldsymbol{\nabla} \times \boldsymbol{B} - \frac{1}{c^2}\frac{\partial \boldsymbol{E}}{\partial t} = \mu_0 \boldsymbol{J} \tag{2.19}$$

which is the Ampère-Maxwell Law. The differential form, tensor and vector formulations of the inhomogeneous laws of standard model classical electrodynamics are summarized as follows:

$$d \wedge \widetilde{F} = \frac{J}{\epsilon_0} \rightarrow \partial_\mu F^{\mu\nu} = \frac{J^\nu}{\epsilon_0} \rightarrow \boldsymbol{\nabla} \cdot \boldsymbol{E} = \frac{\rho}{\epsilon_0}, \tag{2.20}$$

$$\boldsymbol{\nabla} \times \boldsymbol{B} - \frac{1}{c^2}\frac{\partial \boldsymbol{E}}{\partial t} = \mu_0 \boldsymbol{J}.$$

2.12 Appendix 3: Some examples of Hodge duals in Minkowski space-time

In Minkowski space-time the Hodge dual of a rank two anti-symmetric tensor (two-form) in four dimensions is defined by:

$$\widetilde{F}_{\mu\nu} = \frac{1}{2}\epsilon_{\mu\nu\rho\sigma}F^{\rho\sigma}. \tag{2.1}$$

For example, the B(3) field is defined by:

$$F^{\mu\nu} = \begin{bmatrix} 0 & 0 & 0 & 0 \\ 0 & 0 & -cB^{(3)} & 0 \\ 0 & cB^{(3)} & 0 & 0 \\ 0 & 0 & 0 & 0 \end{bmatrix} \tag{2.2}$$

so its Hodge dual is:

$$\widetilde{F}^{\mu\nu} = \begin{bmatrix} 0 & 0 & 0 & cB^{(3)} \\ 0 & 0 & 0 & 0 \\ 0 & 0 & 0 & 0 \\ -cB^{(3)} & 0 & 0 & 0 \end{bmatrix}. \tag{2.3}$$

It can be seen that the Hodge dual of the B(3) field does not imply the existence of an E(3) field, it is a re-arrangement of matrix elements. There appears to be no experimental evidence for the existence of a radiated E(3) field. In other words there is no electric equivalent of the inverse Faraday effect, and there is no electric equivalent of the Faraday effect.

The radiated B(3) field is generated by the spin connection, the static magnetic field of the standard model is defined without the spin connection as follows:

$$\boldsymbol{B} = \boldsymbol{\nabla} \times \boldsymbol{A}. \tag{2.4}$$

In tensor form the static magnetic field is:

$$F^{\mu\nu} = \begin{bmatrix} 0 & 0 & 0 & 0 \\ 0 & 0 & -cB_Z & cB_Y \\ 0 & cB_Z & 0 & -cB_X \\ 0 & -cB_Y & cB_X & 0 \end{bmatrix} \tag{2.5}$$

whose Hodge dual is:

$$\widetilde{F}^{\mu\nu} = \begin{bmatrix} 0 & cB_X & cB_Y & cB_Z \\ -cB_X & 0 & 0 & 0 \\ -cB_Y & 0 & 0 & 0 \\ -cB_Z & 0 & 0 & 0 \end{bmatrix}. \tag{2.6}$$

Again, the Hodge dual does not generate an electric field. In ECE theory the magnetic field in vector notation always includes the spin connection vector as follows:

$$\boldsymbol{B} = \boldsymbol{\nabla} \times \boldsymbol{A} - \boldsymbol{\omega} \times \boldsymbol{A} \tag{2.7}$$

and this is true for all types of magnetic field.

2.13 Appendix 4: Standard tensorial formulation of the homogeneous Maxwell-Heaviside field equations

The standard tensorial formulation developed in this appendix is:

$$\partial_\mu \widetilde{F}^{\mu\nu} = \partial^\mu \widetilde{F}_{\mu\nu} = 0 \tag{2.1}$$

and is needed as a baseline for the development of ECE theory. The field tensor is defined as:

$$F^{\mu\nu} = \begin{bmatrix} 0 & cB^1 & cB^2 & cB^3 \\ -cB^1 & 0 & -E^3 & E^2 \\ -cB^2 & E^3 & 0 & -E^1 \\ -cB^3 & -E^2 & E^1 & 0 \end{bmatrix}. \tag{2.2}$$

where, in standard covariant - contravariant notation and in S.I. units:

$$\partial_\mu = \left(\frac{1}{c}\frac{\partial}{\partial t}, \frac{\partial}{\partial X}, \frac{\partial}{\partial Y}, \frac{\partial}{\partial Z}\right), \tag{2.3}$$

$$\partial^\mu = \left(\frac{1}{c}\frac{\partial}{\partial t}, -\frac{\partial}{\partial X}, -\frac{\partial}{\partial Y}, -\frac{\partial}{\partial Z}\right), \tag{2.4}$$

$$x^\mu = (ct, X, Y, Z), \tag{2.5}$$

$$x_\mu = (ct, -X, -Y, -Z). \tag{2.6}$$

The metric and inverse metric tensors in Minkowski space-time are equal, and are given by:

$$g_{\mu\nu} = g^{\mu\nu} = \begin{bmatrix} 1 & 0 & 0 & 0 \\ 0 & -1 & 0 & 0 \\ 0 & 0 & -1 & 0 \\ 0 & 0 & 0 & -1 \end{bmatrix}. \tag{2.7}$$

Indices are raised and lowered with the metric, for example:

$$\widetilde{F}_{\mu\nu} = g_{\mu\rho}\, g_{\nu\sigma}\, \widetilde{F}^{\rho\sigma} \tag{2.8}$$

where

$$g_{00} = 1, g_{11} = g_{22} = g_{33} = -1 \tag{2.9}$$

and so on. Therefore:

$$\widetilde{F}_{01} = g_{00}\, g_{11} F^{01} = -\widetilde{F}^{01},\ \widetilde{F}_{02} = -\widetilde{F}^{02}, \widetilde{F}_{03} = -\widetilde{F}^{03} \tag{2.10}$$

and so on. Therefore:

$$\begin{gathered} \widetilde{F}^{\mu\nu} = \begin{bmatrix} 0 & cB_X & cB_Y & cB_Z \\ -cB_x & 0 & -E_Z & E_Y \\ -cB_Y & E_Z & 0 & -E_X \\ -cB_Z & -E_Y & E_X & 0 \end{bmatrix}, \\ \widetilde{F}_{\mu\nu} = \begin{bmatrix} 0 & -cB_X & -cB_Y & -cB_Z \\ cB_X & 0 & -E_Z & E_Y \\ cB_Y & E_Z & 0 & -E_X \\ cB_Z & -E_Y & E_X & 0 \end{bmatrix}. \end{gathered} \tag{2.11}$$

If the field tensor is defined with raised indices then the Gauss law is given by:

$$\partial_1 \widetilde{F}^{10} + \partial_2 \widetilde{F}^{20} + \partial_3 \widetilde{F}^{30} = 0 \tag{2.12}$$

i.e.:

$$-\nabla \cdot \boldsymbol{B} = 0 \tag{2.13}$$

and the Faraday law of induction is given by

$$\partial_0 \widetilde{F}^{01} + \partial_2 \widetilde{F}^{21} + \partial_3 \widetilde{F}^{31} = 0 \tag{2.14}$$

$$\partial_0 \widetilde{F}^{02} + \partial_1 \widetilde{F}^{12} + \partial_3 \widetilde{F}^{32} = 0 \tag{2.15}$$

$$\partial_0 \widetilde{F}^{03} + \partial_1 \widetilde{F}^{13} + \partial_2 \widetilde{F}^{23} = 0 \tag{2.16}$$

i.e.

$$\nabla \times \boldsymbol{E} + \frac{\partial \boldsymbol{B}}{\partial t} = \mathbf{0}. \tag{2.17}$$

In almost all textbooks the Gauss law is written as:

$$\nabla \cdot \boldsymbol{B} = 0, \tag{2.18}$$

but the above is the rigorously correct result.

Similarly if the field tensor is written with lowered indices, : i.e.:

$$\partial^{\mu} \widetilde{F}_{\mu\nu} = 0 \tag{2.19}$$

the rigorously correct result is:

$$\begin{aligned} -\nabla \cdot \boldsymbol{B} &= 0 \\ -\left(\nabla \times \boldsymbol{E} + \frac{\partial \boldsymbol{B}}{\partial t}\right) &= \mathbf{0} \end{aligned} \tag{2.20}$$

The minus signs are always omitted in textbook material.

If the field tensor is defined with indices raised:

$$\partial_{\mu} F^{\mu\nu} = \frac{J^{\nu}}{\epsilon_0} \tag{2.21}$$

where:

$$F^{\mu\nu} = \frac{1}{2} \epsilon^{\mu\nu\rho\sigma} \widetilde{F}_{\rho\sigma}. \tag{2.22}$$

The totally anti-symmetric unit tensor in four dimensions has elements:

$$\begin{aligned} \epsilon^{0123} &= -\epsilon^{1230} &= \epsilon^{2301} &= -\epsilon^{3012} &= 1 \\ \epsilon^{1023} &= -\epsilon^{2130} &= \epsilon^{3201} &= -\epsilon^{0312} &= -1 \\ \epsilon^{1032} &= -\epsilon^{2103} &= \epsilon^{3210} &= -\epsilon^{0321} &= 1 \\ \epsilon^{1302} &= -\epsilon^{2013} &= \epsilon^{3120} &= -\epsilon^{0231} &= -1 \end{aligned} \tag{2.23}$$

So for example:

$$
\begin{aligned}
F^{01} &= \frac{1}{2}\left(\epsilon^{0123}\widetilde{F}_{23} + \epsilon^{0132}\widetilde{F}_{32}\right) = \widetilde{F}_{23} = -E_X \\
F^{02} &= \frac{1}{2}\left(\epsilon^{0231}\widetilde{F}_{31} + \epsilon^{0213}\widetilde{F}_{13}\right) = \widetilde{F}_{31} = -E_Y \\
F^{03} &= \frac{1}{2}\left(\epsilon^{0312}\widetilde{F}_{12} + \epsilon^{0321}\widetilde{F}_{21}\right) = \widetilde{F}_{12} = -E_Z \\
F^{23} &= \frac{1}{2}\left(\epsilon^{2301}\widetilde{F}_{01} + \epsilon^{2310}\widetilde{F}_{10}\right) = \widetilde{F}_{01} = -cB_X \\
F^{13} &= \frac{1}{2}\left(\epsilon^{1302}\widetilde{F}_{02} + \epsilon^{1320}\widetilde{F}_{20}\right) = -\widetilde{F}_{02} = cB_Y \\
F^{12} &= \frac{1}{2}\left(\epsilon^{1230}\widetilde{F}_{30} + \epsilon^{1203}\widetilde{F}_{03}\right) = \widetilde{F}_{03} = -cB_Z
\end{aligned}
\tag{2.24}
$$

Therefore:

$$
F^{\mu\nu} = \begin{bmatrix} 0 & -E_X & E_Y & -E_Z \\ E_X & 0 & -cB_Z & cB_Y \\ E_Y & cB_Z & 0 & -cB_X \\ E_Z & -cB_Y & cB_X & 0 \end{bmatrix} = \begin{bmatrix} 0 & -E^1 & -E^2 & -E^3 \\ E^1 & 0 & -cB^3 & cB^2 \\ E^2 & cB^3 & 0 & -cB^1 \\ E^3 & -cB^2 & cB^1 & 0 \end{bmatrix}. \tag{2.25}
$$

The charge current density is:

$$
J^\nu = \left(\rho, \frac{J}{c}\right). \tag{2.26}
$$

The Coulomb law is:

$$
\partial_1 F^{10} + \partial_2 F^{20} + \partial_3 F^{30} = \frac{1}{\epsilon_0} J^0 = \frac{\rho}{\epsilon_0} \tag{2.27}
$$

which in vector notation is:

$$
\boldsymbol{\nabla} \cdot \boldsymbol{E} = \frac{\rho}{\epsilon_0}. \tag{2.28}
$$

The Ampère-Maxwell law is:

$$
\partial_0 F^{01} + \partial_2 F^{21} + \partial_3 F^{31} = J^1/\epsilon_0 \tag{2.29}
$$

$$
\partial_0 F^{02} + \partial_1 F^{12} + \partial_3 F^{32} = J^2/\epsilon_0 \tag{2.30}
$$

$$
\partial_0 F^{03} + \partial_1 F^{13} + \partial_2 F^{23} = J^3/\epsilon_0 \tag{2.31}
$$

i.e.:

$$
-\frac{1}{c}\frac{\partial E_X}{\partial t} + c\left(\frac{\partial B_Z}{\partial Y} - \frac{\partial B_Y}{\partial Z}\right) = \frac{1}{\epsilon_0} J_X \tag{2.32}
$$

$$
-\frac{1}{c}\frac{\partial E_Y}{\partial t} + c\left(\frac{\partial B_X}{\partial Z} - \frac{\partial B_Z}{\partial X}\right) = \frac{1}{\epsilon_0} J_Y \tag{2.33}
$$

$$
-\frac{1}{c}\frac{\partial E_Z}{\partial t} + c\left(\frac{\partial B_Y}{\partial X} - \frac{\partial B_X}{\partial Y}\right) = \frac{1}{\epsilon_0} J_Z \tag{2.34}
$$

which is:

$$
\boldsymbol{\nabla} \times \boldsymbol{B} - \frac{1}{c^2}\frac{\partial \boldsymbol{E}}{\partial t} = \mu_0 \, \boldsymbol{J}. \tag{2.35}
$$

Therefore the standard adopted is:

$$\partial_\mu F^{\mu\nu} = \frac{1}{\epsilon_0} J^\nu \rightarrow \nabla \cdot \boldsymbol{E} = \rho / \epsilon_0 \tag{2.36}$$

$$\nabla \times \boldsymbol{B} - \frac{1}{c^2}\frac{\partial \boldsymbol{E}}{\partial t} = \mu_0 \, \boldsymbol{J}.$$

To be precisely correct therefore, the tensorial formulation of the four laws of electrodynamics is:

$$\partial_\mu F^{\mu\nu} = \frac{1}{\epsilon_0} J^\nu \tag{2.37}$$

$$-\partial^\mu \widetilde{F}_{\mu\nu} = 0 \tag{2.38}$$

where:

$$F^{\mu\nu} = \begin{bmatrix} 0 & -E^1 & -E^2 & -E^3 \\ E^1 & 0 & -cB^3 & cB^2 \\ E^2 & cB^3 & 0 & -cB^1 \\ E^3 & -cB^2 & cB^1 & 0 \end{bmatrix} \tag{2.39}$$

and

$$\widetilde{F}^{\mu\nu} = \begin{bmatrix} 0 & -cB^1 & -cB^2 & -cB^3 \\ cB^1 & 0 & -E^3 & E^2 \\ cB^2 & E^3 & 0 & -E^1 \\ -cB^3 & -E^2 & E^1 & 0 \end{bmatrix}. \tag{2.40}$$

In free space:

$$\partial_\mu F^{\mu\nu} = 0, \tag{2.41}$$

$$-\partial^\mu \widetilde{F}_{\mu\nu} = 0. \tag{2.42}$$

The free space equations are duality invariant under:

$$F^{\mu\nu} \leftrightarrow \widetilde{F}_{\mu\nu} \tag{2.43}$$

i.e.:

$$E_X \leftrightarrow cB_X, E_Y \leftrightarrow cB_Y, E_Z \leftrightarrow cB_Z. \tag{2.44}$$

The Hodge dual transform is:

$$F^{\mu\nu} = \frac{1}{2}\epsilon^{\mu\nu\rho\sigma}\widetilde{F}_{\rho\sigma} \tag{2.45}$$

and can be summarized as:

The presence of matter and charge-current density breaks the duality symmetry, or duality invariance.

2.14 Appendix 5: Illustrating the meaning of the connection with rotation in a plane

Consider the clockwise rotation in a plane of a vector V^1 to V^2 as in Fig. 2.1. This rotation is carried out by moving the vector and keeping the frame of reference fixed. This process is equivalent to keeping the vector fixed and rotating the frame of reference anti-clockwise through an equal angle θ. In Cartesian coordinates (Fig. 2.1):

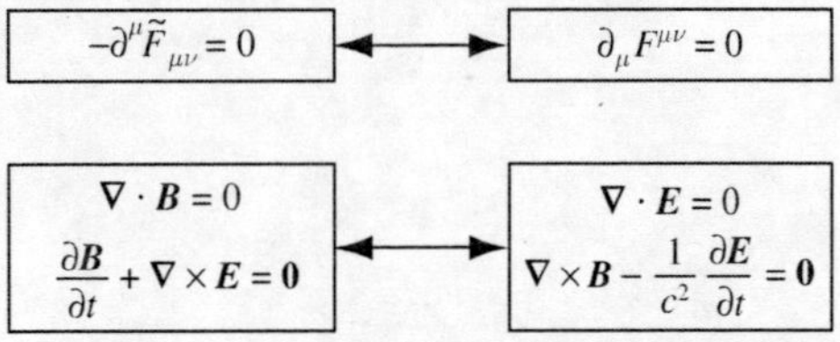

Figure 2.3: Homogeneous ECE Field Equation.

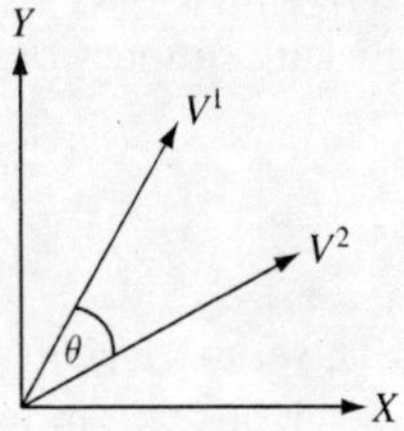

Figure 2.4: Rotation of a Vector in a Plane.

$$\boldsymbol{V}^1 = V_X^1\boldsymbol{i} + V_Y^1\boldsymbol{j} \tag{2.1}$$
$$\boldsymbol{V}^2 = V_X^2\boldsymbol{i} + V_Y^2\boldsymbol{j} \tag{2.2}$$

where:

$$|\boldsymbol{V}^1| = |\boldsymbol{V}^2|, \tag{2.3}$$

$$|\boldsymbol{V}^1| = (V_X^{1\ 2} + V_Y^{1\ 2})^{\frac{1}{2}}, \tag{2.4}$$

$$|\boldsymbol{V}^2| = (V_X^{2\ 2} + V_Y^{2\ 2})^{\frac{1}{2}}. \tag{2.5}$$

This is a rotation in which the frame is fixed, i.e. the Cartesian unit vectors i and j do not change. The rotation could equally well be represented by:

$$\boldsymbol{V}^1 = V_X\boldsymbol{i}_1 + V_Y\boldsymbol{j}_1, \tag{2.6}$$

$$\boldsymbol{V}^2 = V_X\boldsymbol{i}_2 + V_Y\boldsymbol{j}_2, \tag{2.7}$$

and in this case the vector is fixed and the frame rotated anti-clockwise. We now have:

$$|\boldsymbol{V}^1| = |\boldsymbol{V}^2| = (V_X^2 + V_Y^2)^{\frac{1}{2}} \tag{2.8}$$

because:

$$\left.\begin{array}{r} \boldsymbol{i}_1\cdot\boldsymbol{i}_1 = \boldsymbol{i}_2\cdot\boldsymbol{i}_2 = 1 \\ \boldsymbol{j}_1\cdot\boldsymbol{j}_1 = \boldsymbol{j}_2\cdot\boldsymbol{j}_2 = 1 \end{array}\right\}. \tag{2.9}$$

The invariance under rotation of the complete vector field is true in both cases:

$$\begin{array}{ll} \text{a)} & V^{12} = V_X^{12} + V_Y^{12} = V_X^{22} + V_Y^{22} = V^{22}, \\ \text{b)} & V^{12} = V_X^2 + V_Y^2 = V^{22}. \end{array} \tag{2.10}$$

The rotation can also be represented by:

$$\begin{bmatrix} V_X^1 \\ V_Y^1 \\ V_Z^1 \end{bmatrix} = \begin{bmatrix} \cos\theta & \sin\theta & 0 \\ -\sin\theta & \cos\theta & 0 \\ 0 & 0 & 1 \end{bmatrix} \begin{bmatrix} V_X^2 \\ V_Y^2 \\ V_Z^2 \end{bmatrix} \tag{2.11}$$

i.e.:

$$V_X^1 = V_X^2 \cos\theta + V_Y^2 \sin\theta \tag{2.12}$$

$$V_Y^1 = -V_X^2 \sin\theta + V_Y^2 \cos\theta \tag{2.13}$$

$$V_Z^1 = V_Z^2. \tag{2.14}$$

These equations are usually interpreted as the vector rotated clockwise with fixed frame. However they are also true for a fixed vector and frame rotated anti-clockwise. So this is an example of the frame itself moving. Therefore a connection can be defined because the connection determines how the frame itself moves. The general rule for covariant derivative is:

$$D_\nu V^\mu = \partial_\nu V^\mu + \Gamma^\mu{}_{\lambda\nu} V^\lambda. \tag{2.15}$$

This equation means that D_ν acting on V^μ is the four derivative ∂_ν plus the term $\Gamma^\mu{}_{\lambda\nu}V^\lambda$. The three index symbol is referred to as "the connection", and describes the movement of the frame itself. The latter produces, for a given ν:

$$U^\mu = \Gamma^\mu{}_\lambda V^\lambda. \tag{2.16}$$

It is seen that Eq. (2.11) is an example of Eq. (2.16) in three dimensions, *X*, *Y*, and *Z*. So for a rotation of the frame anti-clockwise in three dimensions about the Z axis the matrix is the rotation matrix:

$$\Gamma^\mu{}_\lambda = \begin{bmatrix} \cos\theta & \sin\theta & 0 \\ -\sin\theta & \cos\theta & 0 \\ 0 & 0 & 1 \end{bmatrix}. \tag{2.17}$$

Thus:

$$\left.\begin{aligned} \Gamma^1{}_1 = \cos\theta, \Gamma^1{}_2 = \sin\theta, \Gamma^1{}_3 = 0, \\ \Gamma^2{}_1 = -\sin\theta, \Gamma^2{}_2 = \cos\theta, \Gamma^2{}_3 = 0, \\ \Gamma^3{}_1 = 0, \Gamma^3{}_2 = 0, \Gamma^3{}_3 = 1 \end{aligned}\right\}. \tag{2.18}$$

for each ν. Summation over repeated indices is used in Eq. (2.16) so:

$$\left.\begin{aligned} U^1 = \Gamma^1{}_1 V^1 + \Gamma^1{}_2 V^2 + \Gamma^1{}_3 V^3, \\ U^2 = \Gamma^2{}_1 V^1 + \Gamma^2{}_2 V^2 + \Gamma^2{}_3 V^3, \\ U^3 = \Gamma^3{}_1 V^1 + \Gamma^3{}_2 V^2 + \Gamma^3{}_3 V^3, \end{aligned}\right\} \tag{2.19}$$

for each ν. These equations (2.19) are the same as Eqs. (2.12) to (2.14).

The covariant derivative of Eq. (2.15) in this case is therefore:

$$D_\nu V^\mu = (\partial + \Gamma^\mu{}_\lambda)_\nu V^\lambda. \tag{2.20}$$

For example:

$$\begin{aligned} D_\nu V^1 &= (\partial + \Gamma^1{}_1)_\nu V^1 + \Gamma^1{}_{2\nu} V^2 \\ D_\nu V^1 &= (\partial + \cos\theta)_\nu V^1 + (\sin\theta)_\nu V^2 \\ D_\nu V^1 &= \partial_\nu V^1 + (\cos\theta)_\nu V^1 + (\sin\theta)_\nu V^2 \end{aligned} \tag{2.21}$$

Thus:

$$\Gamma^1{}_{1\nu} = (\cos\theta)_\nu, \Gamma^1{}_{2\nu} = (\sin\theta)_\nu. \tag{2.22}$$

These connections must have the units of inverse meters and must operate in the same way as the four derivative ∂_ν. So it is reasonable to assume:

$$\Gamma^1_{1\nu} = \frac{1}{2}\cos\theta\ \partial_\nu,\ \Gamma^1_{2\nu} = \frac{1}{2}\sin\theta\ \partial_\nu \tag{2.23}$$

and

$$D_\nu V^1 = \frac{1}{2}((1+\cos\theta)\partial_\nu V^1 + \sin\theta\partial_\nu V^2) \tag{2.24}$$

If there is no frame rotation:

$$\theta = 0 \tag{2.25}$$

and

$$D_\nu V^1 = \partial_\nu V^1. \tag{2.26}$$

This method regards the connection as an operator. It is well known that the set is a basis set in Riemann geometry. Others possibilities consistent with the correct dimensions of the connection are

$$(\cos\theta)_\nu = \frac{\cos\theta}{r},\ (\sin\theta)_\nu = \frac{\sin\theta}{r}. \tag{2.27}$$

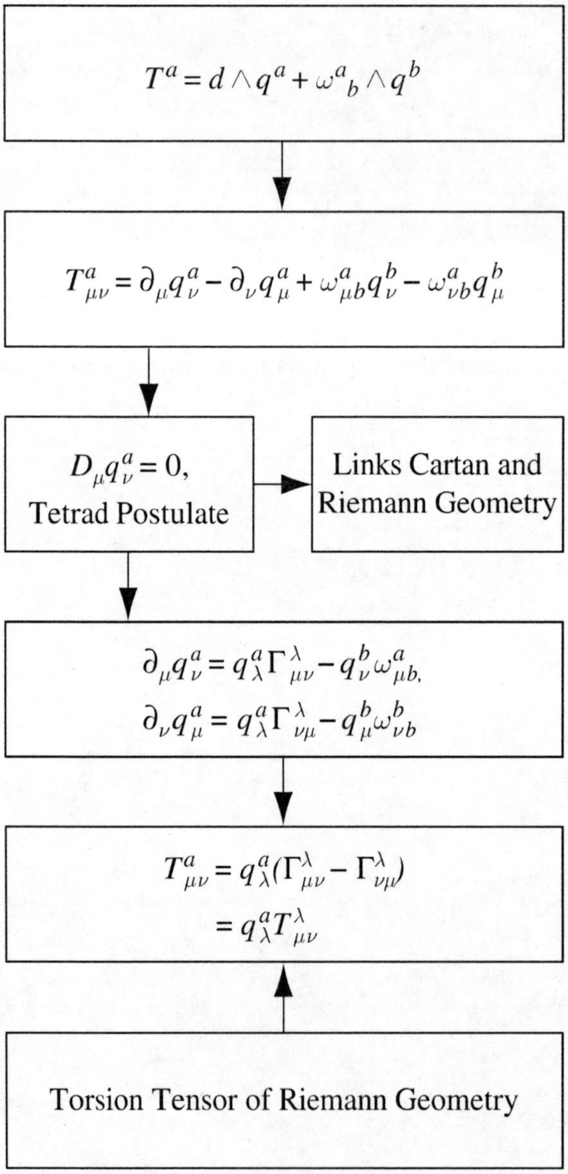

Flowchart 2.1: First Cartan Structure Equation.

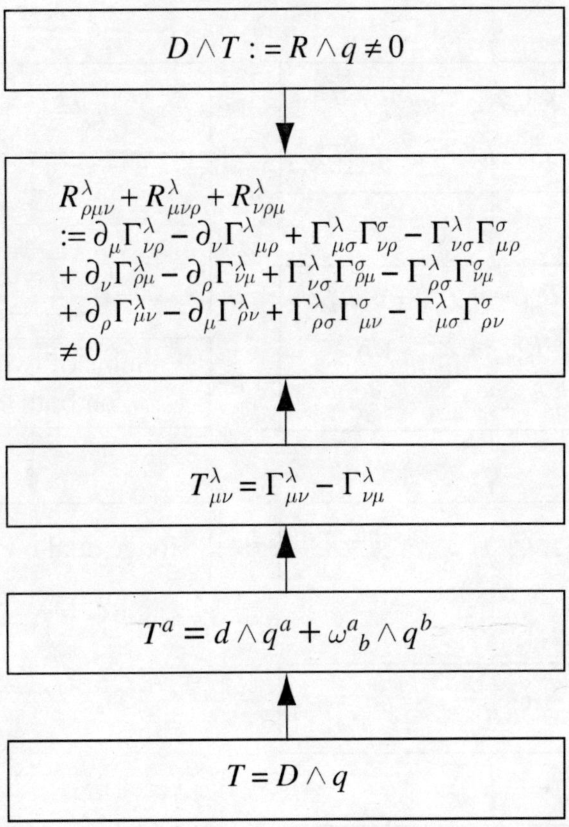

Flowchart 2.2: The Bianchi Identity.

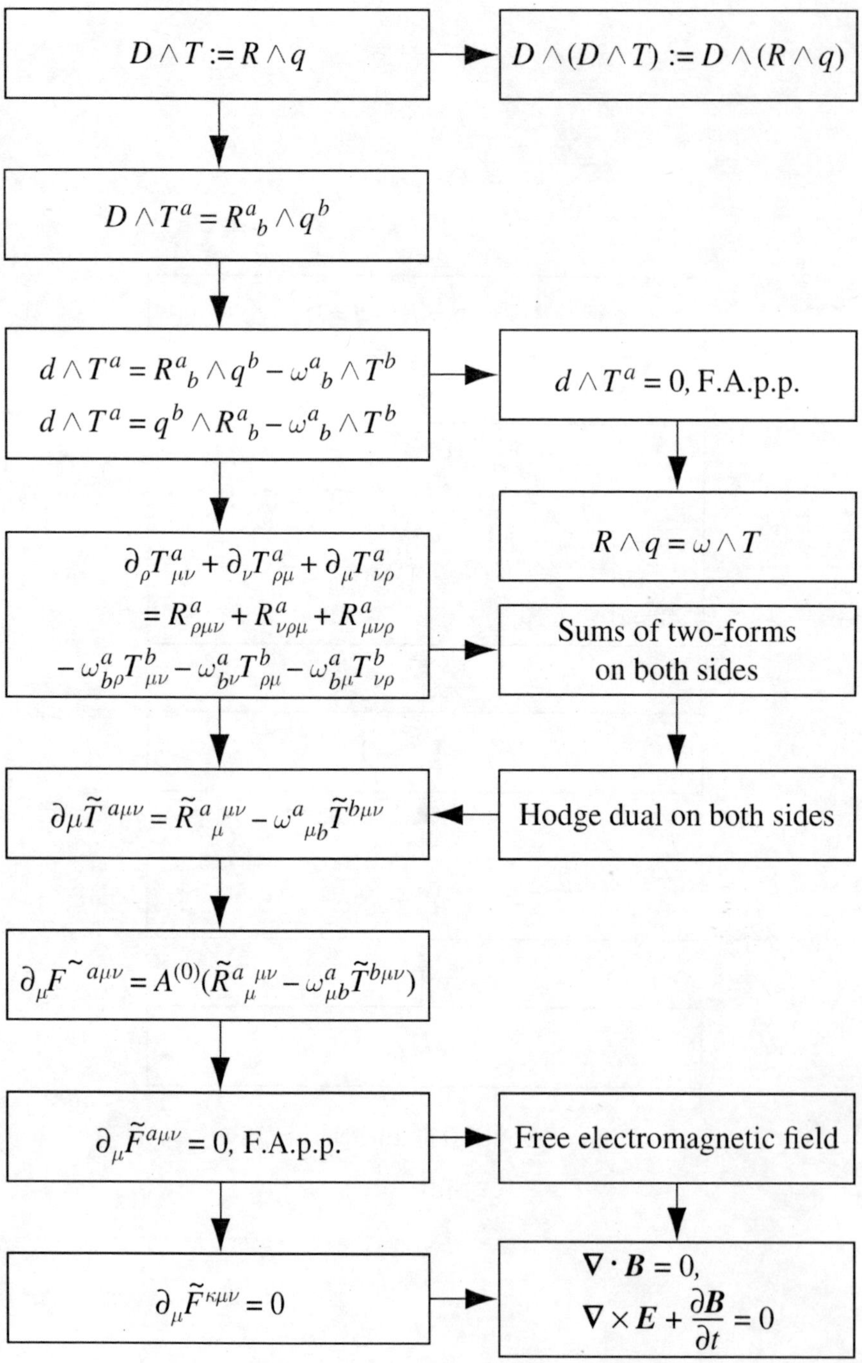

Flowchart 2.3: Homogeneous ECE Field Equation.

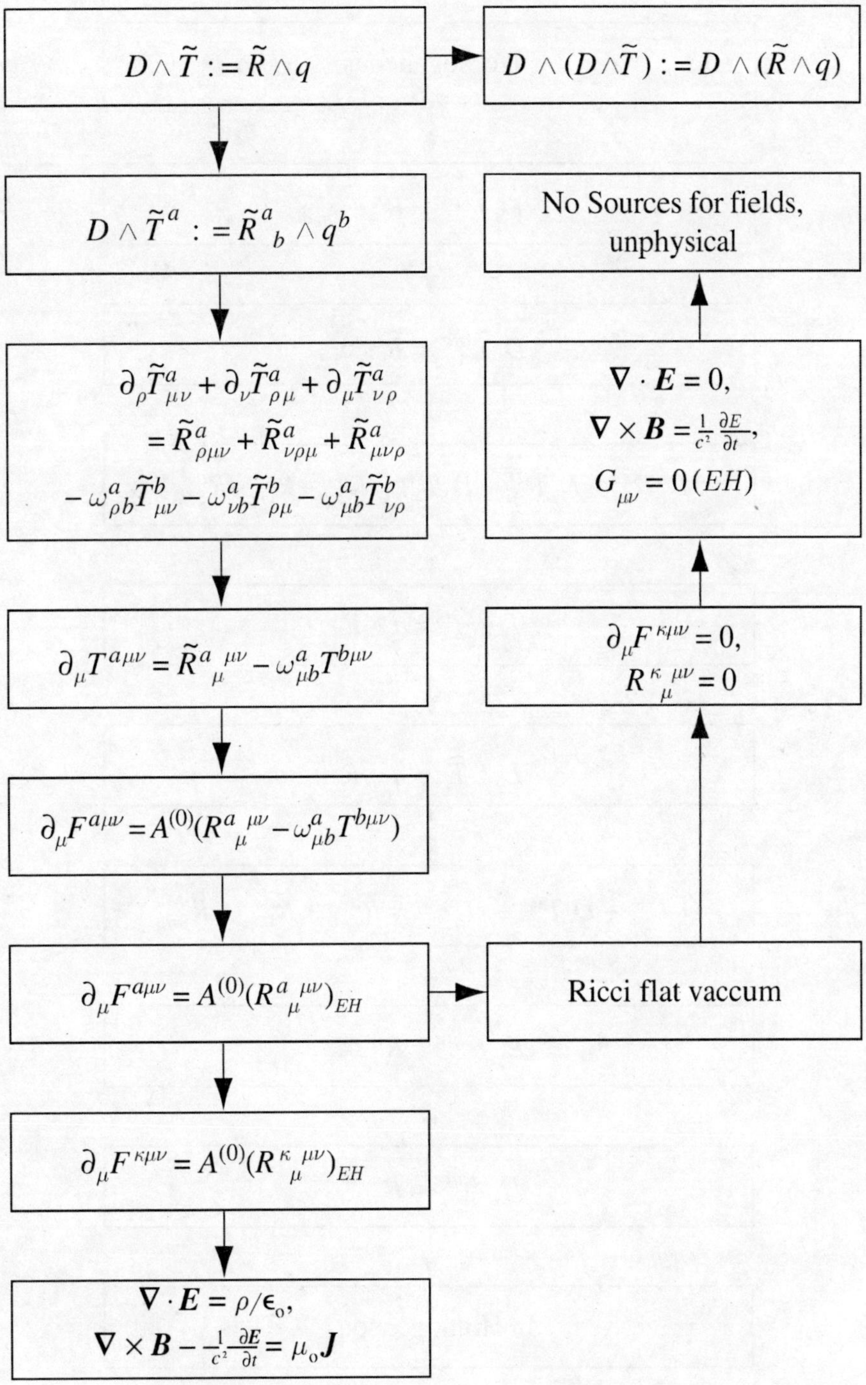

Flowchart 2.4: Inhomogeneous ECE Field Equation.

Homogeneous

↓

$$D_\mu \tilde{T}^{\kappa\mu\nu} = \tilde{R}^{\kappa\ \ \mu\nu}_{\ \mu}$$

↓

$$D_\mu \tilde{T}^{a\mu\nu} = \tilde{R}^{a\ \ \mu\nu}_{\ \mu}$$

↓

$$D_\rho T^a_{\ \mu\nu} + D_\nu T^a_{\ \rho\mu} + D_\mu T^a_{\ \nu\rho} = R^a_{\ \rho\mu\nu} + R^a_{\ \nu\rho\mu} + R^a_{\ \mu\nu\rho}$$

↓

$$D \wedge T := q \wedge R$$

↓

$$D \wedge \tilde{T} := q \wedge \tilde{R}$$

↓

$$D_\rho \tilde{T}^a_{\ \mu\nu} + D_\nu \tilde{T}^a_{\ \rho\mu} + D_\mu \tilde{T}^a_{\ \nu\rho} = \tilde{R}^a_{\ \rho\mu\nu} + \tilde{R}^a_{\ \nu\rho\mu} + \tilde{R}^a_{\ \mu\nu\rho}$$

↓

$$D_\mu T^{a\mu\nu} = R^{a\ \ \mu\nu}_{\ \mu}$$

↓

$$D_\mu T^{\kappa\mu\nu} = R^{\kappa\ \ \mu\nu}_{\ \mu}$$

↓

In Homogeneous

Flowchart 2.5: The Basic Field Equations.

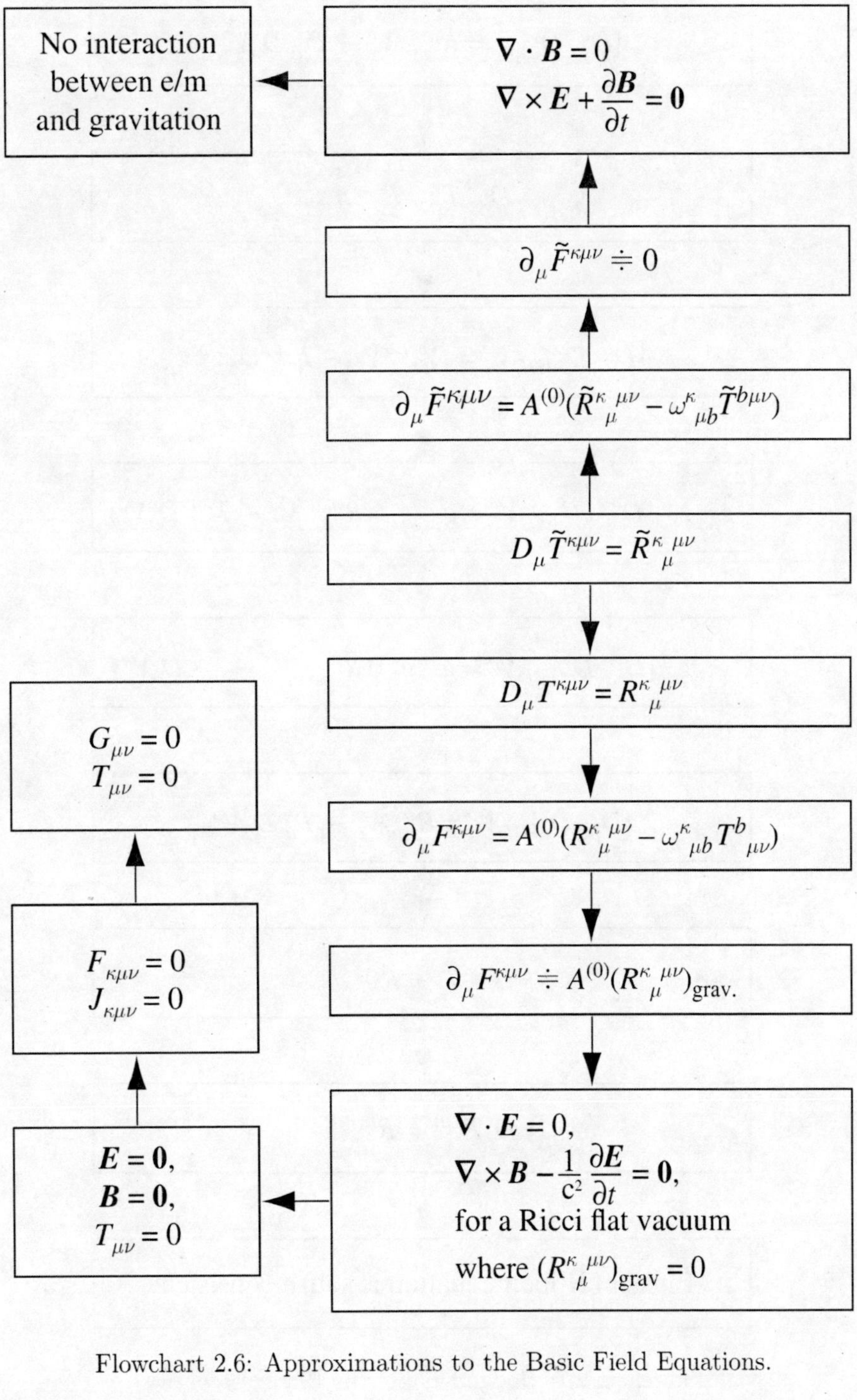

Flowchart 2.6: Approximations to the Basic Field Equations.

$$[D_\mu, D_\nu]V^\rho = R^\rho{}_{\sigma\mu\nu}V^\sigma - T^\lambda_{\mu\nu}D_\lambda V^\rho$$

↓

$$D \wedge T := R \wedge q$$

↓

$$[D^\mu, D^\nu]_{HD}V^\rho = \tilde{R}^\rho{}_\sigma{}^{\mu\nu}V^\sigma - \tilde{T}^{\lambda\mu\nu}D_\lambda V^\rho$$

↓

$$g_{\mu\alpha}g_{\nu\beta}[D^\alpha, D^\beta]_{HD}V^\rho = g_{\mu\alpha}g_{\nu\beta}(\tilde{R}^\rho{}_\sigma{}^{\mu\nu}V^\sigma - \tilde{T}^{\lambda\mu\nu}D_\lambda V^\rho)$$

↓

$$g^{\mu\alpha}g^{\nu\beta}[D_\alpha, D_\beta]_{HD}V^\rho = g^{\mu\alpha}g^{\mu\beta}(\tilde{R}^\rho{}_{\sigma\alpha\beta}V^\sigma - \tilde{T}^\lambda_{\alpha\beta}D_\lambda V^\rho)$$

↓

$$[D_\mu, D_\nu]_{HD}V^\rho = \tilde{R}^\rho_{\sigma\mu\nu}V^\rho - \tilde{T}^\lambda_{\mu\nu}D_\lambda V^\rho$$

↓

$$D \wedge \tilde{T} := \tilde{R} \wedge q$$

↓

$$D_\mu T^{\kappa\mu\nu} = R^\kappa{}_\mu{}^{\mu\nu}$$

↓

Einstein Hilbert equation is self-inconsistent

Flowchart 2.7: Hodge Dual of the Bianchi Identity.

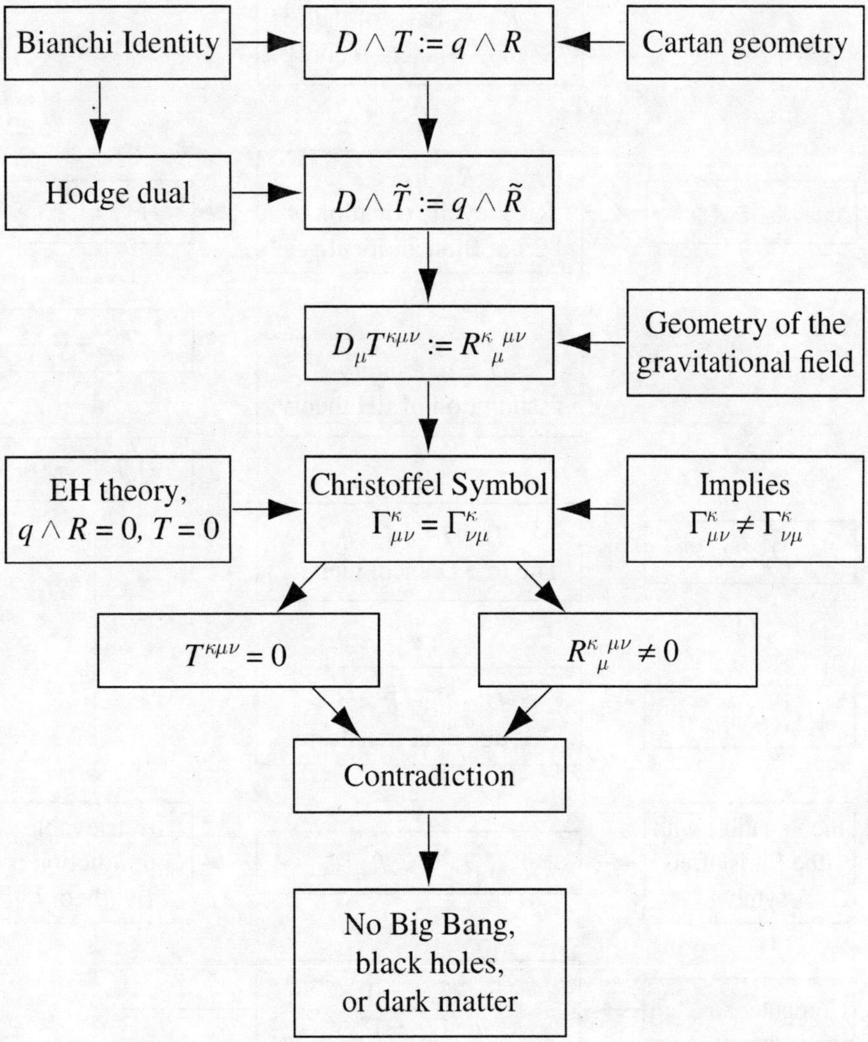

Flowchart 2.8: Self Inconsistency of General Relativity.

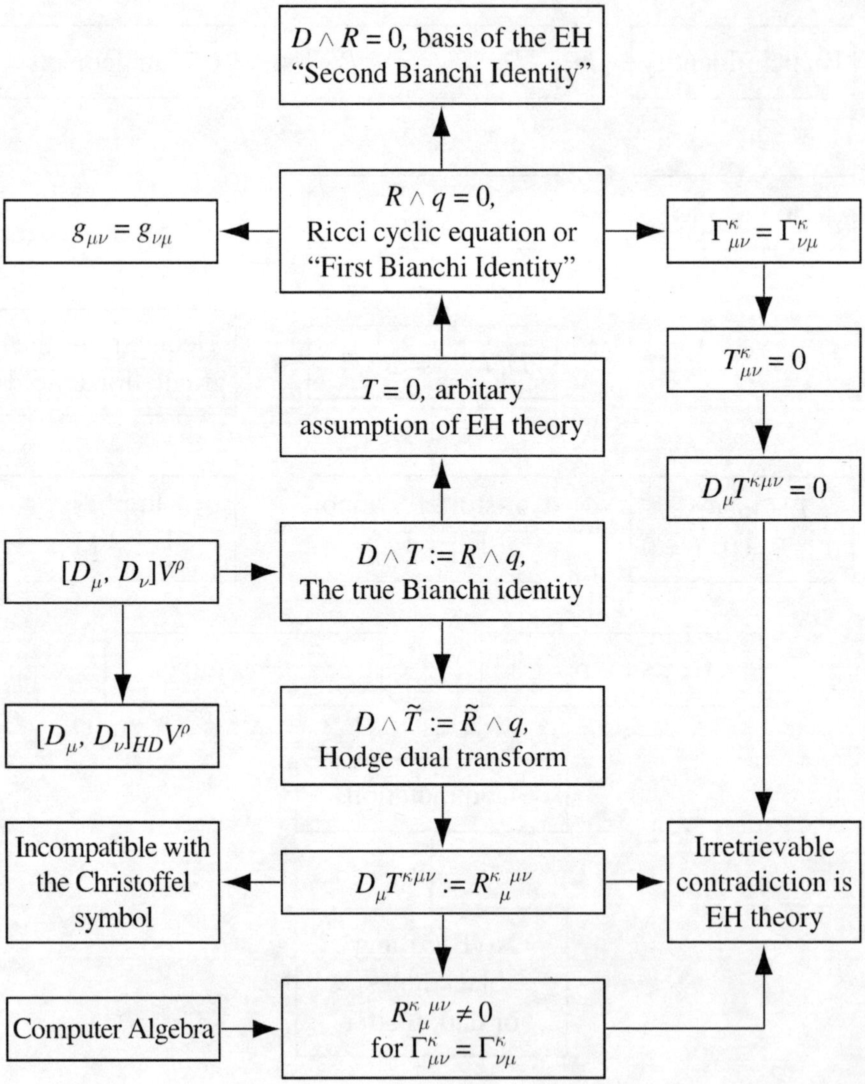

Flowchart 2.9: Irretrievable Flaws in the Geometry of the Einstein-Hilbert Field Theory.

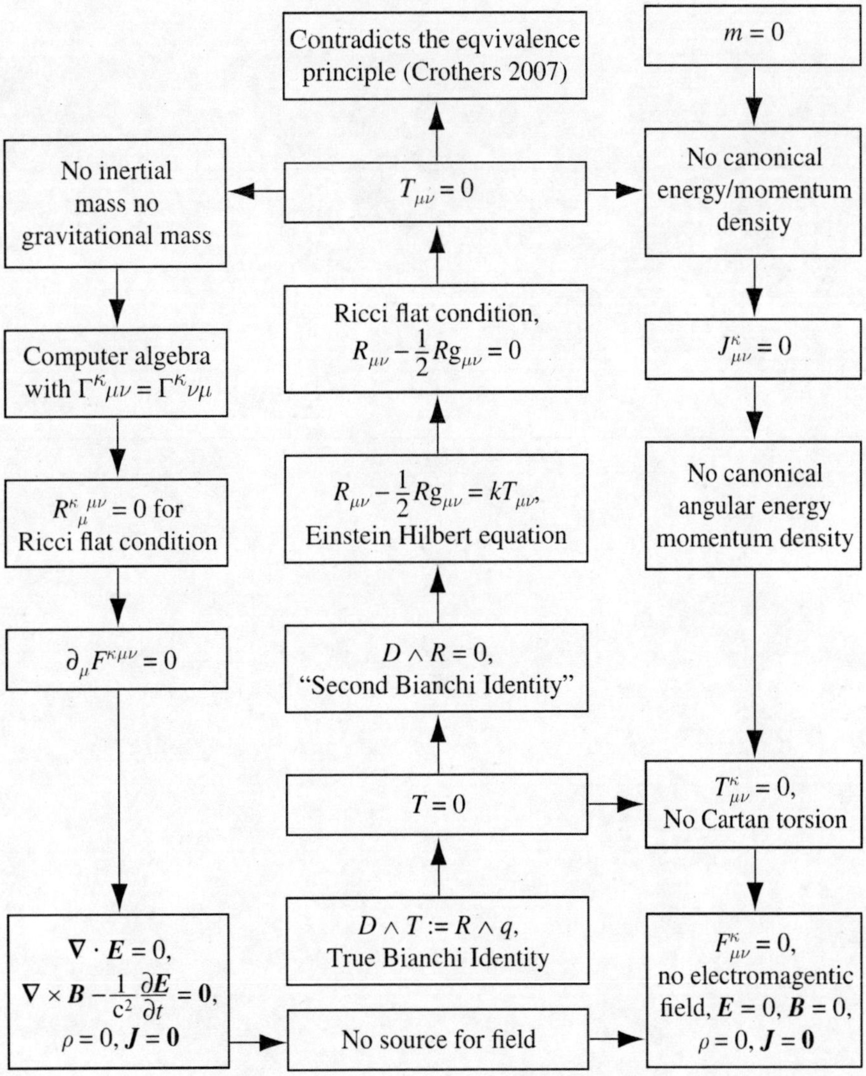

Flowchart 2.10: Irretrievable Contradiction in the Ricci Flat Condition.

Bibliography

[1] M. W. Evans, "Generally Covariant Unified Field Theory: the Geometrization of Physics" (Abramis Academic, 2005 to present), vols. 1 to 4, vol. 5 in prep. (Papers 71 to 93 on www.aias.us).

[2] L. Felker, "The Evans Equations of Unified Field Theory" (Abramis Academic, 2007).

[3] K. Pendergast, "Crystal Spheres" (preprint on www.aias.us, Abramis to be published).

[4] Omnia Opera Section of www.aias.us.

[5] H. Eckardt, L. Felker, D. Indranu and S. Crothers, articles on www.aias.us.

[6] M. W. Evans, (ed.), "Modern Non-linear Optics", a special topical issue of I. Prigogine and S. A. Rice, "Advances in Chemical Physics" (Wiley Interscience, New York, 2001, second edition), vols. 119(1) to 119(3).

[7] M. W. Evans and S. Kielich (eds.), ibid., first edition (Wiley Interscience, New York, 1992, 1993, 1997), vols. 85(1) to 85(3).

[8] M. W. Evans and L. D. Crowell, "Classical and Quantum Electrodynamics and the B(3) Field" (World Scientific, 2001).

[9] M. W. Evans and J.-P. Vigier, "The Enigmatic Photon" (Kluwer, 1994 to 2002), in five volumes.

[10] M. W. Evans and A. A. Hasanein, "The Photomagneton in Quantum Field Theory" (World Scientific, 1994).

[11] M. W. Evans, "The Photon's Magnetic Field, Optical NMR" (World Scientific, 1992).

[12] M. W. Evans, Physica B, **182**, 227 and 237 (1992).

[13] S. P. Carroll, "Spacetime and Geometry, an Introduction to General Relativity" (Addison Wesley, New York, 2004), chapter three.

[14] M. W. Evans, Omnia Opera Section of www.aias.us (1992 to present).

[15] J. B. Marion and S. T. Thornton, "Classical Dynamics of Particles and Systems" (Harcourt Brace College Publishers, 1988, third edition).

[16] See for example a review by Zawodny in ref. (7).

[17] The first papers on the B(3) field are ref. (12), available in the Omnia Opera section of www.aias.us.

[18] L. H. Ryder, "Quantum Field Theory" (Cambridge University Press, 1996, 2nd ed.).

[19] S. Crothers, papers and references therein on the www.aias.us site and papers 93 of the ECE series on www.aias.us.

[20] P. W. Atkins, "Molecular Quantum Mechanics" (Oxford University Press, 1983, 2nd edition and subsequent editions).

[21] J. R. Croca, "Towards a Non-linear Quantum Physics" (World Scientific, 2001).

[22] This claim is made in ref. (18) and corrected in ref. (1).

[23] E. G. Milewski (Chief Editor), "The Vector Analysis Problem Solver" (Research and Education Association, New York, 1987, revised printing).

[24] J. D. Jackson, "Classical Electrodynamics" (Wiley, New York, 1999, third edition).

[25] M. W. Evans, Acta Phys. Polonica, **38**, 2211 (2007).

[26] M. W. Evans and H. Eckardt, Physica B, **400**, 175 (2007).

[27] M. W. Evans, Physica B, **403**, 517 (2008).

[28] Michael Krause (Director), "All About Tesla" (a film available on DVD that premiered in 2007).

[29] P. W. Atkins, frontispiece of ref. (20).

Chapter 3

Fundamental Errors in the General Theory of Relativity

by

Stephen J. Crothers[1],

Alpha Institute for Advance Study (AIAS)
(www.aias.us, www.atomicprecision.com)

3.1 Introduction

The so-called 'Schwarzschild solution' is not Schwarzschild's solution, but a corruption of the Schwarzschild/Droste solutions. In the so-called 'Schwarzschild solution' the quantity m is alleged to be the mass of the source of a gravitational field and the quantity r is alleged to be able to go down to zero (although no valid proof of this claim has ever been advanced), so that there are two alleged 'singularities', one at $r = 2m$ and another at $r = 0$. It is routinely asserted that $r = 2m$ is a 'coordinate' or 'removable' singularity which denotes the so-called 'Schwarzschild radius' (event horizon) and that a 'physical' singularity is at $r = 0$. The quantity r in the 'Schwarzschild solution' has never been rightly identified by the physicists, who, although proposing many and varied concepts for what r therein denotes, effectively treat it as a radial distance from the claimed source of the gravitational field at the 'origin of coordinates'. The consequence of this is that the intrinsic geometry of the metric manifold has been violated. It is easily proven that the said quantity r is in fact the inverse square root of the Gaussian curvature of the spherically symmetric geodesic surface in the spatial section of the 'Schwarzschild solution' and so does not in itself define any distance whatsoever in that manifold. Thus the 'Schwarzschild radius' is not a distance of any sort. With the correct identification of the associated Gaussian curvature it is also easily proven that there is only one singularity associated with all Schwarzschild metrics, of which there is an infinite number that are equivalent. Thus, the standard removal of the singularity at $r = 2m$ is erroneous, as the alleged singularity at $r = 0$ does not exist, very simply demonstrated herein. This has major implications for the localisation of gravitational energy, i.e. gravitational waves.

It is demonstrated herein that Special Relativity forbids infinite density and in consequence of this General Relativity necessarily forbids infinite density, and so the infinitely dense point-mass singularity of the alleged black hole is forbidden by the Theory of Relativity. It is also shown that neither Einstein's Principle of Equivalence not his Laws of Special Relativity can manifest in a space-time that by construction contains no matter, and therefore $Ric = 0$ violates the requirement that both the said Principle and Special Relativity manifest

[1]e-mail: thenarmis@gmail.com

in Einstein's gravitational field. The immediate implication of this is that the total gravitational energy of Einstein's gravitational field is always zero, so that the energy-momentum tensor and the Einstein tensor must vanish identically. Attempts to preserve the usual conservation of energy and momentum by means of Einstein's pseudo-tensor are fatally flawed owing to the fact that the pseudo-tensor implies the existence of a first-order intrinsic differential invariant, dependent solely upon the components of the metric tensor and their first derivatives, an invariant which however does not exist, proven by the pure mathematicians G. Ricci-Curbastro and T. Levi-Civita, in 1900. Although it is standard method to utilise the Kretschmann scalar to justify infinite Schwarzschild space-time curvature at the point-mass singularity, it is demonstrated that the Kretschmann scalar is not an independent curvature invariant, being in fact a function of the Gaussian curvature of the spherically symmetric geodesic surface in the spatial section, and therefore constrained by the limitations set on the said Gaussian curvature by the geometric ground-form of the line-element itself. Since it is easily proven that the said Gaussian curvature cannot become unbounded in Schwarzschild space-time, the Kretschmann scalar is necessarily finite everywhere in the Schwarzschild manifold.

3.2 Schwarzschild space-time

It is reported almost invariably in the literature that Schwarzschild's solution for $Ric = R_{\mu\nu} = 0$ is (using $c = 1$, $G = 1$),

$$ds^2 = \left(1 - \frac{2m}{r}\right) dt^2 - \left(1 - \frac{2m}{r}\right)^{-1} dr^2 - r^2 \left(d\theta^2 + \sin^2\theta d\varphi^2\right), \tag{3.1}$$

wherein it is asserted by inspection that r can go down to zero in some way, producing an infinitely dense point-mass singularity there, with an event horizon at the 'Schwarzschild radius' at $r = 2m$: a black hole. Contrast this metric with that actually obtained by K. Schwarzschild in 1915 (published January 1916),

$$ds^2 = \left(1 - \frac{\alpha}{R}\right) dt^2 - \left(1 - \frac{\alpha}{R}\right)^{-1} dR^2 - R^2 \left(d\theta^2 + \sin^2\theta d\varphi^2\right), \tag{3.2}$$

$$R = R(r) = \left(r^3 + \alpha^3\right)^{\frac{1}{3}}, \quad 0 < r < \infty,$$

wherein α is an undetermined constant. There is only *one* singularity in Schwarz-schild's solution, at $r = 0$, to which his solution is constructed. Contrary to the usual claims made by the astrophysical scientists, Schwarzschild did not set $\alpha = 2m$ where m is mass; he did not breathe a single word about the bizarre object that is called a black hole; he did not allege the so-called 'Schwarzschild radius'; he did not claim that there is an 'event horizon' (by any other name); and his solution clearly forbids the black hole because when Schwarzschild's $r = 0$, his $R = \alpha$, and so there is no possibility for his R to be less than α, let alone take the value $R = 0$. All this can be easily verified by simply reading Schwarzschild's original paper [1], in which he *constructs* his solution so that the singularity occurs at the *"origin"* of coordinates. Thus, Eq. (3.1) for $0 < r < 2m$ is *inconsistent* with Schwarzschild's true solution, Eq. (3.2). It is also inconsistent with the intrinsic geometry of the line-element, whereas Eq. (3.2) is geometrically consistent, as demonstrated herein. Thus, Eq. (3.1) is meaningless for $0 \leq r < 2m$.

In the usual interpretation of Hilbert's [2–4] version, Eq. (3.1), of Schwarz-schild's solution, the quantity r therein has ***never*** been properly identified. It has been variously and vaguely called a "*distance*" [5, 6], "*the radius*" [6–19,78,79], the "*radius of a 2-sphere*" [20], the "*coordinate radius*" [21], the "*radial coordinate*" [22–25,78,79], the "*radial space coordinate*" [26], the "*areal radius*" [21, 24, 27, 28], the "*reduced circumference*" [25], and even "*a gauge choice: it defines the coordinate r*" [29]. In the particular case of $r = 2m = 2GM/c^2$ it is almost invariably referred to as the "*Schwarzschild radius*" or the "*gravitational radius*". However, none of these various and vague concepts of what r is are correct because the irrefutable geometrical fact is that r, in the spatial section of Hilbert's version of the Schwarzschild/Droste line-element, is the inverse square root of the Gaussian curvature of a spherically symmetric geodesic surface in the spatial section [30–32], and as such it ***does not*** of itself determine the geodesic radial distance from the centre of spherical symmetry located at an arbitrary point in the related pseudo-Riemannian metric manifold. It does not of itself determine any distance at all in the spherically symmetric metric manifold. It is the radius of Gaussian curvature merely by virtue of its formal geometric relationship to the Gaussian curvature. It must also be emphasized that a geometry is completely determined by the ***form*** of its line-element [33].

Since r in Eq. (3.1) can be replaced by *any* analytic function $R_c(r)$ [4,30,32,34] without disturbing spherical symmetry and without violation of the field equations $R_{\mu\nu}=0$, which is very easily verified, satisfaction of the field equations is a necessary but insufficient condition for a solution for Einstein's static vacuum 'gravitational' field. Moreover, the admissible *form* of $R_c(r)$ must be determined in such a way that an infinite number of equivalent metrics is generated thereby [32,34]. In addition, the identification of the centre of spherical symmetry, origin of coordinates and the properties of points must also be clarified in relation to the non-Euclidean geometry of Einstein's gravitational field. In relation to Eq. (3.1) it has been routinely presumed that geometric points in the spatial section (which is non-Euclidean) must have the very same properties of points in the spatial section (Euclidean) of Minkowski space-time. However, it is easily proven that the non-Euclidean geometric points in the spatial section of Schwarzschild space-time do not possess the same characteristics of the Euclidean geometric points in the spatial section of Minkowski space-time [32,35]. This should not be surprising, since the indefinite metric of Einstein's Theory of Relativity admits of other geometrical oddities, such as null vectors, i.e. non-zero vectors that have zero magnitude and which are orthogonal to themselves [36].

3.3 Spherical symmetry

Recall that the squared differential element of arc of a curve in a surface is given by **the first fundamental quadratic form** for a surface,

$$ds^2 = E\,du^2 + 2F\,du\,dv + G\,dv^2,$$

wherein u and v are curvilinear coordinates. If either u or v is constant the resulting line-elements are called *parametric curves* in the surface. The differential element of surface area is given by,

$$dA = \left|\sqrt{EG - F^2}\,du\,dv\right|.$$

An expression which depends only on E, F, G and their first and second derivatives is called a bending invariant. It is an intrinsic (or absolute) property of a surface. The Gaussian (or Total) curvature of a surface is an important intrinsic property of a surface.

The **'Theorema Egregium'** of **Gauss**

> *The Gaussian curvature K at any point P of a surface depends only on the values at P of the coefficients in the First Fundamental Form and their first and second derivatives.* [37–39]

And so,

> *"The Gaussian curvature of a surface is a bending invariant."* [38]

The plane has a constant Gaussian curvature of $K = 0$. *"A surface of positive constant Gaussian curvature is called a spherical surface."* [39]

Now a line-element, or squared differential element of arc-length, in spherical coordinates, for 3-dimensional Euclidean space is,

$$ds^2 = dr^2 + r^2\left(d\theta^2 + \sin^2\theta\,d\varphi^2\right), \qquad (3.3)$$
$$0 \le r < \infty.$$

The scalar r can be construed, verified by calculation, as the magnitude of the radius vector $\vec{r}$ from the origin of the coordinate system, the said origin coincident with the centre of the associated sphere. All the components of the metric tensor are well-defined and related geometrical quantities are fixed by the form of the line-element. Indeed, the radius R_p of the associated sphere ($\theta = const.$, $\varphi = const.$) is given by,

$$R_p = \int_0^r dr = r,$$

the length of the geodesic C_p (a parametric curve; $r =$ *const.*, $\theta = \pi/2$) in an associated surface is given by,

$$C_p = r\int_0^{2\pi} d\varphi = 2\pi r,$$

the area A_p of an associated spherically symmetric surface ($r =$ *const.*) is,

$$A_p = r^2\int_0^{\pi} \sin\theta d\theta \int_0^{2\pi} d\varphi = 4\pi r^2,$$

and the volume V_p of the sphere is,

$$V_p = \int_0^r r^2 dr \int_0^{\pi} \sin\theta d\theta \int_0^{2\pi} d\varphi = \frac{4}{3}\pi r^3.$$

Now the point at the centre of spherical symmetry for any problem at hand need not be coincident with the origin of the coordinate system. For example, the equation of a sphere of radius ρ centered at the point C located at the extremity of the fixed vector $\vec{r}_o$ in Euclidean 3-space, is given by

$$(\vec{r} - \vec{r}_o)\cdot(\vec{r} - \vec{r}_o) = \rho^2.$$

If $\vec{r}$ and $\vec{r}_o$ are collinear, the vector notation can be dropped, and this expression becomes,

$$|r - r_o| = \rho,$$

where $r = |\vec{r}|$ and $r_o = |\vec{r}_o|$, and the common direction of $\vec{r}$ and $\vec{r}_o$ becomes entirely immaterial. This scalar expression for a shift of the centre of spherical symmetry away from the origin of the coordinate system plays a significant rôle in the equivalent line-elements for Schwarzschild space-time.

Consider now the generalisation of Eq. (3.3) to a spherically symmetric metric manifold, by the line-element,

$$ds^2 = dR_p^2 + R_c^2\left(d\theta^2 + \sin^2\theta d\varphi^2\right) = \Psi\left(R_c\right)dR_c^2 + R_c^2\left(d\theta^2 + \sin^2\theta d\varphi^2\right), \tag{3.4}$$

$$R_c = R_c(r)$$

$$R_c(0) \leq R_c(r) < \infty,$$

where both $\Psi(R_c)$ and $R_c(r)$ are *a priori* unknown analytic functions. Since neither $\Psi(R_c)$ nor $R_c(r)$ are known, Eq. (3.4) may or may not be well-defined at $R_c(0)$: one cannot know until $\Psi(R_c)$ and $R_c(r)$ are somehow specified. With this proviso, there is a one-to-one point-wise correspondence between the manifolds described by metrics (3) and (4), i.e. a mapping between the auxiliary Euclidean manifold described by metric (3) and the generalised non-Euclidean manifold described by metric (4), as the differential geometers have explained [30]. If R_c is constant, metric (4) reduces to a 2-dimensional spherically symmetric geodesic surface described by the first fundamental quadratic form,

$$ds^2 = R_c^2\left(d\theta^2 + \sin^2\theta d\varphi^2\right). \tag{3.5}$$

If r is constant, Eq. (3.3) reduces to the 2-dimensional spherically symmetric surface described by the first fundamental quadratic form,

$$ds^2 = r^2\left(d\theta^2 + \sin^2\theta d\varphi^2\right). \tag{3.6}$$

Although R_c and r are constants in equations (5) and (6) respectively, they share a definite geometric identity in their respective surfaces: but it is **not** that of being a radial quantity, or of a distance.

A surface is a manifold in its own right. It need not be considered in relation to an embedding space. Therefore, quantities appearing in its line-element must be identified in relation to the surface, not to any embedding space it might be in:

> *"And in any case, if the metric form of a surface is known for a certain system of intrinsic coordinates, then all the results concerning the intrinsic geometry of this surface can be obtained without appealing to the embedding space."* [40]

Note that eqs. (3) and (4) have the same metrical form and that eqs. (5) and (6) have the same metrical form. Metrics of the same form share the same fundamental relations between the components of their respective metric tensors. For example, consider Eq. (3.4) in relation to Eq. (3.3). For Eq. (3.4), the radial geodesic distance (i.e. the proper radius) from the point at the centre of spherical symmetry ($\theta = const.$, $\varphi = const.$) is,

$$R_p = \int_0^{R_p} dR_p = \int_{R_c(0)}^{R_c(r)} \sqrt{\Psi(R_c(r))}dR_c(r) = \int_0^r \sqrt{\Psi(R_c(r))}\frac{dR_c(r)}{dr}dr,$$

the length of the geodesic C_p (a parametric curve; $R_c(r) =$ *const.*, $\theta = \pi/2$) in an associated surface is given by,

$$C_p = R_c(r) \int_0^{2\pi} d\varphi = 2\pi R_c(r),$$

the area A_p of an associated spherically symmetric geodesic surface ($(R_c(r)) =$ *const.*) is,

$$A_p = R_c^2(r) \int_0^{\pi} \sin\theta d\theta \int_0^{2\pi} d\varphi = 4\pi R_c^2(r),$$

and the volume V_p of the geodesic sphere is,

$$V_p = \int_0^{R_p} R_c^2\left(r\right) dR_p \int_0^{\pi} \sin\theta d\theta \int_0^{2\pi} d\varphi = 4\pi \int_{R_c(0)}^{R_c(r)} \sqrt{\Psi\left(R_c\left(r\right)\right)} R_c^2(r) dR_c$$

$$= 4\pi \int_0^r \sqrt{\Psi\left(R_c\left(r\right)\right)} R_c^2(r) \frac{dR_c(r)}{dr} dr.$$

Remarkably, in relation to metric (1), Celotti, Miller and Sciama [11] make the following false assertion:

"The 'mean density' $\overline{\rho}$ of a black hole (its mass M divided by $\frac{4}{3}\pi r_s^3$) is proportional to $1/M^2$"

where r_s is the so-called "Schwarzschild radius". The volume they adduce for a black hole cannot be obtained from metric (1): it is a volume associated with the Euclidean 3-space described by metric (3).

Now in the case of the 2-dimensional metric manifold given by Eq. (3.5) the Riemannian curvature associated with Eq. (3.4) (which depends upon both position and direction) reduces to the Gaussian curvature K (which depends only upon position), and is given by [30,38,39,41–45],

$$K = \frac{R_{1212}}{g}, \tag{3.7}$$

where R_{1212} is a component of the Riemann tensor of the 1st kind and $g = g_{11}g_{22} = g_{\theta\theta}g_{\varphi\varphi}$ (because the metric tensor of Eq. (3.5) is diagonal). Gaussian curvature is an intrinsic geometric property of a surface (Theorema Egregium[2]); independent of any embedding space.

Now recall from elementary differential geometry and tensor analysis that

$$R_{\mu\nu\rho\sigma} = g_{\mu\gamma} R^{\gamma}_{.\,\nu\rho\sigma}$$

$$R^1_{.\,212} = \frac{\partial \Gamma^1_{22}}{\partial x^1} - \frac{\partial \Gamma^1_{21}}{\partial x^2} + \Gamma^k_{22}\Gamma^1_{k1} - \Gamma^k_{21}\Gamma^1_{k2}$$

$$\Gamma^i_{ij} = \Gamma^i_{ji} = \frac{\partial \left(\frac{1}{2}\ln|g_{ii}|\right)}{\partial x^j}$$

$$\Gamma^i_{jj} = -\frac{1}{2g_{ii}}\frac{\partial g_{jj}}{\partial x^i}, \quad (i \neq j) \tag{3.8}$$

[2] i.e. Gauss' Most Excellent Theorem.

and all other Γ^i_{jk} vanish. In the above, $i, j, k = 1, 2$, $x^1 = \theta$, $x^2 = \varphi$. Applying expressions (7) and (8) to expression metric (5) gives,

$$K = \frac{1}{R_c^2} \tag{3.9}$$

so that $R_c(r)$ is the inverse square root of the Gaussian curvature, i.e. the radius of Gaussian curvature, and hence, in Eq. (3.6) the quantity r therein is the radius of Gaussian curvature because this Gaussian curvature is intrinsic to all geometric surfaces having the form of Eq. (3.5) [30], and a geometry is completely determined by the *form* of its line-element [33]. Note that according to Eqs. (3.3), (3.6) and (3.7), the radius calculated for (3) gives the same value as the associated radius of Gaussian curvature of a spherically symmetric surface embedded in the space described by Eq. (3.3). Thus, the Gaussian curvature (and hence the radius of Gaussian curvature) of the spherically symmetric surface embedded in the space of (3) can be associated with the radius calculated from Eq. (3.3). This is a consequence of the Euclidean nature of the space described by metric (3), which also describes the spatial section of Minkowski space-time. However, this is *not* a general relationship. The inverse square root of the Gaussian curvature (the radius of Gaussian curvature) is not a distance at all in Einstein's gravitational manifold but in fact determines the Gaussian curvature of the spherically symmetric geodesic surface through any point in the spatial section of the gravitational manifold, as proven by expression (9). Thus, the quantity r in Eq. (3.1) is the inverse square root of the Gaussian curvature (i.e. the radius of Gaussian curvature) of a spherically symmetric geodesic surface in the spatial section, not the radial geodesic distance from the centre of spherical symmetry of the spatial section, or any other distance.

The platitudinous nature of the concepts "*reduced circumference*" and "*areal radius*" is now plainly evident - neither concept correctly identifies the geometric nature of the quantity r in metric (1). The geodesic C_p in the spherically symmetric geodesic surface in the spatial section of Eq. (3.1) is a function of the curvilinear coordinate φ and the surface area A_p is a function of the curvilinear coordinates θ and φ where, in both cases, r is a constant. However, r therein has a clear and definite geometrical meaning: it is the inverse square root of the Gaussian curvature of the spherically symmetric geodesic surface in the spatial section. The Gaussian curvature K is a positive constant bending invariant of the surface, independent of the values of θ and φ. Thus, neither C_p nor A_p, or the infinite variations of them by means of the integrated values of θ and φ, rightly identify what r is in line-element (1). To illustrate further, when $\theta = constant$, the arc-length in the spherically symmetric geodesic surface is given by:

$$s = s(\varphi) = r \int_0^{\varphi} \sin\theta \, d\varphi = r \sin\theta \, \varphi, \quad 0 \leq \varphi \leq 2\pi,$$

where $r = constant$. This is the equation of a straight line, of gradient $ds/d\varphi = r\sin\theta$. If $\theta = const. = \frac{1}{2}\pi$ then $s = s(\varphi) = r\varphi$, which is the equation of a straight line of gradient $ds/d\varphi = r$. The maximum arc-length of the geodesic $\theta = \; const. \; = \frac{1}{2}\pi$ is therefore $s(2\pi) = 2\pi r = C_p$. Similarly the surface area is:

$$A = A(\varphi, \theta) = r^2 \int_0^{\theta} \int_0^{\varphi} \sin\theta \, d\theta \, d\varphi = r^2 \varphi \left(1 - \cos\theta\right),$$

$$0 \leq \varphi \leq 2\pi, \quad 0 \leq \theta \leq \pi, \quad r = constant.$$

The maximum area (i.e. the area of the entire surface) is $A\left(2\pi, \pi\right) = 4\pi r^2 = A_p$. Clearly, neither s nor A are functions of r, because r is a constant, not a variable. And since r appears in each expression (and so having the same value in each expression), neither s nor A rightly identify the geometrical significance of r in the 1st fundamental form for the spherically symmetric geodesic surface, $ds^2 = r^2\left(d\theta^2 + \sin^2\theta \, d\varphi^2\right)$, because r is **not** a distance in the surface and is **not** the "radius" of the surface. The geometrical significance of r is intrinsic to the surface and is determined from the components of the metric tensor and their derivatives (Gauss' Theorema Egregium): it is the inverse square root of the Gaussian curvature K of the spherically symmetric surface so described (the constant is $K = 1/r^2$). Thus, C_p and A_p are merely platitudinous expressions containing the constant r, and so the "*reduced circumference*" $r = C_p/2\pi$ and the "*areal radius*" $r = \sqrt{A_p/4\pi}$ do not identify the geometric nature of r in either metric (6) or metric (1), the former appearing in the latter. The claims by the astrophysical scientists that the "*areal radius*" and the "*reduced circumference*" each define [21, 25, 48] (in two different ways) the constant r in Eq. (3.1) are entirely false. The "*reduced circumference*" and the "*areal radius*" are in fact one and the same, namely the inverse square root of the Gaussian curvature of the spherically symmetric geodesic surface in the spatial section of Eq. (3.1), as proven above. No proponent of black holes is aware of this simple geometrical fact, which completely subverts all claims made for black holes being predicted by General Relativity.

3.4 Derivation of Schwarzschild space-time

The usual derivation begins with the following metric for Minkowski space-time (using $c=1$),

$$ds^2 = dt^2 - dr^2 - r^2\left(d\theta^2 + \sin^2\theta d\varphi^2\right), \tag{3.10}$$

$$0 \le r < \infty,$$

and proposes a generalisation thereof as, or equivalent to,

$$ds^2 = F(r)dt^2 - G(r)dr^2 - R^2(r)\left(d\theta^2 + \sin^2\theta d\varphi^2\right), \tag{3.11}$$

where $F, G > 0$ and r is that which appears in the metric for Minkowski space-time, making r in Eq. (3.10) a parameter for the components of the metric tensor of Eq. (3.11). The functions $F(r), G(r), R(r)$ are to be determined such that the signature of metric (10) is maintained in metric (11), at $(+,-,-,-)$. The substitution $r^* = R(r)$ is then usually made, to get,

$$ds^2 = W(r^*)dt^2 - M(r^*)dr^{*2} - r^{*2}\left(d\theta^2 + \sin^2\theta d\varphi^2\right), \tag{3.11b}$$

Then the * is simply dropped, and with that it is just *assumed* that $0 \le r < \infty$ can be carried over from Eq. (3.10), to get [5,8,9,21–23,26,30,33,34,36,47–55,79],

$$ds^2 = e^{\lambda}dt^2 - e^{\beta}dr^2 - r^2\left(d\theta^2 + \sin^2\theta d\varphi^2\right), \tag{3.12}$$

$$0 \le r < \infty,$$

the exponential functions being introduced for subsequent ease of mathematical manipulations. It is then required that $e^{\lambda(r)} > 0$ and $e^{\beta(r)} > 0$ be determined so as to satisfy $R_{\mu\nu} = 0$.

Now note that in going from Eq. (3.11b) to Eq. (3.12), it is merely *assumed* that $R(0) = 0$, making $0 \le r^* < \infty$ (and hence in Eq. (3.12), $0 \le r < \infty$), since $r^* = R(r)$: but this cannot be known since $R(r)$ is *a priori* unknown [2,3]. One simply cannot treat r^* in Eq. (3.11b), and hence r in Eqs. (3.12) and (3.1), as the r in Eq. (3.10); contrary to the practice of the astrophysical scientists and their mathematician collaborators. Also note that Eq. (3.12) not only retains the signature -2, but **also retains the signature** $(+,-,-,-)$, because $e^{\lambda} > 0$ and $e^{\beta} > 0$ **by construction**. Thus, neither e^{λ} nor e^{β} can change sign [5,48,55,79]. This is a requirement since there is no possibility for Minkowski space-time (eq. 10) to change signature from $(+,-,-,-)$ to, for example, $(-,+,-,-)$.

The Standard Analysis then obtains the solution given by Eq. (3.1), wherein the constant m is claimed to be the mass generating the alleged associated gravitational field. Then by mere inspection of Eq. (3.1) the Standard Analysis asserts that there are two singularities, one at $r = 2m$ and one at $r = 0$. It is claimed that $r = 2m$ is a removable coordinate singularity, and that $r = 0$ a physical singularity. It is also asserted that $r = 2m$ gives the event horizon (the 'Schwarzschild radius') of a black hole, from which the 'escape velocity' is that of light (in vacuo), and that $r = 0$ is the position of the infinitely dense point-mass singularity of the black hole, produced by irresistible gravitational collapse.

However, these claims cannot be true. First, the construction of Eq. (3.12) to obtain Eq. (3.1) in satisfaction of $R_{\mu\nu} = 0$ is such that neither e^{λ} nor e^{β} can change sign, because $e^{\lambda} > 0$ and $e^{\beta} > 0$. Therefore the claim that r in metric (1) can take values less than $2m$ is false; a contradiction by the very construction of the metric (12) leading to metric (1). Furthermore, since neither e^{λ} nor e^{β} can ever be zero, the claim that $r = 2m$ is a removable coordinate singularity is also false. In addition, the true nature of r in both Eqs. (3.12) and (3.1) is entirely overlooked, and the geometric relations between the components of the metric tensor, fixed by the *form* of the line-element, are not applied, in consequence of which the Standard Analysis fatally falters.

In going from Eq. (3.11) to Eq. (3.12) the Standard Analysis has failed to realise that in Eq. (3.11) all the components of the metric tensor are functions of r by virtue of the fact that all the components of the metric tensor are functions of $R(r)$. Indeed, to illuminate this, consider the metric,

$$ds^2 = B(R)dR^2 + R^2(d\theta^2 + \sin^2\theta d\varphi^2),$$
$$B(R) > 0.$$

This is the most general expression for the metric of a three-dimensional spherically symmetric metric-space [30]. Now if R is a function of some parameter σ, then the metric in terms of σ is,

$$ds^2 = B(R(\sigma))\left(\frac{dR}{d\sigma}\right)^2 d\sigma^2 + R^2(\sigma)(d\theta^2 + \sin^2\theta d\varphi^2),$$
$$B(R(\sigma)) \equiv B(\sigma) > 0.$$

Relabelling the parameter σ with r gives precisely the generalisation of the spatial section of Minkowski space-time. Now Eq. (3.11) is given in terms of the parameter r of Minkowski space-time, not in terms of the function $R(r)$. In Eq. (3.11), set $G(r) = N(R(r))\,(dR/dr)^2$, then Eq. (3.11) becomes,

$$ds^2 = F(R(r))dt^2 - N(R(r))\left(\frac{dR}{dr}\right)^2 dr^2 - R^2(r)\left(d\theta^2 + \sin^2\theta d\varphi^2\right), \tag{3.11c}$$

or simply

$$ds^2 = F(R)dt^2 - N(R)dR^2 - R^2\left(d\theta^2 + \sin^2\theta d\varphi^2\right), \tag{3.11d}$$

wherein $R = R(r)$. Similarly, working backwards from Eq. (3.11b), using $r^* = R(r)$, Eq. (3.11b) becomes,

$$ds^2 = W(R(r))dt^2 - M(R(r))dR^2(r) - R^2(r)\left(d\theta^2 + \sin^2\theta d\varphi^2\right), \tag{3.11e}$$

or simply,

$$ds^2 = W(R)dt^2 - M(R)dR^2 - R^2\left(d\theta^2 + \sin^2\theta d\varphi^2\right),$$

wherein $R = R(r)$; and in terms of the parameter r of Minkowski space-time, this becomes,

$$ds^2 = W(r)dt^2 - M(r)\left(\frac{dR}{dr}\right)^2 dr^2 - R^2(r)\left(d\theta^2 + \sin^2\theta d\varphi^2\right). \tag{3.11f}$$

Writing $W(r) = F(r)$ and $G(r) = M(r)\,(dR/dr)^2$ gives,

$$ds^2 = F(r)dt^2 - G(r)dr^2 - R^2(r)\left(d\theta^2 + \sin^2\theta d\varphi^2\right),$$

which is Eq. (3.11). So Eq. (3.11) is a disguised form of Eq. (3.11d), and so there is no need at all for the 'transformations' applied by the astrophysical scientists to get their Eq. (3.12), from which they get their Eq. (3.1). In other words, what the astrophysical scientists call r in their Eq. (3.1) is actually $R(r)$, for which they have not given any definite admissible form in terms of the parameter r, and they incorrectly treat their $R(r)$, labelled r in Eqs. (3.12) and (3.1), as the r in Eq. (3.10), manifest in the miscarrying over of the range $0 \le r < \infty$ from Eq. (3.10).

Nonetheless, $R(r)$ is still an *a priori* unknown function, and so it cannot be arbitrarily asserted that $R(0) = 0$; contrary to the assertions of the astrophysical scientists. It is now quite plain that the 'transformations' used by the Standard Analysis in going from Eq. (3.11) to Eq. (3.12) are rather pointless, since all the relations are contained in Eq. (3.11) already, and by its pointless procedure the Standard Analysis has confused matters and thereby introduced a major error concerning the range on the quantity r in its expression (12) and hence in its expression (1). One can of course, solve Eq. (3.11d), subject to $R_{\mu\nu} = 0$, in terms of $R(r)$, without determining the admissible form of $R(r)$. However, the range of $R(r)$ **must** be ascertained by means of boundary conditions fixed by the very form of the line-element in which it appears. And if it is required that the parameter r appear explicitly in the solution, by means of a mapping between the manifolds described by Eqs. (3.10) and (3.11), then the admissible form of $R(r)$ must also be ascertained, in which case r in Minkowski space is a parameter, and Minkowski space a parametric space, for the related quantities in Schwarzschild space. To highlight further, rewrite Eq. (3.11) as,

$$ds^2 = A\left(R_c\right) dt^2 - B\left(R_c\right) dR_c^2 - R_c^2\left(d\theta^2 + \sin^2\theta d\varphi^2\right), \tag{3.13}$$

where $A\left(R_c\right), B\left(R_c\right), R_c\left(r\right) > 0$. The solution for $R_{\mu\nu} = 0$ then takes the form,

$$ds^2 = \left(1 + \frac{\kappa}{R_c}\right) dt^2 - \left(1 + \frac{\kappa}{R_c}\right)^{-1} dR_c^2 - R_c^2\left(d\theta^2 + \sin^2\theta d\varphi^2\right),$$

$$R_c = R_c(r),$$

where κ is a constant. There are two cases to consider; $\kappa > 0$ and $\kappa < 0$. In conformity with the astrophysical scientists take $\kappa < 0$, and so set $\kappa = -\alpha$, $\alpha > 0$. Then the solution takes the form,

$$ds^2 = \left(1 - \frac{\alpha}{R_c}\right) dt^2 - \left(1 - \frac{\alpha}{R_c}\right)^{-1} dR_c^2 - R_c^2\left(d\theta^2 + \sin^2\theta d\varphi^2\right), \tag{3.14}$$

$$R_c = R_c(r),$$

where $\alpha > 0$ is a constant. It remains to determine the admissible form of $R_c\left(r\right)$, which, from **Section II**, is the inverse square root of the Gaussian curvature of a spherically symmetric geodesic surface in the spatial section of the manifold associated with Eq. (3.14), owing to the metrical *form* of Eq. (3.14). From **Section II** herein the proper radius associated with metric (14) is,

$$R_p = \int \frac{dR_c}{\sqrt{1 - \frac{\alpha}{R_c}}} = \sqrt{R_c\left(R_c - \alpha\right)} + \alpha \ln\left[\sqrt{R_c} + \sqrt{R_c - \alpha}\right] + k, \tag{3.15}$$

where k is a constant. Now for some r_o, $R_p\left(r_o\right) = 0$. Then by (15) it is required that $R_c\left(r_o\right) = \alpha$ and $k = -\alpha \ln\sqrt{\alpha}$, so

$$R_p\left(r\right) = \sqrt{R_c\left(R_c - \alpha\right)} + \alpha \ln\left[\frac{\sqrt{R_c} + \sqrt{R_c - \alpha}}{\sqrt{\alpha}}\right], \tag{3.16}$$

$$R_c = R_c(r).$$

It is thus also determined that the Gaussian curvature of the spherically symmetric geodesic surface of the spatial section ranges not from ∞ to 0, as it does for Euclidean 3-space, but from α^{-2} to 0. This is an inevitable consequence of the peculiar non-Euclidean geometry described by metric (14).

Schwarzschild's true solution, Eq. (3.2), must be a particular case of the general expression sought for $R_c(r)$. Brillouin's solution [2, 35] must also be a particular case, viz.,

$$ds^2 = \left(1 - \frac{\alpha}{r + \alpha}\right) dt^2 - \left(1 - \frac{\alpha}{r + \alpha}\right)^{-1} dr^2 - \left(r + \alpha\right)^2\left(d\theta^2 + \sin^2\theta d\varphi^2\right), \tag{3.17}$$

$$0 < r < \infty,$$

and Droste's solution [46] must as well be a particular solution, viz.,

$$ds^2 = \left(1 - \frac{\alpha}{r}\right) dt^2 - \left(1 - \frac{\alpha}{r}\right)^{-1} dr^2 - r^2\left(d\theta^2 + \sin^2\theta d\varphi^2\right).$$

$$\alpha < r < \infty. \tag{3.18}$$

All these solutions must be particular cases in an infinite set of equivalent metrics [34]. The only admissible form for $R_c(r)$ is [32],

$$R_c(r) = \left(|r - r_o|^n + \alpha^n\right)^{\frac{1}{n}} = \frac{1}{\sqrt{K(r)}},$$

$$r \in \Re, \quad n \in \Re^+, \quad r \neq r_o, \tag{3.19}$$

where r_o and n are entirely arbitrary constants. So the solution for $R_{\mu\nu} = 0$ is,

$$ds^2 = \left(1 - \frac{\alpha}{R_c}\right) dt^2 - \left(1 - \frac{\alpha}{R_c}\right)^{-1} dR_c^2 - R_c^2 \left(d\theta^2 + \sin^2\theta d\varphi^2\right),$$

$$R_c(r) = \left(|r - r_o|^n + \alpha^n\right)^{\frac{1}{n}} = \frac{1}{\sqrt{K(r)}},$$

$$r \in \Re, \quad n \in \Re^+, \quad r \neq r_o. \tag{3.20}$$

Then if $r_o = 0$, $r > r_o$, $n = 1$, Brillouin's solution Eq. (3.17) results. If $r_o = 0$, $r > r_o$, $n = 3$, then Schwarzschild's actual solution Eq. (3.2) results. If $r_o = \alpha$, $r > r_o$, $n = 1$, then Droste's solution Eq. (3.18) results, which is the correct solution in the particular metric of Eq. (3.1). In addition the required infinite set of equivalent metrics is thereby obtained, all of which are asymptotically Minkowski space-time. Furthermore, if the constant α is set to zero, Eqs. (3.20) reduces to Minkowski space-time, and if in addition r_o is set to zero, then the usual Minkowski metric of Eq. (3.10) is obtained. The significance of the term $|r - r_o|$ was given in **Section II**: it is a shift of the location of the centre of spherical symmetry in the spatial section of the auxiliary manifold away from the origin of coordinates of the auxiliary manifold, along a radial line, to a point at distance r_o from the origin of coordinates. The point r_o in the auxiliary manifold is mapped into the point $R_p(r_o) = 0$ in Schwarzschild space, irrespective of the choice of the parametric point r_o in the auxiliary manifold. Minkowski space-time is the auxiliary manifold for Schwarzschild space-time. Strictly speaking, the parameter r of the auxiliary manifold need not be incorporated into metric (20), in which case the metric is defined only on $\alpha < R_c < \infty$. I have retained the quantity r to fully illustrate its rôle as a parameter and the part played by Minkowski space-time as an auxiliary manifold.

It is clear from expressions (20) that there is only one singularity, at the arbitrary constant r_o, where $R_c(r_o) = \alpha \;\forall\; r_o \;\forall\; n$ and $R_p(r_o) = 0 \;\forall\; r_o \;\forall\; n$, and that **all** components of the metric tensor are affected by the constant α. Hence, the "removal" of the singularity at $r = 2m$ in Eq. (3.1) is fallacious because it is clear from expressions (20), in accordance with the intrinsic geometry of the line-element as given in **Section II**, and the generalisation at Eq. (3.13), that there is no singularity at $r = 0$ in Eq. (3.1) and so $0 \le r \le 2m$ therein is meaningless [1–5,32,41,42,46,57,62]. The Standard claims for Eq. (3.1) violate the geometry fixed by the *form* of its line-element and contradict the generalisations at Eqs. (3.11) and (3.12) from which it has been obtained by the Standard method. There is therefore no black hole associated with Eq. (3.1) since there is no black hole associated with Eq. (3.2) and none with Eq. (3.20), of which Schwarzschild's actual solution, Eq. (3.2), Brillouin's solution, Eq. (3.17), and Droste's solution, Eq. (3.18), are just particular equivalent cases.

In the case of $\kappa > 0$ the proper radius of the line-element is,

$$R_p = \int \frac{dR_c}{\sqrt{1 + \frac{\kappa}{R_c}}} = \sqrt{R_c\left(R_c + \kappa\right)} - \kappa \ln\left[\sqrt{R_c} + \sqrt{R_c + \kappa}\right] + k,$$

$$R_c = R_c(r),$$

where k is a constant. Now for some r_o, $R_p(r_o) = 0$, so it is required that $R_c(r_o) = 0$ and $k = \kappa \ln \sqrt{\kappa}$. The proper radius is then,

$$R_p(r) = \sqrt{R_c\left(R_c + \kappa\right)} - \kappa \ln\left[\frac{\sqrt{R_c} + \sqrt{R_c + \kappa}}{\sqrt{\kappa}}\right],$$

$$R_c = R_c(r).$$

The admissible form of $R_c(r)$ must now be determined. According to Einstein, the metric must be asymptotically Minkowski space-time. Since $\kappa > 0$ by hypothesis, the application of the (spurious) argument for Newtonian approximation used by the astrophysical scientists cannot be applied here. There are no other boundary conditions that provide any means for determining the value of κ, and so it remains indeterminable. The only form that meets the condition $R_c(r_o) = 0$ and the requirement of asymptotic Minkowski space-time is,

$$R_c(r) = |r - r_o| = \frac{1}{\sqrt{K}},$$

$$r \in \Re,$$

where r_o is entirely arbitrary. Then $R_p(r_o) = 0 \ \forall\ r_o$ and $R_c(r_o) = 0 \ \forall\ r_o$, and so, if explicit reference to the auxiliary manifold of Minkowski space-time is not desired, $R_c(r)$ becomes superfluous and can be simply replaced by $R_c(r) = |r - r_o| = \rho$, $0 < \rho < \infty$. Thus, points in the spatial section of this space-time have the very same properties of points in the spatial section of Minkowski space-time. The line-element is again singular at only one point; $\rho = 0$ (i.e. in the case of explicit inclusion of the auxiliary manifold, only at the point $r = r_o$). The signature of this metric is always $(+,-,-,-)$. Clearly there is no possibility for a black hole in this case either.

The usual form of Eq. (3.1) in isotropic coordinates is,

$$ds^2 = \frac{\left(1 - \frac{m}{2r}\right)^2}{\left(1 + \frac{m}{2r}\right)^2} dt^2 - \left(1 + \frac{m}{2r}\right)^4 \left[dr^2 + r^2\left(d\theta^2 + \sin^2\theta d\varphi^2\right)\right],$$

wherein it is again alleged that r can go down to zero. This expression has the very same metrical form as Eq. (3.13) and so shares the very same geometrical character. Now the coefficient of dt^2 is zero when $r = m/2$, which, according to the astrophysical scientists, marks the 'radius' or 'event horizon' of a black hole, and where m is the alleged point-mass of the black hole singularity located at $r = 0$, just as in Eq. (3.1). This further amplifies the fact that the quantity r appearing in both Eq. (3.1) and its isotropic coordinate form is not a distance in the manifold described by these line-elements. Applying the intrinsic geometric relations detailed in **Section II** above it is clear that the inverse square root of the Gaussian curvature of a spherically symmetric geodesic surface in the spatial section of the isotropic coordinate line-element is given by,

$$R_c(r) = r\left(1 + \frac{m}{2r}\right)^2$$

and the proper radius is given by,

$$R_p(r) = r + m \ln\left(\frac{2r}{m}\right) - \frac{m^2}{4r}.$$

Hence, $R_c(m/2) = 2m$, and $R_p(m/2) = 0$, which are scalar invariants necessarily consistent with Eq. (3.20). Furthermore, applying the same geometrical analysis leading to Eq. (3.20), the generalised solution in isotropic coordinates is [57],

$$ds^2 = \frac{\left(1 - \frac{\alpha}{4h}\right)^2}{\left(1 + \frac{\alpha}{4h}\right)^2} dt^2 - \left(1 + \frac{\alpha}{4h}\right)^4 \left[dh^2 + h^2\left(d\theta^2 + \sin^2\theta d\varphi^2\right)\right],$$

$$h = h(r) = \left[|r - r_o|^n + \left(\frac{\alpha}{4}\right)^n\right]^{\frac{1}{n}},$$

$$r \in \Re, \quad n \in \Re^+, \quad r \neq r_o,$$

wherein r_o and n are entirely arbitrary constants. Then,

$$R_c(r) = h(r)\left(1 + \frac{\alpha}{4h(r)}\right)^2 = \frac{1}{\sqrt{K(r)}},$$

$$R_p(r) = h(r) + \frac{\alpha}{2}\ln\left(\frac{4h(r)}{\alpha}\right) - \frac{\alpha^2}{16h(r)},$$

and so

$$R_c(r_o) = \alpha, \quad R_p(r_o) = 0, \quad \forall\, r_o\ \forall\, n,$$

which are scalar invariants, in accordance with Eq. (3.20). Clearly in these isotropic coordinate expressions r does not in itself denote any distance in the manifold, just as it does not in itself denote any distance in Eq. (3.20) of which Eqs. (3.1) and (3.2) are particular cases. It is a parameter for the Gaussian curvature of the spherically symmetric geodesic surface in the spatial section and for the proper radius (i.e. the radial geodesic distance from the point at the centre of spherical symmetry of the spatial section). The 'interior' of the alleged Schwarzschild black hole does not form part of the solution space of the Schwarzschild manifold [2,4,5,32,41,42,57,61–63].

In the same fashion it is easily proven [32,61] that the general expression for the Kerr-Newman geometry is given by,

$$ds^2 = \frac{\Delta}{\rho^2}\left(dt - a\sin^2\theta d\varphi^2\right)^2 - \frac{\sin^2\theta}{\rho^2}\left[\left(R^2 + a^2\right)d\varphi - adt\right]^2 - \frac{\rho^2}{\Delta}dR^2 - \rho^2 d\theta^2$$

$$R = R(r) = \left(|r - r_o|^n + \beta^n\right)^{\frac{1}{n}}, \quad \beta = \frac{\alpha}{2} + \sqrt{\frac{\alpha^2}{4} - \left(q^2 + a^2\cos^2\theta\right)}, \quad a^2 + q^2 < \frac{\alpha^2}{4},$$

$$a = \frac{2L}{\alpha}, \quad \rho^2 = R^2 + a^2\cos^2\theta, \quad \Delta = R^2 - \alpha R + q^2 + a^2,$$

$$r \in \Re, \quad n \in \Re^+, \quad r \neq r_o.$$

The Kruskal-Szekeres coordinates, the Eddington-Finkelstein coordinates, and the Regge-Wheeler coordinates do not take into account the rôle of Gaussian curvature of the spherically symmetric geodesic surface in the spatial section of the Schwarzschild manifold [64], and so they thereby violate the geometric form of the line-element, making them invalid.

The foregoing amplifies the inadmissibility of the introduction of the Newtonian potential into Schwarzschild space-time. The Newtonian potential is a **two-body** concept; it is defined as the work done per unit mass against the gravitational field. There is no meaning to a Newtonian potential for a single mass in an otherwise empty Universe. Newton's theory of gravitation is defined in terms of the interaction of two masses in a space for which the 'Principle of Superposition' applies. In Newton's theory there is no limit set to the number of masses that can be piled up in space, although the analytical relations for the gravitational interactions of many bodies upon one another quickly become intractable. In Einstein's theory matter cannot be piled up in a given space-time because the matter itself determines the structure of the space-time containing the matter. It is clearly impossible for Schwarzschild space-time, which is alleged by the astrophysical scientists to contain one mass in an otherwise totally empty Universe, to reduce to or otherwise contain an expression that is defined in terms of the *a priori* interaction of two masses. This is illustrated even further by writing Eq. (3.1) in terms of c and G explicitly,

$$ds^2 = \left(c^2 - \frac{2Gm}{r}\right)dt^2 - c^2\left(c^2 - \frac{2Gm}{r}\right)^{-1}dr^2 - r^2\left(d\theta^2 + \sin^2\theta d\varphi^2\right).$$

The term $2Gm/r$ is the square of the Newtonian escape velocity from a mass m. And so the astrophysical scientists assert that when the "escape velocity" is that of light in vacuum, there is an event horizon ("Schwarzschild radius") and hence a black hole. But escape velocity is a concept that involves **two bodies** - one body escapes from another body. Even though one mass appears in the expression for escape velocity, it cannot be determined without recourse to a fundamental two-body gravitational interaction. Recall that Newton's Universal Law of Gravitation is,

$$F_g = -G\frac{mM}{r^2},$$

where G is the gravitational constant and r is the distance between the centre of mass of m and the centre of mass of M. A centre of mass is an infinitely dense point-mass, but it is not a physical object; merely a mathematical artifice. Newton's gravitation is clearly defined in terms of the interaction of two bodies. Newton's gravitational potential Φ is defined as,

$$\Phi = \lim_{\sigma\to\infty}\int_\sigma^r -\frac{F_g}{m}dr = -G\frac{M}{r},$$

which is the work done per unit mass in the gravitational field due to masses M and m. This is a two-body concept. The potential energy P of a mass m in the gravitational field due to masses M and m is therefore given by,

$$P = m\Phi = -G\frac{mM}{r},$$

which is clearly a two-body concept.

Similarly, the velocity required by a mass m to escape from the gravitational field due to masses M and m is determined by,

$$F_g = -G\frac{mM}{r^2} = ma = m\frac{dv}{dt} = mv\frac{dv}{dr}.$$

Separating variables and integrating gives,

$$\int_v^0 mv\,dv = \lim_{r_f \to \infty} \int_R^{r_f} -GmM\,\frac{dr}{r^2},$$

whence

$$v = \sqrt{\frac{2GM}{R}},$$

where R is the radius of the mass M. Thus, escape velocity necessarily involves two bodies: m escapes from M. In terms of the conservation of kinetic and potential energies,

$$K_i + P_i = K_f + P_f$$

whence,

$$\frac{1}{2}mv^2 - G\frac{mM}{R} = \frac{1}{2}mv_f^2 - G\frac{mM}{r_f}.$$

Then as $r_f \to \infty$, $v_f \to 0$, and the escape velocity of m from M is,

$$v = \sqrt{\frac{2GM}{R}}.$$

Once again, the relation is derived from a two-body gravitational interaction.

The consequence of all this for black holes and their associated gravitational waves is that there can be no gravitational waves generated by black holes because the latter are fictitious.

3.5 The prohibition of point-mass singularities

The black hole is alleged to contain an infinitely dense point-mass singularity, produced by irresistible gravitational collapse (see for example [17, 24, 77], for the typical claim). According to Hawking [80]:

> *"The work that Roger Penrose and I did between 1965 and 1970 showed that, according to general relativity, there must be a singularity of infinite density, within the black hole."*

The singularity of the alleged Big Bang cosmology is, according to many proponents of the Big Bang, also infinitely dense. Yet according to Special Relativity, infinite densities are forbidden because their existence implies that a material object can acquire the speed of light c in vacuo (or equivalently, the existence of infinite energies), thereby violating the very basis of Special Relativity. Since General Relativity cannot violate Special Relativity, General Relativity must therefore also forbid infinite densities. Point-mass singularities are alleged to be infinitely dense objects. Therefore, point-mass singularities are forbidden by the Theory of Relativity.

Let a cuboid rest-mass m_0 have sides of length L_0. Let m_0 have a relative speed $v < c$ in the direction of one of three mutually orthogonal Cartesian axes attached to an observer of rest-mass M_0. According to the observer M_0, the moving mass m is

$$m = \frac{m_0}{\sqrt{1 - \frac{v^2}{c^2}}},$$

and the volume V thereof is

$$V = L_0^3\sqrt{1 - \frac{v^2}{c^2}}.$$

Thus, the density D is

$$D = \frac{m}{V} = \frac{m_0}{L_0^3\left(1 - \frac{v^2}{c^2}\right)},$$

and so $v \to c \;\Rightarrow\; D \to \infty$. Since, according to Special Relativity, no material object can acquire the speed c (this would require an infinite energy), infinite densities are forbidden by Special Relativity, and so point-mass singularities are forbidden. Since General Relativity cannot violate Special Relativity, it too must thereby forbid infinite densities and hence forbid point-mass singularities. It does not matter how it is alleged that a point-mass singularity is generated by General Relativity because the infinitely dense point-mass cannot be reconciled with Special Relativity. Point-charges too are therefore forbidden by the Theory of Relativity since there can be no charge without mass.

It is nowadays routinely claimed that many black holes have been found. The signatures of the black hole are (a) an infinitely dense 'point-mass' singularity and (b) an event horizon. Nobody has ever found an infinitely dense 'point-mass' singularity and nobody has ever found an event horizon, so nobody has ever assuredly found a black hole. It takes an infinite amount of observer time to verify a black hole event horizon [24,28,36,48,54,56,71]. Nobody has been around and nobody will be around for an infinite amount of time and so no observer can ever verify the presence of an event horizon, and hence a black hole, in principle, and so the notion is irrelevant to physics. All reports of black holes being found are patently false; the product of wishful thinking.

3.6 Laplace's alleged black hole

It has been claimed by the astrophysical scientists that a black hole has an escape velocity c (or $\geq c$, the speed of light in vacuo) [6,12–14,16,18,19,24,28,76,78,80–82]. Chandrasekhar [24] remarked,

> *"Let me be more precise as to what one means by a black hole. One says that a black hole is formed when the gravitational forces on the surface become so strong that light cannot escape from it.*
>
> *... A trapped surface is one from which light cannot escape to infinity."*

According to Hawking,

> *"Eventually when a star has shrunk to a certain critical radius, the gravitational field at the surface becomes so strong that the light cones are bent inward so much that the light can no longer escape. According to the theory of relativity, nothing can travel faster than light. Thus, if light cannot escape, neither can anything else. Everything is dragged back by the gravitational field. So one has a set of events, a region of space-time from which it is not possible to escape to reach a distant observer. Its boundary is called the event horizon. It coincides with the paths of the light rays that just fail to escape from the black hole."*

However, according to the alleged properties of a black hole, nothing at all can even **leave** the black hole. In the very same paper Chandrasekhar made the following quite typical contradictory assertion propounded by the astrophysical scientists:

> *"The problem we now consider is that of the gravitational collapse of a body to a volume so small that a trapped surface forms around it; as we have stated, from such a surface no light can emerge."*

Hughes [28] reiterates,

> *"Things can go into the horizon (from $r > 2M$ to $r < 2M$), but they cannot get out; once inside, all causal trajectories (timelike or null) take us inexorably into the classical singularity at $r = 0$.*
>
> *"The defining property of black holes is their event horizon. Rather than a true surface, black holes have a 'one-way membrane' through which stuff can go in but cannot come out."*

Taylor and Wheeler [25] assert,

> *"... Einstein predicts that nothing, not even light, can be successfully launched outward from the horizon ... and that light launched outward EXACTLY at the horizon will never increase its radial position by so much as a millimeter."*

In the Dictionary of Geophysics, Astrophysics and Astronomy [78], one finds the following assertions:

> "**black hole** *A region of space-time from which the escape velocity exceeds the velocity of light. In Newtonian gravity the escape velocity from the gravitational pull of a spherical star of mass M and radius R is*
>
> $$v_{esc} = \sqrt{\frac{2GM}{R}},$$
>
> *where G is Newton's constant. Adding mass to the star (increasing M), or compressing the star (reducing R) increases v_{esc}. When the escape velocity exceeds the speed of light c, even light cannot escape, and the star becomes a* black hole. *The required radius R_{BH} follows from setting v_{esc} equal to c:*
>
> $$R_{BH} = \frac{2GM}{c^2}.$$
>
> ... *"In General Relativity for spherical black holes (Schwarzschild black holes), exactly the same expression R_{BH} holds for the surface of a black hole. The surface of a black hole at R_{BH} is a null surface, consisting of those photon trajectories (null rays) which just do not escape to infinity. This surface is also called the black hole horizon."*

Now if its escape velocity is really that of light in vacuo, then by definition of escape velocity, light would escape from a black hole, and therefore be seen by all observers. If the escape velocity of the black hole is greater than that of light in vacuo, then light could emerge but not escape, and so there would always be a class of observers that could see it. Not only that, if the black hole had an escape velocity, then material objects with an initial velocity less than the alleged escape velocity, could leave the black hole, and therefore be seen by a class of observers, but not escape (just go out, come to a stop and then fall back), even if the escape velocity is $\geq c$. Escape velocity does not mean that objects cannot leave; it only means they cannot escape if they have an initial velocity less than the escape velocity. So on the one hand it is claimed that black holes have an escape velocity c, but on the other hand that nothing, not even light, can even leave the black hole. The claims are contradictory - nothing but a meaningless play on the words "escape velocity" [67, 68]. Furthermore, as demonstrated in **Section III**, escape velocity is a **two-body** concept, whereas the black hole is derived not from a two-body gravitational interaction, but from a one-body concept. The black hole has no escape velocity.

It is also routinely asserted that the theoretical Michell-Laplace (M-L) dark body of Newton's theory, which has an escape velocity $\geq c$, is a kind of black hole [6, 11, 14, 24, 78, 80] or that Newton's theory somehow predicts *"the radius of a black hole"* [25]. Hawking remarks,

> *"On this assumption a Cambridge don, John Michell, wrote a paper in 1783 in the* Philosophical *Transactions of the Royal Society of London. In it, he pointed out that a star that was sufficiently massive and compact would have such a strong gravitational field that light could not escape. Any light emitted from the surface of the star would be dragged back by the star's gravitational attraction before it could get very far. Michell suggested that there might be a large number of stars like this. Although we would not be able to see them because light from them would not reach us, we could still feel their gravitational attraction. Such objects are what we now call black holes, because that is what they are – black voids in space."*

But the M-L dark body is not a black hole. The M-L dark body possesses an escape velocity, whereas the black hole has no escape velocity; objects can leave the M-L dark body, but nothing can leave the black hole; it does not require irresistible gravitational collapse, whereas the black hole does; it has no infinitely dense point-mass singularity, whereas the black hole does; it has no event horizon, whereas the black hole does; there is always a class of observers that can see the M-L dark body [67,68], but there is no class of observers that can see the black hole; the M-L dark body can persist in a space which contains other matter and interact with that matter, but the space-time of the "Schwarzschild" black hole (and variants thereof) is devoid of matter by construction and so it cannot interact with anything. Thus the M-L dark body does not possess the characteristics of the alleged black hole and so it is not a black hole.

3.7 Black hole interactions and gravitational collapse

The literature abounds with claims that black holes can interact in such situations as binary systems, mergers, collisions, and with surrounding matter generally. According to Chandrasekhar [24], for example, who also cites S. Hawking,

> *"From what I have said, collapse of the kind I have described must be of frequent occurrence in the Galaxy; and black-holes must be present in numbers comparable to, if not exceeding, those of the pulsars. While the black-holes will not be visible to external observers, they can nevertheless interact with one another and with the outside world through their external fields.*
>
> *"In considering the energy that could be released by interactions with black holes, a theorem of Hawking is useful. Hawking's theorem states that* **in the interactions involving black holes, the total surface area of the boundaries of the black holes can never decrease***; it can at best remain unchanged (if the conditions are stationary).*
>
> *"Another example illustrating Hawking's theorem (and considered by him) is the following. Imagine two spherical (Schwarzschild) black holes, each of mass $\frac{1}{2}M$, coalescing to form a single black hole; and let the black hole that is eventually left be, again, spherical and have a mass $\overline{M}$. Then Hawking's theorem requires that*
>
> $$16\pi\overline{M}^2 \geq 16\pi\left[2\left(\frac{1}{2}M\right)^2\right] = 8\pi M^2$$
>
> *or*
>
> $$\overline{M} \geq M/\sqrt{2}.$$
>
> *Hence the maximum amount of energy that can be released in such a coalescence is*
>
> $$M\left(1 - 1/\sqrt{2}\right)c^2 = 0.293Mc^2."$$

Hawking [80] says,

> *"Also, suppose two black holes collided and merged together to form a single black hole. Then the area of the event horizon of the final black hole would be greater than the sum of the areas of the event horizons of the original black holes."*

According to Schutz [48],

> *"... Hawking's area theorem: in any physical process involving a horizon, the area of the horizon cannot decrease in time. ... This fundamental theorem has the result that, while two black holes can collide and coalesce, a single black hole can never bifurcate spontaneously into two smaller ones.*
>
> *"Black holes produced by supernovae would be much harder to observe unless they were part of a binary system which survived the explosion and in which the other star was not so highly evolved."*

Townsend [56] also arbitrarily applies the 'Principle of Superposition' to obtain charged black hole (Reissner-Nordström) interactions as follows:

> "*The extreme RN in isotropic coordinates is*
>
> $$ds^2 = V^{-2}dt^2 + V^2\left(d\rho^2 + \rho^2 d\Omega^2\right)$$
>
> *where*
>
> $$V = 1 + \frac{M}{\rho}$$
>
> *This is a special case of the multi black hole solution*
>
> $$ds^2 = V^{-2}dt^2 + V^2 d\vec{x} \cdot d\vec{x}$$
>
> *where $d\vec{x} \cdot d\vec{x}$ is the Euclidean 3-metric and V is any solution of $\nabla^2 V = 0$. In particular*
>
> $$V = 1 + \sum_{i=1}^{N} \frac{M_i}{\left|\vec{x} - \vec{x}_i\right|}$$
>
> *yields the metric for N extreme black holes of masses M_i at positions $\overline{x}_i$.*

Now Einstein's field equations are non-linear, so the 'Principle of Superposition' **does not apply** [51,67,79]. Therefore, before one can talk of black hole binary systems and the like it must first be proven that the two-body system is theoretically well-defined by General Relativity. This can be done in only two ways:

(a) Derivation of an exact solution to Einstein's field equations for the two-body configuration of matter; or

(b) Proof of an existence theorem.

There are no known solutions to Einstein's field equations for the interaction of two (or more) masses (charged or not), so option (a) has never been fulfilled. No existence theorem has ever been proven, by which Einstein's field equations can even be said to admit of latent solutions for such configurations of matter, and so option (b) has never been fulfilled. The "Schwarzschild" black hole is allegedly obtained from a line-element satisfying $Ric = 0$. For the sake of argument, assuming that black holes are predicted by General Relativity as alleged in relation to metric (1), since $Ric = 0$ is a statement that there is no matter in the Universe, one cannot simply insert a second black hole into the space-time of $Ric = 0$ of a given black hole so that the resulting two black holes (each obtained separately from $Ric = 0$) mutually persist in and mutually interact in a mutual space-time that **by construction contains no matter**! One cannot simply assert by an analogy with Newton's theory that two black holes can be components of binary systems, collide or merge [51, 67, 68], because the 'Principle of Superposition' does not apply in Einstein's theory. Moreover, General Relativity has to date been unable to account for the simple experimental fact that two fixed bodies will approach one another upon release. Thus, black hole binaries, collisions, mergers, black holes from supernovae, and other black hole interactions are all invalid concepts.

Much of the justification for the notion of irresistible gravitational collapse into an infinitely dense point-mass singularity, and hence the formation of a black hole, is given to the analysis due to Oppenheimer and Snyder [69]. Hughes [28] relates it as follows;

> "*In an idealized but illustrative calculation, Oppenheimer and Snyder ... showed in 1939 that a black hole in fact does form in the collapse of ordinary matter. They considered a 'star' constructed out of a pressureless 'dustball'. By Birkhof's Theorem, the entire exterior of this dustball is given by the Schwarzschild metric ... . Due to the self-gravity of this 'star', it immediately begins to collapse. Each mass element of the pressureless star follows a geodesic trajectory toward the star's center; as the collapse proceeds, the star's density increases and more of the space-time is described by the Schwarzschild metric. Eventually, the surface passes through $r = 2M$. At this point, the Schwarzschild*

exterior includes an event horizon: A black hole has formed. Meanwhile, the matter which formerly constituted the star continues collapsing to ever smaller radii. In short order, all of the original matter reaches $r = 0$ and is compressed (classically!) into a singularity[4].

[4]*"Since all of the matter is squashed into a point of zero size, this classical singularity must be modified in a a complete, quantum description. However, since all the singular nastiness is hidden behind an event horizon where it is causally disconnected from us, we need not worry about it (at least for astrophysical black holes)."*

Note that the 'Principle of Superposition' has again been arbitrarily applied by Oppenheimer and Snyder, from the outset. They first **assume** a relativistic universe in which there are multiple mass elements present *a priori*, where the 'Principle of Superposition' however, does not apply, and despite there being no solution or existence theorem for such configurations of matter in General Relativity. Then all these mass elements "collapse" into a central point (zero volume; infinite density). Such a collapse has however not been given any specific general relativistic mechanism in this argument; it is simply asserted that the "collapse" is due to self-gravity. But the "collapse" cannot be due to Newtonian gravitation, given the resulting black hole, which does not occur in Newton's theory of gravitation. And a Newtonian universe cannot "collapse" into a non-Newtonian universe. Moreover, the black hole so formed is in an empty universe, since the "Schwarzschild black hole" relates to $Ric = 0$, a space-time that by construction contains no matter. Nonetheless, Oppenheimer and Snyder permit, within the context of General Relativity, the presence of and the gravitational interaction of many mass elements, which coalesce and collapse into a point of zero volume to form an infinitely dense point-mass singularity, when there is no demonstrated general relativistic mechanism by which any of this can occur.

Furthermore, nobody has ever observed a celestial body undergo irresistible gravitational collapse and there is no laboratory evidence whatsoever for such a phenomenon.

3.8 Further consequences for gravitational waves

The question of the localisation of gravitational energy is related to the validity of the field equations $R_{\mu\nu} = 0$, for according to Einstein, matter is the cause of the gravitational field and the causative matter is described in his theory by a mathematical object called the energy-momentum tensor, which is coupled to geometry (i.e. space-time) by his field equations, so that matter causes space-time curvature (his gravitational field). Einstein's field equations,

> *"... couple the gravitational field (contained in the curvature of space-time) with its sources."* [36]

> *"Since gravitation is determined by the matter present, the same must then be postulated for geometry, too. The geometry of space is not given a priori, but is only determined by matter."* [53]

> *"Again, just as the electric field, for its part, depends upon the charges and is instrumental in producing mechanical interaction between the charges, so we must assume here that* **the metrical field** *(or, in mathematical language, the tensor with components g_{ik})* **is related to the material filling the world."** [5]

> *"... we have, in following the ideas set out just above, to discover the* **invariant law of gravitation, according to which matter determines the components $\Gamma^{\alpha}_{\beta\iota}$ of the gravitational field**, *and which replaces the Newtonian law of attraction in Einstein's Theory."* [5]

> *"Thus the equations of the gravitational field also contain the equations for the matter (material particles and electromagnetic fields) which produces this field."* [51]

> *"Clearly, the mass density, or equivalently, energy density $\varrho(\vec{x}, t)$ must play the role as a source. However, it is the 00 component of a tensor $T_{\mu\nu}(x)$, the mass-energy-momentum distribution of matter. So, this tensor must act as the source of the gravitational field."* [10]

"In general relativity, the stress-energy or energy-momentum tensor T^{ab} acts as the source of the gravitational field. It is related to the Einstein tensor and hence to the curvature of space-time via the Einstein equation". [79]

Qualitatively Einstein's field equations are:

Spacetime geometry = -κ × *causative matter* (i.e. material sources)

where *causative matter* is described by the energy-momentum tensor and κ is a constant. The space-time geometry is described by a mathematical object called Einstein's tensor, $G_{\mu\nu}$, $(\mu, \nu = 0, 1, 2, 3)$ and the energy-momentum tensor is $T_{\mu\nu}$. So Einstein's full field equations are[3]:

$$G_{\mu\nu} = R_{\mu\nu} - \frac{1}{2}Rg_{\mu\nu} = -\kappa T_{\mu\nu}. \tag{3.21}$$

Einstein asserted that his 'Principle of Equivalence' and his laws of Special Relativity must hold in a sufficiently small region of his gravitational field. Here is what Einstein [52] himself said in 1954, the year before his death:

"Let now K be an inertial system. Masses which are sufficiently far from each other and from other bodies are then, with respect to K, free from acceleration. We shall also refer these masses to a system of co-ordinates K',uniformly accelerated with respect to K. Relatively to K' all the masses have equal and parallel accelerations; with respect to K' they behave just as if a gravitational field were present and K' were unaccelerated. Overlooking for the present the question as to the 'cause' of such a gravitational field, which will occupy us later, there is nothing to prevent our conceiving this gravitational field as real, that is, the conception that K' is 'at rest' and a gravitational field is present we may consider as equivalent to the conception that only K is an 'allowable' system of co-ordinates and no gravitational field is present. The assumption of the complete physical equivalence of the systems of coordinates, K and K', we call the 'principle of equivalence'; this principle is evidently intimately connected with the law of the equality between the inert and the gravitational mass, and signifies an extension of the principle of relativity to co-ordinate systems which are in non-uniform motion relatively to each other. In fact, through this conception we arrive at the unity of the nature of inertia and gravitation. For, according to our way of looking at it, the same masses may appear to be either under the action of inertia alone (with respect to K) or under the combined action of inertia and gravitation (with respect to K').

"Stated more exactly, there are finite regions, where, with respect to a suitably chosen space of reference, material particles move freely without acceleration, and in which the laws of special relativity, which have been developed above, hold with remarkable accuracy."

In their textbook, Foster and Nightingale [36] succinctly state the 'Principle of Equivalence' thus:

"We may incorporate these ideas into the principle of equivalence, which is this: In a freely falling (nonrotating) laboratory occupying a small region of space-time, the laws of physics are the laws of special relativity."

According to Pauli [53],

"We can think of the physical realization of the local coordinate system K_o in terms of a freely floating, sufficiently small, box which is not subjected to any external forces apart from gravity, and which is falling under the influence of the latter. ... "It is evidently natural to assume that the special theory of relativity should remain valid in K_o."

Taylor and Wheeler state in their book [25],

[3]The so-called "cosmological constant" is not included.

"General Relativity requires more than one free-float frame."

In the Dictionary of Geophysics, Astrophysics and Astronomy [78],

> *"Near every event in space-time, in a sufficiently small neighborhood, in every freely falling reference frame all phenomena (including gravitational ones) are exactly as they are in the absence of external gravitational sources."*

Note that the 'Principle of Equivalence' involves the *a priori* presence of multiple arbitrarily large finite masses. Similarly, the laws of Special Relativity involve the *a priori* presence of at least two arbitrarily large finite masses; for otherwise relative motion between two bodies cannot manifest. The postulates of Special Relativity are themselves couched in terms of inertial systems, which are in turn defined in terms of mass via Newton's First Law of motion.

In the space of Newton's theory of gravitation, one can simply put in as many masses as one pleases. Although solving for the gravitational interaction of these masses rapidly becomes beyond our capacity, there is nothing to prevent us inserting masses conceptually. This is essentially the 'Principle of Superposition'. However, one cannot do this in General Relativity, because Einstein's field equations are non-linear. In General Relativity, each and every configuration of matter must be described by a corresponding energy-momentum tensor and the field equations solved separately for each and every such configuration, because matter and geometry are coupled, as Eq. (3.21) describes. Not so in Newton's theory where geometry is independent of matter. The 'Principle of Superposition' does not apply in General Relativity:

> *"In a gravitational field, the distribution and motion of the matter producing it cannot at all be assigned arbitrarily — on the contrary it must be determined (by solving the field equations for given initial conditions) simultaneously with the field produced by the same matter."* [51]

> *"An important characteristic of gravity within the framework of general relativity is that the theory is nonlinear. Mathematically, this means that if g_{ab} and γ_{ab} are two solutions of the field equations, then $ag_{ab} + b\gamma_{ab}$ (where a, b are scalars) may not be a solution. This fact manifests itself physically in two ways. First, since a linear combination may not be a solution, we cannot take the overall gravitational field of the two bodies to be the summation of the individual gravitational fields of each body."* [79]

Now Einstein and the relevant physicists assert that the gravitational field "outside" a mass contains no matter, and so they assert that $T_{\mu\nu} = 0$, and that there is only one mass in the whole Universe with this particular problem statement. But setting the energy-momentum tensor to zero means that there is no matter present by which the gravitational field can be caused! Nonetheless, it is so claimed, and it is also claimed that the field equations then reduce to the much simpler form,

$$R_{\mu\nu} = 0. \tag{3.22}$$

So this is a clear statement that space-time is devoid of matter.

> *"Black holes were first discovered as purely mathematical solutions of Einstein's field equations. This solution, the Schwarzschild black hole, is a nonlinear solution of the Einstein equations of General Relativity. It contains no matter, and exists forever in an asymptotically flat space-time."* [78]

However, since this is a space-time that **by construction** contains no matter, Einstein's 'Principle of Equivalence' and his laws of Special Relativity cannot manifest, thus violating the physical requirements of the gravitational field that Einstein himself laid down. It has never been proven that Einstein's 'Principle of Equivalence' and his laws of Special Relativity, both of which are defined in terms of the *a priori* presence of multiple arbitrary large finite masses, can manifest in a space-time that by construction contains no matter. Indeed, it is a contradiction; so $R_{\mu\nu} = 0$ fails. Now Eq. (3.1) relates to Eq. (3.22). However, there is allegedly mass present, denoted by m in Eq. (3.1). This mass is not described by an energy-momentum tensor. That m is actually responsible for the alleged gravitational field associated with Eq. (3.1) is confirmed by the fact that if $m = 0$, Eq.

(3.1) reduces to Minkowski space-time, and hence no gravitational field. So if not for the presence of the alleged mass m in Eq. (3.1) there is no gravitational field. But this contradicts Einstein's relation between geometry and matter, since m is introduced into Eq. (3.1) *post hoc*, not via an energy-momentum tensor describing the matter causing the associated gravitational field. The components of the metric tensor are functions of only one another, and reduce to functions of only one component of the metric tensor. None of the components of the metric tensor contain matter, because the energy-momentum tensor is zero. There is no transformation of matter in Minkowski space-time into Schwarzschild space-time, and so the laws of Special Relativity are not transformed into a gravitational field by $Ric = 0$. The transformation is merely from a pseudo-Euclidean geometry containing no matter into a pseudo-Riemannian (non-Euclidean) geometry containing no matter. Matter is introduced into the space-time of $Ric = 0$ by means of a vicious circle, as follows. It is stated at the outset that $Ric = 0$ describes space-time "outside a body". The words "outside a body" introduce matter, contrary to the energy-momentum tensor, $T_{\mu\nu} = 0$, that describes the causative matter as being absent. There is no matter involved in the transformation of Minkowski space-time into Schwarzschild space-time via $Ric = 0$, since the energy-momentum tensor is zero, making the components of the resulting metric tensor functions solely of one another, and reducible to functions of just one component of the metric tensor. To satisfy the initial claim that $Ric = 0$ describes space-time "outside a body", so that the resulting space-time is caused by the alleged mass present, the alleged causative mass is **inserted** into the resulting metric *ad hoc*, by means of a contrived analogy with Newton's theory, thus closing the vicious circle. Here is how Chandrasekhar [24] presents the vicious circle:

> *"That such a contingency can arise was surmised already by Laplace in 1798. Laplace argued as follows. For a particle to escape from the surface of a spherical body of mass M and radius R, it must be projected with a velocity v such that* $\frac{1}{2}v^2 > GM/R$*; and it cannot escape if* $v^2 < 2GM/R$*. On the basis of this last inequality, Laplace concluded that if* $R < 2GM/c^2 = R_s$ *(say) where c denotes the velocity of light, then light will not be able to escape from such a body and we will not be able to see it!*
>
> *"By a curious coincidence, the limit* R_s *discovered by Laplace is* exactly *the same that general relativity gives for the occurrence of the* trapped surface *around a spherical mass."*

But it is not surprising that general relativity gives the same R_s "*discovered by Laplace*" because the Newtonian expression for escape velocity is deliberately inserted *post hoc* by the astrophysical scientists, into the so-called "Schwarzschild solution" in order to make it so. Newton's escape velocity does not drop out of any of the calculations to Schwarzschild space-time. Furthermore, although $Ric = 0$ is claimed to describe space-time "outside a body", the resulting metric (1) is nonetheless used to describe the **interior** of a black hole, since the black hole begins at the alleged "event horizon", not at its infinitely dense point-mass singularity (said to be at $r = 0$ in Eq. (3.1)).

In the case of a totally empty Universe, what would be the relevant energy-momentum tensor? It must be $T_{\mu\nu} = 0$. Indeed, it is also claimed that space-times can be intrinsically curved, i.e. that there are gravitational fields that have no material cause. An example is de Sitter's empty spherical Universe, based upon the following field equations [33, 34]:

$$R_{\mu\nu} = \lambda g_{\mu\nu} \tag{3.23}$$

where λ is the so-called 'cosmological constant'. In the case of metric (1) the field equations are given by expression (22). On the one hand de Sitter's empty world is devoid of matter ($T_{\mu\nu} = 0$) and so has no material cause for the alleged associated gravitational field. On the other hand it is claimed that the space-time described by Eq. (3.22) has a material cause, *post hoc* as m in metric (1), even though $T_{\mu\nu} = 0$ there as well: a contradiction. This is amplified by the so-called "Schwarzschild-de Sitter" line-element,

$$ds^2 = \left(1 - \frac{2m}{r} - \frac{\lambda}{3}r^2\right) dt^2 - \left(1 - \frac{2m}{r} - \frac{\lambda}{3}r^2\right)^{-1} dr^2 - r^2 \left(d\theta^2 + \sin^2\theta d\varphi^2\right), \tag{3.24}$$

which is the standard solution for Eq. (3.23). Once again, m is identified *post hoc* as mass at the centre of spherical symmetry of the manifold, said to be at $r = 0$. The completely empty universe of de Sitter [33, 34] can be obtained by setting $m=0$ in Eq. (3.24) to yield,

$$ds^2 = \left(1 - \frac{\lambda}{3}r^2\right) dt^2 - \left(1 - \frac{\lambda}{3}r^2\right)^{-1} dr^2 - r^2 \left(d\theta^2 + \sin^2\theta d\varphi^2\right), \tag{3.25}$$

Also, if $\lambda=0$, Eq. (3.23) reduces to Eq. (3.22) and Eq. (3.24) reduces to Eq. (3.1). If both $\lambda=0$ and $m=0$, Eqs. (3.24) and (3.25) reduce to Minkowski space-time. Now in Eq. (3.23) the term $\lambda g_{\mu\nu}$ is not an energy-momentum tensor, since according to the astrophysical scientists, expression (25) describes a world devoid of matter [33,34]. The universe described by Eq. (3.25), which also satisfies Eq. (3.23), is completely empty and so its curvature has no material cause; in Eq. (3.23), just as in Eq. (3.22), $T_{\mu\nu}=0$. So Eq. (3.25) is alleged to describe a gravitational field that has no material cause. Furthermore, although in Eq. (3.22), $T_{\mu\nu}=0$, its usual solution, Eq. (3.1), is said to contain a (*post hoc*) material cause, denoted by m therein. Thus for Eq. (3.1) it is claimed that $T_{\mu\nu}=0$ supports a material cause of a gravitational field, but at the same time, for Eq. (3.25), $T_{\mu\nu}=0$ is also claimed to preclude material cause of a gravitational field. So $T_{\mu\nu}=0$ is claimed to include and to exclude material cause. This is not possible. The contradiction is due to the *post hoc* introduction of mass, as m, in Eq. (3.1), by means of the Newtonian expression for gravitational potential. Furthermore, there is no experimental evidence to support the claim that a gravitational field can be generated without a material cause. Material cause is codified theoretically in Eq. (3.21).

Since $R_{\mu\nu}=0$ cannot describe Einstein's gravitational field, Einstein's field equations cannot reduce to $R_{\mu\nu}=0$ when $T_{\mu\nu}=0$. In other words, if $T_{\mu\nu}=0$ (i.e. there is no matter present) then there is no gravitational field. Consequently Einstein's field equations must take the form [58, 59],

$$\frac{G_{\mu\nu}}{\kappa} + T_{\mu\nu} = 0. \tag{3.26}$$

The $G_{\mu\nu}/\kappa$ are the components of a gravitational energy tensor. Thus the total energy of Einstein's gravitational field is *always* zero; the $G_{\mu\nu}/\kappa$ and the $T_{\mu\nu}$ *must vanish identically*; there is *no possibility* for the localization of gravitational energy (i.e. there are no Einstein gravitational waves). This also means that Einstein's gravitational field violates the experimentally well-established usual conservation of energy and momentum [53]. Since there is no experimental evidence that the usual conservation of energy and momentum is invalid, Einstein's General Theory of Relativity violates the experimental evidence, and so it is invalid.

In an attempt to circumvent the foregoing conservation problem, Einstein invented his gravitational pseudo-tensor, the components of which he says are '*the "energy components" of the gravitational field*' [60]. His invention had a two-fold purpose (a) to bring his theory into line with the usual conservation of energy and momentum, (b) to enable him to get gravitational waves that propagate with speed c. First, Einstein's gravitational pseudo-tensor is not a tensor, and is therefore not in keeping with his theory that all equations be tensorial. Second, he constructed his pseudo-tensor in such a way that it behaves like a tensor in one particular situation, that in which he could get gravitational waves with speed c. Now Einstein's pseudo-tensor is claimed to represent the energy and momentum of the gravitational field and it is routinely applied in relation to the localisation of gravitational energy, the conservation of energy and the flow of energy and momentum.

Dirac [54] pointed out that,

> *"It is not possible to obtain an expression for the energy of the gravitational field satisfying both the conditions: (i) when added to other forms of energy the total energy is conserved, and (ii) the energy within a definite (three dimensional) region at a certain time is independent of the coordinate system. Thus, in general, gravitational energy cannot be localized. The best we can do is to use the pseudo-tensor, which satisfies condition (i) but not condition (ii). It gives us approximate information about gravitational energy, which in some special cases can be accurate."*

On gravitational waves Dirac [54] remarked,

> *"Let us consider the energy of these waves. Owing to the pseudo-tensor not being a real tensor, we do not get, in general, a clear result independent of the coordinate system. But there is one special case in which we do get a clear result; namely, when the waves are all moving in the same direction."*

About the propagation of gravitational waves Eddington [34] remarked ($g_{\mu\nu} = \delta_{\mu\nu} + h_{\mu\nu}$),

$$\frac{\partial^2 h_{\mu\nu}}{\partial t^2} - \frac{\partial^2 h_{\mu\nu}}{\partial x^2} - \frac{\partial^2 h_{\mu\nu}}{\partial y^2} - \frac{\partial^2 h_{\mu\nu}}{\partial z^2} = 0,$$

> *"... showing that the deviations of the gravitational potentials are propagated as waves with unit velocity, i.e. the velocity of light. But it must be remembered that this representation of the propagation, though always permissible, is not unique. ... All the coordinate-systems differ from Galilean coordinates by small quantities of the first order. The potentials $g_{\mu\nu}$ pertain not only to the gravitational influence which is objective reality, but also to the coordinate-system which we select arbitrarily. We can 'propagate' coordinate-changes with the speed of thought, and these may be mixed up at will with the more dilatory propagation discussed above. There does not seem to be any way of distinguishing a physical and a conventional part in the changes of the $g_{\mu\nu}$.*
>
> *"The statement that in the relativity theory gravitational waves are propagated with the speed of light has, I believe, been based entirely upon the foregoing investigation; but it will be seen that it is only true in a very conventional sense. If coordinates are chosen so as to satisfy a certain condition which has no very clear geometrical importance, the speed is that of light; if the coordinates are slightly different the speed is altogether different from that of light. The result stands or falls by the choice of coordinates and, so far as can be judged, the coordinates here used were purposely introduced in order to obtain the simplification which results from representing the propagation as occurring with the speed of light. The argument thus follows a vicious circle."*

Now Einstein's pseudo-tensor, $\sqrt{-g}\, t^{\mu}_{\nu}$, is defined by [23, 33, 34, 53, 54, 58, 60],

$$\sqrt{-g}\, t^{\mu}_{\nu} = \frac{1}{2}\left(\delta^{\mu}_{\nu} L - \frac{\partial L}{\partial g^{\sigma\rho}_{,\mu}}\, g^{\sigma\rho}_{,\nu}\right), \tag{3.27}$$

wherein L is given by

$$L = -g^{\alpha\beta}\left(\Gamma^{\gamma}_{\alpha\kappa}\Gamma^{\kappa}_{\beta\gamma} - \Gamma^{\gamma}_{\alpha\beta}\Gamma^{\kappa}_{\gamma\kappa}\right). \tag{3.28}$$

According to Einstein [60] his pseudo-tensor "*expresses the law of conservation of momentum and of energy for the gravitational field.*"

In a remarkable paper published in 1917, T. Levi-Civita [58] provided a clear and rigorous proof that Einstein's pseudo-tensor is meaningless, and therefore any argument relying upon it is fallacious. I repeat Levi-Civita's proof. Contracting Eq. (3.27) produces a linear invariant, thus

$$\sqrt{-g}\, t^{\mu}_{\mu} = \frac{1}{2}\left(4L - \frac{\partial L}{\partial g^{\sigma\rho}_{,\mu}}\, g^{\sigma\rho}_{,\mu}\right). \tag{3.29}$$

Since L is, according to Eq. (3.28), quadratic and homogeneous with respect to the Riemann-Christoffel symbols, and therefore also with respect to $g^{\sigma\rho}_{,\mu}$, one can apply Euler's theorem to obtain (also see [34]),

$$\frac{\partial L}{\partial g^{\sigma\rho}_{,\mu}}\, g^{\sigma\rho}_{,\mu} = 2L. \tag{3.30}$$

Substituting expression (30) into expression (29) yields the linear invariant at L. This is a first-order, intrinsic differential invariant that depends only on the components of the metric tensor and their first derivatives (see expression (28) for L). However, the mathematicians G. Ricci-Curbastro and T. Levi-Civita [65] proved, in 1900, that such invariants **do not exist**. This is sufficient to render Einstein's pseudo-tensor entirely meaningless, and hence all arguments relying on it false. Einstein's conception of the conservation of energy in the gravitational field is erroneous.

Linearisation of Einstein's field equations and associated perturbations have been popular. "*The existence of exact solutions corresponding to a solution to the linearised equations must be investigated before perturbation analysis can be applied with any reliability*" [21]. Unfortunately, the astrophysical scientists have not properly investigated. Indeed, linearisation of the field equations is inadmissible, even though the astrophysical scientists write down linearised equations and proceed as though they are valid, because linearisation of the field equations implies the existence of a tensor which, except for the trivial case of being precisely zero, **does not exist**; proven by Hermann Weyl [66] in 1944.

3.9 Other violations

In writing Eq. (3.1) the Standard Model incorrectly asserts that only the components g_{00} and g_{11} are modified by $R_{\mu\nu} = 0$. However, it is plain by expressions (20) that this is false. All components of the metric tensor are modified by the constant α appearing in Eqs. (3.20), of which metric (3.1) is but a particular case.

The Standard Model asserts in relation to metric (1) that a 'true' singularity must occur where the Riemann tensor scalar curvature invariant (i.e. the Kretschmann scalar) is unbounded [21, 23, 50]. However, it has never been proven that Einstein's field equations require such a curvature condition to be fulfilled: in fact, it is not required by General Relativity. Since the Kretschmann scalar is finite at $r = 2m$ in metric (1), it is also claimed that $r = 2m$ marks a "coordinate singularity" or "removable singularity". However, these assertions violate the intrinsic geometry of the manifold described by metric (1). The Kretschmann scalar depends upon all the components of the metric tensor and all the components of the metric tensor are functions of the Gaussian curvature of the spherically symmetric geodesic surface in the spatial section, owing to the form of the line-element. The Kretschmann scalar is not therefore an independent curvature invariant. Einstein's gravitational field is manifest in the curvature of space-time, a curvature induced by the presence of matter. It should not therefore be unexpected that the Gaussian curvature of a spherically symmetric geodesic surface in the spatial section of the gravitational manifold might also be modified from that of ordinary Euclidean space, and this is indeed the case for Eq. (3.1). Metric (20) gives the modification of the Gaussian curvature fixed by the intrinsic geometry of the line-element and the required boundary conditions specified by Einstein and the astrophysical scientists, in consequence of which the Kretschmann scalar is constrained by the Gaussian curvature of the spherically symmetric geodesic surface in the spatial section. Recall that the Kretschmann scalar f is,

$$f = R_{\alpha\beta\gamma\delta} R^{\alpha\beta\gamma\delta}.$$

Using metric (20) gives,

$$f = 12\alpha^2 K^3 = \frac{12\alpha^2}{R_c^6} = \frac{12\alpha^2}{\left(|r - r_o|^n + \alpha^n\right)^{\frac{6}{n}}},$$

then

$$f\left(r_o\right) = \frac{12}{\alpha^4} \quad \forall\, r_o \;\forall\, n,$$

which is a scalar invariant that corresponds to the scalar invariants $R_p\left(r_o\right) = 0$, $R_c\left(r_o\right) = \alpha$, $K\left(r_o\right) = \alpha^{-2}$. The usual assumption that the Kretschmann scalar must be unbounded (undefined) at a singularity in Schwarzschild space-time is just that, and has no valid physical or mathematical basis. It is evident from the line-element that the Kretschmann scalar is finite everywhere. This is reaffirmed by the Riemannian (or Sectional) curvature K_s of the spatial section of Schwarzschild space-time, given by

$$K_s = \frac{-\frac{\alpha}{2} W_{1212} - \frac{\alpha}{2} W_{1313} \sin^2\theta + \alpha R_c \left(R_c - \alpha\right) W_{2323}}{R_c^3 W_{1212} + R_c^3 W_{1313} \sin^2\theta + R_c^4 \sin^2\theta \left(R_c - \alpha\right) W_{2323}}.$$

$$R_c = \left(|r - r_o|^n + \alpha^n\right)^{\frac{1}{n}}$$

where

$$W_{ijkl} = \begin{vmatrix} U^i & U^j \\ V^i & V^j \end{vmatrix} \begin{vmatrix} U^k & U^l \\ V^k & V^l \end{vmatrix}$$

and $\langle U^i \rangle$ and $\langle V^i \rangle$ are two arbitrary non-zero contravariant vectors at any point in the space. Thus, in general the Riemannian curvature is dependent upon both position and direction (i.e. the directions of the contravariant vectors). Now

$$K_s(r_o) = -\frac{1}{2\alpha^2} = -\frac{1}{2}K(r_o)$$

which is entirely *independent* of the contravariant vectors and is half the negative of the associated Gaussian curvature of the spherically symmetric geodesic surface in the spatial section. This is a scalar invariant that corresponds to $R_c(r_o) = \alpha\ \forall r_o\ \forall n$ and $R_p(r_o) = 0\ \forall r_o\ \forall n$.

Doughty [70] has shown that the radial geodesic acceleration a of a point in a manifold described by a line-element with the form of Eq. (3.13) is given by,

$$a = \frac{\sqrt{-g_{11}}\left(-g^{11}\right)|g_{00,1}|}{2g_{00}}.$$

Using metric (20) once again gives,

$$a = \frac{\alpha}{R_c^{\frac{3}{2}}(r)\sqrt{R_c(r)-\alpha}}.$$

Now,

$$\lim_{r\to r_o^{\pm}} R_p(r) = 0, \qquad \lim_{r\to r_o^{\pm}} R_c(r) = \alpha,$$

and so

$$r \to r_o^{\pm} \Rightarrow a \to \infty \quad \forall\, r_o\ \forall\, n.$$

According to metric (20) there is no possibility for $R_c \le \alpha$.

In the case of Eq. (3.1), for which $r_o = \alpha = 2m$, $n = 1$, $r > \alpha$, the acceleration is,

$$a = \frac{2m}{r^{\frac{3}{2}}\sqrt{r-2m}}.$$

which is infinite at $r = 2m$. But the usual unproven (and invalid) assumption that r in Eq. (3.1) can go down to zero means that there is an infinite acceleration at $r = 2m$ where, according to the Standard Model, **there is no matter**! However, r can't take the values $0 \le r \le r_o = 2m$ in Eq. (3.1), as Eq. (3.20) shows, by virtue of the nature of the Gaussian curvature of spherically symmetric geodesic surfaces in the spatial section associated with the gravitational manifold, and the intrinsic geometry of the line-element.

The proponents of the Standard Model admit that if $0 < r < 2m$ in Eq. (3.1), the rôles of t and r are interchanged. But this violates their construction at Eq. (3.12), which has the fixed signature $(+,-,-,-)$, and is therefore inadmissible. To further illustrate this violation, when $2m < r < \infty$ the signature of Eq. (3.1) is $(+,-,-,-)$; but if $0 < r < 2m$ in Eq. (3.1), then

$$g_{00} = \left(1 - \frac{2m}{r}\right) \text{ is negative, and } \quad g_{11} = -\left(1 - \frac{2m}{r}\right)^{-1} \text{ is positive.}$$

So the signature of metric (1) changes to $(-,+,-,-)$. Thus the rôles of t and r are interchanged. According to Misner, Thorne and Wheeler, who use the space-time signature $(-,+,+,+)$,

> *"The most obvious pathology at $r = 2M$ is the reversal there of the roles of t and r as timelike and spacelike coordinates. In the region $r > 2M$, the t direction, $\partial/\partial t$, is timelike ($g_{tt} < 0$) and the r direction, $\partial/\partial r$, is spacelike ($g_{rr} > 0$); but in the region $r < 2M$, $\partial/\partial t$, is spacelike ($g_{tt} > 0$) and $\partial/\partial r$, is timelike ($g_{rr} < 0$).*
>
> *"What does it mean for r to 'change in character from a spacelike coordinate to a timelike one'? The explorer in his jet-powered spaceship prior to arrival at $r = 2M$ always has the option to turn on his jets and change his motion from decreasing r (infall) to increasing r (escape). Quite the contrary in the situation when he has once allowed himself to fall inside $r = 2M$. Then the further decrease of r represents the passage of time. No command that the traveler can give to his jet engine will*

> *turn back time. That unseen power of the world which drags everyone forward willy-nilly from age twenty to forty and from forty to eighty also drags the rocket in from time coordinate* $r = 2M$ *to the later time coordinate* $r = 0$. *No human act of will, no engine, no rocket, no force (see exercise 31.3) can make time stand still.. As surely as cells die, as surely as the traveler's watch ticks away 'the unforgiving minutes', with equal certainty, and with never one halt along the way,* r *drops from* $2M$ *to* 0.
>
> *"At* $r = 2M$, *where* r *and* t *exchange roles as space and time coordinates,* g_{tt} *vanishes while* g_{rr} *is infinite."*

Chandrasekhar [24] has expounded the same claim as follows,

> *'There is no alternative to the matter collapsing to an infinite density at a singularity once a point of no-return is passed. The reason is that once the event horizon is passed, all time-like trajectories must necessarily get to the singularity: "all the King's horses and all the King's men" cannot prevent it.'*

Carroll [50] also says,

> *"This is worth stressing; not only can you not escape back to region I, you cannot even stop yourself from moving in the direction of decreasing* r, *since this is simply the timelike direction. (This could have been seen in our original coordinate system; for* $r < 2GM$, t *becomes spacelike and* r *becomes timelike.) Thus you can no more stop moving toward the singularity than you can stop getting older."*

Vladmimirov, Mitskiévich and Horský [71] assert,

> *"For* $r < 2GM/c^2$, *however, the component* g_{oo} *becomes negative, and* g_{rr}, *positive, so that in this domain, the role of time-like coordinate is played by* r, *whereas that of space-like coordinate by* t. *Thus in this domain, the gravitational field depends significantly on time* (r) *and does not depend on the coordinate* t.

To amplify this, set $t = r^*$ and $r = t^*$, and so for $0 < r < 2m$, Eq. (3.1) becomes,

$$ds^2 = \left(1 - \frac{2m}{t^*}\right) dr^{*2} - \left(1 - \frac{2m}{t^*}\right)^{-1} dt^{*2} - t^{*2}\left(d\theta^2 + \sin^2\theta d\varphi^2\right),$$

$$0 < t^* < 2m.$$

It now becomes quite clear that this is a **time-dependent metric** since all the components of the metric tensor are functions of the timelike t^*, and so this metric **bears no relationship to the original time-independent problem to be solved** [35, 46]. In other words, this metric is a **non-static solution to a static problem**: *contra hyp*! Thus, in Eq. (3.1), $0 < r < 2m$ is meaningless, as Eqs. (3.20) demonstrate.

Furthermore, if the signature of "Schwarzschild" space-time is permitted to change from $(+, -, -, -)$ to $(-, +, -, -)$ in the fashion claimed for black holes, then there must be for the latter signature a corresponding generalisation of the Minkowski metric, taking the fundamental form

$$ds^2 = -e^{\lambda}dt^2 + e^{\beta}dr^2 - R^2\left(d\theta^2 + \sin^2\theta d\varphi^2\right),$$

where λ, β and R are all unknown real-valued functions of only the real variable r, and where $e^{\lambda} > 0$ and $e^{\beta} > 0$. But this is impossible because the Minkowski space-time metric has the fixed signature $(+, -, -, -)$, since the spatial section of Minkowski space-time is a positive definite quadratic form; and so the foregoing generalised metric is not a generalistion of Minkowski space-time at all.

Also of importance is the fact that Hagihara [72] proved, in 1931, that all geodesics that do not run into the boundary of the "Schwarzschild" metric at $r = 2m$ (i.e. at $R_p(r_o = 2m) = 0$) are complete.

Nobody has ever found a black hole anywhere because nobody has found an infinitely dense point-mass singularity and nobody has found an event horizon.

> *"Unambiguous observational evidence for the existence of astrophysical black holes has not yet been established.* [28]

All claims for detection of black holes are patently false.

It has recently been admitted by astronomers [73] at the Max Planck Institute for Extraterrestrial Physics that,

(a) Nobody has ever found a black hole, despite the numerous claims for their discovery;

(b) The infinitely dense point-mass singularity of the alleged black hole is nonsense;

(c) The alleged black hole has no escape velocity, despite the claims of the astrophysical scientists;

(d) They were until very recently informed, unaware of Schwarzschild's actual solution.

The LIGO project and its international counterparts have not detected gravitational waves [74]. They are destined to detect nothing. Furthermore, the Lense-Thirring or 'frame dragging' effect was not detected by the Gravity Probe B and NASA has terminated further funding of that project [75].

3.10 Three-dimensional spherically symmetric metric manifolds - first principles

To complete the purely mathematical foundations of this paper, the differential geometry expounded in the foregoing is now developed from first principles.

Following the method suggested by Palatini, and developed by Levi-Civita [30], denote ordinary Euclidean 3-space by $\mathbf{E}^3$. Let $\mathbf{M}^3$ be a 3-dimensional metric manifold. Let there be a one-to-one correspondence between all points of $\mathbf{E}^3$ and $\mathbf{M}^3$. Let the point $O \in \mathbf{E}^3$ and the corresponding point in $\mathbf{M}^3$ be O'. Then a point transformation T of $\mathbf{E}^3$ into itself gives rise to a corresponding point transformation of $\mathbf{M}^3$ into itself.

A rigid motion in a metric manifold is a motion that leaves the metric $d\ell'^2$ unchanged. Thus, a rigid motion changes geodesics into geodesics. The metric manifold $\mathbf{M}^3$ possesses spherical symmetry around any one of its points O' if each of the ∞^3 rigid rotations in $\mathbf{E}^3$ around the corresponding arbitrary point O determines a rigid motion in $\mathbf{M}^3$.

The coefficients of $d\ell'^2$ of $\mathbf{M}^3$ constitute a metric tensor and are naturally assumed to be regular in the region around every point in $\mathbf{M}^3$, except possibly at an arbitrary point, the centre of spherical symmetry $O' \in \mathbf{M}^3$.

Let a ray i emanate from an arbitrary point $O \in \mathbf{E}^3$. There is then a corresponding geodesic $i' \in \mathbf{M}^3$ issuing from the corresponding point $O' \in \mathbf{M}^3$. Let P be any point on i other than O. There corresponds a point P' on $i' \in \mathbf{M}^3$ different to O'. Let g' be a geodesic in $\mathbf{M}^3$ that is tangential to i' at P'.

Taking i as the axis of ∞^1 rotations in $\mathbf{E}^3$, there corresponds ∞^1 rigid motions in $\mathbf{M}^3$ that leaves only all the points on i' unchanged. If g' is distinct from i', then the ∞^1 rigid rotations in $\mathbf{E}^3$ about i would cause g' to occupy an infinity of positions in $\mathbf{M}^3$ wherein g' has for each position the property of being tangential to i' at P' in the same direction, which is impossible. Hence, g' coincides with i'.

Thus, given a spherically symmetric surface Σ in $\mathbf{E}^3$ with centre of symmetry at some arbitrary point $O \in \mathbf{E}^3$, there corresponds a spherically symmetric geodesic surface Σ' in $\mathbf{M}^3$ with centre of symmetry at the corresponding point $O' \in \mathbf{M}^3$.

Let Q be a point in $\Sigma \in \mathbf{E}^3$ and Q' the corresponding point in $\Sigma' \in \mathbf{M}^3$. Let $d\sigma^2$ be a generic line element in Σ issuing from Q. The corresponding generic line element $d\sigma'^2 \in \Sigma'$ issues from the point Q'. Let Σ be described in the usual spherical-polar coordinates r, θ, φ. Then

$$d\sigma^2 = r^2(d\theta^2 + \sin^2\theta d\varphi^2), \tag{3.1.1}$$

$$r = |\overline{OQ}|.$$

Clearly, if r, θ, φ are known, Q is determined and hence also Q' in Σ'. Therefore, θ and φ can be considered to be curvilinear coordinates for Q' in Σ' and the line element $d\sigma' \in \Sigma'$ will also be represented by a quadratic form similar to (3.1.1). To determine $d\sigma'$, consider two elementary arcs of equal length, $d\sigma_1$ and $d\sigma_2$ in Σ, drawn from the point Q in different directions. Then the homologous arcs in Σ' will be $d\sigma'_1$ and $d\sigma'_2$, drawn in different directions from the corresponding point Q'. Now $d\sigma_1$ and $d\sigma_2$ can be obtained from one another by a rotation about the axis $\overline{OQ}$ in $\mathbf{E}^3$, and so $d\sigma'_1$ and $d\sigma'_2$ can be obtained from one another by a rigid motion in $\mathbf{M}^3$, and are therefore also of equal length, since the metric is unchanged by such a motion. It therefore follows that the ratio $\frac{d\sigma'}{d\sigma}$ is the same for the two different directions irrespective of $d\theta$ and $d\varphi$, and so the foregoing ratio is a function of position, i.e. of r, θ, φ. But Q is an arbitrary point in Σ, and so $\frac{d\sigma'}{d\sigma}$ must have the same ratio for any corresponding points Q and Q'. Therefore, $\frac{d\sigma'}{d\sigma}$ is a function of r alone, thus

$$\frac{d\sigma'}{d\sigma} = H(r),$$

and so

$$d\sigma'^2 = H^2(r)d\sigma^2 = H^2(r)r^2(d\theta^2 + \sin^2\theta d\varphi^2), \tag{3.1.2}$$

where $H(r)$ is *a priori* unknown. For convenience set $R_c = R_c(r) = H(r)r$, so that (3.1.2) becomes

$$d\sigma'^2 = R_c^2(d\theta^2 + \sin^2\theta d\varphi^2), \tag{3.1.3}$$

where R_c is a quantity associated with $\mathbf{M}^3$. Comparing (3.1.3) with (3.1.1) it is apparent that R_c is to be rightly interpreted in terms of the Gaussian curvature K at the point Q', i.e. in terms of the relation $K = \frac{1}{R_c^2}$ since the Gaussian curvature of (3.1.1) is $K = \frac{1}{r^2}$. This is an intrinsic property of all line elements of the form (3.1.3) [30]. Accordingly, R_c, the inverse square root of the Gaussian curvature, can be regarded as the radius of Gaussian curvature. Therefore, in (3.1.1) the radius of Gaussian curvature is $R_c = r$. Moreover, owing to spherical symmetry, all points in the corresponding surfaces Σ and Σ' have constant Gaussian curvature relevant to their respective manifolds and centres of symmetry, so that all points in the respective surfaces are umbilics.

Let the element of radial distance from $O \in \mathbf{E}^3$ be dr. Clearly, the radial lines issuing from O cut the surface Σ orthogonally. Combining this with (3.1.1) by the theorem of Pythagoras gives the line element in $\mathbf{E}^3$

$$d\ell^2 = dr^2 + r^2(d\theta^2 + \sin^2\theta d\varphi^2). \tag{3.1.4}$$

Let the corresponding radial geodesic from the point $O' \in \mathbf{M}^3$ be dR_p. Clearly the radial geodesics issuing from O' cut the geodesic surface Σ' orthogonally. Combining this with (3.1.3) by the theorem of Pythagoras gives the line element in $\mathbf{M}^3$ as,

$$d\ell'^2 = dR_p^2 + R_c^2(d\theta^2 + \sin^2\theta d\varphi^2), \tag{3.1.5}$$

where dR_p is, by spherical symmetry, also a function only of R_c. Set $dR_p = \sqrt{B(R_c)}dR_c$, so that (3.1.5) becomes

$$d\ell'^2 = B(R_c)dR_c^2 + R_c^2(d\theta^2 + \sin^2\theta d\varphi^2), \tag{3.1.6}$$

where $B(R_c)$ is an *a priori* unknown function.

Expression (3.1.6) is the most general for a metric manifold $\mathbf{M}^3$ having spherical symmetry about some arbitrary point $O' \in \mathbf{M}^3$.

Considering (3.1.4), the distance $R_p = |\overline{OQ}|$ from the point at the centre of spherical symmetry O to a point $Q \in \Sigma$, is given by

$$R_p = \int_0^r dr = r = R_c.$$

Call R_p the proper radius. Consequently, in the case of $\mathbf{E}^3$, R_p and R_c are identical, and so the Gaussian curvature of the spherically symmetric geodesic surface containing any point in $\mathbf{E}^3$ can be associated with R_p, the radial distance between the centre of spherical symmetry at the point $O \in \mathbf{E}^3$ and the point $Q \in \Sigma$. Thus, in this case, $K = \frac{1}{R_c^2} = \frac{1}{R_p^2} = \frac{1}{r^2}$. However, this is not a general relation, since according to (3.1.5) and (3.1.6), in the case of $\mathbf{M}^3$, the radial geodesic distance from the centre of spherical symmetry at the point $O' \in \mathbf{M}^3$ is not the same as the radius of Gaussian curvature of the associated spherically symmetric geodesic surface, but is given by

$$R_p = \int_0^{R_p} dR_p = \int_{R_c(0)}^{R_c(r)} \sqrt{B(R_c(r))}\, dR_c(r) = \int_0^{r} \sqrt{B(R_c(r))}\, \frac{dR_c(r)}{dr} dr,$$

where $R_c(0)$ is *a priori* unknown owing to the fact that $R_c(r)$ is *a priori* unknown. One cannot simply assume that because $0 \leq r < \infty$ in (3.1.4) that it must follow that in (3.1.5) and (3.1.6) $0 \leq R_c(r) < \infty$. In other words, one cannot simply assume that $R_c(0) = 0$. Furthermore, it is evident from (3.1.5) and (3.1.6) that R_p determines the radial geodesic distance from the centre of spherical symmetry at the arbitrary point O' in $\mathbf{M}^3$ (and correspondingly so from O in $\mathbf{E}^3$) to another point in $\mathbf{M}^3$. Clearly, R_c does not in general render the radial geodesic length from the point at the centre of spherical symmetry to some other point in a metric manifold. Only in the particular case of $\mathbf{E}^3$ does R_c render both the radius of Gaussian curvature of an associated spherically symmetric surface and the radial distance from the point at the centre of spherical symmetry, owing to the fact that R_p and R_c are identical in that special case.

It should also be noted that in writing expressions (3.1.4) and (3.1.5) it is implicit that $O \in \mathbf{E}^3$ is defined as being located at the origin of the coordinate system of (3.1.4), i.e. O is located where $r = 0$, and by correspondence O' is defined as being located at the origin of the coordinate system of (3.1.5) and of (3.1.6); $O' \in \mathbf{M}^3$ is located where $R_p = 0$. Furthermore, since it is well known that a geometry is completely determined by the form of the line element describing it [33], expressions (3.1.4), (3.1.5) and (3.1.6) share the very same fundamental geometry because they are line elements of the same metrical groundform.

Expression (3.1.6) plays an important role in Einstein's alleged gravitational field.

3.11 Conclusions

"Schwarzschild's solution" is not Schwarzschild's solution. Schwarzschild's actual solution forbids black holes. The quantity 'r' appearing in the so-called "Schwarzschild solution" is not a distance of any kind in the associated manifold - it is the inverse square root of the Gaussian curvature of the spherically symmetric geodesic surface in the spatial section. This simple fact completely subverts all claims for black holes.

The generalisation of Minkowski space-time to Schwarzschild space-time, via $Ric = 0$, a space-time that by construction contains no matter, is not a generalisation of Special Relativity. Neither Einstein's 'Principle of Equivalence' nor his laws of Special Relativity can manifest in a space-time that by construction contains no matter.

Despite claims for discovery of black holes, nobody has ever found a black hole; no infinitely dense point-mass singularity and no event horizon have ever been found. There is no physical evidence for the existence of infinitely dense point-masses. The black hole is fictitious. The international search for black holes is destined to find none.

The Michell-Laplace dark body is not a black hole. Newton's theory of gravitation does not predict black holes.

Curved space-times without material cause violate the physical principles of General Relativity. There is no experimental evidence supporting the notion of gravitational fields generated without material cause.

No celestial body has ever been observed to undergo irresistible gravitational collapse. There is no laboratory evidence for irresistible gravitational collapse. Infinitely dense point-mass singularities howsoever formed cannot be reconciled with Special Relativity, i.e. they violate Special Relativity, and therefore violate General Relativity.

The Riemann tensor scalar curvature invariant (the Kretschmann scalar) is not an independent curvature invariant - it is a function of the intrinsic Gaussian curvature of the spherically symmetric geodesic surface in the spatial section.

General Relativity cannot account for the simple experimental fact that two fixed bodies will approach one another upon release. There are no known solutions to Einstein's field equations for two or more masses and there is no existence theorem by which it can even be asserted that his field equations contain latent solutions for such configurations of matter. All claims for black hole interactions are invalid.

Einstein's gravitational waves are fictitious; Einstein's gravitational energy cannot be localised; so the international search for Einstein's gravitational waves is destined to detect nothing. No gravitational waves have been detected. Einstein's pseudo-tensor is meaningless and linearisation of Einstein's field equations inadmissible. And the Lense-Thirring effect was not detected by the Gravity Probe B.

Einstein's field equations violate the experimentally well-established usual conservation of energy and momentum, and therefore violate the experimental evidence.

Dedication

I dedicate this paper to my late brother,

Paul Raymond Crothers

12th May 1968 – 25th December 2008.

Bibliography

[1] Schwarzschild, K. On the gravitational field of a mass point according to Einstein's theory. *Sitzungsber. Preuss. Akad. Wiss., Phys. Math. Kl.*, 189, 1916, www.sjcrothers.plasmaresources.com/schwarzschild.pdf

[2] Abrams, L. S. Black holes: the legacy of Hilbert's error. *Can. J. Phys.*, v. 67, 919, 1989, www.sjcrothers.plasmaresources.com/Abrams1989.pdf

[3] Antoci, S. David Hilbert and the origin of the "Schwarzschild" solution. 2001, www.sjcrothers.plasmaresources.com/hilbert.pdf

[4] Loinger, A. On black holes and gravitational waves. La Goliardica Paves, Pavia, 2002.

[5] Weyl, H. Space Time Matter, Dover Publications Inc., New York, 1952.

[6] Percy, J. University of Toronto, Astronomical Society of the Pacific, 390 Ashton Avenue, San Francisco, CA 94112, www.astrosociety.org/education/publications/tnl/24/24.html

[7] O'Neill, B. www.math.ucla.edu/~bon/kerrhistory.html

[8] Mould, R. A. Basic Relativity, Springer–Verlag New York Inc., New York, 1994.

[9] Dodson, C. T. J. and Poston, T. Tensor Geometry – The Geometric Viewpoint and its Uses, 2nd Ed., Springer–Verlag, 1991.

[10] 't Hooft, G. Introduction to the theory of blackholes, Version February 5, 2009, (Lectures presented at Utrecht University, 2009), Institute for Theoretical Physics Utrecht University, Princetonplein 5, 3584 CC Utrecht, the Netherlands, www.phys.uu.nl/~thooft/lectures/blackholes/BHlecturenotes.pdf

[11] Celloti, A., Miller, J. C., Sciama, D. W. Astrophysical evidence for the existence of black holes, (1999), http://arxiv.org/abs/astro-ph/9912186

[12] Information Leaflet No. 9: 'Blackholes', Science and Engineering Research Council Royal Greenwich Observatory,
http://www.nmm.ac.uk/?PHPSESSID=0093df5b6c376566dfcf1e17a6c4b117

[13] Kunstatter, G. Black holes, Physics Department University of Winnipeg, http://theory.uwinnipeg.ca/users/gabor/blackholes/

[14] Harrison, D. M. Black holes, University of Toronto,
www.upscale.utoronto.ca/GeneralInterest/Harrison/BlackHoles/BlackHoles.html

[15] Nemiroff, R. Virtual Trips to Black Holes and Neutron Stars, Michigan Technological University and NASA Goddard,
http://antwrp.gsfc.nasa.gov/htmltest/rjnbht.html

[16] Cambridge University, UK, www.damtp.cam.ac.uk/user/gr/public/bhhome.html

[17] NASA Goddard Space Flight Center,
http://imagine.gsfc.nasa.gov/docs/science/knowl1/blackholes.html,
http://imagine.gsfc.nasa.gov/docs/science/knowl2/blackholes.html

[18] National Maritime Museum, London, UK,
www.nmm.ac.uk/explore/astronomy-and-time/astronomy-facts/stars/black-holes/*/viewPage/2

[19] http://design.lbl.gov/education/blackholes/index.html

[20] Bruhn, G. W. (private communication), www.sjcrothers.plasmaresources.com/BHLetters.html

[21] Wald, R. M. General Relativity, The University of Chicago Press, Chicago, 1984.

[22] Carroll, B. W. and Ostile, D. A. An Introduction to Modern Astrophysics, Addison–Wesley Publishing Company Inc., 1996.

[23] Misner, C. W., Thorne K. S., Wheeler, J. A. Gravitation, W. H. Freeman and Company, New York, 1970.

[24] Chandrasekhar, S. The increasing rôle of general relativity in astronomy, *The Observatory*, 92, 168, 1972.

[25] Taylor E. F. and Wheeler J. A. Exploring Black Holes — Introduction to General Relativity, Addison Wesley Longman, 2000 (in draft).

[26] Zel'dovich, Ya. B. and Novikov, I. D. Stars and Relativity, Dover Publications Inc., New York, 1996.

[27] Ludvigsen, M. General Relativity – A Geometric Approach, Cambridge University Press, Cambridge, UK, 1999.

[28] Hughes, S. A. Trust but verify: The case for astrophysical black holes, Department of Physics and MIT Kavli Institute, 77 Massachusetts Avenue, Cambridge, MA 02139, SLAC Summer Institute 2005,

[29] 't Hooft, G. (private communication), www.sjcrothers.plasmaresources.com/BHLetters.html

[30] Levi-Civita, T. The Absolute Differential Calculus, Dover Publications Inc., New York, 1977.

[31] Schwarzschild, K. On the gravitational field of a sphere of incompressible fluid according to Einstein's theory. *Sitzungsber. Preuss. Akad. Wiss., Phys. Math. Kl.*, 424, 1916, www.sjcrothers.plasmaresources.com/Schwarzschild2.pdf

[32] Crothers, S. J. On the geometry of the general solution for the vacuum field of the point-mass. *Progress in Physics*, v. 2, 3–14, 2005, www.ptep-online.com/index_files/2005/PP-02-01.PDF

[33] Tolman, R. C. Relativity Thermodynamics and Cosmology, Dover Publications Inc., New York, 1987.

[34] Eddington, A. S. The mathematical theory of relativity, Cambridge University Press, Cambridge, 2nd edition, 1960.

[35] Brillouin, M. The singular points of Einstein's Universe. *Journ Phys. Radium*, v. 23, 43, 1923, www.sjcrothers.plasmaresources.com/brillouin.pdf

[36] Foster, J., Nightingale, J. D. A short course in General Relativity, Springer-Verlag, New York, Inc., 1995.

[37] Wardle, K. L. Differential Geoemtry, Routledge & Kegan Paul Ltd., London, 1965.

[38] Struik, D. J. Lectures on Classical Differential Geometry, 2nd Ed., Dover Publications Inc., New York, 1988.

[39] Kreyszig, E. Differential Geometry, Dover Publications Inc., New York, 1991.

[40] Efimov, N. V. Higher Geometry, Mir Publishers, Moscow, 1980.

[41] Crothers, S. J. Gravitation on a spherically symmetric metric manifold, *Progress in Physics*, v. 2, 68–74, 2007,
www.ptep-online.com/index_files/2007/PP-09-14.PDF

[42] Crothers, S. J. Certain Conceptual Anomalies in Einstein's Theory of Relativity, *Progress in Physics*, v. 1, 52–57, 2008, www.ptep-online.com/index_files/2008/PP-12-11.PDF

[43] Kay, D. C. Tensor Calculus, Schaum's Outline Series, McGraw-Hill Inc., New York, 1998.

[44] McConnell, A. J. Applications of Tensor Analysis, Dover Publications Inc., New York, 1957.

[45] Petrov, A .Z. Einstein Spaces, Pergamon Press Ltd., Oxford, 1969.

[46] Droste, J. The field of a single centre in Einstein's theory of gravitation, and the motion of a particle in that field. *Ned. Acad. Wet., S. A.*, v. 19, 197, 1917, www.sjcrothers.plasmaresources.com/Droste.pdf

[47] Morgan, F. Riemannian Geometry: A Beginner's Guide, 2nd Ed., A.K. Peters Ltd., MA, 1998.

[48] Schutz, B. F. A first course in general relativity, Cambridge University Press, UK, 1990.

[49] D'Inverno, R. Introducing Einstein's Relativity, Oxford University Press Inc., New York, 1992.

[50] Carroll, S. Lecture Notes on General Relativity, http://arxiv.org/abs/gr-qc/9712019

[51] Landau, L., Lifshitz E. The Classical Theory of Fields, Addison-Wesley Publishing Company, Inc., Reading, Massachusettes, 1951.

[52] Einstein, A. The Meaning of Relativity, Science Paperbacks and Methuen & Co. Ltd., 56–57, 1967.

[53] Pauli, W. The Theory of Relativity, Dover Publications, Inc., New York, 1981.

[54] Dirac, P. A. M. General Theory of Relativity. Princeton Landmarks in Physics Series, Princeton University Press, Princeton, New Jersey, 1996.

[55] 't Hooft, G. Introduction to general relativity, Institute for Theoretical Physics Utrecht University, Princetonplein 5, 3584 CC Utrecht, the Netherlands, version 8/4/2002, www.phys.uu.nl/~thooft/lectures/genrel.pdf

[56] Townsend, P. K. Black Holes, lecture notes, DAMTP, University of Cambridge, Silver St., Cambridge, U.K., 4 July, 1977, http://arxiv.org/abs/gr-qc/9707012

[57] Crothers, S. J. On Isotropic Coordinates and Einstein's Gravitational Field, *Progress in Physics*, v. 3, 7–12, 2006, www.ptep-online.com/index_files/2006/PP-06-02.PDF

[58] Levi-Civita, T. Mechanics. - On the analytical expression that must be given to the gravitational tensor in Einstein's theory. *Rendiconti della Reale Accadmeia dei Lincei*, 26, 381, (1917), www.sjcrothers.plasmaresources.com/Levi-Civita.pdf

[59] Lorentz, H. A. *Versl. gewone Vergad. Akad. Amst.*, 25, 468 and 1380, (1916).

[60] Einstein, A. The Foundation of the General Theory of Relativity, *Annalen der Physik*, 49, 1916, The Principle of Relativity (A collection of original memoirs on the special and general theory of relativity), Dover Publications Inc., New York, 1952.

[61] Crothers, S. J. On the Ramifications of the Schwarzschild Space-Time Metric, *Progress in Physics*, v.1, pp. 74-80, 2005,
www.ptep-online.com/index_files/2005/PP-01-10.PDF

[62] Smoller J., Temple B. On the Oppenheimer-Volkoff equations in General Relativity, *Arch. Rational Mech. Anal.*, Springer-Verlag, 142, 177, (1998).

[63] Dunning-Davies, J. Exploding a myth, Conventional wisdom or scientific truth?, Horwood Publishing Limited, Chichester, 2007 (ISBN: 978-1-904275-30-5)

[64] Crothers, S. On the Regge-Wheeler Tortoise and the Kruskal-Szekeres Coordinates, *Progress in Physics*, v.3, 30–34, 2006, www.ptep-online.com/indexfiles/2006/PP-06-06.PDF

[65] Ricci-Curbastro, G. and Levi-Civita, T. Méthodes de calcul différential absolu et leurs applications, *Matematische Annalen*, B. 54, p.162, (1900).

[66] Weyl, H. How far can one get with a linear field theory of gravitation in flat space-time?, *Amer. J. Math.*, 66, 591, (1944), www.sjcrothers.plasmaresources.com/weyl-1.pdf

[67] McVittie, G. C. Laplace's alleged "black hole". *The Observatory*, v.98, 272, 1978, www.sjcrothers.plasmaresources.com/McVittie.pdf

[68] Crothers, S. J. A Brief History of Black Holes. *Progress in Physics*, v.2, 54–57, 2006, www.ptep-online.com/index_files/2006/PP-05-10.PDF

[69] Oppenheimer, J. R., Snyder, H. Phys. Rev. 56, 455 (1939).

[70] Doughty, N., *Am. J. Phys.*, v.49, pp. 720, 1981.

[71] Vladimirov, Yu., Mitskiévich, N., Horský, J. Space Time Gravitation, Mir Publishers, Moscow, 1984.

[72] Hagihara, Y. *Jpn. J. Astron. Geophys.*, 8, 67, (1931).

[73] Gillessen, S., Genzel, R., Eisenhauer, F. (private communication), December 2008, www.sjcrothers.plasmaresources.com/dialogue.pdf

[74] Sintes, A. M., "Report on an all-sky LIGO search for periodic gravitational waves in the S4 data", *Journal of Physics*: Conference Series 122 (2008) 012033, IOP Publishing, www.iop.org/EJ/abstract/1742-6596/122/1/012033/

[75] www.binaryresearchinstitute.org/bri/research/findingit/gravprobeb-update.shtml

[76] National Center for Supercomputing Applications, The Board of Trustees of the University of Illinois, http://archive.ncsa.uiuc.edu/Cyberia/NumRel/BlackHoles.html

[77] National Center for Supercomputing Applications, The Board of Trustees of the University of Illinois, http://archive.ncsa.uiuc.edu/Cyberia/NumRel/BlackHoleFormation.html

[78] DICTIONARY OF GEOPHYSICS, ASTROPHYSICS, and ASTRONOMY, Edited by Richard A. Matzner, CRC Press LLC,, Boca Raton, USA, 2001, http://www.deu.edu.tr/userweb/emre.timur/dosyalar/ Dictionary%20of%20Geophysics,%20Astrophysics%20and%20Astronomy.pdf

[79] McMahon, D. Relativity Demystified, A Self-teaching guide; McGraw-Hill, New York, 2006.

[80] Hawking, S. W. The Theory of Everything, The Origin and Fate of the Universe; New Millennium Press, Beverly Hills, CA., 2002.

[81] Collins Encyclopedia of the Universe, HarperCollins Publishers, London, 2001.

[82] The Penguin Dictionary of Physics, 3rd Ed., Penguin Books Ltd., London, 2000.

Chapter 4

Violation of the Dual Bianchi Identity by Solutions of the Einstein Field Equation

by

Myron W. Evans[1], **Horst Eckardt**[2],

Alpha Institute for Advance Study (AIAS)
(www.aias.us, www.atomicprecision.com)

4.1 Introduction

In chapter two it was shown that solutions of the Einstein field equation violate the Hodge dual of the Bianchi identity of Cartan geometry. In tensor notation the dual identity may be expressed as:

$$D_\mu T^{\kappa\mu\nu} = R^{\kappa}{}_{\mu}{}^{\mu\nu} \tag{4.1}$$

which means that the covariant derivative of the torsion tensor $T^{\kappa\mu\nu}$ is the curvature tensor $R^{\kappa}{}_{\mu}{}^{\mu\nu}$. In this chapter, various classes of exact solutions of the Einstein field equation are tested numerically against equation (4.1), by directly evaluating the curvature tensor. It is found that the Einstein field equation fails the test of Eq. (4.1) because the Einstein field equation is based on a geometry that neglects torsion as discussed in Chapter 2 of this book. In Chapter 3, Crothers shows that the class of vacuum solutions of the Einstein field equation has no physical meaning, because that class of solutions is for the null Einstein tensor of Riemann geometry:

$$G_{\mu\nu} = R_{\mu\nu} - \frac{1}{2} R\, g_{\mu\nu} = 0. \tag{4.2}$$

In Eq. (4.2), the canonical energy momentum density $T_{\mu\nu}$ does not appear, so Eq. (4.2) is one of pure geometry with no physical meaning. It is shown in this chapter that all known solutions of the Einstein field equation:

$$G_{\mu\nu} = k\, T_{\mu\nu} \tag{4.3}$$

violate Eq. (4.1), and therefore violate fundamental geometry. Here k is the Einstein constant, $R_{\mu\nu}$ is the Ricci tensor, R is the Ricci scalar and $g_{\mu\nu}$ is the symmetric metric. These quantities all assume that there is no spacetime torsion, an arbitrary, untenable assumption. The Einstein field equation is not physically meaningful under any circumstances - a fiasco for twentieth century gravitational physics. The ECE gravitational equations [1]- [12] on the other hand rigorously obey Eq. (4.1) and must be used to develop new cosmologies.

[1]e-mail: emyrone@aol.com
[2]e-mail: horsteck@aol.com

All solutions of the Einstein field equation assume the Christoffel or symmetric connection. The use of the symmetric connection means that the torsion tensor vanishes [13]. Therefore for all solutions of the Einstein field equation:

$$T^{\kappa\mu\nu} = 0 \tag{4.4}$$

in Eq. (4.1). It is shown in this chapter that in general:

$$R^{\kappa}{}_{\mu}{}^{\mu\nu} \neq 0 \tag{4.5}$$

for the same solutions of the Einstein field equation that produce Eq. (4.4). So the fundamental geometry of Eq. (4.1) is violated, and the Einstein field equation is mathematically incorrect. The one exception to this result is the class of vacuum solutions, i.e. the purely geometrical solutions of Eq. (4.2). These solutions produce:

$$R^{\kappa}{}_{\mu}{}^{\mu\nu} = 0 \tag{4.6}$$

so that Eq. (4.1) is obeyed fortuitously. However, the class of vacuum solutions by definition assumes that the curvature is zero. Therefore in this case Eq. (4.1) is merely the result of this assumption. Vacuum solutions of the Einstein field equation by definition assume that there is no canonical energy/momentum density present. This concept is analogous to assuming that a field can propagate without a source for that field, a logical self-contradiction. The process of solving Eq. (4.2) without $T_{\mu\nu}$, to give the vacuum solutions, is not consistent with the assumption that the solutions of Eq. (4.2) involve mass M, because mass M is part of $T_{\mu\nu}$, which has already been assumed to be zero. It cannot be first assumed zero and then assumed non-zero. Unfortunately this is the self-inconsistent and meaningless procedure adopted in standard gravitational physics, and it has been criticized by Crothers in chapter 3 of this book.

4.2 Numerical procedure

The procedure adopted was to begin with line elements that are known [14] to be solutions of the Einstein field equation (4.3). There are numerous solutions now known of Eq. (4.3) and are usually classified [14] into groups. For each class of solutions the metric matrix was constructed from the line element. In some classes the metric has off diagonal elements, but in the majority of cases the metric is diagonal. This was used as input parameters for numerical code (see paper 93 of www.aias.us) based on Maxima. The code was rigorously tested with known analytical results and passed this test. The code was then used to evaluate the Christoffel symbols for each line element. It was found that the code correctly reproduced all analytically known Christoffel symbols. The Christoffel symbols were then used to compute all the elements of the Riemann tensor for each line element. Again it was found that the code correctly gave analytically known Riemann elements. The next step was to compute elements of the Ricci tensor from the Riemann tensor elements, and finally to compute the Ricci scalar and Einstein tensor. It was again found that the code correctly reproduced analytically known Ricci tensor elements (for example those in ref [13]).

It is therefore considered that the code is fully accurate and reliable. It was then used to compute the curvature tensor on the right hand side of Eq. (4.1) for selected line elements known to be exact solutions of the Einstein field equation [14]. In general the curvature tensor with raised last two indices is defined by:

$$R^{\kappa}{}_{\mu}{}^{\sigma\rho} = g^{\sigma\alpha} g^{\rho\sigma} R^{\kappa}{}_{\mu\alpha\beta} \tag{4.7}$$

where $g^{\mu\nu}$ are inverse metric elements from the known line elements which are the starting point of the computation. Finally the curvature tensor $R^{\kappa}{}_{\mu}{}^{\mu\nu}$ is defined by summation over repeated indices of the tensor computed in Eq. (4.7):

$$R^{\kappa}{}_{\mu}{}^{\mu\nu} = R^{\kappa}{}_{0}{}^{0\nu} + R^{\kappa}{}_{1}{}^{1\nu} + R^{\kappa}{}_{2}{}^{2\nu} + R^{\kappa}{}_{3}{}^{3\nu}. \tag{4.8}$$

The tensor in Eq. (4.8) was evaluated for:

$$\kappa = 0,\ 1,\ 2,\ 3. \tag{4.9}$$

The results for

$$\kappa = 0 \tag{4.10}$$

were denoted "charge density", and the results for:

$$\kappa = 1,\ 2,\ 3 \tag{4.11}$$

were denoted as elements of the "current density". The reason for this is that they appear as such in the inhomogeneous ECE electro-dynamical equations as explained in chapter 2. For each class of solutions of selected line elements the charge and current densities were evaluated numerically using Maxima. The line element was also tested with the Ricci cyclic equation, which for torsion free situations is:

$$R_{\kappa\mu\nu\sigma} + R_{\kappa\sigma\mu\nu} + R_{\kappa\nu\sigma\mu} = 0 \tag{4.12}$$

and also tested for metric compatibility:

$$D_\mu g_{\nu\rho} = 0. \tag{4.13}$$

Eqs. (4.12) and (4.13) are fundamental requirements which must be satisfied by any line element that claims to be correct, even in the restricted context of torsion-less geometry as used in the Einstein field equation. Line elements that failed the tests of Eqs. (4.12) and (4.13) are mathematically incorrect, and must be discarded. Several well known line elements failed these tests, another fiasco for standard gravitational physics.

Finally the results were graphed and displayed. The computed Christoffel, Riemann, Ricci, Einstein, charge and current density elements and so on can become very intricate, so the computer was used to prepare them in tabular form without any transcription by hand or typesetter. This ensured maximum accuracy.

4.3 Results and discussion

The results are classed into groups as is the custom in standard gravitational physics. The results are given in tabular and graphical format for each class of solutions, and graphed. This procedure was first adopted in paper 93 of www.aias.us and is extended to many known solutions of the Einstein field equation in this chapter. The result is a disaster of standard gravitational physics, because it is shown clearly by computer that the Einstein field equation is mathematically incorrect. This incorrectness is due to the neglect of torsion, and in retrospect it ought to be obvious that an equation that arbitrarily neglects torsion must be incorrect. The Einstein field equation is therefore historically similar to phlogiston and so on.

The fundamental origin of space-time torsion and curvature was briefly discussed in chapter 2, and is presented in a text such as that of Carroll [13]. The origin is as follows:

$$[D_\mu, D_\nu]V^\rho = R^\rho{}_{\sigma\mu\nu}V^\sigma - T^\lambda{}_{\mu\nu}\,D_\lambda V^\rho. \tag{4.14}$$

The commutator of covariant derivatives acting on the vector produces the torsion tensor:

$$T^\lambda{}_{\mu\nu} = \Gamma^\lambda_{\mu\nu} - \Gamma^\lambda_{\nu\mu} \tag{4.15}$$

and the curvature tensor:

$$R^\rho{}_{\sigma\mu\nu} = \partial_\mu\Gamma^\rho_{\nu\sigma} - \partial_\nu\Gamma^\rho_{\mu\sigma} + \Gamma^\rho_{\mu\lambda}\Gamma^\lambda_{\nu\sigma} - \Gamma^\rho_{\nu\lambda}\Gamma^\lambda_{\mu\sigma}. \tag{4.16}$$

This result is true irrespective of any assumption on the connection, and irrespective even of the metric compatibility condition. This is well known and available in a first principles text [13]. The use of a symmetric connection eliminates the torsion tensor:

$$\Gamma^\lambda_{\mu\nu} = \Gamma^\lambda_{\nu\mu} \tag{4.17}$$

and this assumption was used by Einstein in the derivation of the famous field equation. However, this assumption is arbitrary, it is a restriction on a general geometry. The latter is expressed by Cartan through his two well known [1]- [13] structure equations. The first of these defines the Cartan torsion form:

$$T^a = d \wedge q^a + \omega^a{}_b \wedge q^b \tag{4.18}$$

where T^a is the Cartan torsion, q^a is the Cartan tetrad, $\omega^a{}_b$ is the spin connection form, and $d\wedge$ is the exterior derivative. The second structure equation defines the Cartan curvature form:

$$R^a{}_b = d \wedge \omega^a{}_b + \omega^a{}_c \wedge \omega^c{}_b. \tag{4.19}$$

The Bianchi identity:

$$d \wedge T^a + \omega^a{}_b \wedge T^b := R^a{}_b \wedge q^b \tag{4.20}$$

follows from the two Cartan structure equations, as is well known. The dual identity:

$$d \wedge \widetilde{T}^a + \omega^a{}_b \wedge \widetilde{T}^b := \widetilde{R}^a{}_b \wedge q^b \tag{4.21}$$

is an example [1]- [12] of the original identity (4.20), as has been proven recently in several ways. Eq. (4.1) is an example of the dual identity (4.21).

There have been at least two major blunders in the development of standard gravitational physics. These are fundamental errors of geometry. The first is the incorrect elimination of torsion by using a symmetric connection, the second is the use of restricted Bianchi identities instead of the rigorously correct (4.20) and (4.21). The restricted Bianchi identities are known in standard gravitational physics as the first and second Bianchi identities. These are respectively:

$$R^\lambda{}_{\rho\mu\nu} + R^\lambda{}_{\mu\nu\rho} + R^\lambda{}_{\nu\rho\mu} = 0 \tag{4.22}$$

and:

$$D_\mu R^\kappa{}_{\lambda\nu\rho} + D_\rho R^\kappa{}_{\lambda\mu\nu} + D_\nu R^\kappa{}_{\lambda\rho\mu} = 0. \tag{4.23}$$

However, these incorrectly omit torsion and in consequence are not true identities. The equation (4.22) was in fact first derived by Ricci and Levi Civita and is true if and only if the torsion is zero and the connection and metric are symmetric. This has been shown in detail in refs. [1] - [12] and is also shown in ref. [13] for example. Upon translating from the language of differential forms (Eq. (4.20)) to the language of tensors, the true identity that should be used instead of Eq. (4.22) is [1] - [12]:

$$\begin{aligned} R^\lambda{}_{\rho\mu\nu} + R^\lambda{}_{\mu\nu\rho} + R^\lambda{}_{\nu\rho\mu} :=& \\ &\partial_\mu \Gamma^\lambda_{\nu\rho} - \partial_\nu \Gamma^\lambda_{\mu\rho} + \Gamma^\lambda_{\mu\sigma}\Gamma^\sigma_{\nu\rho} - \Gamma^\lambda_{\nu\sigma}\Gamma^\sigma_{\mu\rho} \\ &+\partial_\nu \Gamma^\lambda_{\rho\mu} - \partial_\rho \Gamma^\lambda_{\nu\mu} + \Gamma^\lambda_{\nu\sigma}\Gamma^\sigma_{\rho\mu} - \Gamma^\lambda_{\rho\sigma}\Gamma^\sigma_{\nu\mu} \\ &+\partial_\rho \Gamma^\lambda_{\mu\nu} - \partial_\mu \Gamma^\lambda_{\rho\nu} + \Gamma^\lambda_{\rho\sigma}\Gamma^\sigma_{\mu\nu} - \Gamma^\lambda_{\mu\sigma}\Gamma^\sigma_{\rho\nu} \neq 0 \end{aligned} \tag{4.24}$$

in which the connection is not necessarily symmetric and in which the torsion tensor is not zero. Eq. (4.24) is a rigorously correct mathematical identity, the true Bianchi identity. The curvature tensors appearing in this identity are defined by:

$$R^\lambda{}_{\rho\mu\nu} = \partial_\mu \Gamma^\lambda_{\nu\rho} - \partial_\nu \Gamma^\lambda_{\mu\rho} + \Gamma^\lambda_{\mu\sigma}\Gamma^\sigma_{\nu\rho} - \Gamma^\lambda_{\nu\sigma}\Gamma^\sigma_{\mu\rho} \tag{4.25}$$

and so on in cyclic permutation. However, these curvature tensors are NOT defined in general by a symmetric connection, i.e. the curvature and torsion tensors co-exist as indicated by the very fundamental result (4.14). It is incorrect to claim, as in standard gravitational physics, that the torsion must be zero. Eq. (4.14) is a basic result of geometry, and exists in Riemann geometry as well as in Cartan geometry. The latter is a re-expression of Riemann geometry as is well known [13]. Therefore it is incorrect to claim, as is often done in standard physics, that the Einstein equation is somehow independent of Cartan geometry. This is the same as claiming that the Einstein field equation is independent of geometry, reductio ad absurdum.

The geometry of the Einstein field equation is found by using:

$$T^a = 0 \tag{4.26}$$

in Eqs. (4.18) to (4.21), and as developed in detail [1] - [12] is the geometry:

$$d \wedge q^a = q^b \wedge \omega^a{}_b, \tag{4.27}$$

$$R^a{}_b \wedge q^b = 0. \tag{4.28}$$

Eq. (4.28) is Eq. (4.22) in the notation of differential geometry. Differential forms and tensors are related by the tetrad, as is well known from any good basic textbook [13]. For example the torsion tensor is defined from the torsion form by:

$$T^{\kappa}{}_{\mu\nu} = q^{\kappa}_{a}\, T^{a}{}_{\mu\nu} \tag{4.29}$$

and the curvature tensor is defined from the curvature form by:

$$R^{\kappa}{}_{\mu\nu\sigma} = q^{\kappa}_{a}\, q^{b}_{\mu}\, R^{a}{}_{b\nu\sigma}. \tag{4.30}$$

The way in which Riemann and Cartan geometry inter-relate has been demonstrated recently [1] - [12] in comprehensive detail, not easily found elsewhere. It is incorrect therefore to claim that Cartan and Riemann geometry are somehow "independent". As shown by Eqs. (4.26) to (4.28), the Einstein field equation cannot be independent of Cartan geometry.

In the latter geometry there are no restrictions in general on the connection, because the torsion is in general non-zero. The torsion form is defined by:

$$\begin{aligned} T^{a} &= D \wedge q^{a} \\ &:= d \wedge q^{a} + \omega^{a}{}_{b} \wedge q^{b} \end{aligned} \tag{4.31}$$

and the torsion tensor is defined by:

$$T^{\kappa}{}_{\mu\nu} = \Gamma^{\kappa}_{\mu\nu} - \Gamma^{\kappa}_{\nu\mu}. \tag{4.32}$$

The latter equation may be obtained from the former using the definition:

$$T^{\kappa}{}_{\mu\nu} = q^{\kappa}_{a}\, T^{a}{}_{\mu\nu} \tag{4.33}$$

and the tetrad postulate ([1]- [13]):

$$D_{\mu} q^{a}_{\mu} = 0. \tag{4.34}$$

As may be seen from Eq. (4.14), the torsion tensor (4.32) is the direct result of the commutator of covariant derivatives acting on the vector. In the early days of general relativity, it was assumed just for the sake of ease of calculation that the connection is symmetric:

$$\Gamma^{\kappa}_{\mu\nu} = \Gamma^{\kappa}_{\nu\mu} \tag{4.35}$$

which means from Eq. (4.32) that the torsion is assumed to vanish. This assumption is arbitrary, and the computer results of this chapter show that the assumption leads to a violation of the dual identity. So the assumption of vanishing torsion is not only restrictive, it is fundamentally incorrect. This means that all inferences based on the Einstein field equation must be discarded as obsolete.

As shown in a textbook such as ref. [13], the assumption of symmetric metric leads in turn to the usual

$$\Gamma^{\sigma}_{\mu\nu} = \frac{1}{2} g^{\sigma\rho} \left(\partial_{\mu} g_{\nu\rho} + \partial_{\nu} g_{\rho\mu} - \partial_{\rho} g_{\mu\nu} \right). \tag{4.36}$$

However it is almost never stated in standard physics that this definition depends on assuming that the metric is symmetric:

$$g_{\mu\nu} = g_{\nu\mu} \tag{4.37}$$

and also depends on the assumption of metric compatibility:

$$D_{\mu} g_{\nu\rho} = 0. \tag{4.38}$$

On the other hand, the fundamental Eq. (4.14) makes no such assumptions. Also, the Cartan structure equations make no such assumptions, they depend only on the tetrad postulate:

$$D_{\mu} q^{a}_{\nu} = 0 \tag{4.39}$$

which is the very fundamental requirement that a vector field be independent of its coordinate system. For example a vector field in three dimensional Euclidean space is the same vector field if expressed in the Cartesian system, or any other valid system of coordinates such as the spherical polar or cylindrical polar. The symmetric metric tensor as usually used in standard physics is defined in any good textbook [13] as:

$$g_{\mu\nu} = q^a_\mu \, q^b_\nu \, \eta_{ab} \tag{4.40}$$

where η_{ab} is the Minkowski metric. Eq. (4.40) is only a special case of a more general tensor product:

$$g^{ab}_{\mu\nu} = q^a_\mu \, q^b_\nu \tag{4.41}$$

of two tetrads [13]. It is this special case that is used in Einsteinian gravitational theory.

It has also been shown in detail [1]- [12] that the second Cartan structure equation:

$$\begin{aligned} R^a{}_b &= D \wedge \omega^a{}_b \\ &:= d \wedge \omega^a{}_b + \omega^a{}_c \wedge \omega^c{}_b \end{aligned} \tag{4.42}$$

translates into the definition (4.25) of the curvature tensor by use of the tetrad postulate and the definition:

$$R^a{}_{b\mu\nu} = q^a_\kappa \, q^\rho_b \, R^\kappa{}_{\rho\mu\nu}. \tag{4.43}$$

The two Cartan structure equations are therefore equivalent to the tensor equation (4.14). The Bianchi identity:

$$D \wedge T^a = R^a{}_b \wedge q^b \tag{4.44}$$

and its dual identity:

$$D \wedge \widetilde{T}^a = \widetilde{R}^a{}_b \wedge q^b \tag{4.45}$$

are therefore the results of the two Cartan structure equations and of the tensor Eq. (4.14). It has been indicated already that the tensorial expression of Eq. (4.44) is Eq. (4.24). Similarly, the tensorial expression of Eq. (4.45) is the rigorous identity:

$$\begin{aligned} &\widetilde{R}^\lambda{}_{\rho\mu\nu} + \widetilde{R}^\lambda{}_{\mu\nu\rho} + \widetilde{R}^\lambda{}_{\nu\rho\mu} := \\ &\quad \left(\partial_\mu \Gamma^\lambda_{\nu\rho} - \partial_\nu \Gamma^\lambda_{\mu\rho} + \Gamma^\lambda_{\mu\sigma}\Gamma^\sigma_{\nu\rho} - \Gamma^\lambda_{\nu\sigma}\Gamma^\sigma_{\mu\rho}\right)_{HD} \\ &\quad + \left(\partial_\nu \Gamma^\lambda_{\rho\mu} - \partial_\rho \Gamma^\lambda_{\nu\mu} + \Gamma^\lambda_{\nu\sigma}\Gamma^\sigma_{\rho\mu} - \Gamma^\lambda_{\rho\sigma}\Gamma^\sigma_{\nu\mu}\right)_{HD} \\ &\quad + \left(\partial_\rho \Gamma^\lambda_{\mu\nu} - \partial_\mu \Gamma^\lambda_{\rho\nu} + \Gamma^\lambda_{\rho\sigma}\Gamma^\sigma_{\mu\nu} - \Gamma^\lambda_{\mu\sigma}\Gamma^\sigma_{\rho\nu}\right)_{HD} \neq 0 \end{aligned} \tag{4.46}$$

in which appear the definitions:

$$\widetilde{R}^\lambda{}_{\rho\mu\nu} = \left(\partial_\mu \Gamma^\lambda_{\nu\rho} - \partial_\nu \Gamma^\lambda_{\mu\rho} + \Gamma^\lambda_{\mu\sigma}\Gamma^\sigma_{\nu\rho} - \Gamma^\lambda_{\nu\sigma}\Gamma^\sigma_{\mu\rho}\right)_{HD} \tag{4.47}$$

and so on in cyclic permutation [1]- [12]. These results have been proven in all detail in several ECE papers. It has also been proven that the so called second Bianchi identity of standard physics, Eq. (4.23), should be:

$$D \wedge (D \wedge T^a) := D \wedge \left(R^a{}_b \wedge q^b\right) \tag{4.48}$$

which again includes the torsion.

The Einstein field equation (4.3) was obtained from Eq. (4.23) [1]- [13], which in the language of differential geometry is:

$$D \wedge R^a{}_b = 0, \tag{4.49}$$

an equation in which the torsion is missing incorrectly. The torsionless Bianchi identity (4.23) may be re-expressed [1]- [13] as:

$$D^\mu G_{\mu\nu} = 0. \tag{4.50}$$

The covariant Noether Theorem is:

$$D^\mu T_{\mu\nu} = 0. \tag{4.51}$$

and it was assumed in 1915 by Einstein that:

$$D^\mu G_{\mu\nu} = k\ D^\mu T_{\mu\nu}. \tag{4.52}$$

The Einstein field equation (4.3) is a further assumption from Eq. (4.52).

The Maxima code uses Eq. (4.36) to compute the Christoffel connection elements, and uses Eq. (4.25) to compute the Riemann tensor elements. The Ricci tensor elements are computed from the standard Ricci index contraction:

$$R_{\mu\nu} = R^\lambda{}_{\mu\lambda\nu}. \tag{4.53}$$

and the Ricci scalar is defined by:

$$R = g^{\mu\nu} R_{\mu\nu}. \tag{4.54}$$

All these quantities are computed by the code.

4.4 Exact solutions of the Einstein field equation

Recently the many known exact solutions of the Einstein field equation (4.3) have been collected in a volume [14]. There are several classes of solutions in this volume, and in this chapter examples of the classes of solutions have been tested against the fundamental dual identity (4.1). It is found that all solutions violate the dual identity or are otherwise physically meaningless. The volume is typical of twentieth century standard physics in being abstract and mathematical, losing touch with Baconian physics. For the sake of testing it is sufficient to chose a few examples from the volume, representative of each class. The results of the evaluation of these line elements are given in the tables and graphs of this chapter. Each solution in the book assumes the Christoffel symbol, so each solution incorrectly neglects torsion. The computer algebra shows that this assumption leads to the following violation of eq. (4.1):

$$T^{\kappa\mu\nu} = 0,\ R^\kappa{}_\mu{}^{\mu\nu} \neq 0 \tag{4.55}$$

for each solution in which there is finite energy momentum density. The only exception is the class of solutions described by:

$$G_{\mu\nu} = R_{\mu\nu} - \frac{1}{2} R\ g_{\mu\nu} = 0. \tag{4.56}$$

In this case:

$$T^{\kappa\mu\nu} = 0,\ R^\kappa{}_\mu{}^{\mu\nu} = 0 \tag{4.57}$$

as shown in the Tables. However, the assumption (4.56) precludes the existence of matter, and is merely a geometrical exercise (see chapter 3) again carried out in the early days of general relativity merely for the sake of ease of hand calculation.

The fact that the Einstein field equation is incorrect is a major disaster for standard physics. A glance at the Tables in this chapter shows that the calculation of the Christoffel symbols and Riemann tensor elements is in general an intricately complicated process, essentially impossible by hand for any but the simplest line elements. The key result of the work reported in refs. [1] to [12] is the dual identity, eq. (4.1). Standard physics for the past ninety years has not recognized the existence of this identity, and prior to the emergence of the computer algebra such as Maxima, hand calculations were essentially impossible because of their great complexity. The major problem is that the Einstein field equation has been accepted uncritically. The various claims as to its precision are in fact incorrect. As shown in chapter 2, the correct equations of physics must always contain space-time torsion as a central ingredient.

4.4.1 Minkowski metric with shifted radial coordinate

This form of the spherically symmetric line element shows that Minowski space is invariant against a shift in the r coordinate.

Coordinates

$$\mathbf{x} = \begin{pmatrix} t \\ r \\ \vartheta \\ \varphi \end{pmatrix}$$

Metric

$$g_{\mu\nu} = \begin{pmatrix} -1 & 0 & 0 & 0 \\ 0 & 1 & 0 & 0 \\ 0 & 0 & (r_0 - r)^2 & 0 \\ 0 & 0 & 0 & (r_0 - r)^2 \sin^2\vartheta \end{pmatrix}$$

Contravariant Metric

$$g^{\mu\nu} = \begin{pmatrix} -1 & 0 & 0 & 0 \\ 0 & 1 & 0 & 0 \\ 0 & 0 & \frac{1}{(r-r_0)^2} & 0 \\ 0 & 0 & 0 & \frac{1}{(r-r_0)^2 \sin^2\vartheta} \end{pmatrix}$$

Christoffel Connection

$$\Gamma^1{}_{22} = -(r - r_0)$$

$$\Gamma^1{}_{33} = -(r - r_0)\sin^2\vartheta$$

$$\Gamma^2{}_{12} = \frac{1}{r - r_0}$$

$$\Gamma^2{}_{21} = \Gamma^2{}_{12}$$

$$\Gamma^2{}_{33} = -\cos\vartheta\sin\vartheta$$

$$\Gamma^3{}_{13} = \frac{1}{r - r_0}$$

$$\Gamma^3{}_{23} = \frac{\cos\vartheta}{\sin\vartheta}$$

$$\Gamma^3{}_{31} = \Gamma^3{}_{13}$$

$$\Gamma^3{}_{32} = \Gamma^3{}_{23}$$

Metric Compatibility

——— o.k.

Riemann Tensor

——— all elements zero

Ricci Tensor

——— all elements zero

Ricci Scalar

$$R_{sc} = 0$$

Bianchi identity (Ricci cyclic equation $R^{\kappa}{}_{[\mu\nu\sigma]} = 0$)

——— o.k.

Einstein Tensor

——— all elements zero

Hodge Dual of Bianchi Identity

——— (see charge and current densities)

Scalar Charge Density (-$R^{0}{}_{i}{}^{i0}$)

$$\rho = 0$$

Current Density Class 1 (-$R^{i}{}_{\mu}{}^{\mu j}$)

$$J_1 = 0$$

$$J_2 = 0$$

$$J_3 = 0$$

Current Density Class 2 (-$R^{i}{}_{\mu}{}^{\mu j}$)

$$J_1 = 0$$

$$J_2 = 0$$

$$J_3 = 0$$

Current Density Class 3 ($-R^{i}{}_{\mu}{}^{\mu j}$)

$$J_1 = 0$$

$$J_2 = 0$$

$$J_3 = 0$$

4.4.2 Schwarzschild metric

This is the so-called Schwarzschild metric. The interpretation of the parameters (M: mass, G: Newton's constant of gravitation, c: velocity of light) was added later. The Schwarzschild metric is a true vacuum metric, i.e. Ricci and Einstein tensors vahish.

Coordinates

$$\mathbf{x} = \begin{pmatrix} t \\ r \\ \vartheta \\ \varphi \end{pmatrix}$$

Metric

$$g_{\mu\nu} = \begin{pmatrix} \frac{2\,G\,M}{c^2\,r} - 1 & 0 & 0 & 0 \\ 0 & \frac{1}{1-\frac{2\,G\,M}{c^2\,r}} & 0 & 0 \\ 0 & 0 & r^2 & 0 \\ 0 & 0 & 0 & r^2\,\sin^2\vartheta \end{pmatrix}$$

Contravariant Metric

$$g^{\mu\nu} = \begin{pmatrix} \frac{c^2\,r}{2\,G\,M-c^2\,r} & 0 & 0 & 0 \\ 0 & -\frac{2\,G\,M-c^2\,r}{c^2\,r} & 0 & 0 \\ 0 & 0 & \frac{1}{r^2} & 0 \\ 0 & 0 & 0 & \frac{1}{r^2\,\sin^2\vartheta} \end{pmatrix}$$

Christoffel Connection

$$\Gamma^0{}_{01} = -\frac{G\,M}{r\,\left(2\,G\,M - c^2\,r\right)}$$

$$\Gamma^0{}_{10} = \Gamma^0{}_{01}$$

$$\Gamma^1{}_{00} = -\frac{G\,M\,\left(2\,G\,M - c^2\,r\right)}{c^4\,r^3}$$

$$\Gamma^1{}_{11} = \frac{G\,M}{r\,\left(2\,G\,M - c^2\,r\right)}$$

$$\Gamma^{1}{}_{22} = \frac{2\,G\,M - c^2\,r}{c^2}$$

$$\Gamma^{1}{}_{33} = \frac{\sin^2\vartheta\,\left(2\,G\,M - c^2\,r\right)}{c^2}$$

$$\Gamma^{2}{}_{12} = \frac{1}{r}$$

$$\Gamma^{2}{}_{21} = \Gamma^{2}{}_{12}$$

$$\Gamma^{2}{}_{33} = -\cos\vartheta\,\sin\vartheta$$

$$\Gamma^{3}{}_{13} = \frac{1}{r}$$

$$\Gamma^{3}{}_{23} = \frac{\cos\vartheta}{\sin\vartheta}$$

$$\Gamma^{3}{}_{31} = \Gamma^{3}{}_{13}$$

$$\Gamma^{3}{}_{32} = \Gamma^{3}{}_{23}$$

Metric Compatibility

——— o.k.

Riemann Tensor

$$R^{0}{}_{101} = -\frac{2\,G\,M}{r^2\,(2\,G\,M - c^2\,r)}$$

$$R^{0}{}_{110} = -R^{0}{}_{101}$$

$$R^{0}{}_{202} = -\frac{G\,M}{c^2\,r}$$

$$R^{0}{}_{220} = -R^{0}{}_{202}$$

$$R^{0}{}_{303} = -\frac{\sin^2\vartheta\,G\,M}{c^2\,r}$$

$$R^{0}{}_{330} = -R^{0}{}_{303}$$

$$R^{1}{}_{001} = -\frac{2\,G\,M\,\left(2\,G\,M - c^2\,r\right)}{c^4\,r^4}$$

$$R^{1}{}_{010} = -R^{1}{}_{001}$$

$$R^{1}{}_{212} = -\frac{G\,M}{c^2\,r}$$

$$R^{1}{}_{221} = -R^{1}{}_{212}$$

$$R^{1}{}_{313} = -\frac{\sin^2\vartheta\,G\,M}{c^2\,r}$$

$$R^{1}{}_{331} = -R^{1}{}_{313}$$

$$R^{2}{}_{002} = \frac{G\,M\,\left(2\,G\,M - c^2\,r\right)}{c^4\,r^4}$$

$$R^{2}{}_{020} = -R^{2}{}_{002}$$

$$R^{2}{}_{112} = -\frac{G\,M}{r^2\,(2\,G\,M - c^2\,r)}$$

$$R^{2}{}_{121} = -R^{2}{}_{112}$$

$$R^{2}{}_{323} = \frac{2\,\sin^2\vartheta\,G\,M}{c^2\,r}$$

$$R^{2}{}_{332} = -R^{2}{}_{323}$$

$$R^{3}{}_{003} = \frac{G\,M\,\left(2\,G\,M - c^2\,r\right)}{c^4\,r^4}$$

$$R^{3}{}_{030} = -R^{3}{}_{003}$$

$$R^{3}{}_{113} = -\frac{G\,M}{r^2\,(2\,G\,M - c^2\,r)}$$

$$R^{3}{}_{131} = -R^{3}{}_{113}$$

$$R^{3}{}_{223} = -\frac{2\,G\,M}{c^2\,r}$$

$$R^{3}{}_{232} = -R^{3}{}_{223}$$

Ricci Tensor

——— all elements zero

Ricci Scalar

$$R_{sc} = 0$$

Bianchi identity (Ricci cyclic equation $R^{\kappa}{}_{[\mu\nu\sigma]} = 0$)

——— o.k.

Einstein Tensor

——— all elements zero

Hodge Dual of Bianchi Identity

——— (see charge and current densities)

Scalar Charge Density (-$R^{0}{}_{i}{}^{i0}$)

$\rho = 0$

Current Density Class 1 (-$R^{i}{}_{\mu}{}^{\mu j}$)

$J_1 = 0$

$J_2 = 0$

$J_3 = 0$

Current Density Class 2 (-$R^{i}{}_{\mu}{}^{\mu j}$)

$J_1 = 0$

$J_2 = 0$

$J_3 = 0$

Current Density Class 3 (-$R^{i}{}_{\mu}{}^{\mu j}$)

$J_1 = 0$

$J_2 = 0$

$J_3 = 0$

4.4.3 General Crothers metric

This is a general spherical symmetric metric with a general function $C(r)$. It does not describe a vacuum in general since Ricci and Einstein tensors are different from zero.

Coordinates

$$\mathbf{x} = \begin{pmatrix} t \\ r \\ \vartheta \\ \varphi \end{pmatrix}$$

Metric

$$g_{\mu\nu} = \begin{pmatrix} -A\sqrt{C} & 0 & 0 & 0 \\ 0 & B\sqrt{C} & 0 & 0 \\ 0 & 0 & C & 0 \\ 0 & 0 & 0 & \sin^2\vartheta\, C \end{pmatrix}$$

Contravariant Metric

$$g^{\mu\nu} = \begin{pmatrix} -\frac{1}{A\sqrt{C}} & 0 & 0 & 0 \\ 0 & \frac{1}{B\sqrt{C}} & 0 & 0 \\ 0 & 0 & \frac{1}{C} & 0 \\ 0 & 0 & 0 & \frac{1}{\sin^2\vartheta\, C} \end{pmatrix}$$

Christoffel Connection

$$\Gamma^0{}_{01} = \frac{\frac{d}{d\,r}C}{4\,C}$$

$$\Gamma^0{}_{10} = \Gamma^0{}_{01}$$

$$\Gamma^1{}_{00} = \frac{A\left(\frac{d}{d\,r}C\right)}{4\,B\,C}$$

$$\Gamma^1{}_{11} = \frac{\frac{d}{d\,r}C}{4\,C}$$

$$\Gamma^1{}_{22} = -\frac{\frac{d}{d\,r}C}{2\,B\sqrt{C}}$$

$$\Gamma^1{}_{33} = -\frac{\sin^2\vartheta\left(\frac{d}{d\,r}C\right)}{2\,B\sqrt{C}}$$

$$\Gamma^2{}_{12} = \frac{\frac{d}{d\,r}C}{2\,C}$$

$$\Gamma^2{}_{21} = \Gamma^2{}_{12}$$

$$\Gamma^2{}_{33} = -\cos\vartheta\,\sin\vartheta$$

$$\Gamma^3{}_{13} = \frac{\frac{d}{d\,r}C}{2\,C}$$

$$\Gamma^3{}_{23} = \frac{\cos\vartheta}{\sin\vartheta}$$

$$\Gamma^3{}_{31} = \Gamma^3{}_{13}$$

$$\Gamma^3{}_{32} = \Gamma^3{}_{23}$$

Metric Compatibility

——— o.k.

Riemann Tensor

$$R^0{}_{101} = -\frac{C\left(\frac{d^2}{dr^2}C\right) - \left(\frac{d}{dr}C\right)^2}{4C^2}$$

$$R^0{}_{110} = -R^0{}_{101}$$

$$R^0{}_{202} = -\frac{\left(\frac{d}{dr}C\right)^2}{8BC^{\frac{3}{2}}}$$

$$R^0{}_{220} = -R^0{}_{202}$$

$$R^0{}_{303} = -\frac{\sin^2\vartheta\left(\frac{d}{dr}C\right)^2}{8BC^{\frac{3}{2}}}$$

$$R^0{}_{330} = -R^0{}_{303}$$

$$R^1{}_{001} = -\frac{A\left(C\left(\frac{d^2}{dr^2}C\right) - \left(\frac{d}{dr}C\right)^2\right)}{4BC^2}$$

$$R^1{}_{010} = -R^1{}_{001}$$

$$R^1{}_{212} = -\frac{4C\left(\frac{d^2}{dr^2}C\right) - 3\left(\frac{d}{dr}C\right)^2}{8BC^{\frac{3}{2}}}$$

$$R^1{}_{221} = -R^1{}_{212}$$

$$R^1{}_{313} = -\frac{\sin^2\vartheta\left(4C\left(\frac{d^2}{dr^2}C\right) - 3\left(\frac{d}{dr}C\right)^2\right)}{8BC^{\frac{3}{2}}}$$

$$R^1{}_{331} = -R^1{}_{313}$$

$$R^2{}_{002} = -\frac{A\left(\frac{d}{dr}C\right)^2}{8BC^2}$$

$$R^2{}_{020} = -R^2{}_{002}$$

$$R^2{}_{112} = \frac{4C\left(\frac{d^2}{dr^2}C\right) - 3\left(\frac{d}{dr}C\right)^2}{8C^2}$$

$$R^2{}_{121} = -R^2{}_{112}$$

$$R^2{}_{323} = -\frac{\sin^2\vartheta\left(\left(\frac{d}{dr}C\right)^2 - 4\,B\,C^{\frac{3}{2}}\right)}{4\,B\,C^{\frac{3}{2}}}$$

$$R^2{}_{332} = -R^2{}_{323}$$

$$R^3{}_{003} = -\frac{A\left(\frac{d}{dr}C\right)^2}{8\,B\,C^2}$$

$$R^3{}_{030} = -R^3{}_{003}$$

$$R^3{}_{113} = \frac{4\,C\left(\frac{d^2}{dr^2}C\right) - 3\left(\frac{d}{dr}C\right)^2}{8\,C^2}$$

$$R^3{}_{131} = -R^3{}_{113}$$

$$R^3{}_{223} = \frac{\left(\frac{d}{dr}C\right)^2 - 4\,B\,C^{\frac{3}{2}}}{4\,B\,C^{\frac{3}{2}}}$$

$$R^3{}_{232} = -R^3{}_{223}$$

Ricci Tensor

$$\mathrm{Ric}_{00} = \frac{A\left(\frac{d^2}{dr^2}C\right)}{4\,B\,C}$$

$$\mathrm{Ric}_{11} = -\frac{5\,C\left(\frac{d^2}{dr^2}C\right) - 4\left(\frac{d}{dr}C\right)^2}{4\,C^2}$$

$$\mathrm{Ric}_{22} = -\frac{\frac{d^2}{dr^2}C - 2\,B\,\sqrt{C}}{2\,B\,\sqrt{C}}$$

$$\mathrm{Ric}_{33} = -\frac{\sin^2\vartheta\left(\frac{d^2}{dr^2}C - 2\,B\,\sqrt{C}\right)}{2\,B\,\sqrt{C}}$$

Ricci Scalar

$$R_{sc} = -\frac{5\,C\left(\frac{d^2}{dr^2}C\right) - 2\left(\frac{d}{dr}C\right)^2 - 4\,B\,C^{\frac{3}{2}}}{2\,B\,C^{\frac{5}{2}}}$$

Bianchi identity (Ricci cyclic equation $R^\kappa{}_{[\mu\nu\sigma]} = 0$)

———— o.k.

Einstein Tensor

$$G_{00} = -\frac{A\left(2C\left(\frac{d^2}{dr^2}C\right) - \left(\frac{d}{dr}C\right)^2 - 2BC^{\frac{3}{2}}\right)}{2BC^2}$$

$$G_{11} = \frac{\left(\frac{d}{dr}C\right)^2 - 2BC^{\frac{3}{2}}}{2C^2}$$

$$G_{22} = \frac{3C\left(\frac{d^2}{dr^2}C\right) - 2\left(\frac{d}{dr}C\right)^2}{4BC^{\frac{3}{2}}}$$

$$G_{33} = \frac{\sin^2\vartheta\left(3C\left(\frac{d^2}{dr^2}C\right) - 2\left(\frac{d}{dr}C\right)^2\right)}{4BC^{\frac{3}{2}}}$$

Hodge Dual of Bianchi Identity

——— (see charge and current densities)

Scalar Charge Density (-$R^{0\ \ i0}_{\ \ i}$)

$$\rho = \frac{\frac{d^2}{dr^2}C}{4ABC^2}$$

Current Density Class 1 (-$R^{i\ \ \mu j}_{\ \ \mu}$)

$$J_1 = \frac{5C\left(\frac{d^2}{dr^2}C\right) - 4\left(\frac{d}{dr}C\right)^2}{4B^2C^3}$$

$$J_2 = \frac{\frac{d^2}{dr^2}C - 2B\sqrt{C}}{2BC^{\frac{5}{2}}}$$

$$J_3 = \frac{\frac{d^2}{dr^2}C - 2B\sqrt{C}}{2\sin^2\vartheta\, BC^{\frac{5}{2}}}$$

Current Density Class 2 (-$R^{i\ \ \mu j}_{\ \ \mu}$)

$$J_1 = 0$$

$$J_2 = 0$$

$$J_3 = 0$$

Current Density Class 3 ($-R^{i\ \ \mu j}_{\ \mu}$)

$$J_1 = 0$$

$$J_2 = 0$$

$$J_3 = 0$$

4.4.4 Crothers metric with generalized Schwarzschild parameters

The general Crothers metric has been taken with a special function $C(r)$:

$$C(r) = (|r - r_0|^n + \alpha^n)^{2/n}.$$

Again this is not a vacuum metric.

Coordinates

$$\mathbf{x} = \begin{pmatrix} t \\ r \\ \vartheta \\ \varphi \end{pmatrix}$$

Metric

$$g_{\mu\nu} = \begin{pmatrix} -\sqrt{(|r_0 - r|^n + \alpha^n)^{\frac{2}{n}}}\, A & 0 & 0 & 0 \\ 0 & \sqrt{(|r_0 - r|^n + \alpha^n)^{\frac{2}{n}}}\, B & 0 & 0 \\ 0 & 0 & (|r_0 - r|^n + \alpha^n)^{\frac{2}{n}} & 0 \\ 0 & 0 & 0 & (|r_0 - r|^n + \alpha^n)^{\frac{2}{n}} \sin^2\vartheta \end{pmatrix}$$

Contravariant Metric

$$g^{\mu\nu} = \begin{pmatrix} -\frac{\sqrt{(|r_0-r|^n+\alpha^n)^{\frac{2}{n}}}}{(|r_0-r|^n+\alpha^n)^{\frac{2}{n}} A} & 0 & 0 & 0 \\ 0 & \frac{\sqrt{(|r_0-r|^n+\alpha^n)^{\frac{2}{n}}}}{(|r_0-r|^n+\alpha^n)^{\frac{2}{n}} B} & 0 & 0 \\ 0 & 0 & \frac{1}{(|r_0-r|^n+\alpha^n)^{\frac{2}{n}}} & 0 \\ 0 & 0 & 0 & \frac{1}{(|r_0-r|^n+\alpha^n)^{\frac{2}{n}} \sin^2\vartheta} \end{pmatrix}$$

Christoffel Connection

$$\Gamma^0_{\ 01} = -\frac{|r_0 - r|^n}{2\,(r_0 - r)\,(|r_0 - r|^n + \alpha^n)}$$

$$\Gamma^0_{\ 10} = \Gamma^0_{\ 01}$$

$$\Gamma^1_{\ 00} = -\frac{|r_0 - r|^n\, A}{2\,(r_0 - r)\,(|r_0 - r|^n + \alpha^n)\, B}$$

$$\Gamma^1_{\ 11} = -\frac{|r_0 - r|^n}{2\,(r_0 - r)\,(|r_0 - r|^n + \alpha^n)}$$

$$\Gamma^1_{\ 22} = \frac{|r_0 - r|^n \sqrt{(|r_0 - r|^n + \alpha^n)^{\frac{2}{n}}}}{(r_0 - r)\,(|r_0 - r|^n + \alpha^n)\, B}$$

$$\Gamma^1{}_{33} = \frac{|r_0 - r|^n \sqrt{(|r_0 - r|^n + \alpha^n)^{\frac{2}{n}}} \sin^2\vartheta}{(r_0 - r)\,(|r_0 - r|^n + \alpha^n)\,B}$$

$$\Gamma^2{}_{12} = -\frac{|r_0 - r|^n}{(r_0 - r)\,(|r_0 - r|^n + \alpha^n)}$$

$$\Gamma^2{}_{21} = \Gamma^2{}_{12}$$

$$\Gamma^2{}_{33} = -\cos\vartheta\,\sin\vartheta$$

$$\Gamma^3{}_{13} = -\frac{|r_0 - r|^n}{(r_0 - r)\,(|r_0 - r|^n + \alpha^n)}$$

$$\Gamma^3{}_{23} = \frac{\cos\vartheta}{\sin\vartheta}$$

$$\Gamma^3{}_{31} = \Gamma^3{}_{13}$$

$$\Gamma^3{}_{32} = \Gamma^3{}_{23}$$

Metric Compatibility

——— o.k.

Riemann Tensor

$$R^0{}_{101} = \frac{|r_0 - r|^n\,(|r_0 - r|^n - \alpha^n\,n + \alpha^n)}{2\,(r_0 - r)^2\,(|r_0 - r|^n + \alpha^n)^2}$$

$$R^0{}_{110} = -R^0{}_{101}$$

$$R^0{}_{202} = -\frac{|r_0 - r|^{2\,n} \sqrt{(|r_0 - r|^n + \alpha^n)^{\frac{2}{n}}}}{2\,(r_0 - r)^2\,(|r_0 - r|^n + \alpha^n)^2\,B}$$

$$R^0{}_{220} = -R^0{}_{202}$$

$$R^0{}_{303} = -\frac{|r_0 - r|^{2\,n} \sqrt{(|r_0 - r|^n + \alpha^n)^{\frac{2}{n}}} \sin^2\vartheta}{2\,(r_0 - r)^2\,(|r_0 - r|^n + \alpha^n)^2\,B}$$

$$R^0{}_{330} = -R^0{}_{303}$$

$$R^1{}_{001} = \frac{|r_0 - r|^n\,(|r_0 - r|^n - \alpha^n\,n + \alpha^n)\,A}{2\,(r_0 - r)^2\,(|r_0 - r|^n + \alpha^n)^2\,B}$$

$$R^1{}_{010} = -R^1{}_{001}$$

$$R^1{}_{212} = \frac{|r_0 - r|^n\,(|r_0 - r|^n + \alpha^n)^{\frac{2}{n}-2}\,(|r_0 - r|^n - 2\,\alpha^n\,n + 2\,\alpha^n)}{2\,(r_0 - r)^2 \sqrt{(|r_0 - r|^n + \alpha^n)^{\frac{2}{n}}}\,B}$$

$$R^1{}_{221} = -R^1{}_{212}$$

$$R^1{}_{313} = \frac{|r_0 - r|^n\,(|r_0 - r|^n + \alpha^n)^{\frac{2}{n}-2}\,(|r_0 - r|^n - 2\,\alpha^n\,n + 2\,\alpha^n)\,\sin^2\vartheta}{2\,(r_0 - r)^2 \sqrt{(|r_0 - r|^n + \alpha^n)^{\frac{2}{n}}}\,B}$$

$$R^1{}_{331} = -R^1{}_{313}$$

$$R^2{}_{002} = -\frac{|r_0 - r|^{2n} A}{2 (r_0 - r)^2 (|r_0 - r|^n + \alpha^n)^2 B}$$

$$R^2{}_{020} = -R^2{}_{002}$$

$$R^2{}_{112} = -\frac{|r_0 - r|^n (|r_0 - r|^n - 2\alpha^n n + 2\alpha^n)}{2 (r_0 - r)^2 (|r_0 - r|^n + \alpha^n)^2}$$

$$R^2{}_{121} = -R^2{}_{112}$$

$$R^2{}_{323} = \frac{\sin^2\vartheta \left(r_0{}^2 |r_0 - r|^{2n} B - 2 r r_0 |r_0 - r|^{2n} B + r^2 |r_0 - r|^{2n} B + 2\alpha^n r_0{}^2 |r_0 - r|^n B - 4\alpha^n r r_0 |r_0 - r|^n B + \ldots\right)}{(r_0 - r)^2 (|r_0 - r|^n + \alpha^n)^2 B}$$

$$R^2{}_{332} = -R^2{}_{323}$$

$$R^3{}_{003} = -\frac{|r_0 - r|^{2n} A}{2 (r_0 - r)^2 (|r_0 - r|^n + \alpha^n)^2 B}$$

$$R^3{}_{030} = -R^3{}_{003}$$

$$R^3{}_{113} = -\frac{|r_0 - r|^n (|r_0 - r|^n - 2\alpha^n n + 2\alpha^n)}{2 (r_0 - r)^2 (|r_0 - r|^n + \alpha^n)^2}$$

$$R^3{}_{131} = -R^3{}_{113}$$

$$R^3{}_{223} = -\frac{r_0{}^2 |r_0 - r|^{2n} B - 2 r r_0 |r_0 - r|^{2n} B + r^2 |r_0 - r|^{2n} B + 2\alpha^n r_0{}^2 |r_0 - r|^n B - 4\alpha^n r r_0 |r_0 - r|^n B + \ldots}{(r_0 - r)^2 (|r_0 - r|^n + \alpha^n)^2 B}$$

$$R^3{}_{232} = -R^3{}_{223}$$

Ricci Tensor

$$\mathrm{Ric}_{00} = \frac{|r_0 - r|^n (|r_0 - r|^n + \alpha^n n - \alpha^n) A}{2 (r_0 - r)^2 (|r_0 - r|^n + \alpha^n)^2 B}$$

$$\mathrm{Ric}_{11} = \frac{|r_0 - r|^n (3 |r_0 - r|^n - 5\alpha^n n + 5\alpha^n)}{2 (r_0 - r)^2 (|r_0 - r|^n + \alpha^n)^2}$$

$$\mathrm{Ric}_{22} = \frac{r_0{}^2 |r_0 - r|^{2n} \sqrt{(|r_0 - r|^n + \alpha^n)^{\frac{2}{n}}} B - 2 r r_0 |r_0 - r|^{2n} \sqrt{(|r_0 - r|^n + \alpha^n)^{\frac{2}{n}}} B + r^2 |r_0 - r|^{2n} \sqrt{(|r_0 - r|^n + \alpha^n)^{\frac{2}{n}}} B + \ldots}{(r_0 - r)^2 (|r_0 - r|^n + \alpha^n)^2 \sqrt{(|r_0 - r|^n + \alpha^n)^{\frac{2}{n}}} B}$$

$$\mathrm{Ric}_{33} = \frac{\sin^2\vartheta \left(r_0{}^2 |r_0 - r|^{2n} \sqrt{(|r_0 - r|^n + \alpha^n)^{\frac{2}{n}}} B - 2 r r_0 |r_0 - r|^{2n} \sqrt{(|r_0 - r|^n + \alpha^n)^{\frac{2}{n}}} B + r^2 |r_0 - r|^{2n} \ldots + \ldots\right)}{(r_0 - r)^2 (|r_0 - r|^n + \alpha^n)^2 \sqrt{(|r_0 - r|^n + \alpha^n)^{\frac{2}{n}}} B}$$

Ricci Scalar

$$R_{sc} = \frac{(|r_0 - r|^n + \alpha^n)^{-\frac{2}{n}-2} \left(2 r_0{}^2 |r_0 - r|^{2n} \sqrt{(|r_0 - r|^n + \alpha^n)^{\frac{2}{n}}} B - 4 r r_0 |r_0 - r|^{2n} \sqrt{(|r_0 - r|^n + \alpha^n)^{\frac{2}{n}}} B + \ldots\right)}{(r_0 - r)^2 \sqrt{(|r_0 - r|^n + \alpha^n)^{\frac{2}{n}}} B}$$

Bianchi identity (Ricci cyclic equation $R^{\kappa}{}_{[\mu\nu\sigma]} = 0$)

——— o.k.

Einstein Tensor

$$G_{00} = \frac{(|r_0 - r|^n + \alpha^n)^{-\frac{2}{n}-2} \, A \left({r_0}^2 \, |r_0 - r|^{2\,n} \sqrt{(|r_0 - r|^n + \alpha^n)^{\frac{2}{n}}} \, B - 2\, r\, r_0 \, |r_0 - r|^{2\,n} \sqrt{(|r_0 - r|^n + \alpha^n)^{\frac{2}{n}}} \, B + ... \right)}{(r_0 - r)^2 \, B}$$

$$G_{11} = -\frac{(|r_0 - r|^n + \alpha^n)^{-\frac{2}{n}-2} \left({r_0}^2 \, |r_0 - r|^{2\,n} \sqrt{(|r_0 - r|^n + \alpha^n)^{\frac{2}{n}}} \, B - 2\, r\, r_0 \, |r_0 - r|^{2\,n} \sqrt{(|r_0 - r|^n + \alpha^n)^{\frac{2}{n}}} \, B + ... \right)}{(r_0 - r)^2}$$

$$G_{22} = -\frac{|r_0 - r|^n \sqrt{(|r_0 - r|^n + \alpha^n)^{\frac{2}{n}}} \, (|r_0 - r|^n - 3\,\alpha^n\, n + 3\,\alpha^n)}{2\,(r_0 - r)^2 \,(|r_0 - r|^n + \alpha^n)^2 \, B}$$

$$G_{33} = -\frac{|r_0 - r|^n \sqrt{(|r_0 - r|^n + \alpha^n)^{\frac{2}{n}}} \, (|r_0 - r|^n - 3\,\alpha^n\, n + 3\,\alpha^n) \sin^2\vartheta}{2\,(r_0 - r)^2 \,(|r_0 - r|^n + \alpha^n)^2 \, B}$$

Hodge Dual of Bianchi Identity

——— (see charge and current densities)

Scalar Charge Density (-$R^{0}{}_{i}{}^{i0}$)

$$\rho = \frac{|r_0 - r|^n \,(|r_0 - r|^n + \alpha^n)^{-\frac{2}{n}-2} \,(|r_0 - r|^n + \alpha^n\, n - \alpha^n)}{2\,(r_0 - r)^2 \, A\, B}$$

Current Density Class 1 (-$R^{i}{}_{\mu}{}^{\mu j}$)

$$J_1 = -\frac{|r_0 - r|^n \,(|r_0 - r|^n + \alpha^n)^{-\frac{2}{n}-2} \,(3\,|r_0 - r|^n - 5\,\alpha^n\, n + 5\,\alpha^n)}{2\,(r_0 - r)^2 \, B^2}$$

$$J_2 = -\frac{(|r_0 - r|^n + \alpha^n)^{-\frac{4}{n}-2} \left({r_0}^2 \, |r_0 - r|^{2\,n} \, B - 2\, r\, r_0 \, |r_0 - r|^{2\,n} \, B + r^2 \, |r_0 - r|^{2\,n} \, B + 2\,\alpha^n\, {r_0}^2 \, |r_0 - r|^n \, B - 4\,\alpha^n\, r\, r_0 \, |r_0 - r|^n \, B + ... \right)}{(r_0 - r)^2 \, B}$$

$$J_3 = -\frac{(|r_0 - r|^n + \alpha^n)^{-\frac{4}{n}-2} \left({r_0}^2 \, |r_0 - r|^{2\,n} \, B - 2\, r\, r_0 \, |r_0 - r|^{2\,n} \, B + r^2 \, |r_0 - r|^{2\,n} \, B + 2\,\alpha^n\, {r_0}^2 \, |r_0 - r|^n \, B - 4\,\alpha^n\, r\, r_0 \, |r_0 - r|^n \, B + ... \right)}{(r_0 - r)^2 \, \sin^2\vartheta \, B}$$

Current Density Class 2 (-$R^{i}{}_{\mu}{}^{\mu j}$)

$$J_1 = 0$$

$$J_2 = 0$$

$$J_3 = 0$$

Current Density Class 3 (-$R^{i}{}_{\mu}{}^{\mu j}$)

$$J_1 = 0$$

$$J_2 = 0$$

$$J_3 = 0$$

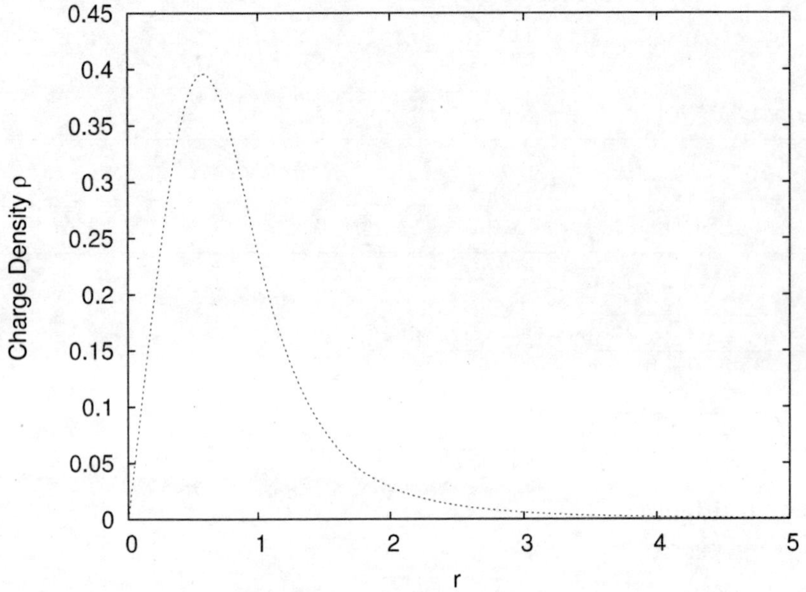

Fig. 4.1: Spherical metric of Crothers, charge density ρ for $r_0 = 0, \alpha = 1, n = 3, A = B = 1$.

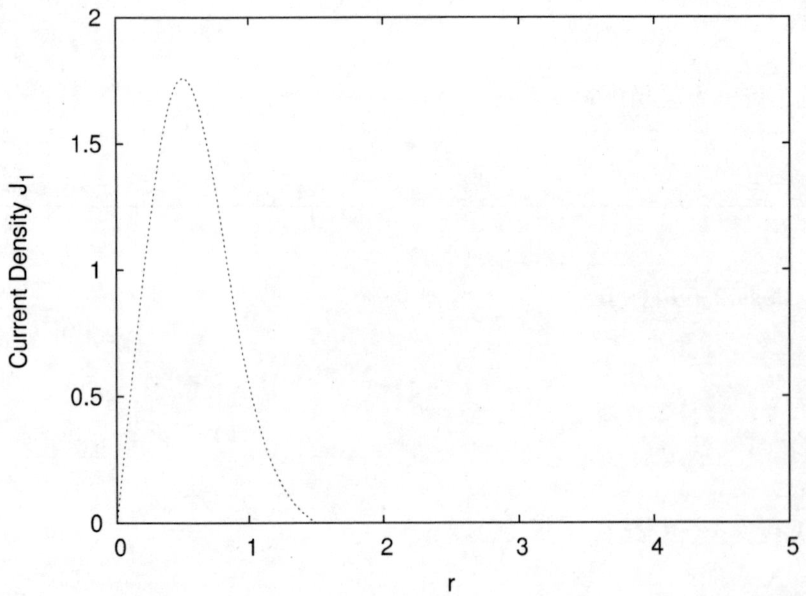

Fig. 4.2: Spherical metric of Crothers, current density J_r for $r_0 = 0, \alpha = 1, n = 3, A = B = 1$.

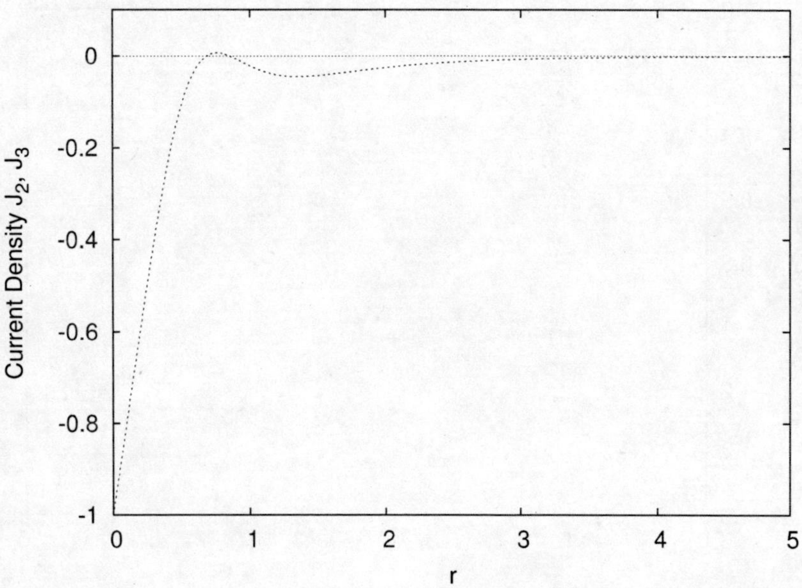

Fig. 4.3: Spherical metric of Crothers, current density J_ϑ, J_φ for $r_0 = 0, \alpha = 1, n = 3, A = B = 1$.

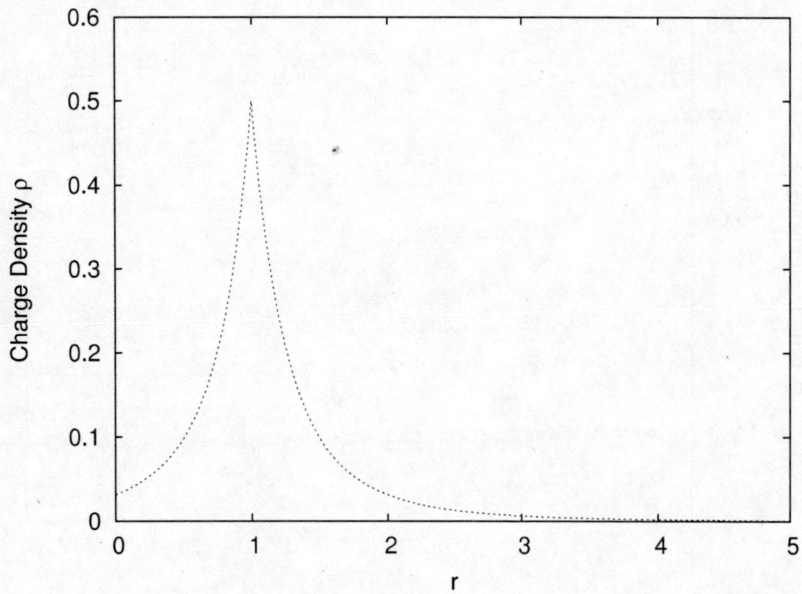

Fig. 4.4: Spherical metric of Crothers, charge density ρ for $r_0 = 1, \alpha = 1, n = 1, A = B = 1$.

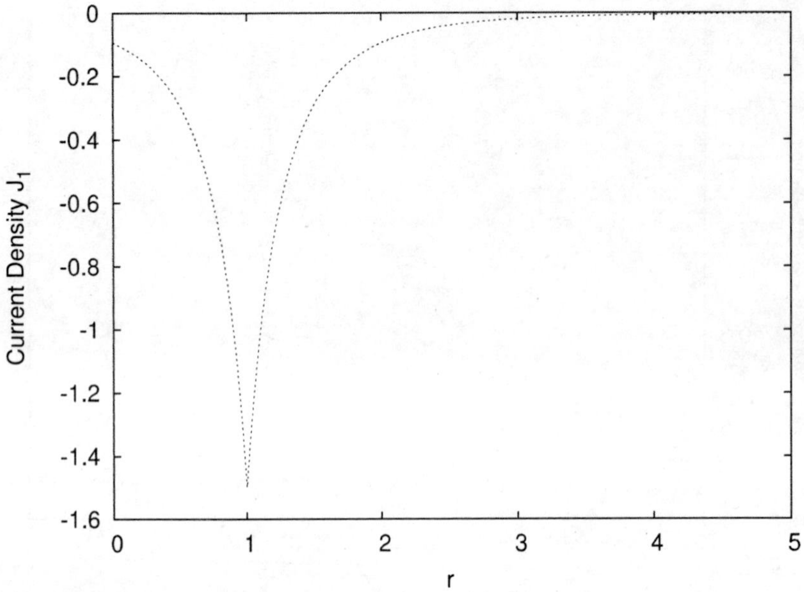

Fig. 4.5: Spherical metric of Crothers, current density J_r for $r_0 = 1, \alpha = 1, n = 1, A = B = 1$.

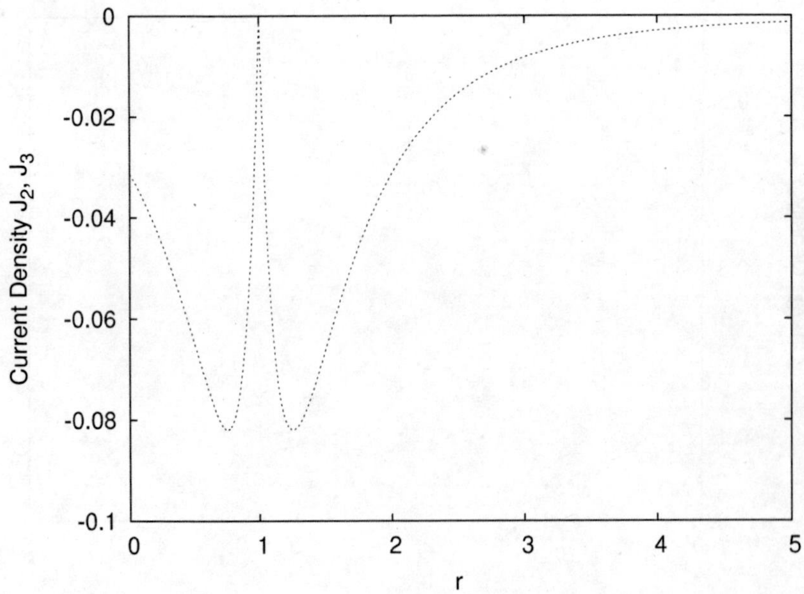

Fig. 4.6: Spherical metric of Crothers, current density J_ϑ, J_φ for $r_0 = 1, \alpha = 0, n = 1, A = B = 1$.

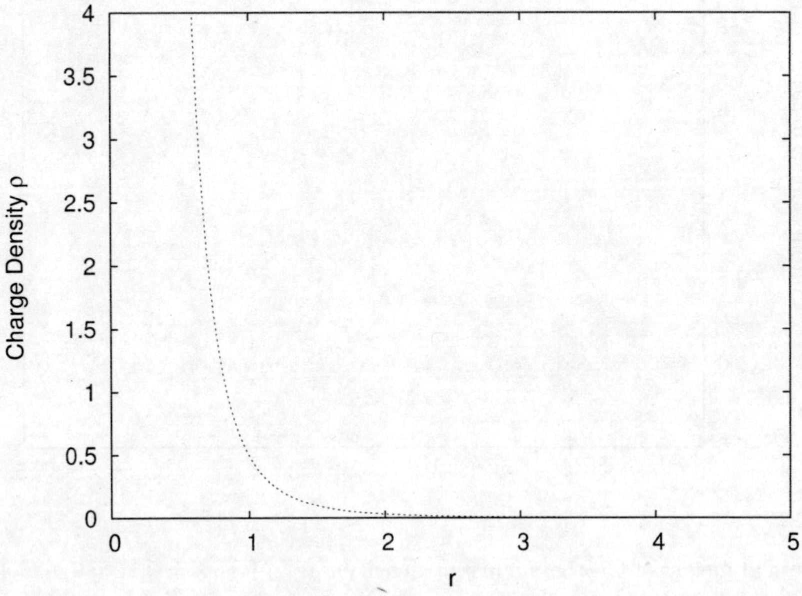

Fig. 4.7: Spherical metric of Crothers, charge density ρ for $r_0 = 0, \alpha = 0, n = 1, A = B = 1$.

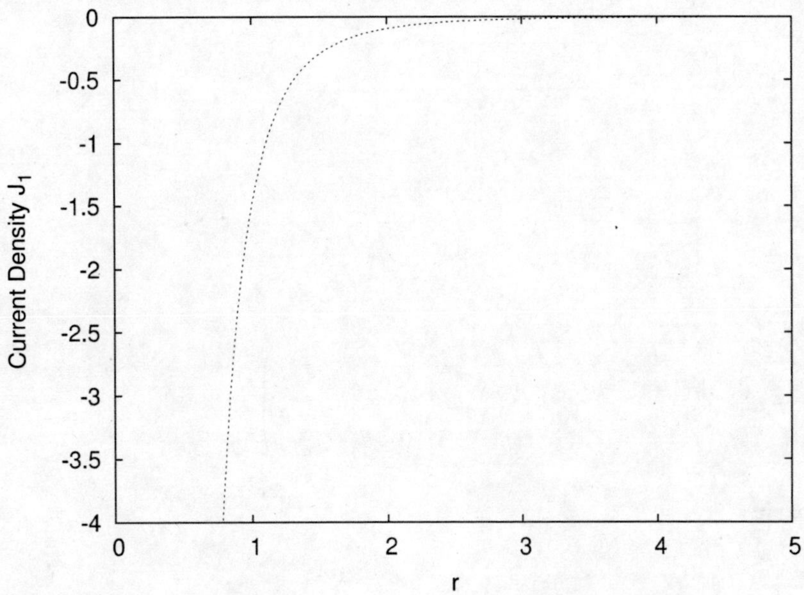

Fig. 4.8: Spherical metric of Crothers, current density J_r for $r_0 = 0, \alpha = 0, n = 1, A = B = 1$.

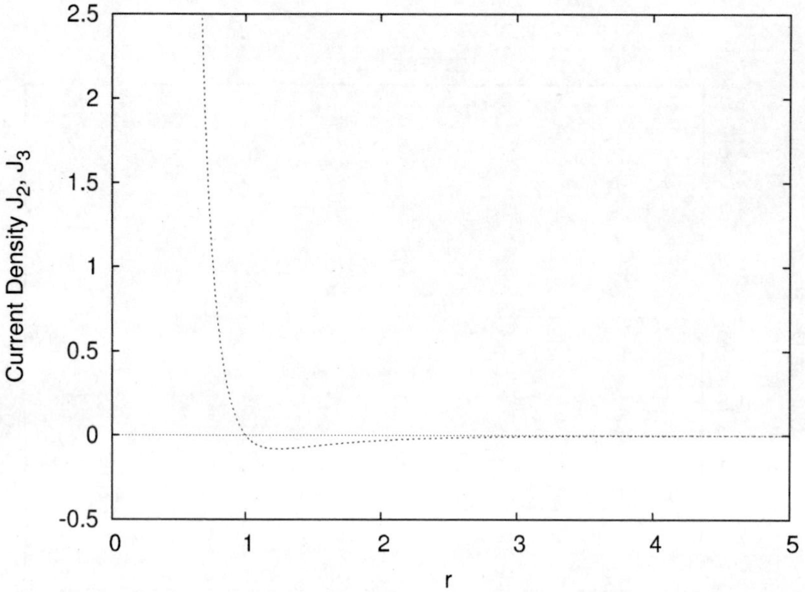

Fig. 4.9: Spherical metric of Crothers, current density J_ϑ, J_φ for $r_0 = 0, \alpha = 0, n = 1, A = B = 1$.

4.4.5 Crothers metric with Schwarzschild parameters

The parameters in the general Crothers metric with:

$$C(r) = (|r - r_0|^n + \alpha^n)^{2/n}$$

have been chosen as $r_0 = 0,\ \alpha = 1,\ n = 3$.

Coordinates

$$\mathbf{x} = \begin{pmatrix} t \\ r \\ \vartheta \\ \varphi \end{pmatrix}$$

Metric

$$g_{\mu\nu} = \begin{pmatrix} -\left(|r|^3+1\right)^{\frac{1}{3}} A & 0 & 0 & 0 \\ 0 & \left(|r|^3+1\right)^{\frac{1}{3}} B & 0 & 0 \\ 0 & 0 & \left(|r|^3+1\right)^{\frac{2}{3}} & 0 \\ 0 & 0 & 0 & \left(|r|^3+1\right)^{\frac{2}{3}} \sin^2\vartheta \end{pmatrix}$$

Contravariant Metric

$$g^{\mu\nu} = \begin{pmatrix} -\frac{1}{\left(|r|^3+1\right)^{\frac{1}{3}} A} & 0 & 0 & 0 \\ 0 & \frac{1}{\left(|r|^3+1\right)^{\frac{1}{3}} B} & 0 & 0 \\ 0 & 0 & \frac{1}{\left(|r|^3+1\right)^{\frac{2}{3}}} & 0 \\ 0 & 0 & 0 & \frac{1}{\left(|r|^3+1\right)^{\frac{2}{3}} \sin^2\vartheta} \end{pmatrix}$$

Christoffel Connection

$$\Gamma^{0}{}_{01} = \frac{r^3}{2\left(|r| + r^4\right)}$$

$$\Gamma^{0}{}_{10} = \Gamma^{0}{}_{01}$$

$$\Gamma^{1}{}_{00} = \frac{r^3 A}{2\left(|r| + r^4\right) B}$$

$$\Gamma^{1}{}_{11} = \frac{r^3}{2\left(|r| + r^4\right)}$$

$$\Gamma^{1}{}_{22} = -\frac{r^3}{|r| \left(|r|^3 + 1\right)^{\frac{2}{3}} B}$$

$$\Gamma^{1}{}_{33} = -\frac{r^3 \sin^2 \vartheta}{|r| \left(|r|^3 + 1\right)^{\frac{2}{3}} B}$$

$$\Gamma^{2}{}_{12} = \frac{r^3}{|r| + r^4}$$

$$\Gamma^{2}{}_{21} = \Gamma^{2}{}_{12}$$

$$\Gamma^{2}{}_{33} = -\cos \vartheta \sin \vartheta$$

$$\Gamma^{3}{}_{13} = \frac{r^3}{|r| + r^4}$$

$$\Gamma^{3}{}_{23} = \frac{\cos \vartheta}{\sin \vartheta}$$

$$\Gamma^{3}{}_{31} = \Gamma^{3}{}_{13}$$

$$\Gamma^{3}{}_{32} = \Gamma^{3}{}_{23}$$

Metric Compatibility

——— o.k.

Riemann Tensor

$$R^{0}{}_{101} = \frac{r^4 \left(r^2 |r| - 2\right)}{2\left(|r|^3 + r^8 |r| + 2 r^6\right)}$$

$$R^{0}{}_{110} = -R^{0}{}_{101}$$

$$R^{0}{}_{202} = -\frac{r^4}{2\left(r^2 |r| + 1\right) \left(|r|^3 + 1\right)^{\frac{2}{3}} B}$$

$$R^{0}{}_{220} = -R^{0}{}_{202}$$

$$R^{0}{}_{303} = -\frac{r^4 \sin^2\vartheta}{2\,(r^2\,|r|+1)\,\left(|r|^3+1\right)^{\frac{2}{3}}\,B}$$

$$R^{0}{}_{330} = -R^{0}{}_{303}$$

$$R^{1}{}_{001} = \frac{r^4\,(r^2\,|r|-2)\,A}{2\,\left(|r|^3+r^8\,|r|+2\,r^6\right)\,B}$$

$$R^{1}{}_{010} = -R^{1}{}_{001}$$

$$R^{1}{}_{212} = -\frac{r^2\,\left(3\,r^4\,|r|^3-4\,r^6\,|r|+4\,|r|+3\,r^4\right)}{2\,\left(|r|^3+1\right)^{\frac{2}{3}}\,\left(|r|^5+r^4\,|r|+r^8+r^2\right)\,B}$$

$$R^{1}{}_{221} = -R^{1}{}_{212}$$

$$R^{1}{}_{313} = -\frac{r^2\,\left(3\,r^4\,|r|^3-4\,r^6\,|r|+4\,|r|+3\,r^4\right)\,\sin^2\vartheta}{2\,\left(|r|^3+1\right)^{\frac{2}{3}}\,\left(|r|^5+r^4\,|r|+r^8+r^2\right)\,B}$$

$$R^{1}{}_{331} = -R^{1}{}_{313}$$

$$R^{2}{}_{002} = -\frac{r^4\,A}{2\,(2\,r^2\,|r|+r^6+1)\,B}$$

$$R^{2}{}_{020} = -R^{2}{}_{002}$$

$$R^{2}{}_{112} = -\frac{r^4\,(r^2\,|r|-4)}{2\,\left(|r|^3+r^8\,|r|+2\,r^6\right)}$$

$$R^{2}{}_{121} = -R^{2}{}_{112}$$

$$R^{2}{}_{323} = \frac{\sin^2\vartheta\,\left(r^2\,|r|\,\left(|r|^3+1\right)^{\frac{2}{3}}\,B+\left(|r|^3+1\right)^{\frac{2}{3}}\,B-r^4\right)}{(r^2\,|r|+1)\,\left(|r|^3+1\right)^{\frac{2}{3}}\,B}$$

$$R^{2}{}_{332} = -R^{2}{}_{323}$$

$$R^{3}{}_{003} = -\frac{r^4\,A}{2\,(2\,r^2\,|r|+r^6+1)\,B}$$

$$R^{3}{}_{030} = -R^{3}{}_{003}$$

$$R^{3}{}_{113} = -\frac{r^4\,(r^2\,|r|-4)}{2\,\left(|r|^3+r^8\,|r|+2\,r^6\right)}$$

$$R^{3}{}_{131} = -R^{3}{}_{113}$$

$$R^{3}{}_{223} = -\frac{r^2\,|r|\,\left(|r|^3+1\right)^{\frac{2}{3}}\,B+\left(|r|^3+1\right)^{\frac{2}{3}}\,B-r^4}{(r^2\,|r|+1)\,\left(|r|^3+1\right)^{\frac{2}{3}}\,B}$$

$$R^{3}{}_{232} = -R^{3}{}_{223}$$

Ricci Tensor

$$\mathrm{Ric}_{00} = \frac{r^4\ (2\,|r|^3 + r^8\,|r| + 3\,r^2\,|r| + 4\,r^6 + 2)\ A}{2\ (r^6\,|r|^3 + |r|^3 + r^{14}\,|r| + 5\,r^8\,|r| + 4\,r^{12} + 4\,r^6)\ B}$$

$$\mathrm{Ric}_{11} = \frac{r^4\ (3\,r^2\,|r| - 10)}{2\ (|r|^3 + r^8\,|r| + 2\,r^6)}$$

$$\mathrm{Ric}_{22} = \frac{2\,|r|^5\ (|r|^3+1)^{\frac{2}{3}}\ B + 2\,r^{10}\,|r|\ (|r|^3+1)^{\frac{2}{3}}\ B + 4\,r^4\,|r|\ (|r|^3+1)^{\frac{2}{3}}\ B + 6\,r^8\ (|r|^3+1)^{\frac{2}{3}}\ B + 2\,r^2\ (|r|^3+1)^{\frac{2}{3}}\ B...}{2\ (|r|^3+1)^{\frac{2}{3}}\ (|r|^5 + r^{10}\,|r| + 2\,r^4\,|r| + 3\,r^8 + r^2)\ B}$$

$$\mathrm{Ric}_{33} = \frac{\sin^2\vartheta\ \left(2\,|r|^5\ (|r|^3+1)^{\frac{2}{3}}\ B + 2\,r^{10}\,|r|\ (|r|^3+1)^{\frac{2}{3}}\ B + 4\,r^4\,|r|\ (|r|^3+1)^{\frac{2}{3}}\ B + 6\,r^8\ (|r|^3+1)^{\frac{2}{3}}\ B + 2\,r^2\ (|r|^3+1)^{\frac{2}{3}}\ B...\right)}{2\ (|r|^3+1)^{\frac{2}{3}}\ (|r|^5 + r^{10}\,|r| + 2\,r^4\,|r| + 3\,r^8 + r^2)\ B}$$

Ricci Scalar

$$R_{sc} = \frac{12\,r^{12}\,|r|^{11}\ B + 14\,r^6\,|r|^{11}\ B + 2\,|r|^{11}\ B + 12\,r^{20}\,|r|^9\ B + 38\,r^{14}\,|r|^9\ B + 20\,r^8\,|r|^9\ B + 2\,r^2\,|r|^9\ B + 26\,r^{22}\,|r|^7\ B + 88\,r^{16}\,|r|^7\ B...}{(|r|^3+1)^{\frac{2}{3}}\ (6\,r^{12}\,|r|^{11} + 7\,r^6\,|r|^{11} + |r|^{11} + 6\,r^{20}\,|r|^9 + 19\,r^{14}\,|r|^9 + 10\,r^8\,|r|^9 + r^2\,|r|^9 + 13\,r^{22}\,|r|^7 + ...)\ B}$$

Bianchi identity (Ricci cyclic equation $R^{\kappa}{}_{[\mu\nu\sigma]} = 0$)

——— o.k.

Einstein Tensor

$$G_{00} = \frac{A\ (48\,r^{20}\,|r|^{11}\ B + 104\,r^{14}\,|r|^{11}\ B + 64\,r^8\,|r|^{11}\ B + 8\,r^2\,|r|^{11}\ B + 48\,r^{28}\,|r|^9\ B + 200\,r^{22}\,|r|^9\ B + 232\,r^{16}\,|r|^9\ B + ...)}{2\,r^2\ (|r|^3+1)^{\frac{1}{3}}\ (24\,r^{18}\,|r|^{11} + 52\,r^{12}\,|r|^{11} + 32\,r^6\,|r|^{11} + 4\,|r|^{11} + 24\,r^{26}\,|r|^9 + 100\,r^{20}\,|r|^9 + 116\,r^{14}\,|r|^9 + 44\,r^8\,|r|^9 + ...)\ B}$$

$$G_{11} = -\frac{24\,r^{14}\,|r|^{11}\ B + 28\,r^8\,|r|^{11}\ B + 4\,r^2\,|r|^{11}\ B + 24\,r^{22}\,|r|^9\ B + 76\,r^{16}\,|r|^9\ B + 40\,r^{10}\,|r|^9\ B + 4\,r^4\,|r|^9\ B + 52\,r^{24}\,|r|^7\ B + ...}{2\,r^2\ (|r|^3+1)^{\frac{1}{3}}\ (12\,r^{12}\,|r|^{11} + 14\,r^6\,|r|^{11} + 2\,|r|^{11} + 12\,r^{20}\,|r|^9 + 38\,r^{14}\,|r|^9 + 20\,r^8\,|r|^9 + 2\,r^2\,|r|^9 + 26\,r^{22}\,|r|^7 + ...)}$$

$$G_{22} = \frac{12\,r^8\,|r|^{19} + 10\,r^2\,|r|^{19} + 8\,r^{16}\,|r|^{17} + 47\,r^{10}\,|r|^{17} + 4\,r^4\,|r|^{17} + 3\,r^{24}\,|r|^{15} + 41\,r^{18}\,|r|^{15} + 73\,r^{12}\,|r|^{15} + 40\,r^6\,|r|^{15} + 4\,|r|^{15} + ...}{2\ (|r|^3+1)^{\frac{2}{3}}\ (18\,r^{18}\,|r|^{11} + 27\,r^{12}\,|r|^{11} + 10\,r^6\,|r|^{11} + |r|^{11} + 18\,r^{26}\,|r|^9 + 63\,r^{20}\,|r|^9 + 49\,r^{14}\,|r|^9 + 13\,r^8\,|r|^9 + ...)\ B}$$

$$G_{33} = \frac{(12\,r^8\,|r|^{19} + 10\,r^2\,|r|^{19} + 8\,r^{16}\,|r|^{17} + 47\,r^{10}\,|r|^{17} + 4\,r^4\,|r|^{17} + 3\,r^{24}\,|r|^{15} + 41\,r^{18}\,|r|^{15} + 73\,r^{12}\,|r|^{15} + 40\,r^6\,|r|^{15} + ...)\ \sin^2\vartheta}{2\ (|r|^3+1)^{\frac{2}{3}}\ (18\,r^{18}\,|r|^{11} + 27\,r^{12}\,|r|^{11} + 10\,r^6\,|r|^{11} + |r|^{11} + 18\,r^{26}\,|r|^9 + 63\,r^{20}\,|r|^9 + 49\,r^{14}\,|r|^9 + 13\,r^8\,|r|^9 + r^2\,|r|^9 + ...)\ B}$$

Hodge Dual of Bianchi Identity

——— (see charge and current densities)

Scalar Charge Density (-$R^{0}{}_{i}{}^{i0}$)

$$\rho = \frac{r^4\ (|r|^3+1)^{\frac{1}{3}}\ (4\,|r|^3 + r^8\,|r| + r^2\,|r| + 4\,r^6 + 2)}{2\ (|r|^9 + r^8\,|r|^7 + r^2\,|r|^7 + r^{10}\,|r|^5 + 2\,r^{12}\,|r|^3 + 5\,r^6\,|r|^3 + |r|^3 + r^{14}\,|r| + 3\,r^8\,|r| + r^{18} + 10\,r^{12} + 5\,r^6)\ A\,B}$$

Current Density Class 1 ($-R^{i}{}_{\mu}{}^{\mu j}$)

$$J_1 = \frac{r^4\,\left(|r|^3+1\right)^{\frac{1}{3}}\,\left(2\,|r|^5+14\,r^8\,|r|^3+8\,r^2\,|r|^3-13\,r^{10}\,|r|+17\,r^4\,|r|-3\,r^{14}+21\,r^8+10\,r^2\right)}{2\,\left(|r|^{11}+2\,r^8\,|r|^9+r^2\,|r|^9+r^{16}\,|r|^7+r^{10}\,|r|^7+r^4\,|r|^7+r^{12}\,|r|^5+2\,r^6\,|r|^5+8\,r^{14}\,|r|^3+\ldots\right)\,B^2}$$

$$J_2 = -\frac{4\,r^{14}\,|r|^7\,\left(|r|^3+1\right)^{\frac{2}{3}}\,B+6\,r^8\,|r|^7\,\left(|r|^3+1\right)^{\frac{2}{3}}\,B+2\,r^2\,|r|^7\,\left(|r|^3+1\right)^{\frac{2}{3}}\,B+2\,r^{16}\,|r|^5\,\left(|r|^3+1\right)^{\frac{2}{3}}\,B+10\,r^{10}\,|r|^5\,\left(|r|^3+1\right)^{\frac{2}{3}}\,B+\ldots}{2\,\left(2\,r^{14}\,|r|^{13}+3\,r^8\,|r|^{13}+r^2\,|r|^{13}+r^{16}\,|r|^{11}+5\,r^{10}\,|r|^{11}+2\,r^4\,|r|^{11}+6\,r^{12}\,|r|^9+6\,r^6\,|r|^9+2\,|r|^9+\ldots\right)\,B}$$

$$J_3 = -\frac{4\,r^{14}\,|r|^7\,\left(|r|^3+1\right)^{\frac{2}{3}}\,B+6\,r^8\,|r|^7\,\left(|r|^3+1\right)^{\frac{2}{3}}\,B+2\,r^2\,|r|^7\,\left(|r|^3+1\right)^{\frac{2}{3}}\,B+2\,r^{16}\,|r|^5\,\left(|r|^3+1\right)^{\frac{2}{3}}\,B+10\,r^{10}\,|r|^5\,\left(|r|^3+1\right)^{\frac{2}{3}}\,B+\ldots}{2\,\left(2\,r^{14}\,|r|^{13}+3\,r^8\,|r|^{13}+r^2\,|r|^{13}+r^{16}\,|r|^{11}+5\,r^{10}\,|r|^{11}+2\,r^4\,|r|^{11}+6\,r^{12}\,|r|^9+6\,r^6\,|r|^9+2\,|r|^9+\ldots\right)\,\sin^2\vartheta\,B}$$

Current Density Class 2 ($-R^{i}{}_{\mu}{}^{\mu j}$)

$$J_1 = 0$$

$$J_2 = 0$$

$$J_3 = 0$$

Current Density Class 3 ($-R^{i}{}_{\mu}{}^{\mu j}$)

$$J_1 = 0$$

$$J_2 = 0$$

$$J_3 = 0$$

4.4.6 General spherical metric

The general spherically symmetric metric contains exponentials of functions α and β which in turn are functions of t and r.

Coordinates

$$\mathbf{x} = \begin{pmatrix} t \\ r \\ \vartheta \\ \varphi \end{pmatrix}$$

Metric

$$g_{\mu\nu} = \begin{pmatrix} -e^{2\,\alpha} & 0 & 0 & 0 \\ 0 & e^{2\,\beta} & 0 & 0 \\ 0 & 0 & r^2 & 0 \\ 0 & 0 & 0 & r^2\,\sin^2\vartheta \end{pmatrix}$$

Contravariant Metric

$$g^{\mu\nu} = \begin{pmatrix} -e^{-2\,\alpha} & 0 & 0 & 0 \\ 0 & e^{-2\,\beta} & 0 & 0 \\ 0 & 0 & \frac{1}{r^2} & 0 \\ 0 & 0 & 0 & \frac{1}{r^2\,\sin^2\vartheta} \end{pmatrix}$$

Christoffel Connection

$$\Gamma^0{}_{00} = \frac{d}{dt}\alpha$$

$$\Gamma^0{}_{01} = \frac{d}{dr}\alpha$$

$$\Gamma^0{}_{10} = \Gamma^0{}_{01}$$

$$\Gamma^0{}_{11} = e^{2\beta-2\alpha}\left(\frac{d}{dt}\beta\right)$$

$$\Gamma^1{}_{00} = \frac{d}{dr}\alpha\, e^{2\alpha-2\beta}$$

$$\Gamma^1{}_{01} = \frac{d}{dt}\beta$$

$$\Gamma^1{}_{10} = \Gamma^1{}_{01}$$

$$\Gamma^1{}_{11} = \frac{d}{dr}\beta$$

$$\Gamma^1{}_{22} = -e^{-2\beta}\, r$$

$$\Gamma^1{}_{33} = -e^{-2\beta}\, r\, \sin^2\vartheta$$

$$\Gamma^2{}_{12} = \frac{1}{r}$$

$$\Gamma^2{}_{21} = \Gamma^2{}_{12}$$

$$\Gamma^2{}_{33} = -\cos\vartheta\, \sin\vartheta$$

$$\Gamma^3{}_{13} = \frac{1}{r}$$

$$\Gamma^3{}_{23} = \frac{\cos\vartheta}{\sin\vartheta}$$

$$\Gamma^3{}_{31} = \Gamma^3{}_{13}$$

$$\Gamma^3{}_{32} = \Gamma^3{}_{23}$$

Metric Compatibility

——— o.k.

Riemann Tensor

$$R^0{}_{101} = e^{-2\alpha}\left(e^{2\beta}\left(\frac{d^2}{dt^2}\beta\right) + e^{2\beta}\left(\frac{d}{dt}\beta\right)^2 - \frac{d}{dt}\alpha\, e^{2\beta}\left(\frac{d}{dt}\beta\right) + e^{2\alpha}\left(\frac{d}{dr}\alpha\right)\left(\frac{d}{dr}\beta\right) - e^{2\alpha}\left(\frac{d^2}{dr^2}\alpha\right) - e^{2\alpha}\left(\frac{d}{dr}\alpha\right)^2\right)$$

$$R^0{}_{110} = -R^0{}_{101}$$

$$R^0{}_{202} = -\frac{d}{dr}\alpha\, e^{-2\beta}\, r$$

$$R^0{}_{212} = -e^{-2\alpha}\left(\frac{d}{dt}\beta\right) r$$

$$R^0{}_{220} = -R^0{}_{202}$$

$$R^0{}_{221} = -R^0{}_{212}$$

$$R^0{}_{303} = -\frac{d}{dr}\alpha\, e^{-2\beta}\, r\, \sin^2\vartheta$$

$$R^0{}_{313} = -e^{-2\alpha}\left(\frac{d}{dt}\beta\right) r\, \sin^2\vartheta$$

$$R^0{}_{330} = -R^0{}_{303}$$

$$R^0{}_{331} = -R^0{}_{313}$$

$$R^1{}_{001} = e^{-2\beta}\left(e^{2\beta}\left(\frac{d^2}{dt^2}\beta\right) + e^{2\beta}\left(\frac{d}{dt}\beta\right)^2 - \frac{d}{dt}\alpha\, e^{2\beta}\left(\frac{d}{dt}\beta\right) + e^{2\alpha}\left(\frac{d}{dr}\alpha\right)\left(\frac{d}{dr}\beta\right) - e^{2\alpha}\left(\frac{d^2}{dr^2}\alpha\right) - e^{2\alpha}\left(\frac{d}{dr}\alpha\right)^2\right)$$

$$R^1{}_{010} = -R^1{}_{001}$$

$$R^1{}_{202} = e^{-2\beta}\left(\frac{d}{dt}\beta\right) r$$

$$R^1{}_{212} = e^{-2\beta}\left(\frac{d}{dr}\beta\right) r$$

$$R^1{}_{220} = -R^1{}_{202}$$

$$R^1{}_{221} = -R^1{}_{212}$$

$$R^1{}_{303} = e^{-2\beta}\left(\frac{d}{dt}\beta\right) r\, \sin^2\vartheta$$

$$R^1{}_{313} = e^{-2\beta}\left(\frac{d}{dr}\beta\right) r\, \sin^2\vartheta$$

$$R^1{}_{330} = -R^1{}_{303}$$

$$R^1{}_{331} = -R^1{}_{313}$$

$$R^2{}_{002} = -\frac{\frac{d}{dr}\alpha\, e^{2\alpha-2\beta}}{r}$$

$$R^2{}_{012} = -\frac{\frac{d}{dt}\beta}{r}$$

$$R^2{}_{020} = -R^2{}_{002}$$

$$R^2{}_{021} = -R^2{}_{012}$$

$$R^2{}_{102} = -\frac{\frac{d}{dt}\beta}{r}$$

$$R^2{}_{112} = -\frac{\frac{d}{dr}\beta}{r}$$

$$R^2{}_{120} = -R^2{}_{102}$$

$$R^2{}_{121} = -R^2{}_{112}$$

$$R^2{}_{323} = e^{-2\beta}\left(e^{\beta}-1\right)\left(e^{\beta}+1\right)\sin^2\vartheta$$

$$R^2{}_{332} = -R^2{}_{323}$$

$$R^3{}_{003} = -\frac{\frac{d}{dr}\alpha\, e^{2\alpha-2\beta}}{r}$$

$$R^3{}_{013} = -\frac{\frac{d}{dt}\beta}{r}$$

$$R^3{}_{030} = -R^3{}_{003}$$

$$R^3{}_{031} = -R^3{}_{013}$$

$$R^3{}_{103} = -\frac{\frac{d}{dt}\beta}{r}$$

$$R^3{}_{113} = -\frac{\frac{d}{dr}\beta}{r}$$

$$R^3{}_{130} = -R^3{}_{103}$$

$$R^3{}_{131} = -R^3{}_{113}$$

$$R^3{}_{223} = -e^{-2\beta}\left(e^{\beta}-1\right)\left(e^{\beta}+1\right)$$

$$R^3{}_{232} = -R^3{}_{223}$$

Ricci Tensor

$$\mathrm{Ric}_{00} = -\frac{e^{-2\beta}\left(e^{2\beta}\left(\frac{d^2}{dt^2}\beta\right) r + e^{2\beta}\left(\frac{d}{dt}\beta\right)^2 r - \frac{d}{dt}\alpha\, e^{2\beta}\left(\frac{d}{dt}\beta\right) r + e^{2\alpha}\left(\frac{d}{dr}\alpha\right)\left(\frac{d}{dr}\beta\right) r - e^{2\alpha}\left(\frac{d^2}{dr^2}\alpha\right) r - \ldots\right)}{r}$$

$$\mathrm{Ric}_{01} = \frac{2\left(\frac{d}{dt}\beta\right)}{r}$$

$$\mathrm{Ric}_{10} = \mathrm{Ric}_{01}$$

$$\mathrm{Ric}_{11} = \frac{e^{-2\alpha}\left(e^{2\beta}\left(\frac{d^2}{dt^2}\beta\right) r + e^{2\beta}\left(\frac{d}{dt}\beta\right)^2 r - \frac{d}{dt}\alpha\, e^{2\beta}\left(\frac{d}{dt}\beta\right) r + e^{2\alpha}\left(\frac{d}{dr}\alpha\right)\left(\frac{d}{dr}\beta\right) r - e^{2\alpha}\left(\frac{d^2}{dr^2}\alpha\right) r - e^{2\alpha}\left(\frac{d}{dr}\alpha\right)^2 r + \ldots\right)}{r}$$

$$\mathrm{Ric}_{22} = e^{-2\beta}\left(\frac{d}{dr}\beta\, r - \frac{d}{dr}\alpha\, r + e^{2\beta} - 1\right)$$

$$\mathrm{Ric}_{33} = e^{-2\beta}\left(\frac{d}{dr}\beta\, r - \frac{d}{dr}\alpha\, r + e^{2\beta} - 1\right)\sin^2\vartheta$$

Ricci Scalar

$$R_{sc} = 2\,e^{-2\beta-2\alpha}\left(\frac{e^{2\beta}\left(\frac{d^2}{dt^2}\beta\right) r^2}{r^2} + \frac{e^{2\beta}\left(\frac{d}{dt}\beta\right)^2 r^2}{r^2} - \frac{\frac{d}{dt}\alpha\, e^{2\beta}\left(\frac{d}{dt}\beta\right) r^2}{r^2} + \frac{e^{2\alpha}\left(\frac{d}{dr}\alpha\right)\left(\frac{d}{dr}\beta\right) r^2}{r^2} - \frac{e^{2\alpha}\left(\frac{d^2}{dr^2}\alpha\right) r^2}{r^2}\right.$$
$$\left. - \frac{e^{2\alpha}\left(\frac{d}{dr}\alpha\right)^2 r^2}{r^2} + \frac{2\,e^{2\alpha}\left(\frac{d}{dr}\beta\right) r}{r^2} - \frac{2\,e^{2\alpha}\left(\frac{d}{dr}\alpha\right) r}{r^2} + \frac{e^{2\beta+2\alpha}}{r^2} - \frac{e^{2\alpha}}{r^2}\right)$$

Bianchi identity (Ricci cyclic equation ${R^{\kappa}}_{[\mu\nu\sigma]} = 0$)

——— o.k.

Einstein Tensor

$$G_{00} = \frac{e^{2\alpha-2\beta}\left(2\left(\frac{d}{dr}\beta\right) r + e^{2\beta} - 1\right)}{r^2}$$

$$G_{01} = \frac{2\left(\frac{d}{dt}\beta\right)}{r}$$

$$G_{10} = G_{01}$$

$$G_{11} = \frac{2\left(\frac{d}{dr}\alpha\right) r - e^{2\beta} + 1}{r^2}$$

$$G_{22} = -e^{-2\beta-2\alpha}\, r\left(e^{2\beta}\left(\frac{d^2}{dt^2}\beta\right) r + e^{2\beta}\left(\frac{d}{dt}\beta\right)^2 r - \frac{d}{dt}\alpha\, e^{2\beta}\left(\frac{d}{dt}\beta\right) r + e^{2\alpha}\left(\frac{d}{dr}\alpha\right)\left(\frac{d}{dr}\beta\right) r - e^{2\alpha}\left(\frac{d^2}{dr^2}\alpha\right) r - e^{2\alpha}\left(\frac{d}{dr}\alpha\right)^2 r + \ldots\right)$$

$$G_{33} = -e^{-2\beta-2\alpha}\, r\left(e^{2\beta}\left(\frac{d^2}{dt^2}\beta\right) r + e^{2\beta}\left(\frac{d}{dt}\beta\right)^2 r - \frac{d}{dt}\alpha\, e^{2\beta}\left(\frac{d}{dt}\beta\right) r + e^{2\alpha}\left(\frac{d}{dr}\alpha\right)\left(\frac{d}{dr}\beta\right) r - e^{2\alpha}\left(\frac{d^2}{dr^2}\alpha\right) r - \ldots\right)\sin^2\vartheta$$

Hodge Dual of Bianchi Identity

——— (see charge and current densities)

Scalar Charge Density (-$R^{0\ i0}_{\ i}$)

$$\rho = -\frac{e^{-2\beta-4\alpha}\left(e^{2\beta}\left(\frac{d^2}{dt^2}\beta\right)r + e^{2\beta}\left(\frac{d}{dt}\beta\right)^2 r - \frac{d}{dt}\alpha\, e^{2\beta}\left(\frac{d}{dt}\beta\right)r + e^{2\alpha}\left(\frac{d}{dr}\alpha\right)\left(\frac{d}{dr}\beta\right)r - e^{2\alpha}\left(\frac{d^2}{dr^2}\alpha\right)r - e^{2\alpha}\left(\frac{d}{dr}\alpha\right)^2 r - ...\right)}{r}$$

Current Density Class 1 (-$R^{i\ \mu j}_{\ \mu}$)

$$J_1 = -\frac{e^{-4\beta-2\alpha}\left(e^{2\beta}\left(\frac{d^2}{dt^2}\beta\right)r + e^{2\beta}\left(\frac{d}{dt}\beta\right)^2 r - \frac{d}{dt}\alpha\, e^{2\beta}\left(\frac{d}{dt}\beta\right)r + e^{2\alpha}\left(\frac{d}{dr}\alpha\right)\left(\frac{d}{dr}\beta\right)r - e^{2\alpha}\left(\frac{d^2}{dr^2}\alpha\right)r - e^{2\alpha}\left(\frac{d}{dr}\alpha\right)^2 r + ...\right)}{r}$$

$$J_2 = -\frac{e^{-2\beta}\left(\frac{d}{dr}\beta\, r - \frac{d}{dr}\alpha\, r + e^{2\beta} - 1\right)}{r^4}$$

$$J_3 = -\frac{e^{-2\beta}\left(\frac{d}{dr}\beta\, r - \frac{d}{dr}\alpha\, r + e^{2\beta} - 1\right)}{r^4\sin^2\vartheta}$$

Current Density Class 2 (-$R^{i\ \mu j}_{\ \mu}$)

$$J_1 = 0$$

$$J_2 = 0$$

$$J_3 = 0$$

Current Density Class 3 (-$R^{i\ \mu j}_{\ \mu}$)

$$J_1 = 0$$

$$J_2 = 0$$

$$J_3 = 0$$

4.4.7 Spherically symmetric metric with perturbation a/r

This spherically symmetric metric contains an additional perturbation a/r.

Coordinates

$$\mathbf{x} = \begin{pmatrix} t \\ r \\ \vartheta \\ \varphi \end{pmatrix}$$

Metric

$$g_{\mu\nu} = \begin{pmatrix} -\frac{\frac{2GM}{c^2}+\frac{a}{r}}{r} - 1 & 0 & 0 & 0 \\ 0 & \frac{1}{\frac{\frac{2GM}{c^2}+\frac{a}{r}}{r}+1} & 0 & 0 \\ 0 & 0 & r^2 & 0 \\ 0 & 0 & 0 & r^2\sin^2\vartheta \end{pmatrix}$$

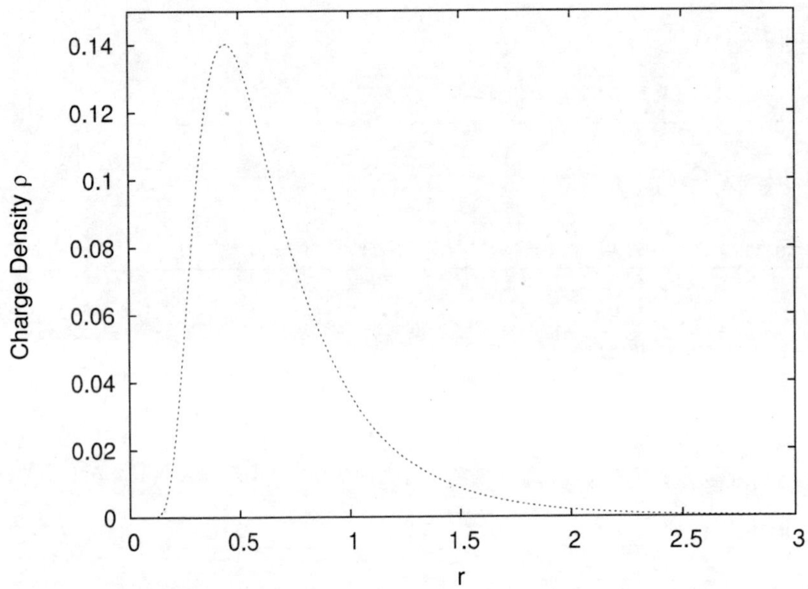

Fig. 4.10: General spherical metric, charge density ρ for $\alpha = 1/r, \beta = r$.

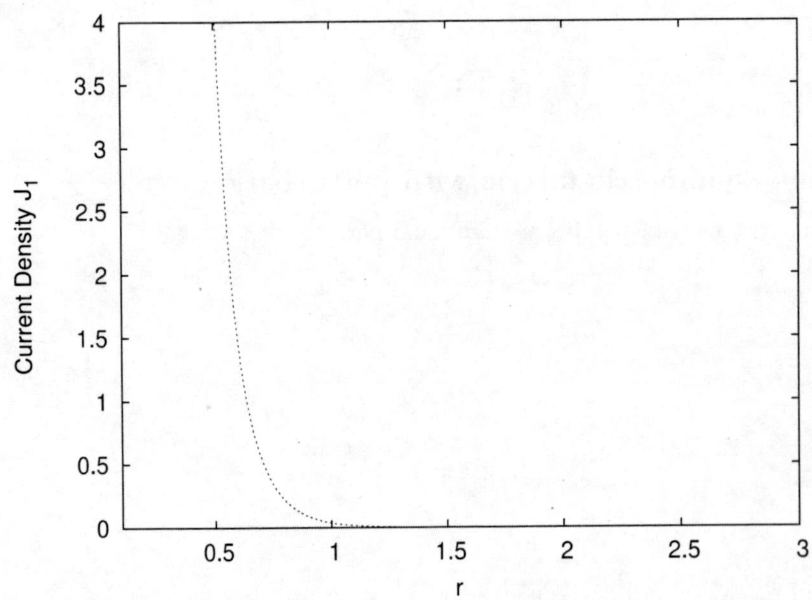

Fig. 4.11: General spherical metric, current density J_r for $\alpha = 1/r, \beta = r$.

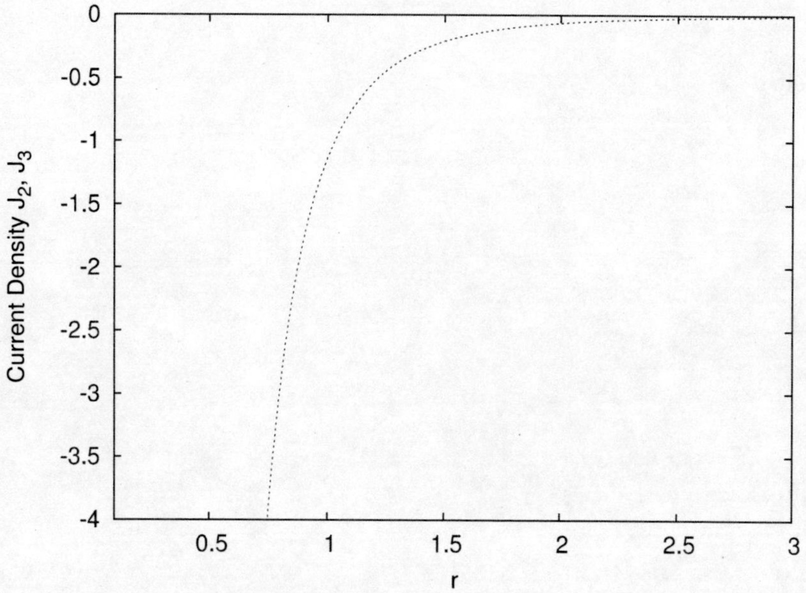

Fig. 4.12: General spherical metric, current density J_θ, J_φ for $\alpha = 1/r, \beta = r$.

Contravariant Metric

$$g^{\mu\nu} = \begin{pmatrix} -\frac{c^2 r^2}{2rGM+c^2r^2+ac^2} & 0 & 0 & 0 \\ 0 & \frac{2rGM+c^2r^2+ac^2}{c^2r^2} & 0 & 0 \\ 0 & 0 & \frac{1}{r^2} & 0 \\ 0 & 0 & 0 & \frac{1}{r^2\sin^2\vartheta} \end{pmatrix}$$

Christoffel Connection

$$\Gamma^0{}_{01} = -\frac{r\,G\,M + a\,c^2}{r\,(2\,r\,G\,M + c^2\,r^2 + a\,c^2)}$$

$$\Gamma^0{}_{10} = \Gamma^0{}_{01}$$

$$\Gamma^1{}_{00} = -\frac{\left(r\,G\,M + a\,c^2\right)\,\left(2\,r\,G\,M + c^2\,r^2 + a\,c^2\right)}{c^4\,r^5}$$

$$\Gamma^1{}_{11} = \frac{r\,G\,M + a\,c^2}{r\,(2\,r\,G\,M + c^2\,r^2 + a\,c^2)}$$

$$\Gamma^1{}_{22} = -\frac{2\,r\,G\,M + c^2\,r^2 + a\,c^2}{c^2\,r}$$

$$\Gamma^1{}_{33} = -\frac{\sin^2\vartheta\,\left(2\,r\,G\,M + c^2\,r^2 + a\,c^2\right)}{c^2\,r}$$

$$\Gamma^2{}_{12} = \frac{1}{r}$$

$$\Gamma^2{}_{21} = \Gamma^2{}_{12}$$

$$\Gamma^2{}_{33} = -\cos\vartheta\,\sin\vartheta$$

$$\Gamma^3{}_{13} = \frac{1}{r}$$

$$\Gamma^3{}_{23} = \frac{\cos\vartheta}{\sin\vartheta}$$

$$\Gamma^3{}_{31} = \Gamma^3{}_{13}$$

$$\Gamma^3{}_{32} = \Gamma^3{}_{23}$$

Metric Compatibility

——— o.k.

Riemann Tensor

$$R^0{}_{101} = -\frac{2\,r\,G\,M + 3\,a\,c^2}{r^2\,(2\,r\,G\,M + c^2\,r^2 + a\,c^2)}$$

$$R^0{}_{110} = -R^0{}_{101}$$

$$R^0{}_{202} = \frac{r\,G\,M + a\,c^2}{c^2\,r^2}$$

$$R^0{}_{220} = -R^0{}_{202}$$

$$R^0{}_{303} = \frac{\sin^2\vartheta\,\left(r\,G\,M + a\,c^2\right)}{c^2\,r^2}$$

$$R^0{}_{330} = -R^0{}_{303}$$

$$R^1{}_{001} = -\frac{\left(2\,r\,G\,M + 3\,a\,c^2\right)\left(2\,r\,G\,M + c^2\,r^2 + a\,c^2\right)}{c^4\,r^6}$$

$$R^1{}_{010} = -R^1{}_{001}$$

$$R^1{}_{212} = \frac{r\,G\,M + a\,c^2}{c^2\,r^2}$$

$$R^1{}_{221} = -R^1{}_{212}$$

$$R^1{}_{313} = \frac{\sin^2\vartheta\,\left(r\,G\,M + a\,c^2\right)}{c^2\,r^2}$$

$$R^1{}_{331} = -R^1{}_{313}$$

$$R^2{}_{002} = \frac{\left(r\,G\,M + a\,c^2\right)\left(2\,r\,G\,M + c^2\,r^2 + a\,c^2\right)}{c^4\,r^6}$$

$$R^2{}_{020} = -R^2{}_{002}$$

$$R^2{}_{112} = -\frac{r\,G\,M + a\,c^2}{r^2\,(2\,r\,G\,M + c^2\,r^2 + a\,c^2)}$$

$$R^2{}_{121} = -R^2{}_{112}$$

$$R^2{}_{323} = -\frac{\sin^2\vartheta\,\left(2\,r\,G\,M + a\,c^2\right)}{c^2\,r^2}$$

$$R^2{}_{332} = -R^2{}_{323}$$

$$R^3{}_{003} = \frac{\left(r\,G\,M + a\,c^2\right)\left(2\,r\,G\,M + c^2\,r^2 + a\,c^2\right)}{c^4\,r^6}$$

$$R^3{}_{030} = -R^3{}_{003}$$

$$R^3{}_{113} = -\frac{r\,G\,M + a\,c^2}{r^2\,(2\,r\,G\,M + c^2\,r^2 + a\,c^2)}$$

$$R^3{}_{131} = -R^3{}_{113}$$

$$R^3{}_{223} = \frac{2\,r\,G\,M + a\,c^2}{c^2\,r^2}$$

$$R^3{}_{232} = -R^3{}_{223}$$

Ricci Tensor

$$\mathrm{Ric}_{00} = \frac{a\,\left(2\,r\,G\,M + c^2\,r^2 + a\,c^2\right)}{c^2\,r^6}$$

$$\mathrm{Ric}_{11} = -\frac{a\,c^2}{r^2\,(2\,r\,G\,M + c^2\,r^2 + a\,c^2)}$$

$$\mathrm{Ric}_{22} = \frac{a}{r^2}$$

$$\mathrm{Ric}_{33} = \frac{a\,\sin^2\vartheta}{r^2}$$

Ricci Scalar

$$R_{sc} = 0$$

Bianchi identity (Ricci cyclic equation $R^{\kappa}{}_{[\mu\nu\sigma]} = 0$)

——— o.k.

Einstein Tensor

$$G_{00} = \frac{a\,(2\,r\,G\,M + c^2\,r^2 + a\,c^2)}{c^2\,r^6}$$

$$G_{11} = -\frac{a\,c^2}{r^2\,(2\,r\,G\,M + c^2\,r^2 + a\,c^2)}$$

$$G_{22} = \frac{a}{r^2}$$

$$G_{33} = \frac{a\,\sin^2\vartheta}{r^2}$$

Hodge Dual of Bianchi Identity

——— (see charge and current densities)

Scalar Charge Density ($-R^{0}{}_{i}{}^{i0}$)

$$\rho = \frac{a\,c^2}{r^2\,(2\,r\,G\,M + c^2\,r^2 + a\,c^2)}$$

Current Density Class 1 ($-R^{i}{}_{\mu}{}^{\mu j}$)

$$J_1 = \frac{a\,(2\,r\,G\,M + c^2\,r^2 + a\,c^2)}{c^2\,r^6}$$

$$J_2 = -\frac{a}{r^6}$$

$$J_3 = -\frac{a}{r^6\,\sin^2\vartheta}$$

Current Density Class 2 ($-R^{i}{}_{\mu}{}^{\mu j}$)

$$J_1 = 0$$

$$J_2 = 0$$

$$J_3 = 0$$

Current Density Class 3 (-$R^{i}{}_{\mu}{}^{\mu j}$)

$$J_1 = 0$$

$$J_2 = 0$$

$$J_3 = 0$$

4.4.8 Spherically symmetric metric with general $\mu(r)$

Spherically symmetric line element with a generalized dependence $\mu(r)$.

Coordinates

$$\mathbf{x} = \begin{pmatrix} t \\ r \\ \vartheta \\ \varphi \end{pmatrix}$$

Metric

$$g_{\mu\nu} = \begin{pmatrix} -\frac{\mu}{r} - 1 & 0 & 0 & 0 \\ 0 & \frac{1}{\frac{\mu}{r}+1} & 0 & 0 \\ 0 & 0 & r^2 & 0 \\ 0 & 0 & 0 & r^2 \sin^2\vartheta \end{pmatrix}$$

Contravariant Metric

$$g^{\mu\nu} = \begin{pmatrix} -\frac{r}{r+\mu} & 0 & 0 & 0 \\ 0 & \frac{r+\mu}{r} & 0 & 0 \\ 0 & 0 & \frac{1}{r^2} & 0 \\ 0 & 0 & 0 & \frac{1}{r^2 \sin^2\vartheta} \end{pmatrix}$$

Christoffel Connection

$$\Gamma^{0}{}_{01} = \frac{\frac{d}{d\,r}\,\mu\,r - \mu}{2\,r\,(r+\mu)}$$

$$\Gamma^{0}{}_{10} = \Gamma^{0}{}_{01}$$

$$\Gamma^{1}{}_{00} = \frac{(r+\mu)\left(\frac{d}{d\,r}\,\mu\,r - \mu\right)}{2\,r^3}$$

$$\Gamma^{1}{}_{11} = -\frac{\frac{d}{d\,r}\,\mu\,r - \mu}{2\,r\,(r+\mu)}$$

$$\Gamma^{1}{}_{22} = -(r+\mu)$$

$$\Gamma^1{}_{33} = -(r+\mu)\sin^2\vartheta$$

$$\Gamma^2{}_{12} = \frac{1}{r}$$

$$\Gamma^2{}_{21} = \Gamma^2{}_{12}$$

$$\Gamma^2{}_{33} = -\cos\vartheta\,\sin\vartheta$$

$$\Gamma^3{}_{13} = \frac{1}{r}$$

$$\Gamma^3{}_{23} = \frac{\cos\vartheta}{\sin\vartheta}$$

$$\Gamma^3{}_{31} = \Gamma^3{}_{13}$$

$$\Gamma^3{}_{32} = \Gamma^3{}_{23}$$

Metric Compatibility

——— o.k.

Riemann Tensor

$$R^0{}_{101} = -\frac{\frac{d^2}{d\,r^2}\,\mu\,r^2 - 2\left(\frac{d}{d\,r}\,\mu\right)r + 2\,\mu}{2\,r^2\,(r+\mu)}$$

$$R^0{}_{110} = -R^0{}_{101}$$

$$R^0{}_{202} = -\frac{\frac{d}{d\,r}\,\mu\,r - \mu}{2\,r}$$

$$R^0{}_{220} = -R^0{}_{202}$$

$$R^0{}_{303} = -\frac{\left(\frac{d}{d\,r}\,\mu\,r - \mu\right)\sin^2\vartheta}{2\,r}$$

$$R^0{}_{330} = -R^0{}_{303}$$

$$R^1{}_{001} = -\frac{(r+\mu)\left(\frac{d^2}{d\,r^2}\,\mu\,r^2 - 2\left(\frac{d}{d\,r}\,\mu\right)r + 2\,\mu\right)}{2\,r^4}$$

$$R^1{}_{010} = -R^1{}_{001}$$

$$R^1{}_{212} = -\frac{\frac{d}{d\,r}\,\mu\,r - \mu}{2\,r}$$

$$R^1{}_{221} = -R^1{}_{212}$$

$$R^1{}_{313} = -\frac{\left(\frac{d}{d\,r}\,\mu\,r - \mu\right)\,\sin^2\vartheta}{2\,r}$$

$$R^1{}_{331} = -R^1{}_{313}$$

$$R^2{}_{002} = -\frac{(r+\mu)\,\left(\frac{d}{d\,r}\,\mu\,r - \mu\right)}{2\,r^4}$$

$$R^2{}_{020} = -R^2{}_{002}$$

$$R^2{}_{112} = \frac{\frac{d}{d\,r}\,\mu\,r - \mu}{2\,r^2\,(r+\mu)}$$

$$R^2{}_{121} = -R^2{}_{112}$$

$$R^2{}_{323} = -\frac{\mu\,\sin^2\vartheta}{r}$$

$$R^2{}_{332} = -R^2{}_{323}$$

$$R^3{}_{003} = -\frac{(r+\mu)\,\left(\frac{d}{d\,r}\,\mu\,\dot{r} - \mu\right)}{2\,r^4}$$

$$R^3{}_{030} = -R^3{}_{003}$$

$$R^3{}_{113} = \frac{\frac{d}{d\,r}\,\mu\,r - \mu}{2\,r^2\,(r+\mu)}$$

$$R^3{}_{131} = -R^3{}_{113}$$

$$R^3{}_{223} = \frac{\mu}{r}$$

$$R^3{}_{232} = -R^3{}_{223}$$

Ricci Tensor

$$\mathrm{Ric}_{00} = \frac{\frac{d^2}{d\,r^2}\,\mu\,(r+\mu)}{2\,r^2}$$

$$\mathrm{Ric}_{11} = -\frac{\frac{d^2}{d\,r^2}\,\mu}{2\,(r+\mu)}$$

$$\mathrm{Ric}_{22} = -\frac{d}{d\,r}\,\mu$$

$$\mathrm{Ric}_{33} = -\frac{d}{d\,r}\,\mu\,\sin^2\vartheta$$

Ricci Scalar

$$R_{sc} = -\frac{\frac{d^2}{d\,r^2}\,\mu\,r + 2\,\left(\frac{d}{d\,r}\,\mu\right)}{r^2}$$

Bianchi identity (Ricci cyclic equation $R^{\kappa}{}_{[\mu\nu\sigma]} = 0$)

——— o.k.

Einstein Tensor

$$G_{00} = -\frac{\frac{d}{d\,r}\,\mu\,(r+\mu)}{r^3}$$

$$G_{11} = \frac{\frac{d}{d\,r}\,\mu}{r\,(r+\mu)}$$

$$G_{22} = \frac{\frac{d^2}{d\,r^2}\,\mu\,r}{2}$$

$$G_{33} = \frac{\frac{d^2}{d\,r^2}\,\mu\,r\,\sin^2\vartheta}{2}$$

Hodge Dual of Bianchi Identity

——— (see charge and current densities)

Scalar Charge Density ($-R^{0}{}_{i}{}^{i0}$)

$$\rho = \frac{\frac{d^2}{d\,r^2}\,\mu}{2\,(r+\mu)}$$

Current Density Class 1 ($-R^{i}{}_{\mu}{}^{\mu j}$)

$$J_1 = \frac{\frac{d^2}{d\,r^2}\,\mu\,(r+\mu)}{2\,r^2}$$

$$J_2 = \frac{\frac{d}{d\,r}\,\mu}{r^4}$$

$$J_3 = \frac{\frac{d}{d\,r}\,\mu}{r^4\,\sin^2\vartheta}$$

Current Density Class 2 ($-R^{i\ \mu j}_{\ \mu}$)

$$J_1 = 0$$

$$J_2 = 0$$

$$J_3 = 0$$

Current Density Class 3 ($-R^{i\ \mu j}_{\ \mu}$)

$$J_1 = 0$$

$$J_2 = 0$$

$$J_3 = 0$$

4.4.9 Spherically symmetric metric with off-diagonal elements

This version of the spherically symmetric line element is a precursor form of the diagonal metric. The functions A, B, C, D depend on t and r.

Coordinates

$$\mathbf{x} = \begin{pmatrix} t \\ r \\ \vartheta \\ \varphi \end{pmatrix}$$

Metric

$$g_{\mu\nu} = \begin{pmatrix} A & -B & 0 & 0 \\ -B & -C & 0 & 0 \\ 0 & 0 & -D & 0 \\ 0 & 0 & 0 & -\sin^2\vartheta\, D \end{pmatrix}$$

Contravariant Metric

$$g^{\mu\nu} = \begin{pmatrix} \frac{C}{A\,C+B^2} & -\frac{B}{A\,C+B^2} & 0 & 0 \\ -\frac{B}{A\,C+B^2} & -\frac{A}{A\,C+B^2} & 0 & 0 \\ 0 & 0 & -\frac{1}{D} & 0 \\ 0 & 0 & 0 & -\frac{1}{\sin^2\vartheta\, D} \end{pmatrix}$$

Christoffel Connection

$$\Gamma^0_{\ 00} = \frac{\frac{d}{d\,t}A\,C + 2\,B\,\left(\frac{d}{d\,t}B\right) + \frac{d}{d\,r}A\,B}{2\,(A\,C + B^2)}$$

$$\Gamma^0_{\ 01} = \frac{B\,\left(\frac{d}{d\,t}C\right) + \frac{d}{d\,r}A\,C}{2\,(A\,C + B^2)}$$

$$\Gamma^0_{\ 10} = \Gamma^0_{\ 01}$$

$$\Gamma^0_{\ 11} = \frac{C\,\left(\frac{d}{d\,t}C\right) + B\,\left(\frac{d}{d\,r}C\right) - 2\,\left(\frac{d}{d\,r}B\right)\,C}{2\,(A\,C + B^2)}$$

$$\Gamma^0{}_{22} = \frac{C\,(\frac{d}{dt}\,D) - B\,(\frac{d}{dr}\,D)}{2\,(A\,C + B^2)}$$

$$\Gamma^0{}_{33} = \frac{\sin^2\vartheta\,(C\,(\frac{d}{dt}\,D) - B\,(\frac{d}{dr}\,D))}{2\,(A\,C + B^2)}$$

$$\Gamma^1{}_{00} = \frac{2\,A\,(\frac{d}{dt}\,B) - \frac{d}{dt}\,A\,B + A\,(\frac{d}{dr}\,A)}{2\,(A\,C + B^2)}$$

$$\Gamma^1{}_{01} = \frac{A\,(\frac{d}{dt}\,C) - \frac{d}{dr}\,A\,B}{2\,(A\,C + B^2)}$$

$$\Gamma^1{}_{10} = \Gamma^1{}_{01}$$

$$\Gamma^1{}_{11} = -\frac{B\,(\frac{d}{dt}\,C) - A\,(\frac{d}{dr}\,C) - 2\,B\,(\frac{d}{dr}\,B)}{2\,(A\,C + B^2)}$$

$$\Gamma^1{}_{22} = -\frac{B\,(\frac{d}{dt}\,D) + A\,(\frac{d}{dr}\,D)}{2\,(A\,C + B^2)}$$

$$\Gamma^1{}_{33} = -\frac{\sin^2\vartheta\,(B\,(\frac{d}{dt}\,D) + A\,(\frac{d}{dr}\,D))}{2\,(A\,C + B^2)}$$

$$\Gamma^2{}_{02} = \frac{\frac{d}{dt}\,D}{2\,D}$$

$$\Gamma^2{}_{12} = \frac{\frac{d}{dr}\,D}{2\,D}$$

$$\Gamma^2{}_{20} = \Gamma^2{}_{02}$$

$$\Gamma^2{}_{21} = \Gamma^2{}_{12}$$

$$\Gamma^2{}_{33} = -\cos\vartheta\,\sin\vartheta$$

$$\Gamma^3{}_{03} = \frac{\frac{d}{dt}\,D}{2\,D}$$

$$\Gamma^3{}_{13} = \frac{\frac{d}{dr}\,D}{2\,D}$$

$$\Gamma^3{}_{23} = \frac{\cos\vartheta}{\sin\vartheta}$$

$$\Gamma^3{}_{30} = \Gamma^3{}_{03}$$

$$\Gamma^3{}_{31} = \Gamma^3{}_{13}$$

$$\Gamma^3{}_{32} = \Gamma^3{}_{23}$$

Metric Compatibility

——— o.k.

Riemann Tensor

$$R^{0}{}_{001} = \frac{B\left(2\,A\,C\left(\frac{d^2}{d\,t^2}\,C\right) + 2\,B^2\left(\frac{d^2}{d\,t^2}\,C\right) - A\left(\frac{d}{d\,t}\,C\right)^2 - \frac{d}{d\,t}\,A\,C\left(\frac{d}{d\,t}\,C\right) - 2\,B\left(\frac{d}{d\,t}\,B\right)\left(\frac{d}{d\,t}\,C\right) + \frac{d}{d\,r}\,A\,B\left(\frac{d}{d\,t}\,C\right) + 2\,A\left(\frac{d}{d\,t}\,B\right)\left(\frac{d}{d\,r}\,C\right) - \ldots\right)}{4\left(A\,C + B^2\right)^2}$$

$$R^{0}{}_{010} = -R^{0}{}_{001}$$

$$R^{0}{}_{101} = \frac{C\left(2\,A\,C\left(\frac{d^2}{d\,t^2}\,C\right) + 2\,B^2\left(\frac{d^2}{d\,t^2}\,C\right) - A\left(\frac{d}{d\,t}\,C\right)^2 - \frac{d}{d\,t}\,A\,C\left(\frac{d}{d\,t}\,C\right) - 2\,B\left(\frac{d}{d\,t}\,B\right)\left(\frac{d}{d\,t}\,C\right) + \frac{d}{d\,r}\,A\,B\left(\frac{d}{d\,t}\,C\right) + 2\,A\left(\frac{d}{d\,t}\,B\right)\left(\frac{d}{d\,r}\,C\right) - \ldots\right)}{4\left(A\,C + B^2\right)^2}$$

$$R^{0}{}_{110} = -R^{0}{}_{101}$$

$$R^{0}{}_{202} = \frac{2\,A\,C^2\,D\left(\frac{d^2}{d\,t^2}\,D\right) + 2\,B^2\,C\,D\left(\frac{d^2}{d\,t^2}\,D\right) - A\,C^2\left(\frac{d}{d\,t}\,D\right)^2 - B^2\,C\left(\frac{d}{d\,t}\,D\right)^2 + A\,B\,C\left(\frac{d}{d\,r}\,D\right)\left(\frac{d}{d\,t}\,D\right) + B^3\left(\frac{d}{d\,r}\,D\right)\left(\frac{d}{d\,t}\,D\right) + \ldots}{4\,(A\,C + B^2)^2\,D}$$

$$R^{0}{}_{212} = -\frac{A\,C^2\left(\frac{d}{d\,r}\,D\right)\left(\frac{d}{d\,t}\,D\right) + B^2\,C\left(\frac{d}{d\,r}\,D\right)\left(\frac{d}{d\,t}\,D\right) - B^2\left(\frac{d}{d\,r}\,C\right)D\left(\frac{d}{d\,t}\,D\right) + \frac{d}{d\,r}\,A\,C^2\,D\left(\frac{d}{d\,t}\,D\right) + 2\,B\left(\frac{d}{d\,r}\,B\right)C\,D\left(\frac{d}{d\,t}\,D\right) + \ldots}{4\,(A\,C + B^2)^2\,D}$$

$$R^{0}{}_{220} = -R^{0}{}_{202}$$

$$R^{0}{}_{221} = -R^{0}{}_{212}$$

$$R^{0}{}_{303} = \frac{\sin^2\vartheta\left(2\,A\,C^2\,D\left(\frac{d^2}{d\,t^2}\,D\right) + 2\,B^2\,C\,D\left(\frac{d^2}{d\,t^2}\,D\right) - A\,C^2\left(\frac{d}{d\,t}\,D\right)^2 - B^2\,C\left(\frac{d}{d\,t}\,D\right)^2 + A\,B\,C\left(\frac{d}{d\,r}\,D\right)\left(\frac{d}{d\,t}\,D\right) + B^3\left(\frac{d}{d\,r}\,D\right)\left(\frac{d}{d\,t}\,D\right) + \ldots\right)}{4\left(A\,C + B^2\right)^2\,D}$$

$$R^{0}{}_{313} = -\frac{\sin^2\vartheta\left(A\,C^2\left(\frac{d}{d\,r}\,D\right)\left(\frac{d}{d\,t}\,D\right) + B^2\,C\left(\frac{d}{d\,r}\,D\right)\left(\frac{d}{d\,t}\,D\right) - B^2\left(\frac{d}{d\,r}\,C\right)D\left(\frac{d}{d\,t}\,D\right) + \frac{d}{d\,r}\,A\,C^2\,D\left(\frac{d}{d\,t}\,D\right) + 2\,B\left(\frac{d}{d\,r}\,B\right)C\,D\left(\frac{d}{d\,t}\,D\right) + \ldots\right)}{4\left(A\,C + B^2\right)^2\,D}$$

$$R^{0}{}_{330} = -R^{0}{}_{303}$$

$$R^{0}{}_{331} = -R^{0}{}_{313}$$

$$R^{1}{}_{001} = \frac{A\left(2\,A\,C\left(\frac{d^2}{d\,t^2}\,C\right) + 2\,B^2\left(\frac{d^2}{d\,t^2}\,C\right) - A\left(\frac{d}{d\,t}\,C\right)^2 - \frac{d}{d\,t}\,A\,C\left(\frac{d}{d\,t}\,C\right) - 2\,B\left(\frac{d}{d\,t}\,B\right)\left(\frac{d}{d\,t}\,C\right) + \frac{d}{d\,r}\,A\,B\left(\frac{d}{d\,t}\,C\right) + 2\,A\left(\frac{d}{d\,t}\,B\right)\left(\frac{d}{d\,r}\,C\right) - \ldots\right)}{4\left(A\,C + B^2\right)^2}$$

$$R^{1}{}_{010} = -R^{1}{}_{001}$$

$$R^{1}{}_{101} = -\frac{B\left(2\,A\,C\left(\frac{d^2}{d\,t^2}\,C\right) + 2\,B^2\left(\frac{d^2}{d\,t^2}\,C\right) - A\left(\frac{d}{d\,t}\,C\right)^2 - \frac{d}{d\,t}\,A\,C\left(\frac{d}{d\,t}\,C\right) - 2\,B\left(\frac{d}{d\,t}\,B\right)\left(\frac{d}{d\,t}\,C\right) + \frac{d}{d\,r}\,A\,B\left(\frac{d}{d\,t}\,C\right) + 2\,A\left(\frac{d}{d\,t}\,B\right)\left(\frac{d}{d\,r}\,C\right) - \ldots\right)}{4\left(A\,C + B^2\right)^2}$$

$$R^1{}_{110} = -R^1{}_{101}$$

$$R^1{}_{202} = -\frac{2\,A\,B\,C\,D\left(\frac{d^2}{d\,t^2}\,D\right) + 2\,B^3\,D\left(\frac{d^2}{d\,t^2}\,D\right) - A\,B\,C\,\left(\frac{d}{d\,t}\,D\right)^2 - B^3\,\left(\frac{d}{d\,t}\,D\right)^2 - A^2\,C\,\left(\frac{d}{d\,r}\,D\right)\left(\frac{d}{d\,t}\,D\right) - A\,B^2\,\left(\frac{d}{d\,r}\,D\right)\left(\frac{d}{d\,t}\,D\right) - \ldots}{4\,(A\,C + B^2)^2\,D}$$

$$R^1{}_{212} = \frac{A\,B\,C\,\left(\frac{d}{d\,r}\,D\right)\left(\frac{d}{d\,t}\,D\right) + B^3\,\left(\frac{d}{d\,r}\,D\right)\left(\frac{d}{d\,t}\,D\right) + A\,C\,\left(\frac{d}{d\,t}\,C\right)\,D\,\left(\frac{d}{d\,t}\,D\right) + B^2\,\left(\frac{d}{d\,t}\,C\right)\,D\,\left(\frac{d}{d\,t}\,D\right) + A\,B\,\left(\frac{d}{d\,r}\,C\right)\,D\,\left(\frac{d}{d\,t}\,D\right) - \ldots}{4\,(A\,C + B^2)^2\,D}$$

$$R^1{}_{220} = -R^1{}_{202}$$

$$R^1{}_{221} = -R^1{}_{212}$$

$$R^1{}_{303} = -\frac{\sin^2\vartheta\,\left(2\,A\,B\,C\,D\left(\frac{d^2}{d\,t^2}\,D\right) + 2\,B^3\,D\left(\frac{d^2}{d\,t^2}\,D\right) - A\,B\,C\,\left(\frac{d}{d\,t}\,D\right)^2 - B^3\,\left(\frac{d}{d\,t}\,D\right)^2 - A^2\,C\,\left(\frac{d}{d\,r}\,D\right)\left(\frac{d}{d\,t}\,D\right) - A\,B^2\,\left(\frac{d}{d\,r}\,D\right)\left(\frac{d}{d\,t}\,D\right) - \ldots\right)}{4\,\left(A\,C + B^2\right)^2\,D}$$

$$R^1{}_{313} = \frac{\sin^2\vartheta\,\left(A\,B\,C\,\left(\frac{d}{d\,r}\,D\right)\left(\frac{d}{d\,t}\,D\right) + B^3\,\left(\frac{d}{d\,r}\,D\right)\left(\frac{d}{d\,t}\,D\right) + A\,C\,\left(\frac{d}{d\,t}\,C\right)\,D\,\left(\frac{d}{d\,t}\,D\right) + B^2\,\left(\frac{d}{d\,t}\,C\right)\,D\,\left(\frac{d}{d\,t}\,D\right) + A\,B\,\left(\frac{d}{d\,r}\,C\right)\,D\,\left(\frac{d}{d\,t}\,D\right) - \ldots\right)}{4\,\left(A\,C + B^2\right)^2\,D}$$

$$R^1{}_{330} = -R^1{}_{303}$$

$$R^1{}_{331} = -R^1{}_{313}$$

$$R^2{}_{002} = \frac{2\,A\,C\,D\left(\frac{d^2}{d\,t^2}\,D\right) + 2\,B^2\,D\left(\frac{d^2}{d\,t^2}\,D\right) - A\,C\,\left(\frac{d}{d\,t}\,D\right)^2 - B^2\,\left(\frac{d}{d\,t}\,D\right)^2 - \frac{d}{d\,t}\,A\,C\,D\,\left(\frac{d}{d\,t}\,D\right) - 2\,B\,\left(\frac{d}{d\,t}\,B\right)\,D\,\left(\frac{d}{d\,t}\,D\right) - \frac{d}{d\,r}\,A\,B\,D\,\left(\frac{d}{d\,t}\,D\right) - \ldots}{4\,\left(A\,C + B^2\right)\,D^2}$$

$$R^2{}_{012} = -\frac{A\,C\,\left(\frac{d}{d\,r}\,D\right)\left(\frac{d}{d\,t}\,D\right) + B^2\,\left(\frac{d}{d\,r}\,D\right)\left(\frac{d}{d\,t}\,D\right) + B\,\left(\frac{d}{d\,t}\,C\right)\,D\,\left(\frac{d}{d\,t}\,D\right) + \frac{d}{d\,r}\,A\,C\,D\,\left(\frac{d}{d\,t}\,D\right) - 2\,A\,C\,D\,\left(\frac{d^2}{d\,r\,d\,t}\,D\right) - 2\,B^2\,D\,\left(\frac{d^2}{d\,r\,d\,t}\,D\right) + \ldots}{4\,\left(A\,C + B^2\right)\,D^2}$$

$$R^2{}_{020} = -R^2{}_{002}$$

$$R^2{}_{021} = -R^2{}_{012}$$

$$R^2{}_{102} = -\frac{A\,C\,\left(\frac{d}{d\,r}\,D\right)\left(\frac{d}{d\,t}\,D\right) + B^2\,\left(\frac{d}{d\,r}\,D\right)\left(\frac{d}{d\,t}\,D\right) + B\,\left(\frac{d}{d\,t}\,C\right)\,D\,\left(\frac{d}{d\,t}\,D\right) + \frac{d}{d\,r}\,A\,C\,D\,\left(\frac{d}{d\,t}\,D\right) - 2\,A\,C\,D\,\left(\frac{d^2}{d\,r\,d\,t}\,D\right) - 2\,B^2\,D\,\left(\frac{d^2}{d\,r\,d\,t}\,D\right) + \ldots}{4\,\left(A\,C + B^2\right)\,D^2}$$

$$R^2{}_{112} = -\frac{C\,\left(\frac{d}{d\,t}\,C\right)\,D\,\left(\frac{d}{d\,t}\,D\right) + B\,\left(\frac{d}{d\,r}\,C\right)\,D\,\left(\frac{d}{d\,t}\,D\right) - 2\,\left(\frac{d}{d\,r}\,B\right)\,C\,D\,\left(\frac{d}{d\,t}\,D\right) - 2\,A\,C\,D\,\left(\frac{d^2}{d\,r^2}\,D\right) - 2\,B^2\,D\,\left(\frac{d^2}{d\,r^2}\,D\right) + A\,C\,\left(\frac{d}{d\,r}\,D\right)^2 + \ldots}{4\,\left(A\,C + B^2\right)\,D^2}$$

$$R^{2}{}_{120} = -R^{2}{}_{102}$$

$$R^{2}{}_{121} = -R^{2}{}_{112}$$

$$R^{2}{}_{323} = \frac{\sin^2\vartheta \left(C\,\left(\frac{d}{d\,t}\,D\right)^2 - 2\,B\,\left(\frac{d}{d\,r}\,D\right)\,\left(\frac{d}{d\,t}\,D\right) - A\,\left(\frac{d}{d\,r}\,D\right)^2 + 4\,A\,C\,D + 4\,B^2\,D\right)}{4\,(A\,C + B^2)\,D}$$

$$R^{2}{}_{332} = -R^{2}{}_{323}$$

$$R^{3}{}_{003} = \frac{2\,A\,C\,D\,\left(\frac{d^2}{d\,t^2}\,D\right) + 2\,B^2\,D\,\left(\frac{d^2}{d\,t^2}\,D\right) - A\,C\,\left(\frac{d}{d\,t}\,D\right)^2 - B^2\,\left(\frac{d}{d\,t}\,D\right)^2 - \frac{d}{d\,t}\,A\,C\,D\,\left(\frac{d}{d\,t}\,D\right) - 2\,B\,\left(\frac{d}{d\,t}\,B\right)\,D\,\left(\frac{d}{d\,t}\,D\right) - \frac{d}{d\,r}\,A\,B\,D\,\left(\frac{d}{d\,t}\,D\right) - \ldots}{4\,\left(A\,C + B^2\right)\,D^2}$$

$$R^{3}{}_{013} = -\frac{A\,C\,\left(\frac{d}{d\,r}\,D\right)\,\left(\frac{d}{d\,t}\,D\right) + B^2\,\left(\frac{d}{d\,r}\,D\right)\,\left(\frac{d}{d\,t}\,D\right) + B\,\left(\frac{d}{d\,t}\,C\right)\,D\,\left(\frac{d}{d\,t}\,D\right) + \frac{d}{d\,r}\,A\,C\,D\,\left(\frac{d}{d\,t}\,D\right) - 2\,A\,C\,D\,\left(\frac{d^2}{d\,r\,d\,t}\,D\right) - 2\,B^2\,D\,\left(\frac{d^2}{d\,r\,d\,t}\,D\right) + \ldots}{4\,\left(A\,C + B^2\right)\,D^2}$$

$$R^{3}{}_{030} = -R^{3}{}_{003}$$

$$R^{3}{}_{031} = -R^{3}{}_{013}$$

$$R^{3}{}_{103} = -\frac{A\,C\,\left(\frac{d}{d\,r}\,D\right)\,\left(\frac{d}{d\,t}\,D\right) + B^2\,\left(\frac{d}{d\,r}\,D\right)\,\left(\frac{d}{d\,t}\,D\right) + B\,\left(\frac{d}{d\,t}\,C\right)\,D\,\left(\frac{d}{d\,t}\,D\right) + \frac{d}{d\,r}\,A\,C\,D\,\left(\frac{d}{d\,t}\,D\right) - 2\,A\,C\,D\,\left(\frac{d^2}{d\,r\,d\,t}\,D\right) - 2\,B^2\,D\,\left(\frac{d^2}{d\,r\,d\,t}\,D\right) + \ldots}{4\,\left(A\,C + B^2\right)\,D^2}$$

$$R^{3}{}_{113} = -\frac{C\,\left(\frac{d}{d\,t}\,C\right)\,D\,\left(\frac{d}{d\,t}\,D\right) + B\,\left(\frac{d}{d\,r}\,C\right)\,D\,\left(\frac{d}{d\,t}\,D\right) - 2\,\left(\frac{d}{d\,r}\,B\right)\,C\,D\,\left(\frac{d}{d\,t}\,D\right) - 2\,A\,C\,D\,\left(\frac{d^2}{d\,r^2}\,D\right) - 2\,B^2\,D\,\left(\frac{d^2}{d\,r^2}\,D\right) + A\,C\,\left(\frac{d}{d\,r}\,D\right)^2 + \ldots}{4\,\left(A\,C + B^2\right)\,D^2}$$

$$R^{3}{}_{130} = -R^{3}{}_{103}$$

$$R^{3}{}_{131} = -R^{3}{}_{113}$$

$$R^{3}{}_{223} = -\frac{C\,\left(\frac{d}{d\,t}\,D\right)^2 - 2\,B\,\left(\frac{d}{d\,r}\,D\right)\,\left(\frac{d}{d\,t}\,D\right) - A\,\left(\frac{d}{d\,r}\,D\right)^2 + 4\,A\,C\,D + 4\,B^2\,D}{4\,(A\,C + B^2)\,D}$$

$$R^{3}{}_{232} = -R^{3}{}_{223}$$

Ricci Tensor

$$\mathrm{Ric}_{00} = -\frac{4\,A^2\,C^2\,D\left(\frac{d^2}{d\,t^2}\,D\right) + 8\,A\,B^2\,C\,D\left(\frac{d^2}{d\,t^2}\,D\right) + 4\,B^4\,D\left(\frac{d^2}{d\,t^2}\,D\right) - 2\,A^2\,C^2\left(\frac{d}{d\,t}\,D\right)^2 - 4\,A\,B^2\,C\left(\frac{d}{d\,t}\,D\right)^2 - 2\,B^4\left(\frac{d}{d\,t}\,D\right)^2 - \ldots}{4\left(A\,C + B^2\right)^2\,D^2}$$

$$\mathrm{Ric}_{01} = \frac{2\,A^2\,C^2\left(\frac{d}{d\,r}\,D\right)\left(\frac{d}{d\,t}\,D\right) + 4\,A\,B^2\,C\left(\frac{d}{d\,r}\,D\right)\left(\frac{d}{d\,t}\,D\right) + 2\,B^4\left(\frac{d}{d\,r}\,D\right)\left(\frac{d}{d\,t}\,D\right) + 2\,A\,B\,C\left(\frac{d}{d\,t}\,C\right)\,D\left(\frac{d}{d\,t}\,D\right) + 2\,B^3\left(\frac{d}{d\,t}\,C\right)\,D\left(\frac{d}{d\,t}\,D\right) + \ldots}{4\left(A\,C + B^2\right)^2\,D^2}$$

$$\mathrm{Ric}_{10} = \mathrm{Ric}_{01}$$

$$\mathrm{Ric}_{11} = \frac{2\,A\,C^2\left(\frac{d}{d\,t}\,C\right)\,D\left(\frac{d}{d\,t}\,D\right) + 2\,B^2\,C\left(\frac{d}{d\,t}\,C\right)\,D\left(\frac{d}{d\,t}\,D\right) + 2\,A\,B\,C\left(\frac{d}{d\,r}\,C\right)\,D\left(\frac{d}{d\,t}\,D\right) + 2\,B^3\left(\frac{d}{d\,r}\,C\right)\,D\left(\frac{d}{d\,t}\,D\right) - 4\,A\left(\frac{d}{d\,r}\,B\right)\,C^2\,D\left(\frac{d}{d\,t}\,D\right) - \ldots}{4\left(A\,C + B^2\right)^2\,D^2}$$

$$\mathrm{Ric}_{22} = \frac{2\,A\,C^2\left(\frac{d^2}{d\,t^2}\,D\right) + 2\,B^2\,C\left(\frac{d^2}{d\,t^2}\,D\right) + A\,C\left(\frac{d}{d\,t}\,C\right)\left(\frac{d}{d\,t}\,D\right) + 2\,B^2\left(\frac{d}{d\,t}\,C\right)\left(\frac{d}{d\,t}\,D\right) + A\,B\left(\frac{d}{d\,r}\,C\right)\left(\frac{d}{d\,t}\,D\right) - \frac{d}{d\,t}\,A\,C^2\left(\frac{d}{d\,t}\,D\right) - \ldots}{4\left(A\,C + B^2\right)^2}$$

$$\mathrm{Ric}_{33} = \frac{\sin^2\vartheta\left(2\,A\,C^2\left(\frac{d^2}{d\,t^2}\,D\right) + 2\,B^2\,C\left(\frac{d^2}{d\,t^2}\,D\right) + A\,C\left(\frac{d}{d\,t}\,C\right)\left(\frac{d}{d\,t}\,D\right) + 2\,B^2\left(\frac{d}{d\,t}\,C\right)\left(\frac{d}{d\,t}\,D\right) + A\,B\left(\frac{d}{d\,r}\,C\right)\left(\frac{d}{d\,t}\,D\right) - \frac{d}{d\,t}\,A\,C^2\left(\frac{d}{d\,t}\,D\right) - \ldots\right)}{4\left(A\,C + B^2\right)^2}$$

Ricci Scalar

$$R_{sc} = -\frac{4\,A\,C^2\,D\left(\frac{d^2}{d\,t^2}\,D\right)}{2\,(A\,C + B^2)^2\,D^2} + \frac{4\,B^2\,C\,D\left(\frac{d^2}{d\,t^2}\,D\right)}{r^2} - \frac{A\,C^2\,\left(\frac{d}{d\,t}\,D\right)^2}{2\,(A\,C + B^2)^2\,D^2} - \frac{B^2\,C\,\left(\frac{d}{d\,t}\,D\right)^2}{2\,(A\,C + B^2)^2\,D^2} + \frac{2\,A\,B\,C\,\left(\frac{d}{d\,r}\,D\right)\,\left(\frac{d}{d\,t}\,D\right)}{r^2}$$

$$+\frac{2\,B^3\,\left(\frac{d}{d\,r}\,D\right)\,\left(\frac{d}{d\,t}\,D\right)}{2\,(A\,C + B^2)^2\,D^2} + \frac{2\,A\,C\,\left(\frac{d}{d\,t}\,C\right)\,D\,\left(\frac{d}{d\,t}\,D\right)}{2\,(A\,C + B^2)^2\,D^2} + \frac{4\,B^2\,\left(\frac{d}{d\,t}\,C\right)\,D\,\left(\frac{d}{d\,t}\,D\right)}{2\,(A\,C + B^2)^2\,D^2} + \frac{2\,A\,B\,\left(\frac{d}{d\,r}\,C\right)\,D\,\left(\frac{d}{d\,t}\,D\right)}{2\,(A\,C + B^2)^2\,D^2} - \frac{2\,\left(\frac{d}{d\,t}\,A\right)\,C^2\,D\,\left(\frac{d}{d\,t}\,D\right)}{2\,(A\,C + B^2)^2\,D^2}$$

$$-\frac{4\,B\,\left(\frac{d}{d\,t}\,B\right)\,C\,D\,\left(\frac{d}{d\,t}\,D\right)}{2\,(A\,C + B^2)^2\,D^2} - \frac{4\,A\,\left(\frac{d}{d\,r}\,B\right)\,C\,D\,\left(\frac{d}{d\,t}\,D\right)}{2\,(A\,C + B^2)^2\,D^2} + \frac{2\,\left(\frac{d}{d\,r}\,A\right)\,B\,C\,D\,\left(\frac{d}{d\,t}\,D\right)}{2\,(A\,C + B^2)^2\,D^2} - \frac{4\,A^2\,C\,D\left(\frac{d^2}{d\,r^2}\,D\right)}{2\,(A\,C + B^2)^2\,D^2} - \frac{4\,A\,B^2\,D\left(\frac{d^2}{d\,r^2}\,D\right)}{2\,(A\,C + B^2)^2\,D^2}$$

$$-\frac{8\,A\,B\,C\,D\left(\frac{d^2}{d\,r\,d\,t}\,D\right)}{2\,(A\,C + B^2)^2\,D^2} - \frac{8\,B^3\,D\left(\frac{d^2}{d\,r\,d\,t}\,D\right)}{2\,(A\,C + B^2)^2\,D^2} + \frac{A^2\,C\,\left(\frac{d}{d\,r}\,D\right)^2}{2\,(A\,C + B^2)^2\,D^2} + \frac{A\,B^2\,\left(\frac{d}{d\,r}\,D\right)^2}{2\,(A\,C + B^2)^2\,D^2} + \frac{2\,A\,B\,\left(\frac{d}{d\,t}\,C\right)\,D\,\left(\frac{d}{d\,r}\,D\right)}{r^2}$$

$$+\frac{2\,A^2\,\left(\frac{d}{d\,r}\,C\right)\,D\,\left(\frac{d}{d\,r}\,D\right)}{2\,(A\,C + B^2)^2\,D^2} - \frac{4\,A\,\left(\frac{d}{d\,t}\,B\right)\,C\,D\,\left(\frac{d}{d\,r}\,D\right)1}{2\,(A\,C + B^2)^2\,D^2} + \frac{2\,\left(\frac{d}{d\,t}\,A\right)\,B\,C\,D\,\left(\frac{d}{d\,r}\,D\right)}{2\,(A\,C + B^2)^2\,D^2} - \frac{2\,A\,\left(\frac{d}{d\,r}\,A\right)\,C\,D\,\left(\frac{d}{d\,r}\,D\right)}{2\,(A\,C + B^2)^2\,D^2} + \frac{4\,A\,B\,\left(\frac{d}{d\,r}\,B\right)\,D\,\left(\frac{d}{d\,r}\,D\right)}{2\,(A\,C + B^2)^2\,D^2}$$

$$-\frac{4\,\left(\frac{d}{d\,r}\,A\right)\,B^2\,D\,\left(\frac{d}{d\,r}\,D\right)}{2\,(A\,C + B^2)^2\,D^2} + \frac{2\,A\,C\left(\frac{d^2}{d\,t^2}\,C\right)\,D^2}{2\,(A\,C + B^2)^2\,D^2} + \frac{2\,B^2\left(\frac{d^2}{d\,t^2}\,C\right)\,D^2}{r^2} - \frac{A\,\left(\frac{d}{d\,t}\,C\right)^2\,D^2}{2\,(A\,C + B^2)^2\,D^2} - \frac{\frac{d}{d\,t}\,A\,C\,\left(\frac{d}{d\,t}\,C\right)\,D^2}{2\,(A\,C + B^2)^2\,D^2}$$

$$-\frac{2\,B\,\left(\frac{d}{d\,t}\,B\right)\,\left(\frac{d}{d\,t}\,C\right)\,D^2}{2\,(A\,C + B^2)^2\,D^2} + \frac{\frac{d}{d\,r}\,A\,B\,\left(\frac{d}{d\,t}\,C\right)\,D^2}{2\,(A\,C + B^2)^2\,D^2} + \frac{2\,A\,\left(\frac{d}{d\,t}\,B\right)\,\left(\frac{d}{d\,r}\,C\right)\,D^2}{2\,(A\,C + B^2)^2\,D^2} - \frac{\frac{d}{d\,t}\,A\,B\,\left(\frac{d}{d\,r}\,C\right)\,D^2}{2\,(A\,C + B^2)^2\,D^2}$$

$$+\frac{A\,\left(\frac{d}{d\,r}\,A\right)\,\left(\frac{d}{d\,r}\,C\right)\,D^2}{2\,(A\,C + B^2)^2\,D^2} - \frac{4\,A\left(\frac{d^2}{d\,r\,d\,t}\,B\right)\,C\,D^2}{2\,(A\,C + B^2)^2\,D^2} + \frac{2\,\left(\frac{d}{d\,t}\,A\right)\,\left(\frac{d}{d\,r}\,B\right)\,C\,D^2}{2\,(A\,C + B^2)^2\,D^2} - \frac{2\,A\left(\frac{d^2}{d\,r^2}\,A\right)\,C\,D^2}{2\,(A\,C + B^2)^2\,D^2} + \frac{\left(\frac{d}{d\,r}\,A\right)^2\,C\,D^2}{2\,(A\,C + B^2)^2\,D^2}$$

$$+\frac{4\,B\,\left(\frac{d}{d\,r}\,B\right)\,\left(\frac{d}{d\,t}\,B\right)\,D^2}{2\,(A\,C + B^2)^2\,D^2} - \frac{4\,B^2\left(\frac{d^2}{d\,r\,d\,t}\,B\right)\,D^2}{2\,(A\,C + B^2)^2\,D^2} + \frac{4\,A^2\,C^2\,D}{2\,(A\,C + B^2)^2\,D^2} + \frac{8\,A\,B^2\,C\,D}{2\,(A\,C + B^2)^2\,D^2} + \frac{4\,B^4\,D}{2\,(A\,C + B^2)^2\,D^2}$$

Bianchi identity (Ricci cyclic equation $R^{\kappa}{}_{[\mu\nu\sigma]} = 0$)

——— o.k.

Einstein Tensor

$$G_{01} = -\frac{4\,A\,B\,C^2\,D\left(\frac{d^2}{d\,t^2}\,D\right) + 4\,B^3\,C\,D\left(\frac{d^2}{d\,t^2}\,D\right) - A\,B\,C^2\left(\frac{d}{d\,t}\,D\right)^2 - B^3\,C\left(\frac{d}{d\,t}\,D\right)^2 - 2\,A^2\,C^2\left(\frac{d}{d\,r}\,D\right)\left(\frac{d}{d\,t}\,D\right) - \ldots}{4\,(A\,C + B^2)^2\,D^2}$$

$$G_{10} = G_{01}$$

$$G_{11} = -\frac{4\,A\,C^3\,D\left(\frac{d^2}{d\,t^2}\,D\right) + 4\,B^2\,C^2\,D\left(\frac{d^2}{d\,t^2}\,D\right) - A\,C^3\left(\frac{d}{d\,t}\,D\right)^2 - B^2\,C^2\left(\frac{d}{d\,t}\,D\right)^2 + 2\,A\,B\,C^2\left(\frac{d}{d\,r}\,D\right)\left(\frac{d}{d\,t}\,D\right) + 2\,B^3\,C\left(\frac{d}{d\,r}\,D\right)\left(\frac{d}{d\,t}\,D\right) + \ldots}{4\left(A\,C + B^2\right)^2\,D^2}$$

$$G_{22} = -\frac{2\,A\,C^2\,D\left(\frac{d^2}{d\,t^2}\,D\right) + 2\,B^2\,C\,D\left(\frac{d^2}{d\,t^2}\,D\right) - A\,C^2\left(\frac{d}{d\,t}\,D\right)^2 - B^2\,C\left(\frac{d}{d\,t}\,D\right)^2 + 2\,A\,B\,C\left(\frac{d}{d\,r}\,D\right)\left(\frac{d}{d\,t}\,D\right) + 2\,B^3\left(\frac{d}{d\,r}\,D\right)\left(\frac{d}{d\,t}\,D\right) + \ldots}{4\left(A\,C + B^2\right)^2\,D}$$

$$G_{33} = -\frac{\sin^2\vartheta\left(2\,A\,C^2\,D\left(\frac{d^2}{d\,t^2}\,D\right) + 2\,B^2\,C\,D\left(\frac{d^2}{d\,t^2}\,D\right) - A\,C^2\left(\frac{d}{d\,t}\,D\right)^2 - B^2\,C\left(\frac{d}{d\,t}\,D\right)^2 + 2\,A\,B\,C\left(\frac{d}{d\,r}\,D\right)\left(\frac{d}{d\,t}\,D\right) + 2\,B^3\left(\frac{d}{d\,r}\,D\right)\left(\frac{d}{d\,t}\,D\right) + \ldots\right)}{4\left(A\,C + B^2\right)^2\,D}$$

Hodge Dual of Bianchi Identity

——— (see charge and current densities)

Scalar Charge Density (-$R^{0}{}_{i}{}^{i0}$)

$$\rho = -\frac{4\,A\,C^3\,D\left(\frac{d^2}{d\,t^2}\,D\right) + 4\,B^2\,C^2\,D\left(\frac{d^2}{d\,t^2}\,D\right) - 2\,A\,C^3\left(\frac{d}{d\,t}\,D\right)^2 - 2\,B^2\,C^2\left(\frac{d}{d\,t}\,D\right)^2 + 4\,A\,B\,C^2\left(\frac{d}{d\,r}\,D\right)\left(\frac{d}{d\,t}\,D\right) + 4\,B^3\,C\left(\frac{d}{d\,r}\,D\right)\left(\frac{d}{d\,t}\,D\right) + \ldots}{4\left(A\,C + B^2\right)^3\,D^2}$$

Current Density Class 1 (-$R^{i}{}_{\mu}{}^{\mu j}$)

$$J_1 = \frac{4\,A\,B^2\,C\,D\left(\frac{d^2}{d\,t^2}\,D\right) + 4\,B^4\,D\left(\frac{d^2}{d\,t^2}\,D\right) - 2\,A\,B^2\,C\left(\frac{d}{d\,t}\,D\right)^2 - 2\,B^4\left(\frac{d}{d\,t}\,D\right)^2 - 4\,A^2\,B\,C\left(\frac{d}{d\,r}\,D\right)\left(\frac{d}{d\,t}\,D\right) - 4\,A\,B^3\left(\frac{d}{d\,r}\,D\right)\left(\frac{d}{d\,t}\,D\right) - \ldots}{4\left(A\,C + B^2\right)^3\,D^2}$$

$$J_2 = -\frac{2\,A\,C^2\left(\frac{d^2}{d\,t^2}\,D\right) + 2\,B^2\,C\left(\frac{d^2}{d\,t^2}\,D\right) + A\,C\left(\frac{d}{d\,t}\,C\right)\left(\frac{d}{d\,t}\,D\right) + 2\,B^2\left(\frac{d}{d\,t}\,C\right)\left(\frac{d}{d\,t}\,D\right) + A\,B\left(\frac{d}{d\,r}\,C\right)\left(\frac{d}{d\,t}\,D\right) - \frac{d}{d\,t}\,A\,C^2\left(\frac{d}{d\,t}\,D\right) - \ldots}{4\left(A\,C + B^2\right)^2\,D^2}$$

$$J_3 = -\frac{2\,A\,C^2\left(\frac{d^2}{d\,t^2}\,D\right) + 2\,B^2\,C\left(\frac{d^2}{d\,t^2}\,D\right) + A\,C\left(\frac{d}{d\,t}\,C\right)\left(\frac{d}{d\,t}\,D\right) + 2\,B^2\left(\frac{d}{d\,t}\,C\right)\left(\frac{d}{d\,t}\,D\right) + A\,B\left(\frac{d}{d\,r}\,C\right)\left(\frac{d}{d\,t}\,D\right) - \frac{d}{d\,t}\,A\,C^2\left(\frac{d}{d\,t}\,D\right) - \ldots}{4\sin^2\vartheta\left(A\,C + B^2\right)^2\,D^2}$$

Current Density Class 2 (-$R^{i}{}_{\mu}{}^{\mu j}$)

$$J_1 = 0$$

$$J_2 = 0$$

$$J_3 = 0$$

Current Density Class 3 ($-R^{i}{}_{\mu}{}^{\mu j}$)

$$J_1 = 0$$

$$J_2 = 0$$

$$J_3 = 0$$

4.4.10 Reissner-Nordstrom metric

This is a metric of a charged mass. M is a mass parameter, Q a charge parameter. Cosmological charge and current densities do exist.

Coordinates

$$\mathbf{x} = \begin{pmatrix} t \\ r \\ \vartheta \\ \varphi \end{pmatrix}$$

Metric

$$g_{\mu\nu} = \begin{pmatrix} -\frac{Q^2}{r^2} + \frac{2M}{r} - 1 & 0 & 0 & 0 \\ 0 & \frac{1}{\frac{Q^2}{r^2} - \frac{2M}{r} + 1} & 0 & 0 \\ 0 & 0 & r^2 & 0 \\ 0 & 0 & 0 & r^2 \sin^2\vartheta \end{pmatrix}$$

Contravariant Metric

$$g^{\mu\nu} = \begin{pmatrix} -\frac{r^2}{Q^2 - 2rM + r^2} & 0 & 0 & 0 \\ 0 & \frac{Q^2 - 2rM + r^2}{r^2} & 0 & 0 \\ 0 & 0 & \frac{1}{r^2} & 0 \\ 0 & 0 & 0 & \frac{1}{r^2 \sin^2\vartheta} \end{pmatrix}$$

Christoffel Connection

$$\Gamma^{0}{}_{01} = -\frac{Q^2 - rM}{r\,(Q^2 - 2rM + r^2)}$$

$$\Gamma^{0}{}_{10} = \Gamma^{0}{}_{01}$$

$$\Gamma^{1}{}_{00} = -\frac{\left(Q^2 - 2rM + r^2\right)\left(Q^2 - rM\right)}{r^5}$$

$$\Gamma^{1}{}_{11} = \frac{Q^2 - rM}{r\,(Q^2 - 2rM + r^2)}$$

$$\Gamma^{1}{}_{22} = -\frac{Q^2 - 2rM + r^2}{r}$$

$$\Gamma^1{}_{33} = -\frac{\sin^2\vartheta\,\left(Q^2 - 2\,r\,M + r^2\right)}{r}$$

$$\Gamma^2{}_{12} = \frac{1}{r}$$

$$\Gamma^2{}_{21} = \Gamma^2{}_{12}$$

$$\Gamma^2{}_{33} = -\cos\vartheta\,\sin\vartheta$$

$$\Gamma^3{}_{13} = \frac{1}{r}$$

$$\Gamma^3{}_{23} = \frac{\cos\vartheta}{\sin\vartheta}$$

$$\Gamma^3{}_{31} = \Gamma^3{}_{13}$$

$$\Gamma^3{}_{32} = \Gamma^3{}_{23}$$

Metric Compatibility

———— o.k.

Riemann Tensor

$$R^0{}_{101} = -\frac{3\,Q^2 - 2\,r\,M}{r^2\,\left(Q^2 - 2\,r\,M + r^2\right)}$$

$$R^0{}_{110} = -R^0{}_{101}$$

$$R^0{}_{202} = \frac{Q^2 - r\,M}{r^2}$$

$$R^0{}_{220} = -R^0{}_{202}$$

$$R^0{}_{303} = \frac{\sin^2\vartheta\,\left(Q^2 - r\,M\right)}{r^2}$$

$$R^0{}_{330} = -R^0{}_{303}$$

$$R^1{}_{001} = -\frac{\left(Q^2 - 2\,r\,M + r^2\right)\,\left(3\,Q^2 - 2\,r\,M\right)}{r^6}$$

$$R^1{}_{010} = -R^1{}_{001}$$

$$R^1{}_{212} = \frac{Q^2 - r\,M}{r^2}$$

$$R^1{}_{221} = -R^1{}_{212}$$

$$R^1{}_{313} = \frac{\sin^2\vartheta\,(Q^2 - r\,M)}{r^2}$$

$$R^1{}_{331} = -R^1{}_{313}$$

$$R^2{}_{002} = \frac{(Q^2 - 2\,r\,M + r^2)\,(Q^2 - r\,M)}{r^6}$$

$$R^2{}_{020} = -R^2{}_{002}$$

$$R^2{}_{112} = -\frac{Q^2 - r\,M}{r^2\,(Q^2 - 2\,r\,M + r^2)}$$

$$R^2{}_{121} = -R^2{}_{112}$$

$$R^2{}_{323} = -\frac{\sin^2\vartheta\,(Q^2 - 2\,r\,M)}{r^2}$$

$$R^2{}_{332} = -R^2{}_{323}$$

$$R^3{}_{003} = \frac{(Q^2 - 2\,r\,M + r^2)\,(Q^2 - r\,M)}{r^6}$$

$$R^3{}_{030} = -R^3{}_{003}$$

$$R^3{}_{113} = -\frac{Q^2 - r\,M}{r^2\,(Q^2 - 2\,r\,M + r^2)}$$

$$R^3{}_{131} = -R^3{}_{113}$$

$$R^3{}_{223} = \frac{Q^2 - 2\,r\,M}{r^2}$$

$$R^3{}_{232} = -R^3{}_{223}$$

Ricci Tensor

$$\mathrm{Ric}_{00} = \frac{Q^2\,(Q^2 - 2\,r\,M + r^2)}{r^6}$$

$$\mathrm{Ric}_{11} = -\frac{Q^2}{r^2\,(Q^2 - 2\,r\,M + r^2)}$$

$$\mathrm{Ric}_{22} = \frac{Q^2}{r^2}$$

$$\mathrm{Ric}_{33} = \frac{\sin^2\vartheta\,Q^2}{r^2}$$

Ricci Scalar

$$R_{sc} = 0$$

Bianchi identity (Ricci cyclic equation $R^{\kappa}{}_{[\mu\nu\sigma]} = 0$)

———— o.k.

Einstein Tensor

$$G_{00} = \frac{Q^2\,(Q^2 - 2\,r\,M + r^2)}{r^6}$$

$$G_{11} = -\frac{Q^2}{r^2\,(Q^2 - 2\,r\,M + r^2)}$$

$$G_{22} = \frac{Q^2}{r^2}$$

$$G_{33} = \frac{\sin^2\vartheta\,Q^2}{r^2}$$

Hodge Dual of Bianchi Identity

———— (see charge and current densities)

Scalar Charge Density (-$R^{0}{}_{i}{}^{i0}$)

$$\rho = \frac{Q^2}{r^2\,(Q^2 - 2\,r\,M + r^2)}$$

Current Density Class 1 (-$R^{i}{}_{\mu}{}^{\mu j}$)

$$J_1 = \frac{Q^2\,(Q^2 - 2\,r\,M + r^2)}{r^6}$$

$$J_2 = -\frac{Q^2}{r^6}$$

$$J_3 = -\frac{Q^2}{r^6\,\sin^2\vartheta}$$

Current Density Class 2 (-$R^{i}{}_{\mu}{}^{\mu j}$)

$$J_1 = 0$$

$$J_2 = 0$$

$$J_3 = 0$$

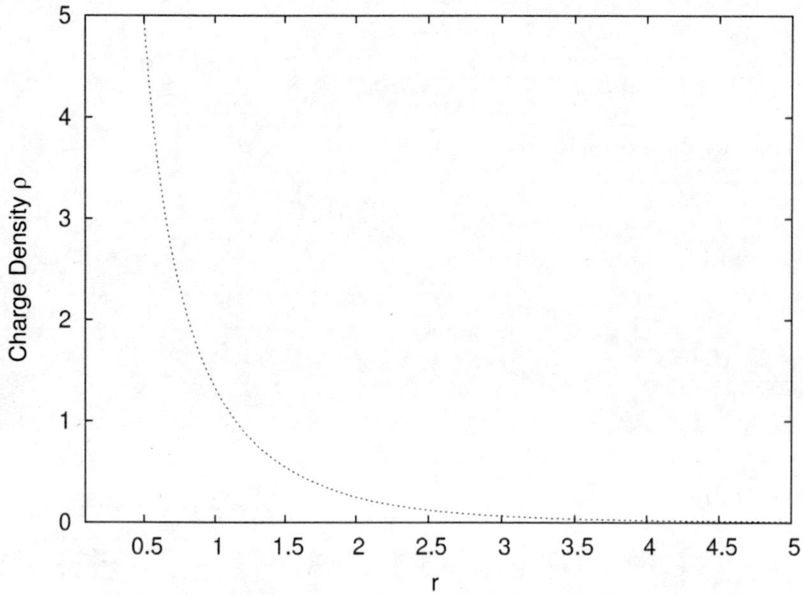

Fig. 4.13: Reissner-Nordstrom metric, charge density ρ for M=1, Q=2.

Current Density Class 3 (-$R^{i}{}_{\mu}{}^{\mu j}$)

$$J_1 = 0$$

$$J_2 = 0$$

$$J_3 = 0$$

4.4.11 Extended Reissner-Weyl metric

General solution of Einstein-Maxwell field theory for electromagnetism unified with gravitation. This metric assumes a vacuum outside of the center, therefore it should be a vacuum metric, but it isn't (only the Ricci scalar is zero). A, B, C, and κ are parameters.

This metric is identical with the Reissner-Nordstrom metric. The parameter C was introduced experimentally to see differences to the Reissner-Nordstrom metric.

Coordinates

$$\mathbf{x} = \begin{pmatrix} t \\ r \\ \vartheta \\ \varphi \end{pmatrix}$$

Metric

$$g_{\mu\nu} = \begin{pmatrix} -\frac{C}{r^3} - \frac{\kappa B^2}{2r^2} - \frac{A}{r} + 1 & 0 & 0 & 0 \\ 0 & -\frac{1}{-\frac{C}{r^3} - \frac{\kappa B^2}{2r^2} - \frac{A}{r} + 1} & 0 & 0 \\ 0 & 0 & r^2 & 0 \\ 0 & 0 & 0 & r^2 \sin^2\vartheta \end{pmatrix}$$

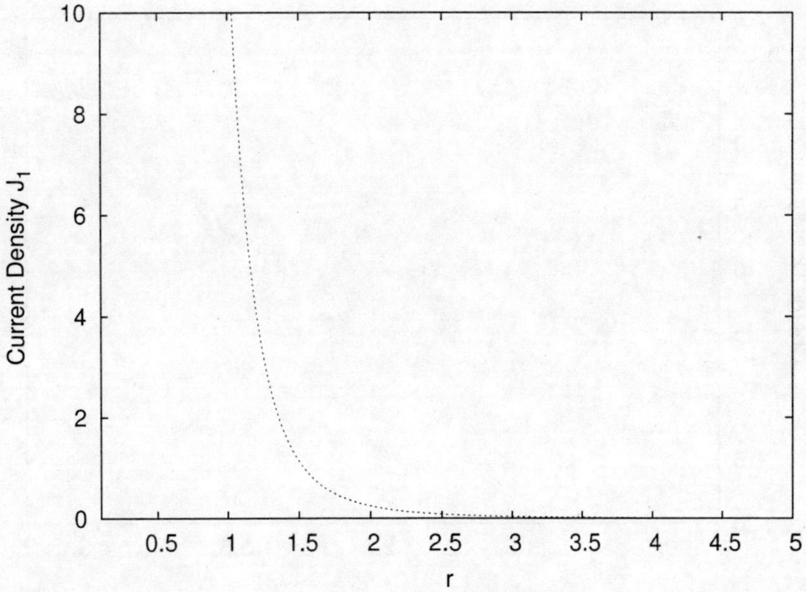

Fig. 4.14: Reissner-Nordstrom metric, current density J_r for M=1, Q=2.

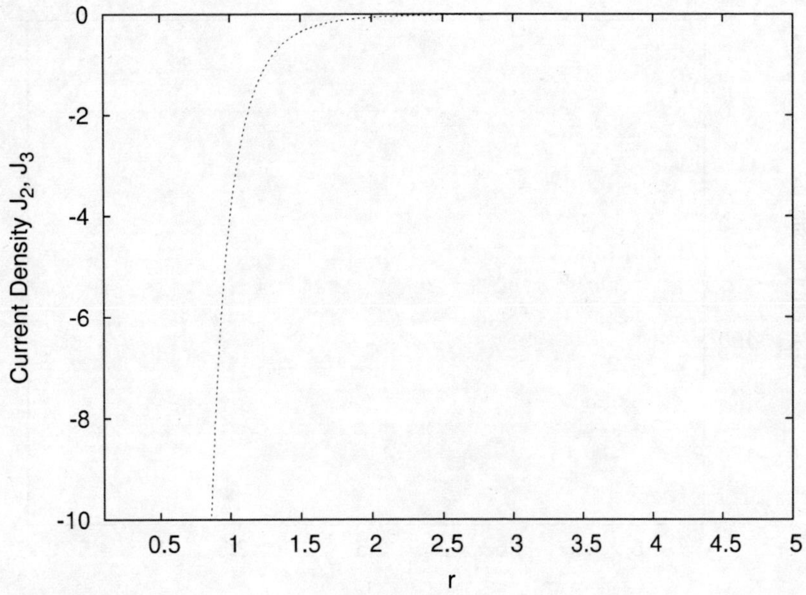

Fig. 4.15: Reissner-Nordstrom metric, current density J_θ, J_φ for M=1, Q=2.

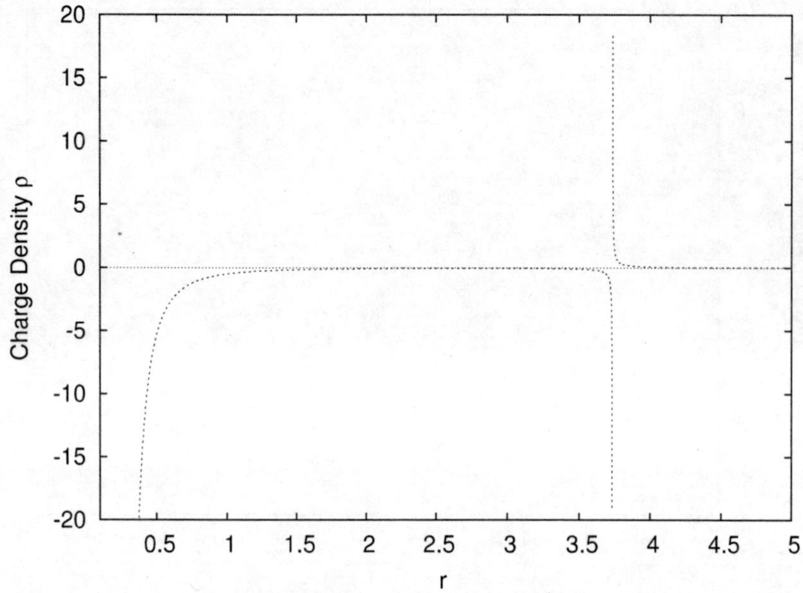

Fig. 4.16: Reissner-Nordstrom metric, charge density ρ for M=2, Q=1.

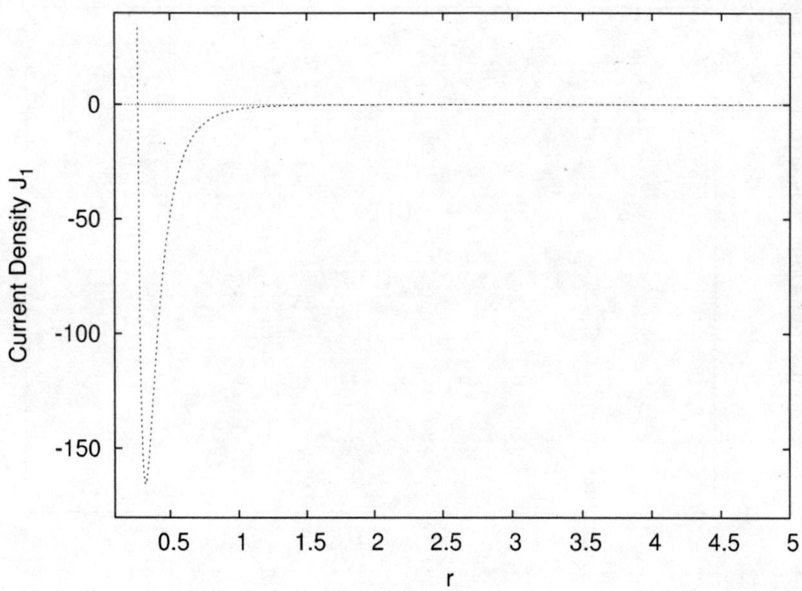

Fig. 4.17: Reissner-Nordstrom metric, current density J_r for M=2, Q=1.

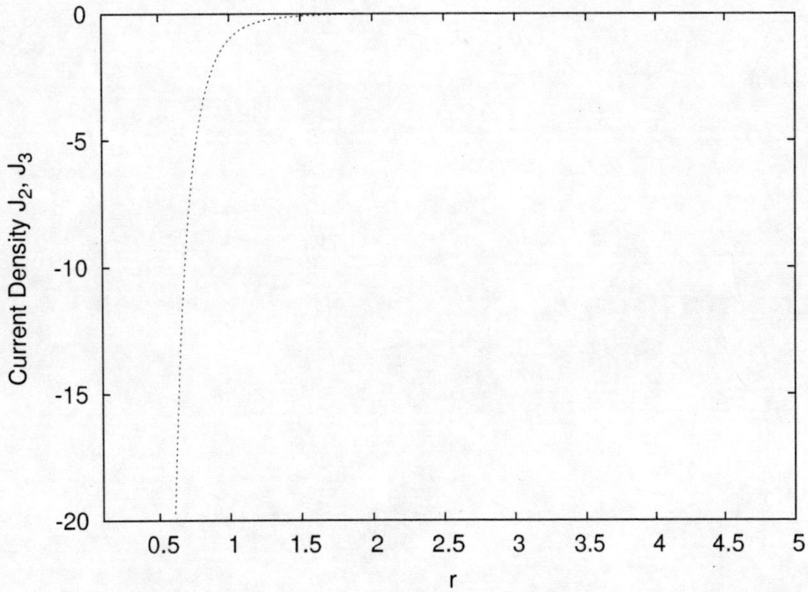

Fig. 4.18: Reissner-Nordstrom metric, current density J_θ, J_φ for M=2, Q=1.

Contravariant Metric

$$g^{\mu\nu} = \begin{pmatrix} -\frac{2\,r^3}{2\,C+\kappa\,r\,B^2+2\,r^2\,A-2\,r^3} & 0 & 0 & 0 \\ 0 & \frac{2\,C+\kappa\,r\,B^2+2\,r^2\,A-2\,r^3}{2\,r^3} & 0 & 0 \\ 0 & 0 & \frac{1}{r^2} & 0 \\ 0 & 0 & 0 & \frac{1}{r^2\,\sin^2\vartheta} \end{pmatrix}$$

Christoffel Connection

$$\Gamma^0{}_{01} = -\frac{3\,C+\kappa\,r\,B^2+r^2\,A}{r\,\left(2\,C+\kappa\,r\,B^2+2\,r^2\,A-2\,r^3\right)}$$

$$\Gamma^0{}_{10} = \Gamma^0{}_{01}$$

$$\Gamma^1{}_{00} = -\frac{\left(2\,C+\kappa\,r\,B^2+2\,r^2\,A-2\,r^3\right)\,\left(3\,C+\kappa\,r\,B^2+r^2\,A\right)}{4\,r^7}$$

$$\Gamma^1{}_{11} = \frac{3\,C+\kappa\,r\,B^2+r^2\,A}{r\,\left(2\,C+\kappa\,r\,B^2+2\,r^2\,A-2\,r^3\right)}$$

$$\Gamma^1{}_{22} = -\frac{2\,C+\kappa\,r\,B^2+2\,r^2\,A-2\,r^3}{2\,r^2}$$

$$\Gamma^1{}_{33} = -\frac{\sin^2\vartheta\,\left(2\,C+\kappa\,r\,B^2+2\,r^2\,A-2\,r^3\right)}{2\,r^2}$$

$$\Gamma^2{}_{12} = \frac{1}{r}$$

$$\Gamma^2{}_{21} = \Gamma^2{}_{12}$$

$$\Gamma^2{}_{33} = -\cos\vartheta \sin\vartheta$$

$$\Gamma^3{}_{13} = \frac{1}{r}$$

$$\Gamma^3{}_{23} = \frac{\cos\vartheta}{\sin\vartheta}$$

$$\Gamma^3{}_{31} = \Gamma^3{}_{13}$$

$$\Gamma^3{}_{32} = \Gamma^3{}_{23}$$

Metric Compatibility

——— o.k.

Riemann Tensor

$$R^0{}_{101} = -\frac{12\,C + 3\,\kappa\,r\,B^2 + 2\,r^2\,A}{r^2\,(2\,C + \kappa\,r\,B^2 + 2\,r^2\,A - 2\,r^3)}$$

$$R^0{}_{110} = -R^0{}_{101}$$

$$R^0{}_{202} = \frac{3\,C + \kappa\,r\,B^2 + r^2\,A}{2\,r^3}$$

$$R^0{}_{220} = -R^0{}_{202}$$

$$R^0{}_{303} = \frac{\sin^2\vartheta\,\left(3\,C + \kappa\,r\,B^2 + r^2\,A\right)}{2\,r^3}$$

$$R^0{}_{330} = -R^0{}_{303}$$

$$R^1{}_{001} = -\frac{\left(2\,C + \kappa\,r\,B^2 + 2\,r^2\,A - 2\,r^3\right)\,\left(12\,C + 3\,\kappa\,r\,B^2 + 2\,r^2\,A\right)}{4\,r^8}$$

$$R^1{}_{010} = -R^1{}_{001}$$

$$R^1{}_{212} = \frac{3\,C + \kappa\,r\,B^2 + r^2\,A}{2\,r^3}$$

$$R^1{}_{221} = -R^1{}_{212}$$

$$R^1{}_{313} = \frac{\sin^2\vartheta\,\left(3\,C + \kappa\,r\,B^2 + r^2\,A\right)}{2\,r^3}$$

$$R^1{}_{331} = -R^1{}_{313}$$

$$R^2{}_{002} = \frac{\left(2\,C + \kappa\,r\,B^2 + 2\,r^2\,A - 2\,r^3\right)\,\left(3\,C + \kappa\,r\,B^2 + r^2\,A\right)}{4\,r^8}$$

$$R^2{}_{020} = -R^2{}_{002}$$

$$R^2{}_{112} = -\frac{3\,C + \kappa\,r\,B^2 + r^2\,A}{r^2\,\left(2\,C + \kappa\,r\,B^2 + 2\,r^2\,A - 2\,r^3\right)}$$

$$R^2{}_{121} = -R^2{}_{112}$$

$$R^2{}_{323} = -\frac{\sin^2\vartheta\,\left(2\,C + \kappa\,r\,B^2 + 2\,r^2\,A - 4\,r^3\right)}{2\,r^3}$$

$$R^2{}_{332} = -R^2{}_{323}$$

$$R^3{}_{003} = \frac{\left(2\,C + \kappa\,r\,B^2 + 2\,r^2\,A - 2\,r^3\right)\,\left(3\,C + \kappa\,r\,B^2 + r^2\,A\right)}{4\,r^8}$$

$$R^3{}_{030} = -R^3{}_{003}$$

$$R^3{}_{113} = -\frac{3\,C + \kappa\,r\,B^2 + r^2\,A}{r^2\,\left(2\,C + \kappa\,r\,B^2 + 2\,r^2\,A - 2\,r^3\right)}$$

$$R^3{}_{131} = -R^3{}_{113}$$

$$R^3{}_{223} = \frac{2\,C + \kappa\,r\,B^2 + 2\,r^2\,A - 4\,r^3}{2\,r^3}$$

$$R^3{}_{232} = -R^3{}_{223}$$

Ricci Tensor

$$\mathrm{Ric}_{00} = \frac{\left(2\,C + \kappa\,r\,B^2 + 2\,r^2\,A - 2\,r^3\right)\,\left(6\,C + \kappa\,r\,B^2\right)}{4\,r^8}$$

$$\mathrm{Ric}_{11} = -\frac{6\,C + \kappa\,r\,B^2}{r^2\,\left(2\,C + \kappa\,r\,B^2 + 2\,r^2\,A - 2\,r^3\right)}$$

$$\mathrm{Ric}_{22} = \frac{4\,C + \kappa\,r\,B^2 + 4\,r^3}{2\,r^3}$$

$$\mathrm{Ric}_{33} = \frac{\sin^2\vartheta\,\left(4\,C + \kappa\,r\,B^2 + 4\,r^3\right)}{2\,r^3}$$

Ricci Scalar

$$R_{sc} = -\frac{2\,(C - 2\,r^3)}{r^5}$$

Bianchi identity (Ricci cyclic equation $R^{\kappa}{}_{[\mu\nu\sigma]} = 0$)

——— o.k.

Einstein Tensor

$$G_{00} = \frac{\left(2\,C + \kappa\,r\,B^2 + 2\,r^2\,A - 2\,r^3\right)\left(4\,C + \kappa\,r\,B^2 + 4\,r^3\right)}{4\,r^8}$$

$$G_{11} = -\frac{4\,C + \kappa\,r\,B^2 + 4\,r^3}{r^2\,(2\,C + \kappa\,r\,B^2 + 2\,r^2\,A - 2\,r^3)}$$

$$G_{22} = \frac{6\,C + \kappa\,r\,B^2}{2\,r^3}$$

$$G_{33} = \frac{\sin^2\vartheta\,(6\,C + \kappa\,r\,B^2)}{2\,r^3}$$

Hodge Dual of Bianchi Identity

——— (see charge and current densities)

Scalar Charge Density (-$R^0{}_i{}^{i0}$)

$$\rho = \frac{6\,C + \kappa\,r\,B^2}{r^2\,(2\,C + \kappa\,r\,B^2 + 2\,r^2\,A - 2\,r^3)}$$

Current Density Class 1 (-$R^i{}_\mu{}^{\mu j}$)

$$J_1 = \frac{\left(2\,C + \kappa\,r\,B^2 + 2\,r^2\,A - 2\,r^3\right)\left(6\,C + \kappa\,r\,B^2\right)}{4\,r^8}$$

$$J_2 = -\frac{4\,C + \kappa\,r\,B^2 + 4\,r^3}{2\,r^7}$$

$$J_3 = -\frac{4\,C + \kappa\,r\,B^2 + 4\,r^3}{2\,r^7\,\sin^2\vartheta}$$

Current Density Class 2 (-$R^i{}_\mu{}^{\mu j}$)

$$J_1 = 0$$

$$J_2 = 0$$

$$J_3 = 0$$

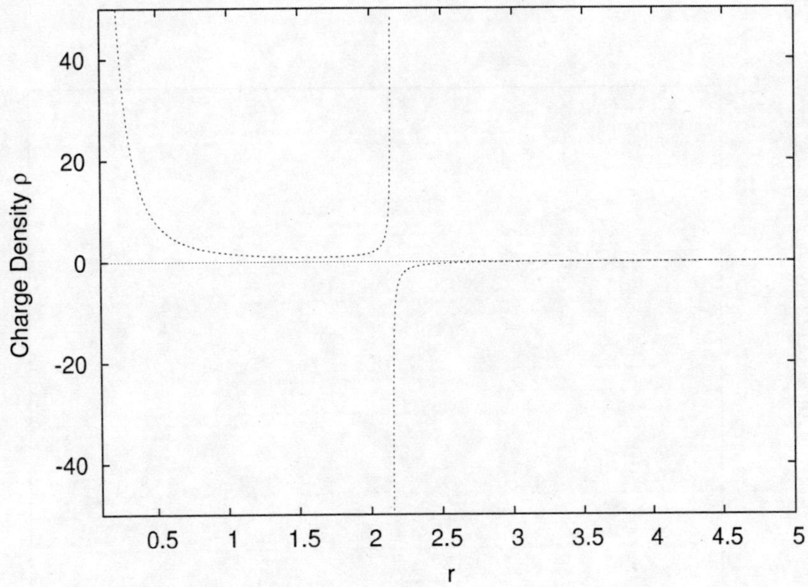

Fig. 4.19: Extended Reissner-Weyl metric, charge density ρ for A=1, B=2, C=1.

Current Density Class 3 (-$R^i{}_\mu{}^{\mu j}$)

$$J_1 = 0$$

$$J_2 = 0$$

$$J_3 = 0$$

4.4.12 Kerr metric

This metric describes a rotating mass without charge. M is the mass parameter, J the parameter of angular momentum. Cosmological charge and current densities do exist.

Coordinates

$$\mathbf{x} = \begin{pmatrix} t \\ r \\ \vartheta \\ \varphi \end{pmatrix}$$

Metric

$$g_{\mu\nu} = \begin{pmatrix} \frac{2M}{r} - 1 & 0 & 0 & -\frac{4\sin^2\vartheta\, J}{r} \\ 0 & \frac{1}{1-\frac{2M}{r}} & 0 & 0 \\ 0 & 0 & r^2 & 0 \\ -\frac{4\sin^2\vartheta\, J}{r} & 0 & 0 & r^2\sin^2\vartheta \end{pmatrix}$$

Contravariant Metric

$$g^{\mu\nu} = \begin{pmatrix} \frac{r^4}{2r^3M-16\sin^2\vartheta\, J^2-r^4} & 0 & 0 & \frac{4rJ}{2r^3M-16\sin^2\vartheta\, J^2-r^4} \\ 0 & -\frac{2M-r}{r} & 0 & 0 \\ 0 & 0 & \frac{1}{r^2} & 0 \\ \frac{4rJ}{2r^3M-16\sin^2\vartheta\, J^2-r^4} & 0 & 0 & \frac{r(2M-r)}{\sin^2\vartheta\,(2r^3M-16\sin^2\vartheta\, J^2-r^4)} \end{pmatrix}$$

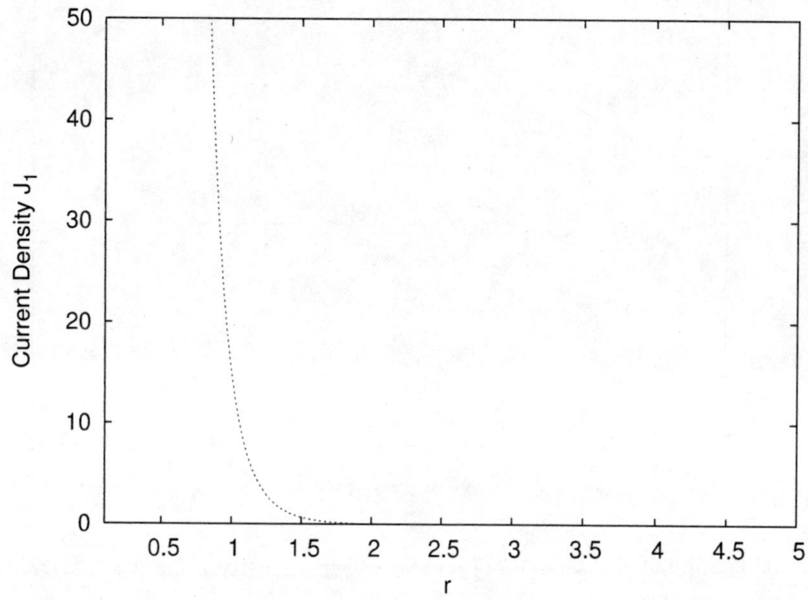

Fig. 4.20: Extended Reissner-Weyl metric, current density J_r for A=1, B=2, C=1.

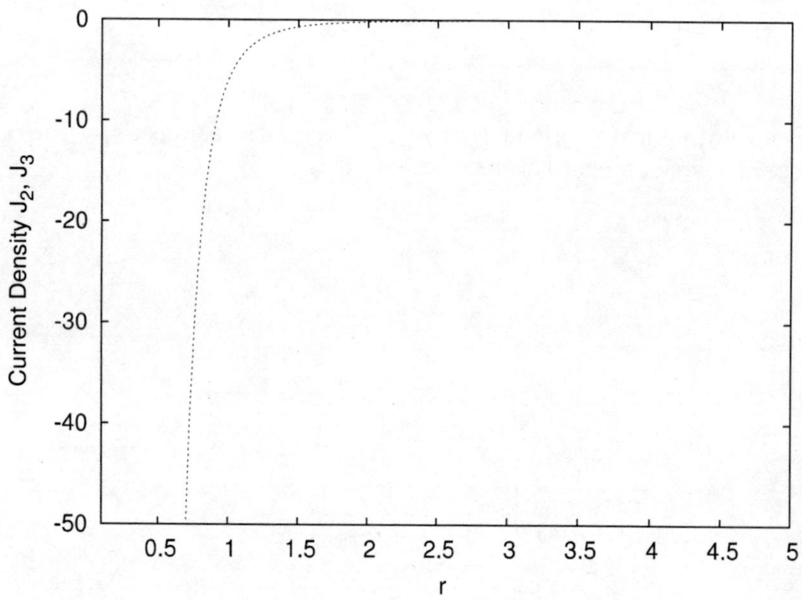

Fig. 4.21: Extended Reissner-Weyl metric, current density J_θ, J_φ for A=1, B=2, C=1.

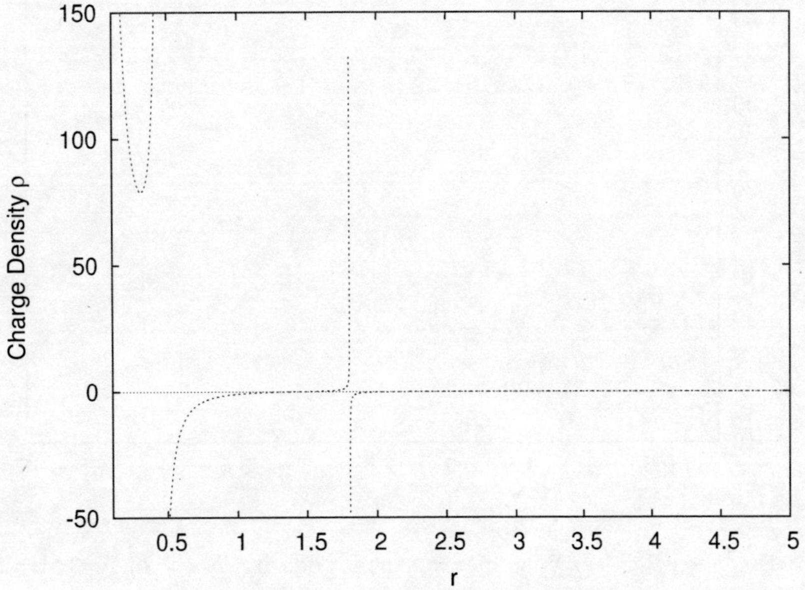

Fig. 4.22: Extended Reissner-Weyl metric, charge density ρ for A=1, B=2, C=-1.

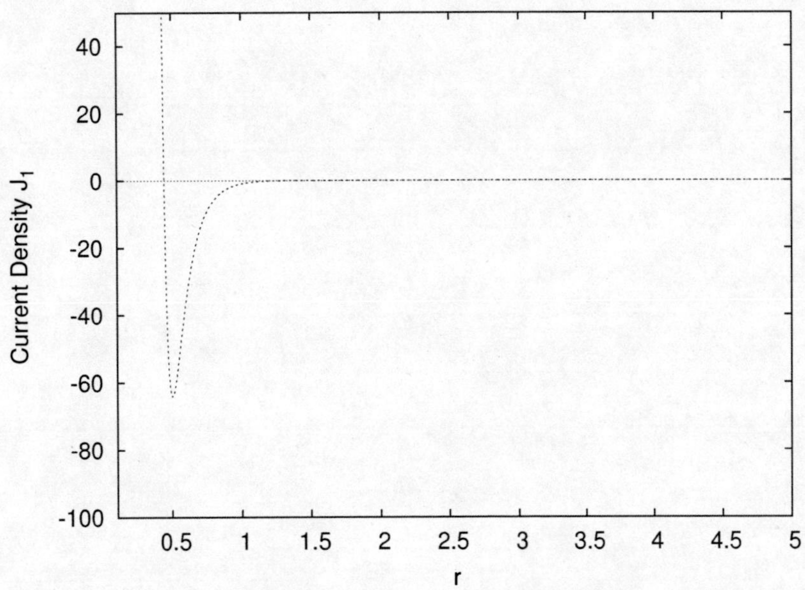

Fig. 4.23: Extended Reissner-Weyl metric, current density J_r for A=1, B=2, C=-1.

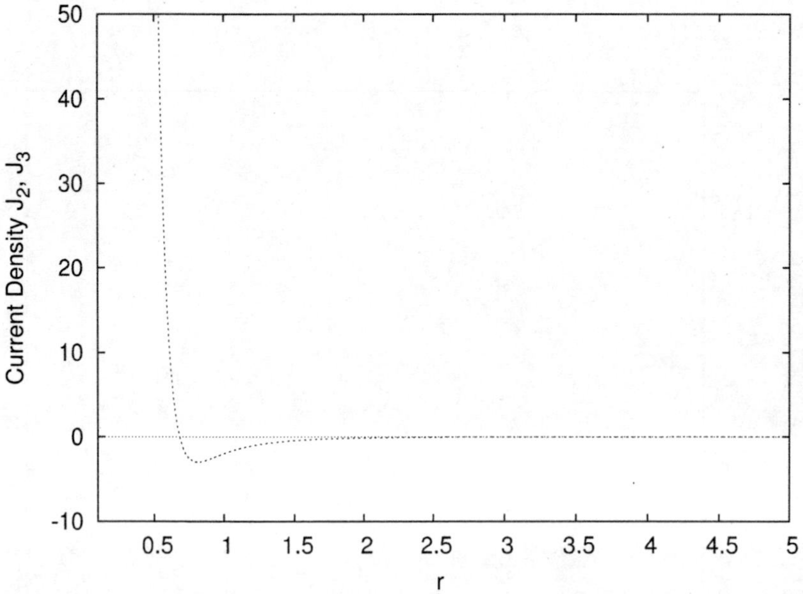

Fig. 4.24: Extended Reissner-Weyl metric, current density J_θ, J_φ for A=1, B=2, C=-1.

Christoffel Connection

$$\Gamma^0{}_{01} = -\frac{r^3\,M - 8\,\sin^2\vartheta\,J^2}{r\,(2\,r^3\,M - 16\,\sin^2\vartheta\,J^2 - r^4)}$$

$$\Gamma^0{}_{02} = -\frac{16\,\cos\vartheta\,\sin\vartheta\,J^2}{2\,r^3\,M - 16\,\sin^2\vartheta\,J^2 - r^4}$$

$$\Gamma^0{}_{10} = \Gamma^0{}_{01}$$

$$\Gamma^0{}_{13} = \frac{6\,r^2\,\sin^2\vartheta\,J}{2\,r^3\,M - 16\,\sin^2\vartheta\,J^2 - r^4}$$

$$\Gamma^0{}_{20} = \Gamma^0{}_{02}$$

$$\Gamma^0{}_{31} = \Gamma^0{}_{13}$$

$$\Gamma^1{}_{00} = -\frac{M\,(2\,M - r)}{r^3}$$

$$\Gamma^1{}_{03} = \frac{2\,\sin^2\vartheta\,J\,(2\,M - r)}{r^3}$$

$$\Gamma^1{}_{11} = \frac{M}{r\,(2\,M - r)}$$

$$\Gamma^1{}_{22} = 2\,M - r$$

$$\Gamma^1{}_{30} = \Gamma^1{}_{03}$$

$$\Gamma^1{}_{33} = \sin^2\vartheta\,(2\,M - r)$$

$$\Gamma^2{}_{03} = \frac{4\cos\vartheta\sin\vartheta\, J}{r^3}$$

$$\Gamma^2{}_{12} = \frac{1}{r}$$

$$\Gamma^2{}_{21} = \Gamma^2{}_{12}$$

$$\Gamma^2{}_{30} = \Gamma^2{}_{03}$$

$$\Gamma^2{}_{33} = -\cos\vartheta\sin\vartheta$$

$$\Gamma^3{}_{01} = -\frac{2J}{2r^3M - 16\sin^2\vartheta\, J^2 - r^4}$$

$$\Gamma^3{}_{02} = -\frac{4\cos\vartheta\, J\,(2M-r)}{\sin\vartheta\,(2r^3M - 16\sin^2\vartheta\, J^2 - r^4)}$$

$$\Gamma^3{}_{10} = \Gamma^3{}_{01}$$

$$\Gamma^3{}_{13} = \frac{2r^3M + 8\sin^2\vartheta\, J^2 - r^4}{r\,(2r^3M - 16\sin^2\vartheta\, J^2 - r^4)}$$

$$\Gamma^3{}_{20} = \Gamma^3{}_{02}$$

$$\Gamma^3{}_{23} = \frac{\cos\vartheta}{\sin\vartheta}$$

$$\Gamma^3{}_{31} = \Gamma^3{}_{13}$$

$$\Gamma^3{}_{32} = \Gamma^3{}_{23}$$

Metric Compatibility

——— o.k.

Riemann Tensor

$$R^0{}_{003} = \frac{4\sin^2\vartheta\, J\left(2r^3M^2 + 8\sin^2\vartheta\, J^2M - r^4M + 12r\sin^2\vartheta\, J^2 - 16rJ^2\right)}{r^4\,(2r^3M - 16\sin^2\vartheta\, J^2 - r^4)}$$

$$R^0{}_{012} = \frac{24r^2\cos\vartheta\sin\vartheta\, J^2\,(2M-r)}{(2r^3M - 16\sin^2\vartheta\, J^2 - r^4)^2}$$

$$R^0{}_{021} = -R^0{}_{012}$$

$$R^0{}_{030} = -R^0{}_{003}$$

$$R^0{}_{101} = -\frac{2\left(4r^6M^3 - 64r^3\sin^2\vartheta\, J^2M^2 - 4r^7M^2 + 256\sin^4\vartheta\, J^4M + 56r^4\sin^2\vartheta\, J^2M + r^8M - 96r\sin^4\vartheta\, J^4 - 14r^5\sin^2\vartheta\, J^2\right)}{r^2\,(2M-r)\,(2r^3M - 16\sin^2\vartheta\, J^2 - r^4)^2}$$

$$R^0{}_{102} = -\frac{16\cos\vartheta\sin\vartheta\, J^2\left(3r^3M - 24\sin^2\vartheta\, J^2 - 2r^4\right)}{r\,(2r^3M - 16\sin^2\vartheta\, J^2 - r^4)^2}$$

$$R^0{}_{110} = -R^0{}_{101}$$

$$R^0{}_{113} = -\frac{6\,r\,\sin^2\vartheta\,J\,\left(4\,r^3\,M^2 + 16\,\sin^2\vartheta\,J^2\,M - 4\,r^4\,M - 16\,r\,\sin^2\vartheta\,J^2 + r^5\right)}{(2\,M - r)\,(2\,r^3\,M - 16\,\sin^2\vartheta\,J^2 - r^4)^2}$$

$$R^0{}_{120} = -R^0{}_{102}$$

$$R^0{}_{123} = \frac{12\,r^2\,\cos\vartheta\,\sin\vartheta\,J\,\left(2\,r^3\,M - 8\,\sin^2\vartheta\,J^2 - r^4\right)}{(2\,r^3\,M - 16\,\sin^2\vartheta\,J^2 - r^4)^2}$$

$$R^0{}_{131} = -R^0{}_{113}$$

$$R^0{}_{132} = -R^0{}_{123}$$

$$R^0{}_{201} = -\frac{8\,\cos\vartheta\,\sin\vartheta\,J^2\,\left(12\,r^3\,M - 48\,\sin^2\vartheta\,J^2 - 7\,r^4\right)}{r\,(2\,r^3\,M - 16\,\sin^2\vartheta\,J^2 - r^4)^2}$$

$$R^0{}_{202} = -\frac{4\,r^6\,M^3 - 64\,r^3\,\sin^2\vartheta\,J^2\,M^2 - 4\,r^7\,M^2 + 256\,\sin^4\vartheta\,J^4\,M + 112\,r^4\,\sin^2\vartheta\,J^2\,M - 32\,r^4\,J^2\,M + r^8\,M - 384\,r\,\sin^4\vartheta\,J^4 - 40\,r^5\,\sin^2\vartheta\,J^2 + \ldots}{r\,\left(2\,r^3\,M - 16\,\sin^2\vartheta\,J^2 - r^4\right)^2}$$

$$R^0{}_{210} = -R^0{}_{201}$$

$$R^0{}_{213} = \frac{6\,r^2\,\cos\vartheta\,\sin\vartheta\,J}{2\,r^3\,M - 16\,\sin^2\vartheta\,J^2 - r^4}$$

$$R^0{}_{220} = -R^0{}_{202}$$

$$R^0{}_{223} = -\frac{6\,r^2\,\sin^2\vartheta\,J\,(2\,M - r)}{2\,r^3\,M - 16\,\sin^2\vartheta\,J^2 - r^4}$$

$$R^0{}_{231} = -R^0{}_{213}$$

$$R^0{}_{232} = -R^0{}_{223}$$

$$R^0{}_{303} = -\frac{\sin^2\vartheta\,(2\,r^3\,M^2 + 8\,\sin^2\vartheta\,J^2\,M - r^4\,M + 12\,r\,\sin^2\vartheta\,J^2 - 16\,r\,J^2)}{r\,(2\,r^3\,M - 16\,\sin^2\vartheta\,J^2 - r^4)}$$

$$R^0{}_{312} = -\frac{6\,r^5\,\cos\vartheta\,\sin\vartheta\,J\,(2\,M - r)}{(2\,r^3\,M - 16\,\sin^2\vartheta\,J^2 - r^4)^2}$$

$$R^0{}_{321} = -R^0{}_{312}$$

$$R^0{}_{330} = -R^0{}_{303}$$

$$R^1{}_{001} = -\frac{2\,\left(4\,r^3\,M^3 - 32\,\sin^2\vartheta\,J^2\,M^2 - 4\,r^4\,M^2 + 16\,r\,\sin^2\vartheta\,J^2\,M + r^5\,M - 2\,r^2\,\sin^2\vartheta\,J^2\right)}{r^4\,(2\,r^3\,M - 16\,\sin^2\vartheta\,J^2 - r^4)}$$

$$R^1{}_{002} = \frac{8\,\cos\vartheta\,\sin\vartheta\,J^2\,(2\,M - r)}{r^2\,(2\,r^3\,M - 16\,\sin^2\vartheta\,J^2 - r^4)}$$

$$R^1{}_{010} = -R^1{}_{001}$$

$$R^1{}_{013} = -\frac{2\,\sin^2\vartheta\,J\,\left(8\,r^3\,M^2 - 64\,\sin^2\vartheta\,J^2\,M - 10\,r^4\,M + 24\,r\,\sin^2\vartheta\,J^2 + 3\,r^5\right)}{r^4\,(2\,r^3\,M - 16\,\sin^2\vartheta\,J^2 - r^4)}$$

$$R^1{}_{020} = -R^1{}_{002}$$

$$R^1{}_{023} = \frac{12\cos\vartheta\,\sin\vartheta\,J\,(2M-r)\,\left(2r^3M - 8\sin^2\vartheta\,J^2 - r^4\right)}{r^3\,(2r^3M - 16\sin^2\vartheta\,J^2 - r^4)}$$

$$R^1{}_{031} = -R^1{}_{013}$$

$$R^1{}_{032} = -R^1{}_{023}$$

$$R^1{}_{203} = \frac{6\cos\vartheta\,\sin\vartheta\,J\,(2M-r)^2}{2r^3M - 16\sin^2\vartheta\,J^2 - r^4}$$

$$R^1{}_{212} = -\frac{M}{r}$$

$$R^1{}_{221} = -R^1{}_{212}$$

$$R^1{}_{230} = -R^1{}_{203}$$

$$R^1{}_{301} = \frac{2\sin^2\vartheta\,J\,\left(8r^3M^2 - 64\sin^2\vartheta\,J^2M - 10r^4M + 24r\sin^2\vartheta\,J^2 + 3r^5\right)}{r^4\,(2r^3M - 16\sin^2\vartheta\,J^2 - r^4)}$$

$$R^1{}_{302} = -\frac{6\cos\vartheta\,\sin\vartheta\,J\,(2M-r)}{r^3}$$

$$R^1{}_{310} = -R^1{}_{301}$$

$$R^1{}_{313} = -\frac{\sin^2\vartheta\,\left(2r^3M^2 + 56\sin^2\vartheta\,J^2M - r^4M - 36r\sin^2\vartheta\,J^2\right)}{r\,(2r^3M - 16\sin^2\vartheta\,J^2 - r^4)}$$

$$R^1{}_{320} = -R^1{}_{302}$$

$$R^1{}_{331} = -R^1{}_{313}$$

$$R^2{}_{001} = -\frac{8\cos\vartheta\,\sin\vartheta\,J^2}{r^3\,(2r^3M - 16\sin^2\vartheta\,J^2 - r^4)}$$

$$R^2{}_{002} = \frac{(2M-r)\,\left(2r^3M^2 - 16\sin^2\vartheta\,J^2M - r^4M + 16r\sin^2\vartheta\,J^2 - 16rJ^2\right)}{r^4\,(2r^3M - 16\sin^2\vartheta\,J^2 - r^4)}$$

$$R^2{}_{010} = -R^2{}_{001}$$

$$R^2{}_{013} = -\frac{6\cos\vartheta\,\sin\vartheta\,J}{r^4}$$

$$R^2{}_{020} = -R^2{}_{002}$$

$$R^2{}_{023} = \frac{2\sin^2\vartheta\,J\,(2M-3r)}{r^4}$$

$$R^2{}_{031} = -R^2{}_{013}$$

$$R^2{}_{032} = -R^2{}_{023}$$

$$R^2{}_{103} = \frac{6 \cos\vartheta \sin\vartheta\, J\, (2\,M - r)}{r\,(2\,r^3\,M - 16 \sin^2\vartheta\, J^2 - r^4)}$$

$$R^2{}_{112} = -\frac{M}{r^2\,(2\,M - r)}$$

$$R^2{}_{121} = -R^2{}_{112}$$

$$R^2{}_{130} = -R^2{}_{103}$$

$$R^2{}_{301} = \frac{12 \cos\vartheta \sin\vartheta\, J\, \left(2\,r^3\,M - 8 \sin^2\vartheta\, J^2 - r^4\right)}{r^4\,(2\,r^3\,M - 16 \sin^2\vartheta\, J^2 - r^4)}$$

$$R^2{}_{302} = -\frac{2 \sin^2\vartheta\, J\, (2\,M - 3\,r)}{r^4}$$

$$R^2{}_{310} = -R^2{}_{301}$$

$$R^2{}_{320} = -R^2{}_{302}$$

$$R^2{}_{323} = \frac{2 \sin^2\vartheta\, M}{r}$$

$$R^2{}_{332} = -R^2{}_{323}$$

$$R^3{}_{003} = \frac{(2\,M - r)\,\left(2\,r^3\,M^2 + 8 \sin^2\vartheta\, J^2\,M - r^4\,M + 12\,r \sin^2\vartheta\, J^2 - 16\,r\,J^2\right)}{r^4\,(2\,r^3\,M - 16 \sin^2\vartheta\, J^2 - r^4)}$$

$$R^3{}_{012} = \frac{6\,r^2 \cos\vartheta\, J\,(2\,M - r)^2}{\sin\vartheta\,(2\,r^3\,M - 16 \sin^2\vartheta\, J^2 - r^4)^2}$$

$$R^3{}_{021} = -R^3{}_{012}$$

$$R^3{}_{030} = -R^3{}_{003}$$

$$R^3{}_{101} = -\frac{2\,J\,\left(12\,r^3\,M^2 - 48 \sin^2\vartheta\, J^2\,M - 12\,r^4\,M + 16\,r \sin^2\vartheta\, J^2 + 3\,r^5\right)}{r\,(2\,M - r)\,(2\,r^3\,M - 16 \sin^2\vartheta\, J^2 - r^4)^2}$$

$$R^3{}_{102} = -\frac{2 \cos\vartheta\, J\,\left(12\,r^3\,M^2 - 96 \sin^2\vartheta\, J^2\,M - 12\,r^4\,M + 32\,r \sin^2\vartheta\, J^2 + 3\,r^5\right)}{r \sin\vartheta\,(2\,r^3\,M - 16 \sin^2\vartheta\, J^2 - r^4)^2}$$

$$R^3{}_{110} = -R^3{}_{101}$$

$$R^3{}_{113} = -\frac{4\,r^6\,M^3 + 176\,r^3 \sin^2\vartheta\, J^2\,M^2 - 4\,r^7\,M^2 - 512 \sin^4\vartheta\, J^4\,M - 208\,r^4 \sin^2\vartheta\, J^2\,M + r^8\,M + 192\,r \sin^4\vartheta\, J^4 + 60\,r^5 \sin^2\vartheta\, J^2}{r^2\,(2\,M - r)\,(2\,r^3\,M - 16 \sin^2\vartheta\, J^2 - r^4)^2}$$

$$R^3{}_{120} = -R^3{}_{102}$$

$$R^3{}_{123} = \frac{48 \cos\vartheta \sin\vartheta\, J^2\,\left(2\,r^3\,M - 8 \sin^2\vartheta\, J^2 - r^4\right)}{r\,(2\,r^3\,M - 16 \sin^2\vartheta\, J^2 - r^4)^2}$$

$$R^3{}_{131} = -R^3{}_{113}$$

$$R^3{}_{132} = -R^3{}_{123}$$

$$R^3{}_{201} = -\frac{4\cos\vartheta\, J\,\left(12\,r^3\,M^2 - 48\,\sin^2\vartheta\,J^2\,M - 12\,r^4\,M + 16\,r\,\sin^2\vartheta\,J^2 + 3\,r^5\right)}{r\,\sin\vartheta\,\left(2\,r^3\,M - 16\,\sin^2\vartheta\,J^2 - r^4\right)^2}$$

$$R^3{}_{202} = -\frac{2\,J\,(2\,M - r)\,\left(6\,r^3\,M - 16\,\sin^2\vartheta\,J^2 - 32\,J^2 - 3\,r^4\right)}{\left(2\,r^3\,M - 16\,\sin^2\vartheta\,J^2 - r^4\right)^2}$$

$$R^3{}_{210} = -R^3{}_{201}$$

$$R^3{}_{213} = \frac{24\cos\vartheta\,\sin\vartheta\,J^2}{r\,\left(2\,r^3\,M - 16\,\sin^2\vartheta\,J^2 - r^4\right)}$$

$$R^3{}_{220} = -R^3{}_{202}$$

$$R^3{}_{223} = -\frac{2\,\left(2\,r^3\,M^2 + 8\,\sin^2\vartheta\,J^2\,M - r^4\,M - 12\,r\,\sin^2\vartheta\,J^2\right)}{r\,\left(2\,r^3\,M - 16\,\sin^2\vartheta\,J^2 - r^4\right)}$$

$$R^3{}_{231} = -R^3{}_{213}$$

$$R^3{}_{232} = -R^3{}_{223}$$

$$R^3{}_{303} = -\frac{4\,\sin^2\vartheta\,J\,\left(2\,r^3\,M^2 + 8\,\sin^2\vartheta\,J^2\,M - r^4\,M + 12\,r\,\sin^2\vartheta\,J^2 - 16\,r\,J^2\right)}{r^4\,\left(2\,r^3\,M - 16\,\sin^2\vartheta\,J^2 - r^4\right)}$$

$$R^3{}_{312} = -\frac{24\,r^2\,\cos\vartheta\,\sin\vartheta\,J^2\,(2\,M - r)}{\left(2\,r^3\,M - 16\,\sin^2\vartheta\,J^2 - r^4\right)^2}$$

$$R^3{}_{321} = -R^3{}_{312}$$

$$R^3{}_{330} = -R^3{}_{303}$$

Ricci Tensor

$$\mathrm{Ric}_{00} = \frac{8\,J^2\,\left(6\,\cos^2\vartheta\,M^2 - 6\,M^2 + 4\,r\,\cos^2\vartheta\,M + 4\,r\,M - 3\,r^2\,\cos^2\vartheta - r^2\right)}{r^4\,\left(2\,r^3\,M - 16\,\sin^2\vartheta\,J^2 - r^4\right)}$$

$$\mathrm{Ric}_{03} = \frac{32\,\sin^2\vartheta\,J^3\,\left(3\,\sin^2\vartheta\,M + 3\,r\,\sin^2\vartheta - 2\,r\right)}{r^4\,\left(2\,r^3\,M - 16\,\sin^2\vartheta\,J^2 - r^4\right)}$$

$$\mathrm{Ric}_{11} = \frac{8\,\sin^2\vartheta\,J^2\,\left(30\,r^3\,M^2 - 96\,\sin^2\vartheta\,J^2\,M - 36\,r^4\,M + 48\,r\,\sin^2\vartheta\,J^2 + 11\,r^5\right)}{r^2\,(2\,M - r)\,\left(2\,r^3\,M - 16\,\sin^2\vartheta\,J^2 - r^4\right)^2}$$

$$\mathrm{Ric}_{12} = -\frac{16\cos\vartheta\,\sin\vartheta\,J^2\,\left(9\,r^3\,M - 48\,\sin^2\vartheta\,J^2 - 5\,r^4\right)}{r\,\left(2\,r^3\,M - 16\,\sin^2\vartheta\,J^2 - r^4\right)^2}$$

$$\mathrm{Ric}_{21} = \mathrm{Ric}_{12}$$

$$\mathrm{Ric}_{22} = \frac{16\,J^2\,\left(6\,r^3\,\sin^2\vartheta\,M^2 - 48\,\sin^4\vartheta\,J^2\,M - 11\,r^4\,\sin^2\vartheta\,M + 2\,r^4\,M + 48\,r\,\sin^4\vartheta\,J^2 + 4\,r^5\,\sin^2\vartheta - r^5\right)}{r\,\left(2\,r^3\,M - 16\,\sin^2\vartheta\,J^2 - r^4\right)^2}$$

$$\mathrm{Ric}_{30} = \mathrm{Ric}_{03}$$

$$\mathrm{Ric}_{33} = -\frac{8\,\sin^2\vartheta\,J^2\,\left(12\,\sin^2\vartheta\,M - 3\,r\,\sin^2\vartheta - 2\,r\right)}{r\,\left(2\,r^3\,M - 16\,\sin^2\vartheta\,J^2 - r^4\right)}$$

Ricci Scalar

$$R_{sc} = -\frac{8\,J^2\,\left(48\,r^3\,\sin^2\vartheta\,M^2 - 96\,\sin^4\vartheta\,J^2\,M - 28\,r^4\,\sin^2\vartheta\,M - 16\,r^4\,M - 144\,r\,\sin^4\vartheta\,J^2 + 64\,r\,\sin^2\vartheta\,J^2 + 3\,r^5\,\sin^2\vartheta + 8\,r^5\right)}{r^3\,\left(2\,r^3\,M - 16\,\sin^2\vartheta\,J^2 - r^4\right)^2}$$

Bianchi identity (Ricci cyclic equation $R^{\kappa}{}_{[\mu\nu\sigma]} = 0$)

——— o.k.

Einstein Tensor

$$G_{00} = \frac{4\,\sin^2\vartheta\,J^2\,\left(72\,r^2\,M^3 - 108\,r^3\,M^2 - 64\,\sin^2\vartheta\,J^2\,M - 128\,J^2\,M + 54\,r^4\,M + 48\,r\,\sin^2\vartheta\,J^2 + 64\,r\,J^2 - 9\,r^5\right)}{r^3\,\left(2\,r^3\,M - 16\,\sin^2\vartheta\,J^2 - r^4\right)^2}$$

$$G_{03} = -\frac{16\,\sin^2\vartheta\,J^3\,\left(36\,r^2\,\sin^2\vartheta\,M^2 - 34\,r^3\,\sin^2\vartheta\,M - 8\,r^3\,M - 48\,\sin^4\vartheta\,J^2 + 9\,r^4\,\sin^2\vartheta + 4\,r^4\right)}{r^3\,\left(2\,r^3\,M - 16\,\sin^2\vartheta\,J^2 - r^4\right)^2}$$

$$G_{11} = \frac{4\,J^2\,\left(12\,r^3\,\sin^2\vartheta\,M^2 - 96\,\sin^4\vartheta\,J^2\,M - 44\,r^4\,\sin^2\vartheta\,M + 16\,r^4\,M + 240\,r\,\sin^4\vartheta\,J^2 - 64\,r\,\sin^2\vartheta\,J^2 + 19\,r^5\,\sin^2\vartheta - 8\,r^5\right)}{r^2\,(2\,M - r)\,\left(2\,r^3\,M - 16\,\sin^2\vartheta\,J^2 - r^4\right)^2}$$

$$G_{12} = -\frac{16\,\cos\vartheta\,\sin\vartheta\,J^2\,\left(9\,r^3\,M - 48\,\sin^2\vartheta\,J^2 - 5\,r^4\right)}{r\,\left(2\,r^3\,M - 16\,\sin^2\vartheta\,J^2 - r^4\right)^2}$$

$$G_{21} = G_{12}$$

$$G_{22} = \frac{4\,J^2\,\left(72\,r^3\,\sin^2\vartheta\,M^2 - 288\,\sin^4\vartheta\,J^2\,M - 72\,r^4\,\sin^2\vartheta\,M - 8\,r^4\,M + 48\,r\,\sin^4\vartheta\,J^2 + 64\,r\,\sin^2\vartheta\,J^2 + 19\,r^5\,\sin^2\vartheta + 4\,r^5\right)}{r\,\left(2\,r^3\,M - 16\,\sin^2\vartheta\,J^2 - r^4\right)^2}$$

$$G_{30} = G_{03}$$

$$G_{33} = \frac{4\,\sin^2\vartheta\,J^2\,\left(288\,\sin^4\vartheta\,J^2\,M + 8\,r^4\,\sin^2\vartheta\,M - 8\,r^4\,M - 240\,r\,\sin^4\vartheta\,J^2 - 3\,r^5\,\sin^2\vartheta + 4\,r^5\right)}{r\,\left(2\,r^3\,M - 16\,\sin^2\vartheta\,J^2 - r^4\right)^2}$$

Hodge Dual of Bianchi Identity

——— (see charge and current densities)

Scalar Charge Density (-$R^{0}{}_{i}{}^{i0}$)

$$\rho = \frac{8\,r\,J^2\,\left(6\,r^3\,\cos^2\vartheta\,M^2 - 6\,r^3\,M^2 - 96\,\cos^4\vartheta\,J^2\,M + 192\,\cos^2\vartheta\,J^2\,M - 96\,J^2\,M + 4\,r^4\,\cos^2\vartheta\,M + 4\,r^4\,M + 144\,r\,\cos^4\vartheta\,J^2 - \ldots\right)}{\left(2\,r^3\,M - 16\,\sin^2\vartheta\,J^2 - r^4\right)^3}$$

Current Density Class 1 (-$R^{i}{}_{\mu}{}^{\mu j}$)

$$J_1 = -\frac{8\,\sin^2\vartheta\,J^2\,(2\,M - r)\,\left(30\,r^3\,M^2 - 96\,\sin^2\vartheta\,J^2\,M - 36\,r^4\,M + 48\,r\,\sin^2\vartheta\,J^2 + 11\,r^5\right)}{r^4\,\left(2\,r^3\,M - 16\,\sin^2\vartheta\,J^2 - r^4\right)^2}$$

$$J_2 = -\frac{16\,J^2\,\left(6\,r^3\,\sin^2\vartheta\,M^2 - 48\,\sin^4\vartheta\,J^2\,M - 11\,r^4\,\sin^2\vartheta\,M + 2\,r^4\,M + 48\,r\,\sin^4\vartheta\,J^2 + 4\,r^5\,\sin^2\vartheta - r^5\right)}{r^5\,\left(2\,r^3\,M - 16\,\sin^2\vartheta\,J^2 - r^4\right)^2}$$

$$J_3 = \frac{8\,J^2\,\left(48\,r^3\,\sin^2\vartheta\,M^3 - 96\,\sin^4\vartheta\,J^2\,M^2 - 60\,r^4\,\sin^2\vartheta\,M^2 - 8\,r^4\,M^2 - 32\,r\,\sin^4\vartheta\,J^2\,M + 24\,r^5\,\sin^2\vartheta\,M + 8\,r^5\,M + 48\,r^2\,\sin^4\vartheta\,J^2 - \ldots\right)}{r^2\,\sin^2\vartheta\,\left(2\,r^3\,M - 16\,\sin^2\vartheta\,J^2 - r^4\right)^3}$$

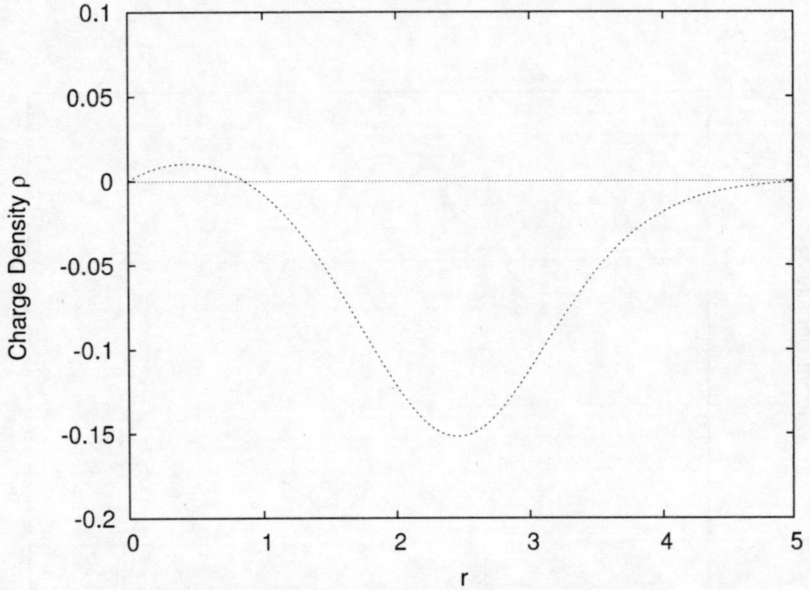

Fig. 4.25: Kerr metric, cosmological charge density ρ for M=1, J=2.

Current Density Class 2 (-$R^{i}{}_{\mu}{}^{\mu j}$)

$$J_1 = 0$$

$$J_2 = -\frac{16 \cos\vartheta \sin\vartheta\, J^2\, (2\,M - r)\, (9\,r^3\,M - 48\,\sin^2\vartheta\, J^2 - 5\,r^4)}{r^4\,(2\,r^3\,M - 16\,\sin^2\vartheta\, J^2 - r^4)^2}$$

$$J_3 = 0$$

Current Density Class 3 (-$R^{i}{}_{\mu}{}^{\mu j}$)

$$J_1 = -\frac{16 \cos\vartheta \sin\vartheta\, J^2\, (2\,M - r)\, (9\,r^3\,M - 48\,\sin^2\vartheta\, J^2 - 5\,r^4)}{r^4\,(2\,r^3\,M - 16\,\sin^2\vartheta\, J^2 - r^4)^2}$$

$$J_2 = 0$$

$$J_3 = 0$$

4.4.13 Kerr-Newman (Charged Kerr metric) with $M = 0, \rho = const.$

Metric of a charged mass with rotation. The quantities of this metric could only be calculated by assuming the following functions to be constant in the Maxima code:

$$\rho = \sqrt{r^2 + a^2\cos^2\theta} \approx const.$$
$$\Delta = r^2 - 2Mr + a^2 + Q^2$$

and further assuming

$$M \approx 0.$$

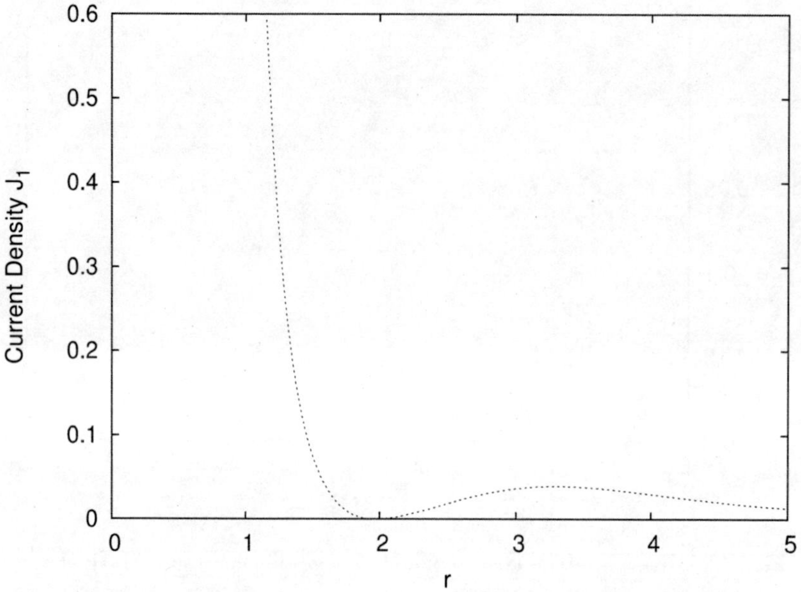

Fig. 4.26: Kerr metric, cosmological current density J_r for M=1, J=2.

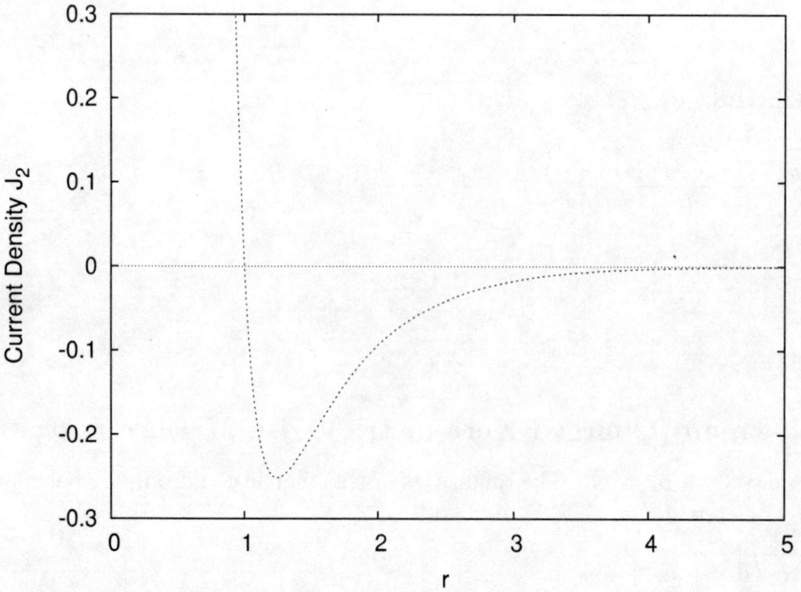

Fig. 4.27: Kerr metric, cosmological current density J_θ for M=1, J=2.

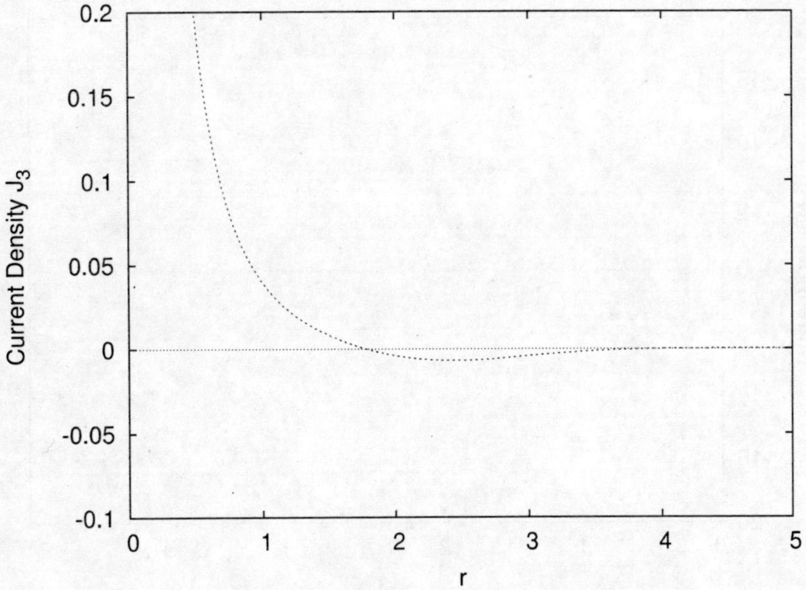

Fig. 4.28: Kerr metric, cosmological current density J_φ for M=1, J=2.

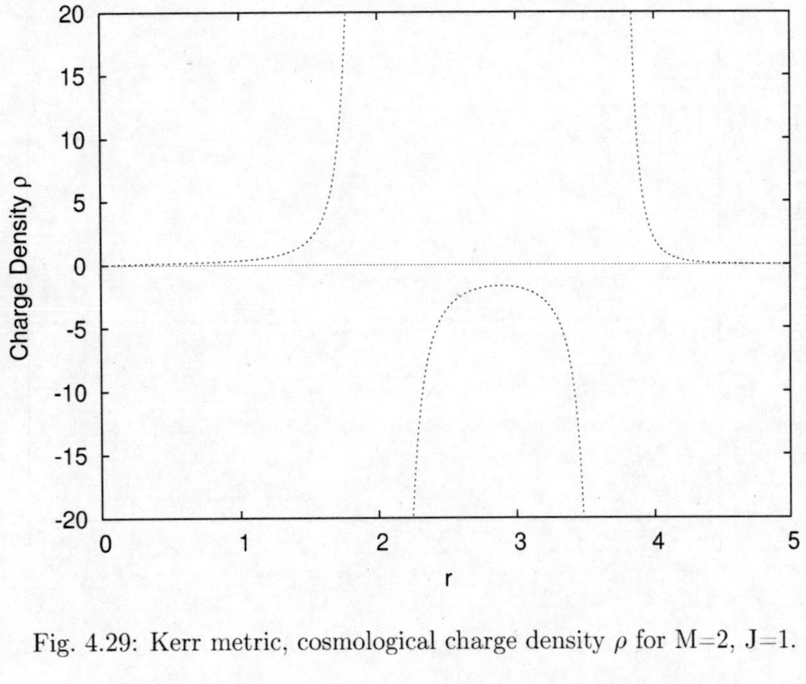

Fig. 4.29: Kerr metric, cosmological charge density ρ for M=2, J=1.

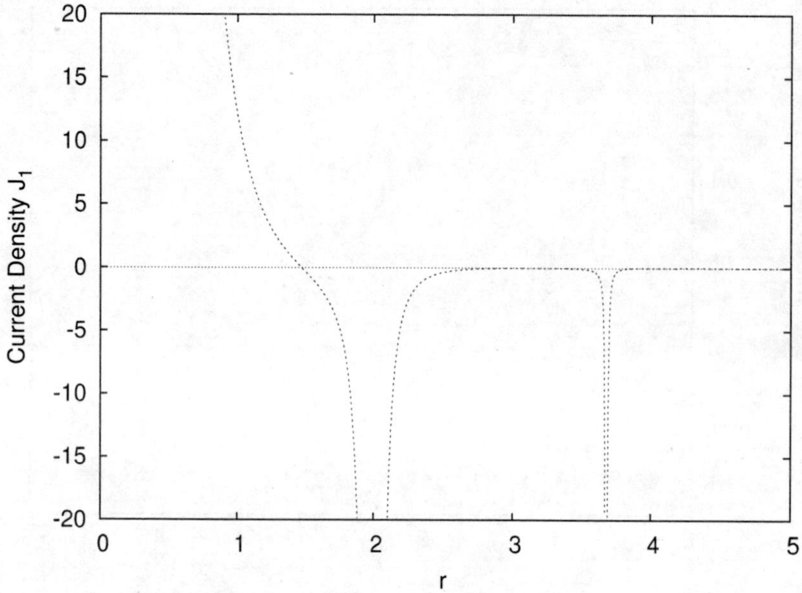

Fig. 4.30: Kerr metric, cosmological current density J_r for M=2, J=1.

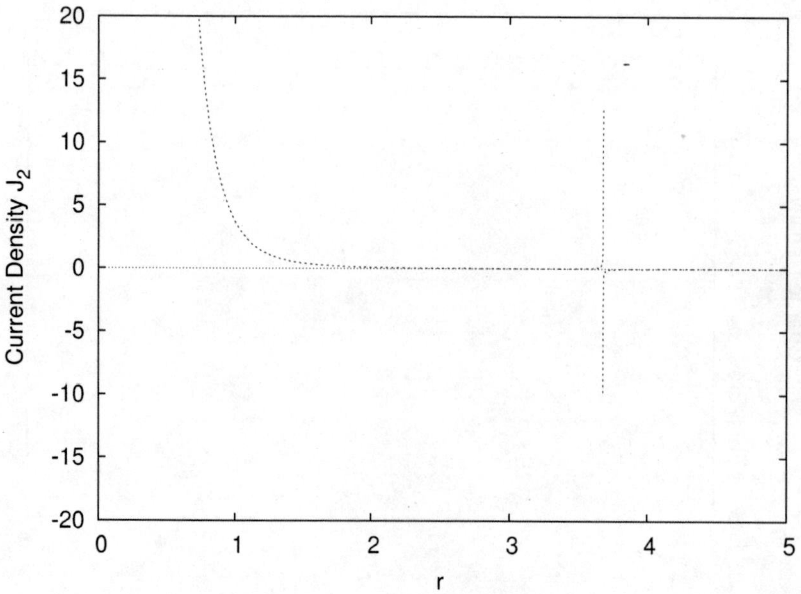

Fig. 4.31: Kerr metric, cosmological current density J_θ for M=2, J=1.

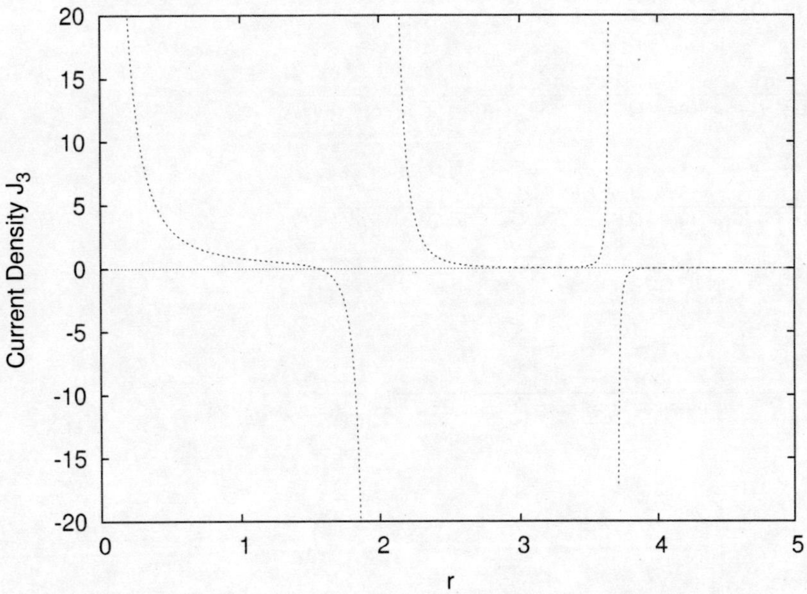

Fig. 4.32: Kerr metric, cosmological current density J_φ for M=2, J=1.

These expressions have to be inserted into the metric. a, M, and Q are parameters:

$$\begin{pmatrix} -(1-(2Mr-Q^2)/\rho^2) & 0 & 0 & -((4Mr-2Q^2)a\sin(\theta)^2/\rho^2) \\ 0 & \rho^2/\Delta & 0 & 0 \\ 0 & 0 & rho^2 & 0 \\ -((4Mr-2Q^2)a\sin(\theta)^2/\rho^2) & 0 & 0 & (r^2+a^2+(2Mr-Q^2)a^2\sin(\theta)^2/\rho^2)\sin(\theta)^2 \end{pmatrix}$$

Results are extremely complicated. Even charge densities of class 2 and 3 exist.

Coordinates

$$\mathbf{x} = \begin{pmatrix} t \\ r \\ \vartheta \\ \varphi \end{pmatrix}$$

Metric

$$g_{\mu\nu} = \begin{pmatrix} -\frac{Q^2}{\rho^2} - 1 & 0 & 0 & \frac{2\,a\,\sin^2\vartheta\,Q^2}{\rho^2} \\ 0 & \frac{\rho^2}{\Delta} & 0 & 0 \\ 0 & 0 & \rho^2 & 0 \\ \frac{2\,a\,\sin^2\vartheta\,Q^2}{\rho^2} & 0 & 0 & \sin^2\vartheta\left(-\frac{a^2\,\sin^2\vartheta\,Q^2}{\rho^2} + r^2 + a^2\right) \end{pmatrix}$$

Contravariant Metric

$$d = 3\,a^2\,\sin^2\vartheta\,Q^4 - a^2\,\rho^2\,\sin^2\vartheta\,Q^2 + r^2\,\rho^2\,Q^2 + a^2\,\rho^2\,Q^2 + r^2\,\rho^4 + a^2\,\rho^4$$

$$g^{\mu\nu} = \begin{pmatrix} \frac{\rho^2\left(a^2\,\sin^2\vartheta\,Q^2 - r^2\,\rho^2 - a^2\,\rho^2\right)}{d} & 0 & 0 & \frac{2\,a\,\rho^2\,Q^2}{d} \\ 0 & \frac{\Delta}{\rho^2} & 0 & 0 \\ 0 & 0 & \frac{1}{\rho^2} & 0 \\ \frac{2\,a\,\rho^2\,Q^2}{d} & 0 & 0 & \frac{\rho^2\left(Q^2+\rho^2\right)}{\sin^2\vartheta\,d} \end{pmatrix}$$

Christoffel Connection

$$\Gamma^0{}_{02} = \frac{4\,a^2\,\cos\vartheta\,\sin\vartheta\,Q^4}{3\,a^2\,\sin^2\vartheta\,Q^4 - a^2\,\rho^2\,\sin^2\vartheta\,Q^2 + r^2\,\rho^2\,Q^2 + a^2\,\rho^2\,Q^2 + r^2\,\rho^4 + a^2\,\rho^4}$$

$$\Gamma^0{}_{13} = \frac{2\,a\,r\,\rho^2\,\sin^2\vartheta\,Q^2}{3\,a^2\,\sin^2\vartheta\,Q^4 - a^2\,\rho^2\,\sin^2\vartheta\,Q^2 + r^2\,\rho^2\,Q^2 + a^2\,\rho^2\,Q^2 + r^2\,\rho^4 + a^2\,\rho^4}$$

$$\Gamma^0{}_{20} = \Gamma^0{}_{02}$$

$$\Gamma^0{}_{23} = -\frac{2\,a^3\,\cos\vartheta\,\sin^3\vartheta\,Q^4}{3\,a^2\,\sin^2\vartheta\,Q^4 - a^2\,\rho^2\,\sin^2\vartheta\,Q^2 + r^2\,\rho^2\,Q^2 + a^2\,\rho^2\,Q^2 + r^2\,\rho^4 + a^2\,\rho^4}$$

$$\Gamma^0{}_{31} = \Gamma^0{}_{13}$$

$$\Gamma^0{}_{32} = \Gamma^0{}_{23}$$

$$\Gamma^1{}_{33} = -\frac{\Delta\,r\,\sin^2\vartheta}{\rho^2}$$

$$\Gamma^2{}_{03} = -\frac{2\,a\,\cos\vartheta\,\sin\vartheta\,Q^2}{\rho^4}$$

$$\Gamma^2{}_{30} = \Gamma^2{}_{03}$$

$$\Gamma^2{}_{33} = \frac{\cos\vartheta\,\sin\vartheta\,\left(2\,a^2\,\sin^2\vartheta\,Q^2 - r^2\,\rho^2 - a^2\,\rho^2\right)}{\rho^4}$$

$$\Gamma^3{}_{02} = -\frac{2\,a\,\cos\vartheta\,Q^2\,\left(Q^2+\rho^2\right)}{\sin\vartheta\,\left(3\,a^2\,\cos^2\vartheta\,Q^4 - 3\,a^2\,Q^4 - a^2\,\rho^2\,\cos^2\vartheta\,Q^2 - r^2\,\rho^2\,Q^2 - r^2\,\rho^4 - a^2\,\rho^4\right)}$$

$$\Gamma^3{}_{13} = -\frac{r\,\rho^2\,\left(Q^2+\rho^2\right)}{3\,a^2\,\cos^2\vartheta\,Q^4 - 3\,a^2\,Q^4 - a^2\,\rho^2\,\cos^2\vartheta\,Q^2 - r^2\,\rho^2\,Q^2 - r^2\,\rho^4 - a^2\,\rho^4}$$

$$\Gamma^3{}_{20} = \Gamma^3{}_{02}$$

$$\Gamma^3{}_{23} = \frac{\cos\vartheta\,\left(2\,a^2\,\sin^2\vartheta\,Q^4 - 2\,a^2\,\rho^2\,\sin^2\vartheta\,Q^2 + r^2\,\rho^2\,Q^2 + a^2\,\rho^2\,Q^2 + r^2\,\rho^4 + a^2\,\rho^4\right)}{\sin\vartheta\,\left(3\,a^2\,\sin^2\vartheta\,Q^4 - a^2\,\rho^2\,\sin^2\vartheta\,Q^2 + r^2\,\rho^2\,Q^2 + a^2\,\rho^2\,Q^2 + r^2\,\rho^4 + a^2\,\rho^4\right)}$$

$$\Gamma^3{}_{31} = \Gamma^3{}_{13}$$

$$\Gamma^3{}_{32} = \Gamma^3{}_{23}$$

Metric Compatibility

——— o.k.

Riemann Tensor

$$R^{0}{}_{003} = \frac{8\,a^3\,\cos^2\vartheta\,\sin^2\vartheta\,Q^6}{\rho^4\,(3\,a^2\,\cos^2\vartheta\,Q^4 - 3\,a^2\,Q^4 - a^2\,\rho^2\,\cos^2\vartheta\,Q^2 - r^2\,\rho^2\,Q^2 - r^2\,\rho^4 - a^2\,\rho^4)}$$

$$R^{0}{}_{012} = -\frac{4\,a^2\,r\,\rho^2\,\cos\vartheta\,\sin\vartheta\,Q^4\,(Q^2+\rho^2)}{(3\,a^2\,\sin^2\vartheta\,Q^4 - a^2\,\rho^2\,\sin^2\vartheta\,Q^2 + r^2\,\rho^2\,Q^2 + a^2\,\rho^2\,Q^2 + r^2\,\rho^4 + a^2\,\rho^4)^2}$$

$$R^{0}{}_{021} = -R^{0}{}_{012}$$

$$R^{0}{}_{030} = -R^{0}{}_{003}$$

$$R^{0}{}_{113} = \frac{2\,a^3\,\rho^2\,\sin^2\vartheta\,Q^2\,(3\,\sin^2\vartheta\,Q^4 - \rho^2\,\sin^2\vartheta\,Q^2 + \rho^2\,Q^2 + \rho^4)}{(3\,a^2\,\sin^2\vartheta\,Q^4 - a^2\,\rho^2\,\sin^2\vartheta\,Q^2 + r^2\,\rho^2\,Q^2 + a^2\,\rho^2\,Q^2 + r^2\,\rho^4 + a^2\,\rho^4)^2}$$

$$R^{0}{}_{123} = \frac{2\,a\,r\,\rho^2\,\cos\vartheta\,\sin\vartheta\,Q^2\,(3\,a^2\,\sin^2\vartheta\,Q^4 - a^2\,\rho^2\,\sin^2\vartheta\,Q^2 + 2\,r^2\,\rho^2\,Q^2 + 2\,a^2\,\rho^2\,Q^2 + 2\,r^2\,\rho^4 + 2\,a^2\,\rho^4)}{(3\,a^2\,\sin^2\vartheta\,Q^4 - a^2\,\rho^2\,\sin^2\vartheta\,Q^2 + r^2\,\rho^2\,Q^2 + a^2\,\rho^2\,Q^2 + r^2\,\rho^4 + a^2\,\rho^4)^2}$$

$$R^{0}{}_{131} = -R^{0}{}_{113}$$

$$R^{0}{}_{132} = -R^{0}{}_{123}$$

$$R^{0}{}_{201} = \frac{4\,a^2\,r\,\rho^2\,\cos\vartheta\,\sin\vartheta\,Q^4\,(Q^2+\rho^2)}{(3\,a^2\,\sin^2\vartheta\,Q^4 - a^2\,\rho^2\,\sin^2\vartheta\,Q^2 + r^2\,\rho^2\,Q^2 + a^2\,\rho^2\,Q^2 + r^2\,\rho^4 + a^2\,\rho^4)^2}$$

$$R^{0}{}_{202} = \frac{4\,a^2\,Q^4\,\left(3\,a^2\,\sin^4\vartheta\,Q^4 - a^2\,\rho^2\,\sin^4\vartheta\,Q^2 + 2\,r^2\,\rho^2\,\sin^2\vartheta\,Q^2 + 2\,a^2\,\rho^2\,\sin^2\vartheta\,Q^2 - r^2\,\rho^2\,Q^2 - a^2\,\rho^2\,Q^2 + 2\,r^2\,\rho^4\,\sin^2\vartheta + 2\,a^2\,\rho^4\,\sin^2\vartheta - \ldots\right)}{\left(3\,a^2\,\sin^2\vartheta\,Q^4 - a^2\,\rho^2\,\sin^2\vartheta\,Q^2 + r^2\,\rho^2\,Q^2 + a^2\,\rho^2\,Q^2 + r^2\,\rho^4 + a^2\,\rho^4\right)^2}$$

$$R^{0}{}_{210} = -R^{0}{}_{201}$$

$$R^{0}{}_{213} = \frac{2\,a\,r\,\rho^2\,\cos\vartheta\,\sin\vartheta\,Q^2\,(4\,a^2\,\sin^2\vartheta\,Q^4 + r^2\,\rho^2\,Q^2 + a^2\,\rho^2\,Q^2 + r^2\,\rho^4 + a^2\,\rho^4)}{(3\,a^2\,\cos^2\vartheta\,Q^4 - 3\,a^2\,Q^4 - a^2\,\rho^2\,\cos^2\vartheta\,Q^2 - r^2\,\rho^2\,Q^2 - r^2\,\rho^4 - a^2\,\rho^4)^2}$$

$$R^{0}{}_{220} = -R^{0}{}_{202}$$

$$R^{0}{}_{223} = \frac{2\,a^3\,\sin^2\vartheta\,Q^4\,\left(12\,a^2\,\sin^4\vartheta\,Q^4 - 9\,a^2\,\sin^2\vartheta\,Q^4 - 4\,a^2\,\rho^2\,\sin^4\vartheta\,Q^2 + 5\,r^2\,\rho^2\,\sin^2\vartheta\,Q^2 + 8\,a^2\,\rho^2\,\sin^2\vartheta\,Q^2 - 4\,r^2\,\rho^2\,Q^2 - 4\,a^2\,\rho^2\,Q^2 + \ldots\right)}{\left(3\,a^2\,\cos^2\vartheta\,Q^4 - 3\,a^2\,Q^4 - a^2\,\rho^2\,\cos^2\vartheta\,Q^2 - r^2\,\rho^2\,Q^2 - r^2\,\rho^4 - a^2\,\rho^4\right)^2}$$

$$R^{0}{}_{231} = -R^{0}{}_{213}$$

$$R^{0}{}_{232} = -R^{0}{}_{223}$$

$$R^{0}{}_{303} = -\frac{4\,a^2\,\cos^2\vartheta\,\sin^2\vartheta\,Q^4\,(a^2\,\sin^2\vartheta\,Q^2 - r^2\,\rho^2 - a^2\,\rho^2)}{\rho^4\,(3\,a^2\,\cos^2\vartheta\,Q^4 - 3\,a^2\,Q^4 - a^2\,\rho^2\,\cos^2\vartheta\,Q^2 - r^2\,\rho^2\,Q^2 - r^2\,\rho^4 - a^2\,\rho^4)}$$

$$R^{0}{}_{312} = \frac{2\,a\,r\,\rho^2\,\cos\vartheta\,\sin\vartheta\,Q^2\,(Q^2+\rho^2)\,(a^2\,\sin^2\vartheta\,Q^2 - r^2\,\rho^2 - a^2\,\rho^2)}{(3\,a^2\,\sin^2\vartheta\,Q^4 - a^2\,\rho^2\,\sin^2\vartheta\,Q^2 + r^2\,\rho^2\,Q^2 + a^2\,\rho^2\,Q^2 + r^2\,\rho^4 + a^2\,\rho^4)^2}$$

$$R^0{}_{321} = -R^0{}_{312}$$

$$R^0{}_{330} = -R^0{}_{303}$$

$$R^1{}_{023} = \frac{2\,a\,\Delta\,r\,\cos\vartheta\,\sin\vartheta\,Q^2\,\left(Q^2+\rho^2\right)}{\rho^2\,\left(3\,a^2\,\sin^2\vartheta\,Q^4 - a^2\,\rho^2\,\sin^2\vartheta\,Q^2 + r^2\,\rho^2\,Q^2 + a^2\,\rho^2\,Q^2 + r^2\,\rho^4 + a^2\,\rho^4\right)}$$

$$R^1{}_{032} = -R^1{}_{023}$$

$$R^1{}_{203} = \frac{2\,a\,\Delta\,r\,\cos\vartheta\,\sin\vartheta\,Q^2\,\left(Q^2+\rho^2\right)}{\rho^2\,\left(3\,a^2\,\sin^2\vartheta\,Q^4 - a^2\,\rho^2\,\sin^2\vartheta\,Q^2 + r^2\,\rho^2\,Q^2 + a^2\,\rho^2\,Q^2 + r^2\,\rho^4 + a^2\,\rho^4\right)}$$

$$R^1{}_{230} = -R^1{}_{203}$$

$$R^1{}_{313} = -\frac{a^2\,\Delta\,\sin^2\vartheta\,\left(3\,\sin^2\vartheta\,Q^4 - \rho^2\,\sin^2\vartheta\,Q^2 + \rho^2\,Q^2 + \rho^4\right)}{\rho^2\,\left(3\,a^2\,\sin^2\vartheta\,Q^4 - a^2\,\rho^2\,\sin^2\vartheta\,Q^2 + r^2\,\rho^2\,Q^2 + a^2\,\rho^2\,Q^2 + r^2\,\rho^4 + a^2\,\rho^4\right)}$$

$$R^1{}_{323} = \frac{\Delta\,r\,\cos\vartheta\,\sin\vartheta\,\left(4\,a^2\,\sin^2\vartheta\,Q^4 + r^2\,\rho^2\,Q^2 + a^2\,\rho^2\,Q^2 + r^2\,\rho^4 + a^2\,\rho^4\right)}{\rho^2\,\left(3\,a^2\,\cos^2\vartheta\,Q^4 - 3\,a^2\,Q^4 - a^2\,\rho^2\,\cos^2\vartheta\,Q^2 - r^2\,\rho^2\,Q^2 - r^2\,\rho^4 - a^2\,\rho^4\right)}$$

$$R^1{}_{331} = -R^1{}_{313}$$

$$R^1{}_{332} = -R^1{}_{323}$$

$$R^2{}_{002} = -\frac{4\,a^2\,\cos^2\vartheta\,Q^4\,\left(Q^2+\rho^2\right)}{\rho^4\,\left(3\,a^2\,\sin^2\vartheta\,Q^4 - a^2\,\rho^2\,\sin^2\vartheta\,Q^2 + r^2\,\rho^2\,Q^2 + a^2\,\rho^2\,Q^2 + r^2\,\rho^4 + a^2\,\rho^4\right)}$$

$$R^2{}_{020} = -R^2{}_{002}$$

$$R^2{}_{023} = \frac{2\,a\,\sin^2\vartheta\,Q^2\,\left(4\,a^2\,\sin^2\vartheta\,Q^4 - a^2\,Q^4 + r^2\,\rho^2\,Q^2 + r^2\,\rho^4 + a^2\,\rho^4\right)}{\rho^4\,\left(3\,a^2\,\sin^2\vartheta\,Q^4 - a^2\,\rho^2\,\sin^2\vartheta\,Q^2 + r^2\,\rho^2\,Q^2 + a^2\,\rho^2\,Q^2 + r^2\,\rho^4 + a^2\,\rho^4\right)}$$

$$R^2{}_{032} = -R^2{}_{023}$$

$$R^2{}_{103} = -\frac{2\,a\,r\,\cos\vartheta\,\sin\vartheta\,Q^2\,\left(Q^2+\rho^2\right)}{\rho^2\,\left(3\,a^2\,\sin^2\vartheta\,Q^4 - a^2\,\rho^2\,\sin^2\vartheta\,Q^2 + r^2\,\rho^2\,Q^2 + a^2\,\rho^2\,Q^2 + r^2\,\rho^4 + a^2\,\rho^4\right)}$$

$$R^2{}_{130} = -R^2{}_{103}$$

$$R^2{}_{301} = -\frac{2\,a\,r\,\cos\vartheta\,\sin\vartheta\,Q^2\,\left(Q^2+\rho^2\right)}{\rho^2\,\left(3\,a^2\,\sin^2\vartheta\,Q^4 - a^2\,\rho^2\,\sin^2\vartheta\,Q^2 + r^2\,\rho^2\,Q^2 + a^2\,\rho^2\,Q^2 + r^2\,\rho^4 + a^2\,\rho^4\right)}$$

$$R^2{}_{302} = \frac{2\,a\,\sin^2\vartheta\,Q^2\,\left(4\,a^2\,\sin^2\vartheta\,Q^4 - a^2\,Q^4 + r^2\,\rho^2\,Q^2 + r^2\,\rho^4 + a^2\,\rho^4\right)}{\rho^4\,\left(3\,a^2\,\cos^2\vartheta\,Q^4 - 3\,a^2\,Q^4 - a^2\,\rho^2\,\cos^2\vartheta\,Q^2 - r^2\,\rho^2\,Q^2 - r^2\,\rho^4 - a^2\,\rho^4\right)}$$

$$R^2{}_{310} = -R^2{}_{301}$$

$$R^2{}_{313} = -\frac{r\,\cos\vartheta\,\sin\vartheta\,\left(4\,a^2\,\sin^2\vartheta\,Q^4 + r^2\,\rho^2\,Q^2 + a^2\,\rho^2\,Q^2 + r^2\,\rho^4 + a^2\,\rho^4\right)}{\rho^2\,\left(3\,a^2\,\sin^2\vartheta\,Q^4 - a^2\,\rho^2\,\sin^2\vartheta\,Q^2 + r^2\,\rho^2\,Q^2 + a^2\,\rho^2\,Q^2 + r^2\,\rho^4 + a^2\,\rho^4\right)}$$

$$R^2{}_{320} = -R^2{}_{302}$$

$$R^2{}_{323} = \frac{\sin^2\vartheta\,\left(16\,a^4\,\sin^4\vartheta\,Q^6 - 10\,a^4\,\sin^2\vartheta\,Q^6 - 4\,a^4\,\rho^2\,\sin^4\vartheta\,Q^4 + 2\,a^2\,r^2\,\rho^2\,\sin^2\vartheta\,Q^4 + 4\,a^4\,\rho^2\,\sin^2\vartheta\,Q^4 - 3\,a^2\,r^2\,\rho^2\,Q^4 - 3\,a^4\,\rho^2\,Q^4 + \ldots\right)}{\rho^4\,\left(3\,a^2\,\cos^2\vartheta\,Q^4 - 3\,a^2\,Q^4 - a^2\,\rho^2\,\cos^2\vartheta\,Q^2 - r^2\,\rho^2\,Q^2 - r^2\,\rho^4 - a^2\,\rho^4\right)}$$

$$R^2{}_{331} = -R^2{}_{313}$$

$$R^2{}_{332} = -R^2{}_{323}$$

$$R^3{}_{003} = -\frac{4\,a^2\,\cos^2\vartheta\,Q^4\,\left(Q^2+\rho^2\right)}{\rho^4\,\left(3\,a^2\,\sin^2\vartheta\,Q^4 - a^2\,\rho^2\,\sin^2\vartheta\,Q^2 + r^2\,\rho^2\,Q^2 + a^2\,\rho^2\,Q^2 + r^2\,\rho^4 + a^2\,\rho^4\right)}$$

$$R^3{}_{012} = -\frac{2\,a\,r\,\rho^2\,\cos\vartheta\,Q^2\,\left(Q^2+\rho^2\right)^2}{\sin\vartheta\,\left(3\,a^2\,\sin^2\vartheta\,Q^4 - a^2\,\rho^2\,\sin^2\vartheta\,Q^2 + r^2\,\rho^2\,Q^2 + a^2\,\rho^2\,Q^2 + r^2\,\rho^4 + a^2\,\rho^4\right)^2}$$

$$R^3{}_{021} = -R^3{}_{012}$$

$$R^3{}_{030} = -R^3{}_{003}$$

$$R^3{}_{113} = \frac{a^2\,\rho^2\,\left(Q^2+\rho^2\right)\,\left(3\,\sin^2\vartheta\,Q^4 - \rho^2\,\sin^2\vartheta\,Q^2 + \rho^2\,Q^2 + \rho^4\right)}{\left(3\,a^2\,\sin^2\vartheta\,Q^4 - a^2\,\rho^2\,\sin^2\vartheta\,Q^2 + r^2\,\rho^2\,Q^2 + a^2\,\rho^2\,Q^2 + r^2\,\rho^4 + a^2\,\rho^4\right)^2}$$

$$R^3{}_{123} = \frac{r\,\left(r^2+a^2\right)\,\rho^4\,\cos\vartheta\,\left(Q^2+\rho^2\right)^2}{\sin\vartheta\,\left(3\,a^2\,\sin^2\vartheta\,Q^4 - a^2\,\rho^2\,\sin^2\vartheta\,Q^2 + r^2\,\rho^2\,Q^2 + a^2\,\rho^2\,Q^2 + r^2\,\rho^4 + a^2\,\rho^4\right)^2}$$

$$R^3{}_{131} = -R^3{}_{113}$$

$$R^3{}_{132} = -R^3{}_{123}$$

$$R^3{}_{201} = \frac{2\,a\,r\,\rho^2\,\cos\vartheta\,Q^2\,\left(Q^2+\rho^2\right)^2}{\sin\vartheta\,\left(3\,a^2\,\sin^2\vartheta\,Q^4 - a^2\,\rho^2\,\sin^2\vartheta\,Q^2 + r^2\,\rho^2\,Q^2 + a^2\,\rho^2\,Q^2 + r^2\,\rho^4 + a^2\,\rho^4\right)^2}$$

$$R^3{}_{202} = -\frac{2\,a\,Q^2\,\left(Q^2+\rho^2\right)\,\left(3\,a^2\,Q^4 + r^2\,\rho^2\,Q^2 + r^2\,\rho^4 + a^2\,\rho^4\right)\,\left(3\,a^2\,\cos^2\vartheta\,Q^4 - 3\,a^2\,Q^4 - a^2\,\rho^2\,\cos^2\vartheta\,Q^2 - r^2\,\rho^2\,Q^2 - r^2\,\rho^4 - a^2\,\rho^4\right)}{\left(3\,a^2\,\sin^2\vartheta\,Q^4 - a^2\,\rho^2\,\sin^2\vartheta\,Q^2 + r^2\,\rho^2\,Q^2 + a^2\,\rho^2\,Q^2 + r^2\,\rho^4 + a^2\,\rho^4\right)^3}$$

$$R^3{}_{210} = -R^3{}_{201}$$

$$R^3{}_{213} = \frac{r\,\rho^2\,\cos\vartheta\,\left(Q^2+\rho^2\right)\,\left(4\,a^2\,\sin^2\vartheta\,Q^4 + r^2\,\rho^2\,Q^2 + a^2\,\rho^2\,Q^2 + r^2\,\rho^4 + a^2\,\rho^4\right)}{\sin\vartheta\,\left(3\,a^2\,\sin^2\vartheta\,Q^4 - a^2\,\rho^2\,\sin^2\vartheta\,Q^2 + r^2\,\rho^2\,Q^2 + a^2\,\rho^2\,Q^2 + r^2\,\rho^4 + a^2\,\rho^4\right)^2}$$

$$R^3{}_{220} = -R^3{}_{202}$$

$$R^3{}_{223} = -\frac{6\,a^4\,\sin^2\vartheta\,Q^8 - 12\,a^4\,\rho^2\,\sin^4\vartheta\,Q^6 + 2\,a^2\,r^2\,\rho^2\,\sin^2\vartheta\,Q^6 + 6\,a^4\,\rho^2\,\sin^2\vartheta\,Q^6 + 3\,a^2\,r^2\,\rho^2\,Q^6 + 3\,a^4\,\rho^2\,Q^6 + 4\,a^4\,\rho^4\,\sin^4\vartheta\,Q^4 - \ldots}{\left(3\,a^2\,\sin^2\vartheta\,Q^4 - a^2\,\rho^2\,\sin^2\vartheta\,Q^2 + r^2\,\rho^2\,Q^2 + a^2\,\rho^2\,Q^2 + r^2\,\rho^4 + a^2\,\rho^4\right)^2}$$

$$R^3{}_{231} = -R^3{}_{213}$$

$$R^3{}_{232} = -R^3{}_{223}$$

$$R^3{}_{303} = -\frac{8\,a^3\,\left(\sin\vartheta - 1\right)\,\sin^2\vartheta\,\left(\sin\vartheta + 1\right)\,Q^6}{\rho^4\,\left(3\,a^2\,\sin^2\vartheta\,Q^4 - a^2\,\rho^2\,\sin^2\vartheta\,Q^2 + r^2\,\rho^2\,Q^2 + a^2\,\rho^2\,Q^2 + r^2\,\rho^4 + a^2\,\rho^4\right)}$$

$$R^3{}_{312} = \frac{4\,a^2\,r\,\rho^2\,\cos\vartheta\,\sin\vartheta\,Q^4\,\left(Q^2+\rho^2\right)}{\left(3\,a^2\,\sin^2\vartheta\,Q^4 - a^2\,\rho^2\,\sin^2\vartheta\,Q^2 + r^2\,\rho^2\,Q^2 + a^2\,\rho^2\,Q^2 + r^2\,\rho^4 + a^2\,\rho^4\right)^2}$$

$$R^3{}_{321} = -R^3{}_{312}$$

$$R^3{}_{330} = -R^3{}_{303}$$

Ricci Tensor

$$\mathrm{Ric}_{00} = -\frac{8\,a^2\,\cos^2\vartheta\,Q^4\,\left(Q^2+\rho^2\right)}{\rho^4\,\left(3\,a^2\,\cos^2\vartheta\,Q^4-3\,a^2\,Q^4-a^2\,\rho^2\,\cos^2\vartheta\,Q^2-r^2\,\rho^2\,Q^2-r^2\,\rho^4-a^2\,\rho^4\right)}$$

$$\mathrm{Ric}_{03} = \frac{2\,a\,\sin^2\vartheta\,Q^2\,\left(8\,a^2\,\sin^2\vartheta\,Q^4-5\,a^2\,Q^4+r^2\,\rho^2\,Q^2+r^2\,\rho^4+a^2\,\rho^4\right)}{\rho^4\,\left(3\,a^2\,\sin^2\vartheta\,Q^4-a^2\,\rho^2\,\sin^2\vartheta\,Q^2+r^2\,\rho^2\,Q^2+a^2\,\rho^2\,Q^2+r^2\,\rho^4+a^2\,\rho^4\right)}$$

$$\mathrm{Ric}_{11} = \frac{a^2\,\rho^2\,\left(Q^2+\rho^2\right)\left(3\,\cos^2\vartheta\,Q^4-3\,Q^4-\rho^2\,\cos^2\vartheta\,Q^2-\rho^4\right)}{\left(3\,a^2\,\cos^2\vartheta\,Q^4-3\,a^2\,Q^4-a^2\,\rho^2\,\cos^2\vartheta\,Q^2-r^2\,\rho^2\,Q^2-r^2\,\rho^4-a^2\,\rho^4\right)^2}$$

$$\mathrm{Ric}_{12} = -\frac{r\,\left(r^2+a^2\right)\,\rho^4\,\cos\vartheta\,\left(Q^2+\rho^2\right)^2}{\sin\vartheta\,\left(3\,a^2\,\cos^2\vartheta\,Q^4-3\,a^2\,Q^4-a^2\,\rho^2\,\cos^2\vartheta\,Q^2-r^2\,\rho^2\,Q^2-r^2\,\rho^4-a^2\,\rho^4\right)^2}$$

$$\mathrm{Ric}_{21} = \mathrm{Ric}_{12}$$

$$\mathrm{Ric}_{22} = \frac{12\,a^4\,\cos^4\vartheta\,Q^8-30\,a^4\,\cos^2\vartheta\,Q^8+18\,a^4\,Q^8-16\,a^4\,\rho^2\,\cos^4\vartheta\,Q^6-10\,a^2\,r^2\,\rho^2\,\cos^2\vartheta\,Q^6+18\,a^4\,\rho^2\,\cos^2\vartheta\,Q^6+9\,a^2\,r^2\,\rho^2\,Q^6-\ldots}{\left(3\,a^2\,\cos^2\vartheta\,Q^4-3\,a^2\,Q^4-a^2\,\rho^2\,\cos^2\vartheta\,Q^2-r^2\,\rho^2\,Q^2-r^2\,\rho^4-a^2\,\rho^4\right)^2}$$

$$\mathrm{Ric}_{30} = \mathrm{Ric}_{03}$$

$$\mathrm{Ric}_{33} = -\frac{\sin^2\vartheta\,\left(20\,a^4\,\sin^4\vartheta\,Q^6-14\,a^4\,\sin^2\vartheta\,Q^6-4\,a^4\,\rho^2\,\sin^4\vartheta\,Q^4-2\,a^2\,r^2\,\rho^2\,\sin^2\vartheta\,Q^4+3\,a^2\,\Delta\,\rho^2\,\sin^2\vartheta\,Q^4+a^2\,r^2\,\rho^2\,Q^4+a^4\,\rho^2\,Q^4+\ldots\right)}{\rho^4\,\left(3\,a^2\,\sin^2\vartheta\,Q^4-a^2\,\rho^2\,\sin^2\vartheta\,Q^2+r^2\,\rho^2\,Q^2+a^2\,\rho^2\,Q^2+r^2\,\rho^4+a^2\,\rho^4\right)}$$

Ricci Scalar

$$R_{sc} = \frac{2\,\left(24\,a^4\,\sin^4\vartheta\,Q^8-6\,a^4\,\sin^2\vartheta\,Q^8-20\,a^4\,\rho^2\,\sin^4\vartheta\,Q^6+14\,a^2\,r^2\,\rho^2\,\sin^2\vartheta\,Q^6-3\,a^2\,\Delta\,\rho^2\,\sin^2\vartheta\,Q^6+22\,a^4\,\rho^2\,\sin^2\vartheta\,Q^6-5\,a^2\,r^2\,\rho^2\,Q^6-\ldots\right)}{\rho^2\,\left(3\,a^2\,\sin^2\vartheta\,Q^4-a^2\,\rho^2\,\sin^2\vartheta\,Q^2+r^2\,\rho^2\,Q^2+a^2\,\rho^2\,Q^2+r^2\,\rho^4+a^2\,\rho^4\right)^2}$$

Bianchi identity (Ricci cyclic equation $R^{\kappa}{}_{[\mu\nu\sigma]} = 0$)

———— o.k.

Einstein Tensor

$$G_{00} = \frac{\left(Q^2+\rho^2\right)\left(18\,a^4\,\sin^2\vartheta\,Q^8-12\,a^4\,\rho^2\,\sin^4\vartheta\,Q^6+6\,a^2\,r^2\,\rho^2\,\sin^2\vartheta\,Q^6-3\,a^2\,\Delta\,\rho^2\,\sin^2\vartheta\,Q^6+6\,a^4\,\rho^2\,\sin^2\vartheta\,Q^6+3\,a^2\,r^2\,\rho^2\,Q^6+3\,a^4\,\rho^2\,Q^6+\ldots\right)}{\rho^4\,\left(3\,a^2\,\sin^2\vartheta\,Q^4-a^2\,\rho^2\,\sin^2\vartheta\,Q^2+r^2\,\rho^2\,Q^2+a^2\,\rho^2\,Q^2+r^2\,\rho^4+a^2\,\rho^4\right)^2}$$

$$G_{03} = -\frac{2\,a^3\,\sin^2\vartheta\,Q^2\,\left(9\,a^2\,\sin^2\vartheta\,Q^8-12\,a^2\,\rho^2\,\sin^4\vartheta\,Q^6+3\,r^2\,\rho^2\,\sin^2\vartheta\,Q^6-3\,\Delta\,\rho^2\,\sin^2\vartheta\,Q^6+9\,a^2\,\rho^2\,\sin^2\vartheta\,Q^6+4\,a^2\,\rho^4\,\sin^4\vartheta\,Q^4-\ldots\right)}{\rho^4\,\left(3\,a^2\,\cos^2\vartheta\,Q^4-3\,a^2\,Q^4-a^2\,\rho^2\,\cos^2\vartheta\,Q^2-r^2\,\rho^2\,Q^2-r^2\,\rho^4-a^2\,\rho^4\right)^2}$$

$$G_{11} = -\frac{24\,a^4\,\sin^4\vartheta\,Q^8-6\,a^4\,\sin^2\vartheta\,Q^8-20\,a^4\,\rho^2\,\sin^4\vartheta\,Q^6+14\,a^2\,r^2\,\rho^2\,\sin^2\vartheta\,Q^6+22\,a^4\,\rho^2\,\sin^2\vartheta\,Q^6-5\,a^2\,r^2\,\rho^2\,Q^6-5\,a^4\,\rho^2\,Q^6+\ldots}{\Delta\,\left(3\,a^2\,\sin^2\vartheta\,Q^4-a^2\,\rho^2\,\sin^2\vartheta\,Q^2+r^2\,\rho^2\,Q^2+a^2\,\rho^2\,Q^2+r^2\,\rho^4+a^2\,\rho^4\right)^2}$$

$$G_{12} = -\frac{r\,\left(r^2+a^2\right)\,\rho^4\,\cos\vartheta\,\left(Q^2+\rho^2\right)^2}{\sin\vartheta\,\left(3\,a^2\,\sin^2\vartheta\,Q^4-a^2\,\rho^2\,\sin^2\vartheta\,Q^2+r^2\,\rho^2\,Q^2+a^2\,\rho^2\,Q^2+r^2\,\rho^4+a^2\,\rho^4\right)^2}$$

$$G_{21} = G_{12}$$

$$G_{22,} = -\frac{a^2\,\left(12\,a^2\,\sin^4\vartheta\,Q^8-12\,a^2\,\sin^2\vartheta\,Q^8-4\,a^2\,\rho^2\,\sin^4\vartheta\,Q^6+4\,r^2\,\rho^2\,\sin^2\vartheta\,Q^6-3\,\Delta\,\rho^2\,\sin^2\vartheta\,Q^6+8\,a^2\,\rho^2\,\sin^2\vartheta\,Q^6-4\,r^2\,\rho^2\,Q^6-\ldots\right)}{\left(3\,a^2\,\sin^2\vartheta\,Q^4-a^2\,\rho^2\,\sin^2\vartheta\,Q^2+r^2\,\rho^2\,Q^2+a^2\,\rho^2\,Q^2+r^2\,\rho^4+a^2\,\rho^4\right)^2}$$

$$G_{30} = G_{03}$$

$$G_{33} = -\frac{4\,a^2\,\sin^2\vartheta\,Q^4\,\left(9\,a^4\,\sin^6\vartheta\,Q^6-9\,a^4\,\sin^4\vartheta\,Q^6-3\,a^4\,\rho^2\,\sin^6\vartheta\,Q^4+6\,a^2\,r^2\,\rho^2\,\sin^4\vartheta\,Q^4+3\,a^2\,\Delta\,\rho^2\,\sin^4\vartheta\,Q^4+9\,a^4\,\rho^2\,\sin^4\vartheta\,Q^4-\ldots\right)}{\rho^4\,\left(3\,a^2\,\cos^2\vartheta\,Q^4-3\,a^2\,Q^4-a^2\,\rho^2\,\cos^2\vartheta\,Q^2-r^2\,\rho^2\,Q^2-r^2\,\rho^4-a^2\,\rho^4\right)^2}$$

Hodge Dual of Bianchi Identity

——— (see charge and current densities)

Scalar Charge Density ($-R^{0\ \ i0}_{\ \ i}$)

$$\rho = -\frac{4\,a^2\,Q^4\,\left(6\,a^4\,\sin^6\vartheta\,Q^6 - 6\,a^4\,\sin^4\vartheta\,Q^6 - 2\,a^4\,\rho^2\,\sin^6\vartheta\,Q^4 + 8\,a^2\,r^2\,\rho^2\,\sin^4\vartheta\,Q^4 + 3\,a^2\,\Delta\,\rho^2\,\sin^4\vartheta\,Q^4 + 10\,a^4\,\rho^2\,\sin^4\vartheta\,Q^4 - ...\right)}{\left(3\,a^2\,\sin^2\vartheta\,Q^4 - a^2\,\rho^2\,\sin^2\vartheta\,Q^2 + r^2\,\rho^2\,Q^2 + a^2\,\rho^2\,Q^2 + r^2\,\rho^4 + a^2\,\rho^4\right)^3}$$

Current Density Class 1 ($-R^{i\ \ \mu j}_{\ \ \mu}$)

$$J_1 = -\frac{a^2\,\Delta^2\,\left(Q^2+\rho^2\right)\,\left(3\,\cos^2\vartheta\,Q^4 - 3\,Q^4 - \rho^2\,\cos^2\vartheta\,Q^2 - \rho^4\right)}{\rho^2\,\left(3\,a^2\,\cos^2\vartheta\,Q^4 - 3\,a^2\,Q^4 - a^2\,\rho^2\,\cos^2\vartheta\,Q^2 - r^2\,\rho^2\,Q^2 - r^2\,\rho^4 - a^2\,\rho^4\right)^2}$$

$$J_2 = -\frac{12\,a^4\,\sin^4\vartheta\,Q^8 + 6\,a^4\,\sin^2\vartheta\,Q^8 - 16\,a^4\,\rho^2\,\sin^4\vartheta\,Q^6 + 10\,a^2\,r^2\,\rho^2\,\sin^2\vartheta\,Q^6 + 14\,a^4\,\rho^2\,\sin^2\vartheta\,Q^6 - a^2\,r^2\,\rho^2\,Q^6 - a^4\,\rho^2\,Q^6 + ...}{\rho^4\,\left(3\,a^2\,\sin^2\vartheta\,Q^4 - a^2\,\rho^2\,\sin^2\vartheta\,Q^2 + r^2\,\rho^2\,Q^2 + a^2\,\rho^2\,Q^2 + r^2\,\rho^4 + a^2\,\rho^4\right)^2}$$

$$J_3 = -\frac{\left(Q^2+\rho^2\right)\,\left(12\,a^4\,\sin^4\vartheta\,Q^8 + 6\,a^4\,\sin^2\vartheta\,Q^8 - 16\,a^4\,\rho^2\,\sin^4\vartheta\,Q^6 + 10\,a^2\,r^2\,\rho^2\,\sin^2\vartheta\,Q^6 - 3\,a^2\,\Delta\,\rho^2\,\sin^2\vartheta\,Q^6 + 14\,a^4\,\rho^2\,\sin^2\vartheta\,Q^6 - ...\right)}{\sin^2\vartheta\,\left(3\,a^2\,\sin^2\vartheta\,Q^4 - a^2\,\rho^2\,\sin^2\vartheta\,Q^2 + r^2\,\rho^2\,Q^2 + a^2\,\rho^2\,Q^2 + r^2\,\rho^4 + a^2\,\rho^4\right)^3}$$

Current Density Class 2 ($-R^{i\ \ \mu j}_{\ \ \mu}$)

$$J_1 = 0$$

$$J_2 = \frac{\Delta\,r\,\left(r^2+a^2\right)\,\cos\vartheta\,\left(Q^2+\rho^2\right)^2}{\sin\vartheta\,\left(3\,a^2\,\sin^2\vartheta\,Q^4 - a^2\,\rho^2\,\sin^2\vartheta\,Q^2 + r^2\,\rho^2\,Q^2 + a^2\,\rho^2\,Q^2 + r^2\,\rho^4 + a^2\,\rho^4\right)^2}$$

$$J_3 = 0$$

Current Density Class 3 ($-R^{i\ \ \mu j}_{\ \ \mu}$)

$$J_1 = \frac{\Delta\,r\,\left(r^2+a^2\right)\,\cos\vartheta\,\left(Q^2+\rho^2\right)^2}{\sin\vartheta\,\left(3\,a^2\,\sin^2\vartheta\,Q^4 - a^2\,\rho^2\,\sin^2\vartheta\,Q^2 + r^2\,\rho^2\,Q^2 + a^2\,\rho^2\,Q^2 + r^2\,\rho^4 + a^2\,\rho^4\right)^2}$$

$$J_2 = 0$$

$$J_3 = 0$$

4.4.14 Kerr-Newman (Charged Kerr metric) with $a = 0$

In this approximation the non-diagonal terms vanish due to $a = 0$.

Coordinates

$$\mathbf{x} = \begin{pmatrix} t \\ r \\ \vartheta \\ \varphi \end{pmatrix}$$

Metric

$$g_{\mu\nu} = \begin{pmatrix} \frac{2\,r\,M-Q^2}{r^2} - 1 & 0 & 0 & 0 \\ 0 & \frac{r^2}{Q^2-2\,r\,M+r^2} & 0 & 0 \\ 0 & 0 & r^2 & 0 \\ 0 & 0 & 0 & r^2\,\sin^2\vartheta \end{pmatrix}$$

Contravariant Metric

$$g^{\mu\nu} = \begin{pmatrix} -\frac{r^2}{Q^2-2\,r\,M+r^2} & 0 & 0 & 0 \\ 0 & \frac{Q^2-2\,r\,M+r^2}{r^2} & 0 & 0 \\ 0 & 0 & \frac{1}{r^2} & 0 \\ 0 & 0 & 0 & \frac{1}{r^2\,\sin^2\vartheta} \end{pmatrix}$$

Christoffel Connection

$$\Gamma^0{}_{01} = -\frac{Q^2 - r\,M}{r\,(Q^2 - 2\,r\,M + r^2)}$$

$$\Gamma^0{}_{10} = \Gamma^0{}_{01}$$

$$\Gamma^1{}_{00} = -\frac{\left(Q^2 - 2\,r\,M + r^2\right)\,\left(Q^2 - r\,M\right)}{r^5}$$

$$\Gamma^1{}_{11} = \frac{Q^2 - r\,M}{r\,(Q^2 - 2\,r\,M + r^2)}$$

$$\Gamma^1{}_{22} = -\frac{Q^2 - 2\,r\,M + r^2}{r}$$

$$\Gamma^1{}_{33} = -\frac{\sin^2\vartheta\,\left(Q^2 - 2\,r\,M + r^2\right)}{r}$$

$$\Gamma^2{}_{12} = \frac{1}{r}$$

$$\Gamma^2{}_{21} = \Gamma^2{}_{12}$$

$$\Gamma^2{}_{33} = -\cos\vartheta\,\sin\vartheta$$

$$\Gamma^3{}_{13} = \frac{1}{r}$$

$$\Gamma^3{}_{23} = \frac{\cos\vartheta}{\sin\vartheta}$$

$$\Gamma^3{}_{31} = \Gamma^3{}_{13}$$

$$\Gamma^3{}_{32} = \Gamma^3{}_{23}$$

Metric Compatibility

——— o.k.

Riemann Tensor

$$R^0{}_{101} = -\frac{3\,Q^2 - 2\,r\,M}{r^2\,(Q^2 - 2\,r\,M + r^2)}$$

$$R^0{}_{110} = -R^0{}_{101}$$

$$R^0{}_{202} = \frac{Q^2 - r\,M}{r^2}$$

$$R^0{}_{220} = -R^0{}_{202}$$

$$R^0{}_{303} = \frac{\sin^2\vartheta\,\left(Q^2 - r\,M\right)}{r^2}$$

$$R^0{}_{330} = -R^0{}_{303}$$

$$R^1{}_{001} = -\frac{\left(Q^2 - 2\,r\,M + r^2\right)\,\left(3\,Q^2 - 2\,r\,M\right)}{r^6}$$

$$R^1{}_{010} = -R^1{}_{001}$$

$$R^1{}_{212} = \frac{Q^2 - r\,M}{r^2}$$

$$R^1{}_{221} = -R^1{}_{212}$$

$$R^1{}_{313} = \frac{\sin^2\vartheta\,\left(Q^2 - r\,M\right)}{r^2}$$

$$R^1{}_{331} = -R^1{}_{313}$$

$$R^2{}_{002} = \frac{\left(Q^2 - 2\,r\,M + r^2\right)\,\left(Q^2 - r\,M\right)}{r^6}$$

$$R^2{}_{020} = -R^2{}_{002}$$

$$R^2{}_{112} = -\frac{Q^2 - r\,M}{r^2\,(Q^2 - 2\,r\,M + r^2)}$$

$$R^2{}_{121} = -R^2{}_{112}$$

$$R^2{}_{323} = -\frac{\sin^2\vartheta\,(Q^2 - 2\,r\,M)}{r^2}$$

$$R^2{}_{332} = -R^2{}_{323}$$

$$R^3{}_{003} = \frac{(Q^2 - 2\,r\,M + r^2)\,(Q^2 - r\,M)}{r^6}$$

$$R^3{}_{030} = -R^3{}_{003}$$

$$R^3{}_{113} = -\frac{Q^2 - r\,M}{r^2\,(Q^2 - 2\,r\,M + r^2)}$$

$$R^3{}_{131} = -R^3{}_{113}$$

$$R^3{}_{223} = \frac{Q^2 - 2\,r\,M}{r^2}$$

$$R^3{}_{232} = -R^3{}_{223}$$

Ricci Tensor

$$\mathrm{Ric}_{00} = \frac{Q^2\,(Q^2 - 2\,r\,M + r^2)}{r^6}$$

$$\mathrm{Ric}_{11} = -\frac{Q^2}{r^2\,(Q^2 - 2\,r\,M + r^2)}$$

$$\mathrm{Ric}_{22} = \frac{Q^2}{r^2}$$

$$\mathrm{Ric}_{33} = \frac{\sin^2\vartheta\,Q^2}{r^2}$$

Ricci Scalar

$$R_{sc} = 0$$

Bianchi identity (Ricci cyclic equation $R^{\kappa}{}_{[\mu\nu\sigma]} = 0$)

——— o.k.

Einstein Tensor

$$G_{00} = \frac{Q^2\left(Q^2 - 2\,r\,M + r^2\right)}{r^6}$$

$$G_{11} = -\frac{Q^2}{r^2\left(Q^2 - 2\,r\,M + r^2\right)}$$

$$G_{22} = \frac{Q^2}{r^2}$$

$$G_{33} = \frac{\sin^2\vartheta\,Q^2}{r^2}$$

Hodge Dual of Bianchi Identity

——— (see charge and current densities)

Scalar Charge Density ($-R^{0\ \,i0}_{\ \,i}$)

$$\rho = \frac{Q^2}{r^2\left(Q^2 - 2\,r\,M + r^2\right)}$$

Current Density Class 1 ($-R^{i\ \,\mu j}_{\ \,\mu}$)

$$J_1 = \frac{Q^2\left(Q^2 - 2\,r\,M + r^2\right)}{r^6}$$

$$J_2 = -\frac{Q^2}{r^6}$$

$$J_3 = -\frac{Q^2}{r^6\,\sin^2\vartheta}$$

Current Density Class 2 ($-R^{i\ \,\mu j}_{\ \,\mu}$)

$$J_1 = 0$$

$$J_2 = 0$$

$$J_3 = 0$$

Current Density Class 3 ($-R^{i\ \,\mu j}_{\ \,\mu}$)

$$J_1 = 0$$

$$J_2 = 0$$

$$J_3 = 0$$

4.4.15 Goedel metric

This is the Goedel metric. ω is a parameter.

Coordinates

$$\mathbf{x} = \begin{pmatrix} t \\ x_1 \\ x_2 \\ x_3 \end{pmatrix}$$

Metric

$$g_{\mu\nu} = \begin{pmatrix} -\frac{1}{2\,\omega^2} & 0 & 0 & 2\,e^{x_1} \\ 0 & \frac{1}{2\,\omega^2} & 0 & 0 \\ 0 & 0 & \frac{1}{2\,\omega^2} & 0 \\ 2\,e^{x_1} & 0 & 0 & -\frac{e^{2\,x_1}}{4\,\omega^2} \end{pmatrix}$$

Contravariant Metric

$$g^{\mu\nu} = \begin{pmatrix} \frac{2\,\omega^2}{32\,\omega^4-1} & 0 & 0 & \frac{16\,\omega^4\,e^{-x_1}}{32\,\omega^4-1} \\ 0 & 2\,\omega^2 & 0 & 0 \\ 0 & 0 & 2\,\omega^2 & 0 \\ \frac{16\,\omega^4\,e^{-x_1}}{32\,\omega^4-1} & 0 & 0 & \frac{4\,\omega^2\,e^{-2\,x_1}}{32\,\omega^4-1} \end{pmatrix}$$

Christoffel Connection

$$\Gamma^0{}_{01} = \frac{16\,\omega^4}{32\,\omega^4 - 1}$$

$$\Gamma^0{}_{10} = \Gamma^0{}_{01}$$

$$\Gamma^0{}_{13} = -\frac{2\,\omega^2\,e^{x_1}}{32\,\omega^4 - 1}$$

$$\Gamma^0{}_{31} = \Gamma^0{}_{13}$$

$$\Gamma^1{}_{03} = -2\,\omega^2\,e^{x_1}$$

$$\Gamma^1{}_{30} = \Gamma^1{}_{03}$$

$$\Gamma^1{}_{33} = \frac{e^{2\,x_1}}{2}$$

$$\Gamma^3{}_{01} = \frac{4\,\omega^2\,e^{-x_1}}{32\,\omega^4 - 1}$$

$$\Gamma^3{}_{10} = \Gamma^3{}_{01}$$

$$\Gamma^3{}_{13} = \frac{(2\,\omega - 1)\,(2\,\omega + 1)\,(4\,\omega^2 + 1)}{32\,\omega^4 - 1}$$

$$\Gamma^3{}_{31} = \Gamma^3{}_{13}$$

Metric Compatibility

——— o.k.

Riemann Tensor

$$R^0{}_{003} = -\frac{32\,\omega^6\,e^{x_I}}{32\,\omega^4 - 1}$$

$$R^0{}_{030} = -R^0{}_{003}$$

$$R^0{}_{101} = -\frac{8\,\omega^4}{32\,\omega^4 - 1}$$

$$R^0{}_{110} = -R^0{}_{101}$$

$$R^0{}_{113} = -\frac{4\,\omega^2\,e^{x_I}}{32\,\omega^4 - 1}$$

$$R^0{}_{131} = -R^0{}_{113}$$

$$R^0{}_{303} = \frac{4\,\omega^4\,e^{2\,x_I}}{32\,\omega^4 - 1}$$

$$R^0{}_{330} = -R^0{}_{303}$$

$$R^1{}_{001} = -\frac{8\,\omega^4}{32\,\omega^4 - 1}$$

$$R^1{}_{010} = -R^1{}_{001}$$

$$R^1{}_{013} = -\frac{32\,\omega^6\,e^{x_I}}{32\,\omega^4 - 1}$$

$$R^1{}_{031} = -R^1{}_{013}$$

$$R^1{}_{301} = \frac{32\,\omega^6\,e^{x_I}}{32\,\omega^4 - 1}$$

$$R^1{}_{310} = -R^1{}_{301}$$

$$R^1{}_{313} = \frac{\left(40\,\omega^4 - 1\right)\,e^{2\,x_I}}{2\,\left(32\,\omega^4 - 1\right)}$$

$$R^1{}_{331} = -R^1{}_{313}$$

$$R^3{}_{003} = -\frac{8\,\omega^4}{32\,\omega^4 - 1}$$

$$R^3{}_{030} = -R^3{}_{003}$$

$$R^3{}_{113} = \frac{8\,\omega^4 - 1}{32\,\omega^4 - 1}$$

$$R^3{}_{131} = -R^3{}_{113}$$

$$R^3{}_{303} = \frac{32\,\omega^6\,e^{x_1}}{32\,\omega^4 - 1}$$

$$R^3{}_{330} = -R^3{}_{303}$$

Ricci Tensor

$$\mathrm{Ric}_{00} = \frac{16\,\omega^4}{32\,\omega^4 - 1}$$

$$\mathrm{Ric}_{03} = -\frac{64\,\omega^6\,e^{x_1}}{32\,\omega^4 - 1}$$

$$\mathrm{Ric}_{11} = -\frac{(2\,\omega - 1)\,(2\,\omega + 1)\,\left(4\,\omega^2 + 1\right)}{32\,\omega^4 - 1}$$

$$\mathrm{Ric}_{30} = \mathrm{Ric}_{03}$$

$$\mathrm{Ric}_{33} = \frac{\left(48\,\omega^4 - 1\right)\,e^{2\,x_1}}{2\,\left(32\,\omega^4 - 1\right)}$$

Ricci Scalar

$$R_{sc} = -\frac{4\,\omega^2\,\left(24\,\omega^4 - 1\right)}{32\,\omega^4 - 1}$$

Bianchi identity (Ricci cyclic equation $R^{\kappa}{}_{[\mu\nu\sigma]} = 0$)

——— o.k.

Einstein Tensor

$$G_{00} = -\frac{8\,\omega^4 - 1}{32\,\omega^4 - 1}$$

$$G_{03} = \frac{4\,\omega^2\,(8\,\omega^4 - 1)\,e^{x_1}}{32\,\omega^4 - 1}$$

$$G_{11} = \frac{8\,\omega^4}{32\,\omega^4 - 1}$$

$$G_{22} = \frac{24\,\omega^4 - 1}{32\,\omega^4 - 1}$$

$$G_{30} = G_{03}$$

$$G_{33} = \frac{12\,\omega^4\,e^{2\,x_1}}{32\,\omega^4 - 1}$$

Hodge Dual of Bianchi Identity

——— (see charge and current densities)

Scalar Charge Density (-$R^{0\ \ i0}_{\ \ i}$)

$$\rho = \frac{64\,\omega^8}{(32\,\omega^4 - 1)^2}$$

Current Density Class 1 (-$R^{i\ \ \mu j}_{\ \ \mu}$)

$$J_1 = \frac{4\,\omega^4\,(2\,\omega - 1)\,(2\,\omega + 1)\,(4\,\omega^2 + 1)}{32\,\omega^4 - 1}$$

$$J_2 = 0$$

$$J_3 = \frac{8\,\omega^4\,(2\,\omega - 1)\,(2\,\omega + 1)\,(4\,\omega^2 + 1)\,e^{-2\,x_1}}{(32\,\omega^4 - 1)^2}$$

Current Density Class 2 (-$R^{i\ \ \mu j}_{\ \ \mu}$)

$$J_1 = 0$$

$$J_2 = 0$$

$$J_3 = 0$$

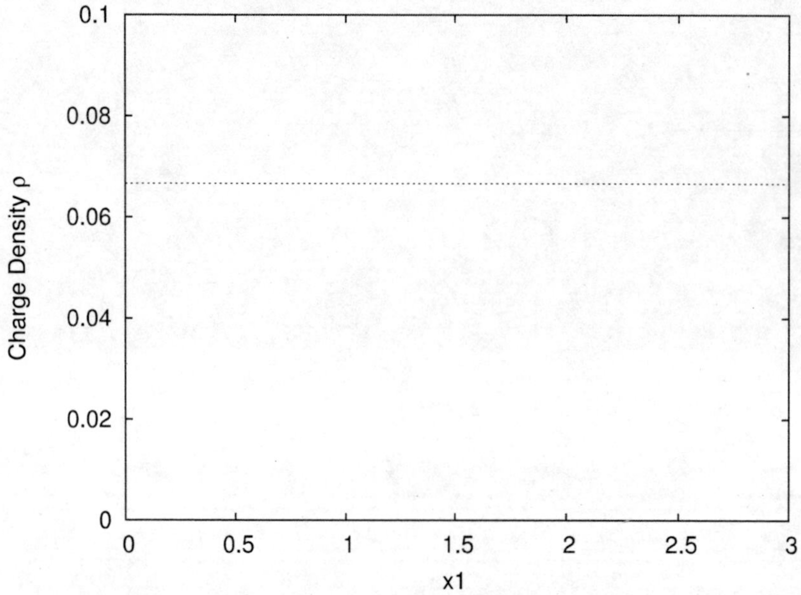

Fig. 4.33: Goedel Metric, charge density ρ for ω=1.

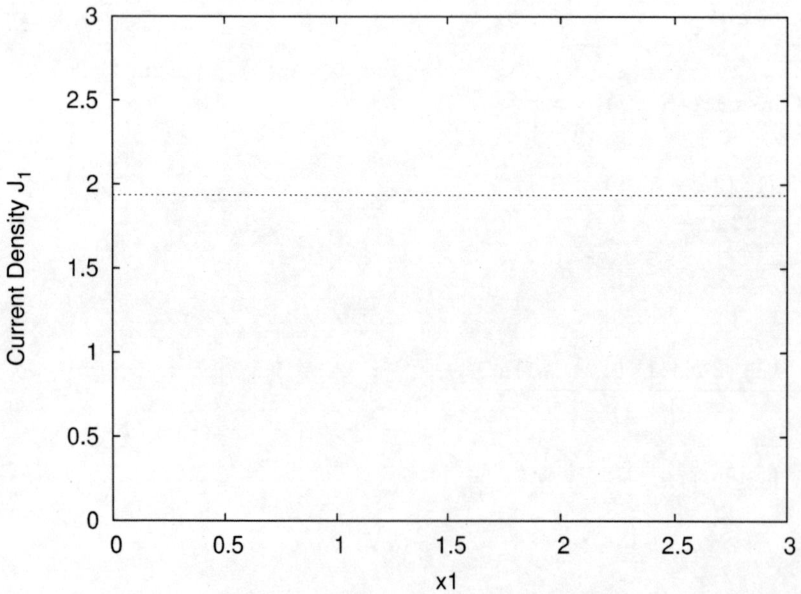

Fig. 4.34: Goedel Metric, current density J_1 for ω=1.

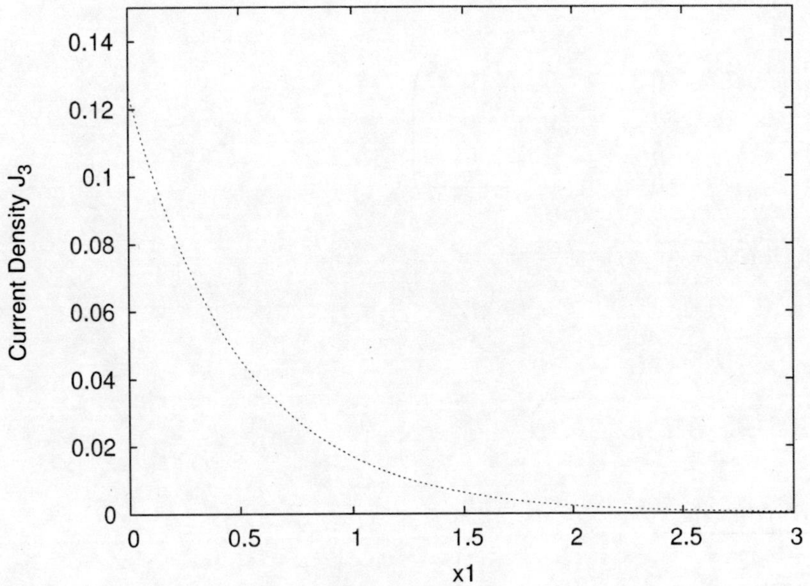

Fig. 4.35: Goedel Metric, current density J_3 for ω=1.

Current Density Class 3 (-$R^{i}{}_{\mu}{}^{\mu j}$)

$$J_1 = 0$$

$$J_2 = 0$$

$$J_3 = 0$$

4.4.16 Static De Sitter metric

This metric describes a universe with a constant scalar curvature. α is a parameter. There is a horizon at $r = \alpha$. This is also visible in the cosmological charge and current densities.

Coordinates

$$\mathbf{x} = \begin{pmatrix} t \\ r \\ \vartheta \\ \varphi \end{pmatrix}$$

Metric

$$g_{\mu\nu} = \begin{pmatrix} \frac{r^2}{\alpha^2} - 1 & 0 & 0 & 0 \\ 0 & \frac{1}{1-\frac{r^2}{\alpha^2}} & 0 & 0 \\ 0 & 0 & r^2 & 0 \\ 0 & 0 & 0 & r^2 \sin^2 \vartheta \end{pmatrix}$$

Contravariant Metric

$$g^{\mu\nu} = \begin{pmatrix} \frac{\alpha^2}{(r-\alpha)\,(r+\alpha)} & 0 & 0 & 0 \\ 0 & -\frac{(r-\alpha)\,(r+\alpha)}{\alpha^2} & 0 & 0 \\ 0 & 0 & \frac{1}{r^2} & 0 \\ 0 & 0 & 0 & \frac{1}{r^2 \sin^2\vartheta} \end{pmatrix}$$

Christoffel Connection

$$\Gamma^0{}_{01} = \frac{r}{(r-\alpha)\,(r+\alpha)}$$

$$\Gamma^0{}_{10} = \Gamma^0{}_{01}$$

$$\Gamma^1{}_{00} = \frac{r\,(r-\alpha)\,(r+\alpha)}{\alpha^4}$$

$$\Gamma^1{}_{11} = -\frac{r}{(r-\alpha)\,(r+\alpha)}$$

$$\Gamma^1{}_{22} = \frac{r\,(r-\alpha)\,(r+\alpha)}{\alpha^2}$$

$$\Gamma^1{}_{33} = \frac{r\,(r-\alpha)\,(r+\alpha)\sin^2\vartheta}{\alpha^2}$$

$$\Gamma^2{}_{12} = \frac{1}{r}$$

$$\Gamma^2{}_{21} = \Gamma^2{}_{12}$$

$$\Gamma^2{}_{33} = -\cos\vartheta\,\sin\vartheta$$

$$\Gamma^3{}_{13} = \frac{1}{r}$$

$$\Gamma^3{}_{23} = \frac{\cos\vartheta}{\sin\vartheta}$$

$$\Gamma^3{}_{31} = \Gamma^3{}_{13}$$

$$\Gamma^3{}_{32} = \Gamma^3{}_{23}$$

Metric Compatibility

———— o.k.

Riemann Tensor

$$R^0{}_{101} = -\frac{1}{(r-\alpha)\,(r+\alpha)}$$

$$R^0{}_{110} = -R^0{}_{101}$$

$$R^0{}_{202} = \frac{r^2}{\alpha^2}$$

$$R^0{}_{220} = -R^0{}_{202}$$

$$R^0{}_{303} = \frac{r^2\sin^2\vartheta}{\alpha^2}$$

$$R^0{}_{330} = -R^0{}_{303}$$

$$R^1{}_{001} = -\frac{(r-\alpha)\,(r+\alpha)}{\alpha^4}$$

$$R^1{}_{010} = -R^1{}_{001}$$

$$R^1{}_{212} = \frac{r^2}{\alpha^2}$$

$$R^1{}_{221} = -R^1{}_{212}$$

$$R^1{}_{313} = \frac{r^2\sin^2\vartheta}{\alpha^2}$$

$$R^1{}_{331} = -R^1{}_{313}$$

$$R^2{}_{002} = -\frac{(r-\alpha)\,(r+\alpha)}{\alpha^4}$$

$$R^2{}_{020} = -R^2{}_{002}$$

$$R^2{}_{112} = \frac{1}{(r-\alpha)\,(r+\alpha)}$$

$$R^2{}_{121} = -R^2{}_{112}$$

$$R^2{}_{323} = \frac{r^2\sin^2\vartheta}{\alpha^2}$$

$$R^2{}_{332} = -R^2{}_{323}$$

$$R^3{}_{003} = -\frac{(r-\alpha)\,(r+\alpha)}{\alpha^4}$$

$$R^3{}_{030} = -R^3{}_{003}$$

$$R^3{}_{113} = \frac{1}{(r-\alpha)\,(r+\alpha)}$$

$$R^3{}_{131} = -R^3{}_{113}$$

$$R^3{}_{223} = -\frac{r^2}{\alpha^2}$$

$$R^3{}_{232} = -R^3{}_{223}$$

Ricci Tensor

$$\mathrm{Ric}_{00} = \frac{3\,(r-\alpha)\,(r+\alpha)}{\alpha^4}$$

$$\mathrm{Ric}_{11} = -\frac{3}{(r-\alpha)\,(r+\alpha)}$$

$$\mathrm{Ric}_{22} = \frac{3\,r^2}{\alpha^2}$$

$$\mathrm{Ric}_{33} = \frac{3\,r^2\,\sin^2\vartheta}{\alpha^2}$$

Ricci Scalar

$$R_{sc} = \frac{12}{\alpha^2}$$

Bianchi identity (Ricci cyclic equation $R^{\kappa}{}_{[\mu\nu\sigma]} = 0$)

———— o.k.

Einstein Tensor

$$G_{00} = -\frac{3\,(r-\alpha)\,(r+\alpha)}{\alpha^4}$$

$$G_{11} = \frac{3}{(r-\alpha)\,(r+\alpha)}$$

$$G_{22} = -\frac{3\,r^2}{\alpha^2}$$

$$G_{33} = -\frac{3\,r^2\,\sin^2\vartheta}{\alpha^2}$$

Hodge Dual of Bianchi Identity

——— (see charge and current densities)

Scalar Charge Density (-$R^{0\ i0}_{\ i}$)

$$\rho = \frac{3}{(r-\alpha)\,(r+\alpha)}$$

Current Density Class 1 (-$R^{i\ \mu j}_{\ \mu}$)

$$J_1 = \frac{3\,(r-\alpha)\,(r+\alpha)}{\alpha^4}$$

$$J_2 = -\frac{3}{\alpha^2\,r^2}$$

$$J_3 = -\frac{3}{\alpha^2\,r^2\,\sin^2\vartheta}$$

Current Density Class 2 (-$R^{i\ \mu j}_{\ \mu}$)

$$J_1 = 0$$

$$J_2 = 0$$

$$J_3 = 0$$

Current Density Class 3 (-$R^{i\ \mu j}_{\ \mu}$)

$$J_1 = 0$$

$$J_2 = 0$$

$$J_3 = 0$$

4.4.17 FLRW metric

The Friedmann-Lemaitre-Robertson-Walker metric is a cosmological metric of a homogeneous and isotropic space. It contains a (normally increasing) time-dependent function $a(t)$ and a constant k which restricts the describable size of the universe by the condition

$$r < \frac{1}{\sqrt{k}}.$$

Although the time function a grows, the charge and current densities go to zero over time. The reverse is true if a decreases in time.

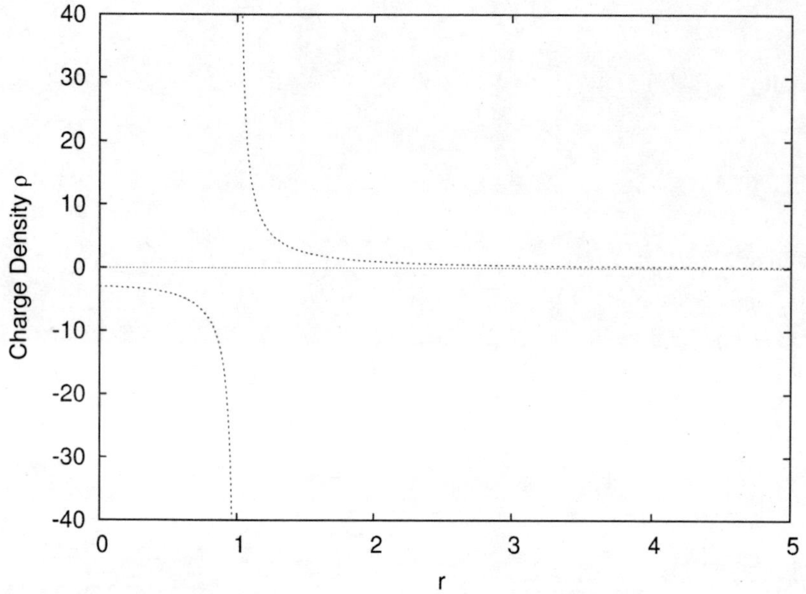

Fig. 4.36: Static De Sitter metric, charge density ρ for $\alpha = 1$.

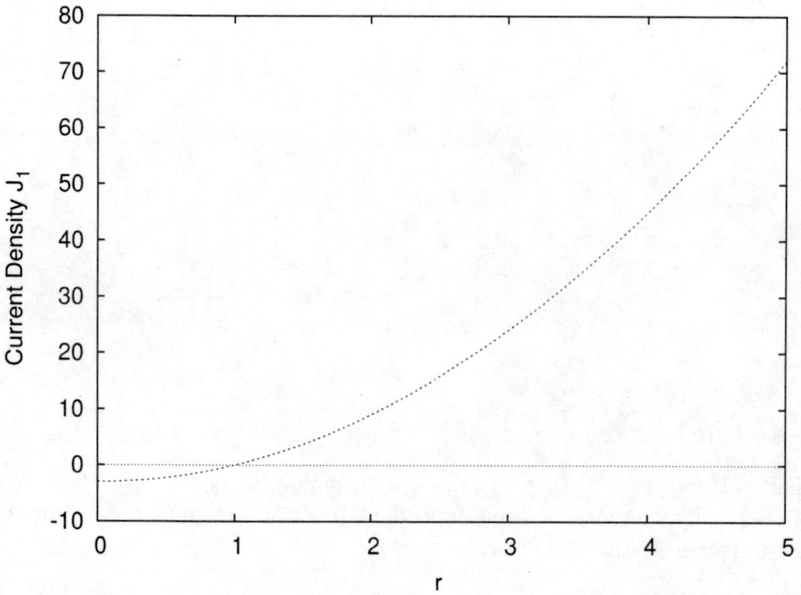

Fig. 4.37: Static De Sitter metric, current density J_r for $\alpha = 1$.

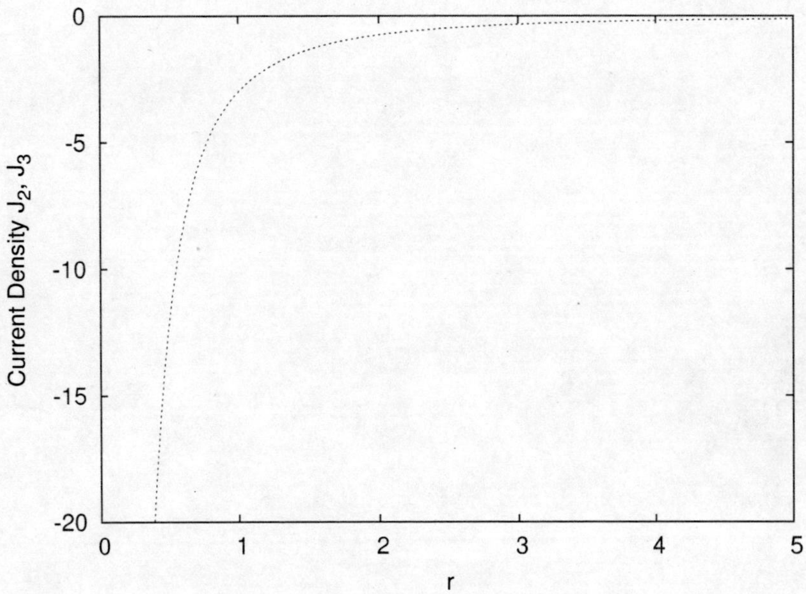

Fig. 4.38: Static De Sitter metric, current density J_ϑ, J_φ for $\alpha = 1$.

Coordinates

$$\mathbf{x} = \begin{pmatrix} t \\ r \\ \vartheta \\ \varphi \end{pmatrix}$$

Metric

$$g_{\mu\nu} = \begin{pmatrix} 1 & 0 & 0 & 0 \\ 0 & -\frac{a^2}{1-k\,r^2} & 0 & 0 \\ 0 & 0 & -a^2\,r^2 & 0 \\ 0 & 0 & 0 & -a^2\,r^2\,\sin^2\vartheta \end{pmatrix}$$

Contravariant Metric

$$g^{\mu\nu} = \begin{pmatrix} 1 & 0 & 0 & 0 \\ 0 & \frac{k\,r^2-1}{a^2} & 0 & 0 \\ 0 & 0 & -\frac{1}{a^2\,r^2} & 0 \\ 0 & 0 & 0 & -\frac{1}{a^2\,r^2\,\sin^2\vartheta} \end{pmatrix}$$

Christoffel Connection

$$\Gamma^0{}_{11} = -\frac{a\,\left(\frac{d}{d\,t}\,a\right)}{k\,r^2-1}$$

$$\Gamma^0{}_{22} = a\,\left(\frac{d}{d\,t}\,a\right)\,r^2$$

$$\Gamma^0{}_{33} = a \left(\frac{d}{dt}\,a\right) r^2 \sin^2\vartheta$$

$$\Gamma^1{}_{01} = \frac{\frac{d}{dt}\,a}{a}$$

$$\Gamma^1{}_{10} = \Gamma^1{}_{01}$$

$$\Gamma^1{}_{11} = -\frac{k\,r}{k\,r^2 - 1}$$

$$\Gamma^1{}_{22} = r\,\left(k\,r^2 - 1\right)$$

$$\Gamma^1{}_{33} = r\,\left(k\,r^2 - 1\right)\,\sin^2\vartheta$$

$$\Gamma^2{}_{02} = \frac{\frac{d}{dt}\,a}{a}$$

$$\Gamma^2{}_{12} = \frac{1}{r}$$

$$\Gamma^2{}_{20} = \Gamma^2{}_{02}$$

$$\Gamma^2{}_{21} = \Gamma^2{}_{12}$$

$$\Gamma^2{}_{33} = -\cos\vartheta\,\sin\vartheta$$

$$\Gamma^3{}_{03} = \frac{\frac{d}{dt}\,a}{a}$$

$$\Gamma^3{}_{13} = \frac{1}{r}$$

$$\Gamma^3{}_{23} = \frac{\cos\vartheta}{\sin\vartheta}$$

$$\Gamma^3{}_{30} = \Gamma^3{}_{03}$$

$$\Gamma^3{}_{31} = \Gamma^3{}_{13}$$

$$\Gamma^3{}_{32} = \Gamma^3{}_{23}$$

Metric Compatibility

——— o.k.

Riemann Tensor

$$R^0{}_{101} = -\frac{a\left(\frac{d^2}{d\,t^2}\,a\right)}{k\,r^2-1}$$

$$R^0{}_{110} = -R^0{}_{101}$$

$$R^0{}_{202} = a\left(\frac{d^2}{d\,t^2}\,a\right)r^2$$

$$R^0{}_{220} = -R^0{}_{202}$$

$$R^0{}_{303} = a\left(\frac{d^2}{d\,t^2}\,a\right)r^2\,\sin^2\vartheta$$

$$R^0{}_{330} = -R^0{}_{303}$$

$$R^1{}_{001} = \frac{\frac{d^2}{d\,t^2}\,a}{a}$$

$$R^1{}_{010} = -R^1{}_{001}$$

$$R^1{}_{212} = \left(k+\left(\frac{d}{d\,t}\,a\right)^2\right)r^2$$

$$R^1{}_{221} = -R^1{}_{212}$$

$$R^1{}_{313} = \left(k+\left(\frac{d}{d\,t}\,a\right)^2\right)r^2\,\sin^2\vartheta$$

$$R^1{}_{331} = -R^1{}_{313}$$

$$R^2{}_{002} = \frac{\frac{d^2}{d\,t^2}\,a}{a}$$

$$R^2{}_{020} = -R^2{}_{002}$$

$$R^2{}_{112} = \frac{k+\left(\frac{d}{d\,t}\,a\right)^2}{k\,r^2-1}$$

$$R^2{}_{121} = -R^2{}_{112}$$

$$R^2{}_{323} = \left(k+\left(\frac{d}{d\,t}\,a\right)^2\right)r^2\,\sin^2\vartheta$$

$$R^2{}_{332} = -R^2{}_{323}$$

$$R^3{}_{003} = \frac{\frac{d^2}{dt^2}\,a}{a}$$

$$R^3{}_{030} = -R^3{}_{003}$$

$$R^3{}_{113} = \frac{k + \left(\frac{d}{dt}\,a\right)^2}{k\,r^2 - 1}$$

$$R^3{}_{131} = -R^3{}_{113}$$

$$R^3{}_{223} = -\left(k + \left(\frac{d}{dt}\,a\right)^2\right)\,r^2$$

$$R^3{}_{232} = -R^3{}_{223}$$

Ricci Tensor

$$\mathrm{Ric}_{00} = -\frac{3\,\left(\frac{d^2}{dt^2}\,a\right)}{a}$$

$$\mathrm{Ric}_{11} = -\frac{2\,k + a\,\left(\frac{d^2}{dt^2}\,a\right) + 2\,\left(\frac{d}{dt}\,a\right)^2}{k\,r^2 - 1}$$

$$\mathrm{Ric}_{22} = \left(2\,k + a\,\left(\frac{d^2}{dt^2}\,a\right) + 2\,\left(\frac{d}{dt}\,a\right)^2\right)\,r^2$$

$$\mathrm{Ric}_{33} = \left(2\,k + a\,\left(\frac{d^2}{dt^2}\,a\right) + 2\,\left(\frac{d}{dt}\,a\right)^2\right)\,r^2\,\sin^2\vartheta$$

Ricci Scalar

$$R_{sc} = -\frac{6\,\left(k + a\,\left(\frac{d^2}{dt^2}\,a\right) + \left(\frac{d}{dt}\,a\right)^2\right)}{a^2}$$

Bianchi identity (Ricci cyclic equation $R^{\kappa}{}_{[\mu\nu\sigma]} = 0$)

——— o.k.

Einstein Tensor

$$G_{00} = \frac{3\left(k + \left(\frac{d}{dt}\,a\right)^2\right)}{a^2}$$

$$G_{11} = \frac{k + 2\,a\left(\frac{d^2}{dt^2}\,a\right) + \left(\frac{d}{dt}\,a\right)^2}{k\,r^2 - 1}$$

$$G_{22} = -\left(k + 2\,a\left(\frac{d^2}{dt^2}\,a\right) + \left(\frac{d}{dt}\,a\right)^2\right) r^2$$

$$G_{33} = -\left(k + 2\,a\left(\frac{d^2}{dt^2}\,a\right) + \left(\frac{d}{dt}\,a\right)^2\right) r^2\,\sin^2\vartheta$$

Hodge Dual of Bianchi Identity

——— (see charge and current densities)

Scalar Charge Density (-$R^{0}{}_{i}{}^{i0}$)

$$\rho = -\frac{3\left(\frac{d^2}{dt^2}\,a\right)}{a}$$

Current Density Class 1 (-$R^{i}{}_{\mu}{}^{\mu j}$)

$$J_1 = \frac{\left(2\,k + a\left(\frac{d^2}{dt^2}\,a\right) + 2\left(\frac{d}{dt}\,a\right)^2\right)\left(k\,r^2 - 1\right)}{a^4}$$

$$J_2 = -\frac{2\,k + a\left(\frac{d^2}{dt^2}\,a\right) + 2\left(\frac{d}{dt}\,a\right)^2}{a^4\,r^2}$$

$$J_3 = -\frac{2\,k + a\left(\frac{d^2}{dt^2}\,a\right) + 2\left(\frac{d}{dt}\,a\right)^2}{a^4\,r^2\,\sin^2\vartheta}$$

Current Density Class 2 (-$R^{i}{}_{\mu}{}^{\mu j}$)

$$J_1 = 0$$

$$J_2 = 0$$

$$J_3 = 0$$

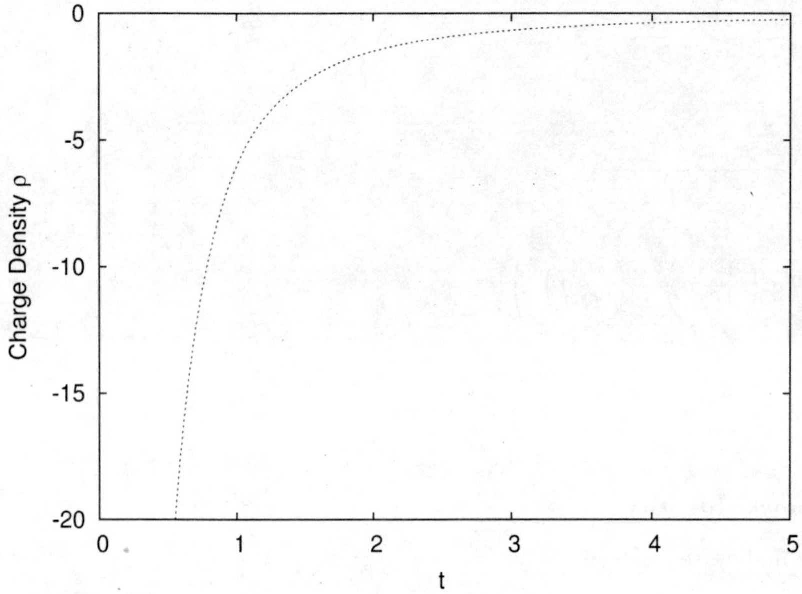

Fig. 4.39: FLRW metric, charge density ρ for $a = t^2, k = .5, r = 1$.

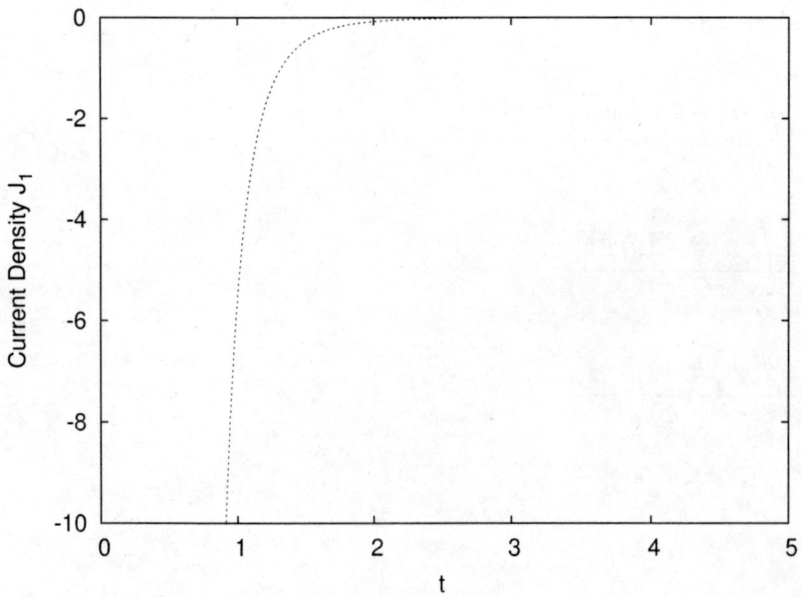

Fig. 4.40: FLRW metric, current density J_r for $a = t^2, k = .5, r = 1$.

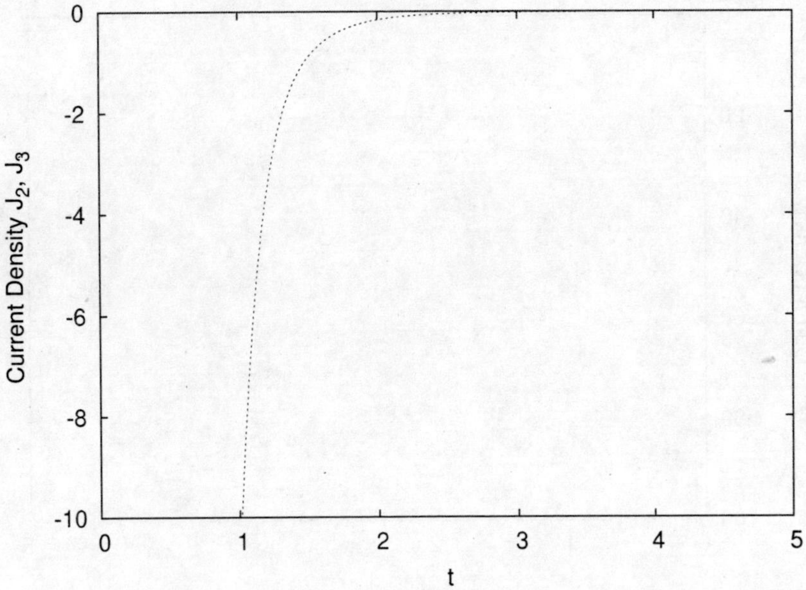

Fig. 4.41: FLRW metric, current density J_ϑ, J_φ for $a = t^2, k = .5, r = 1$.

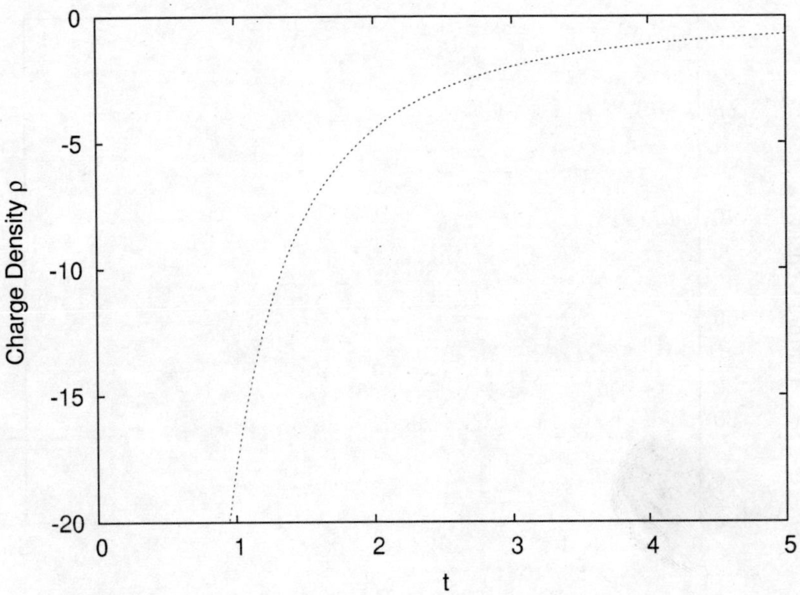

Fig. 4.42: FLRW metric, charge density ρ for $a = t^{-2}, k = .5, r = 1$.

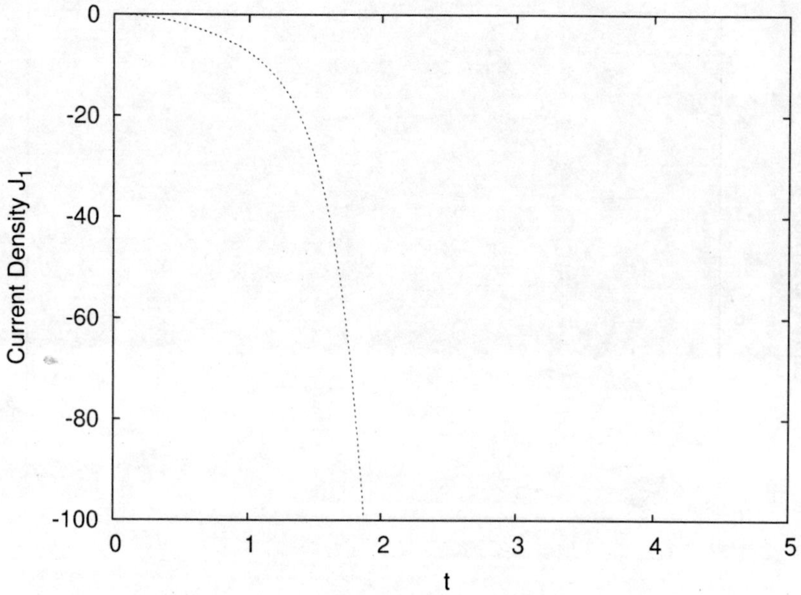

Fig. 4.43: FLRW metric, current density J_r for $a = t^{-2}, k = .5, r = 1$.

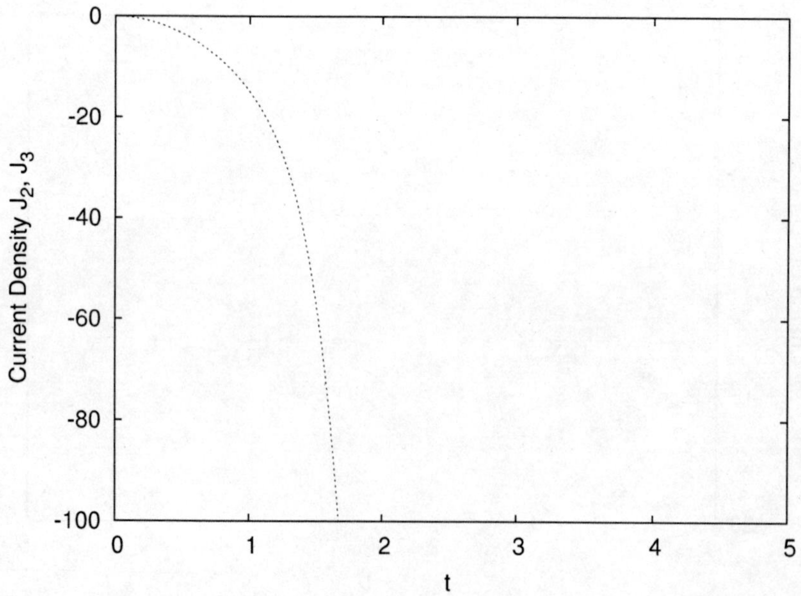

Fig. 4.44: FLRW metric, current density J_ϑ, J_φ for $a = t^{-2}, k = .5, r = 1$.

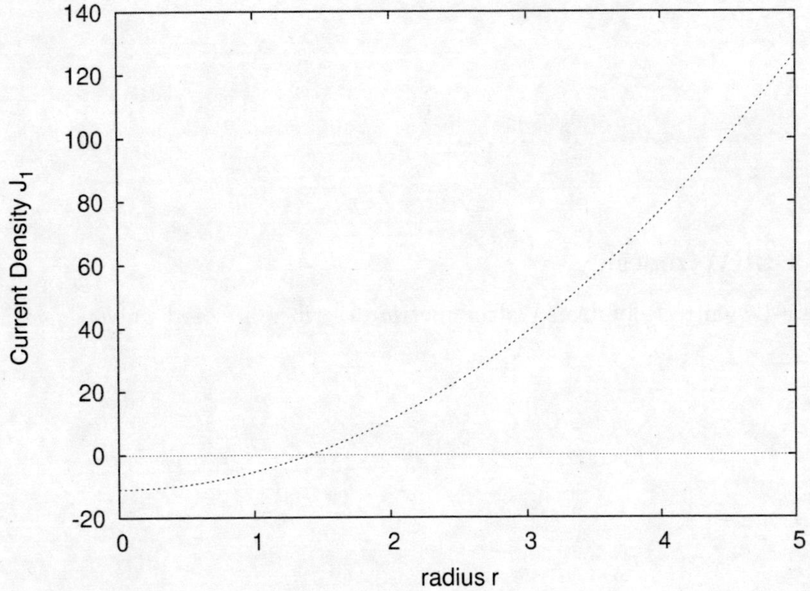

Fig. 4.45: FLRW metric, current density, r dependence of J_r for $a = t^2, t = 1, k = .5$.

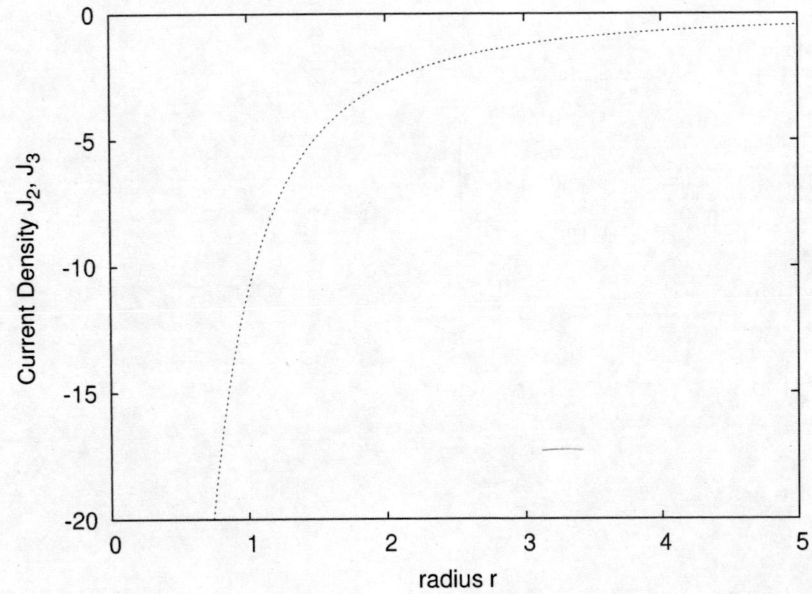

Fig. 4.46: FLRW metric, current density, r dependence of J_ϑ, J_φ for $a = t^2, t = 1, k = .5$.

Current Density Class 3 (-$R^{i}{}_{\mu}{}^{\mu j}$)

$$J_1 = 0$$

$$J_2 = 0$$

$$J_3 = 0$$

4.4.18 Closed FLRW metric

The closed Friedmann-Lemaitre-Robertson-Walker metric describes a closed universe. a is a time dependent function.

Coordinates

$$\mathbf{x} = \begin{pmatrix} t \\ \chi \\ \vartheta \\ \varphi \end{pmatrix}$$

Metric

$$g_{\mu\nu} = \begin{pmatrix} 1 & 0 & 0 & 0 \\ 0 & -a^2 & 0 & 0 \\ 0 & 0 & -a^2 \sin^2\chi & 0 \\ 0 & 0 & 0 & -a^2 \sin^2\chi \sin^2\vartheta \end{pmatrix}$$

Contravariant Metric

$$g^{\mu\nu} = \begin{pmatrix} 1 & 0 & 0 & 0 \\ 0 & -\frac{1}{a^2} & 0 & 0 \\ 0 & 0 & -\frac{1}{a^2 \sin^2\chi} & 0 \\ 0 & 0 & 0 & -\frac{1}{a^2 \sin^2\chi \sin^2\vartheta} \end{pmatrix}$$

Christoffel Connection

$$\Gamma^{0}{}_{11} = a \left(\frac{d}{dt} a\right)$$

$$\Gamma^{0}{}_{22} = a \left(\frac{d}{dt} a\right) \sin^2\chi$$

$$\Gamma^{0}{}_{33} = a \left(\frac{d}{dt} a\right) \sin^2\chi \sin^2\vartheta$$

$$\Gamma^{1}{}_{01} = \frac{\frac{d}{dt} a}{a}$$

$$\Gamma^1{}_{10} = \Gamma^1{}_{01}$$

$$\Gamma^1{}_{22} = -\cos\chi\,\sin\chi$$

$$\Gamma^1{}_{33} = -\cos\chi\,\sin\chi\,\sin^2\vartheta$$

$$\Gamma^2{}_{02} = \frac{\frac{d}{dt}\,a}{a}$$

$$\Gamma^2{}_{12} = \frac{\cos\chi}{\sin\chi}$$

$$\Gamma^2{}_{20} = \Gamma^2{}_{02}$$

$$\Gamma^2{}_{21} = \Gamma^2{}_{12}$$

$$\Gamma^2{}_{33} = -\cos\vartheta\,\sin\vartheta$$

$$\Gamma^3{}_{03} = \frac{\frac{d}{dt}\,a}{a}$$

$$\Gamma^3{}_{13} = \frac{\cos\chi}{\sin\chi}$$

$$\Gamma^3{}_{23} = \frac{\cos\vartheta}{\sin\vartheta}$$

$$\Gamma^3{}_{30} = \Gamma^3{}_{03}$$

$$\Gamma^3{}_{31} = \Gamma^3{}_{13}$$

$$\Gamma^3{}_{32} = \Gamma^3{}_{23}$$

Metric Compatibility

——— o.k.

Riemann Tensor

$$R^0{}_{101} = a\left(\frac{d^2}{dt^2}\,a\right)$$

$$R^0{}_{110} = -R^0{}_{101}$$

$$R^0{}_{202} = a\left(\frac{d^2}{dt^2}\,a\right)\sin^2\chi$$

$$R^0{}_{220} = -R^0{}_{202}$$

$$R^0{}_{303} = a\left(\frac{d^2}{d\,t^2}\,a\right)\sin^2\chi\,\sin^2\vartheta$$

$$R^0{}_{330} = -R^0{}_{303}$$

$$R^1{}_{001} = \frac{\frac{d^2}{d\,t^2}\,a}{a}$$

$$R^1{}_{010} = -R^1{}_{001}$$

$$R^1{}_{212} = \left(\left(\frac{d}{d\,t}\,a\right)^2 + 1\right)\sin^2\chi$$

$$R^1{}_{221} = -R^1{}_{212}$$

$$R^1{}_{313} = \left(\left(\frac{d}{d\,t}\,a\right)^2 + 1\right)\sin^2\chi\,\sin^2\vartheta$$

$$R^1{}_{331} = -R^1{}_{313}$$

$$R^2{}_{002} = \frac{\frac{d^2}{d\,t^2}\,a}{a}$$

$$R^2{}_{020} = -R^2{}_{002}$$

$$R^2{}_{112} = -\left(\left(\frac{d}{d\,t}\,a\right)^2 + 1\right)$$

$$R^2{}_{121} = -R^2{}_{112}$$

$$R^2{}_{323} = \left(\left(\frac{d}{d\,t}\,a\right)^2 + 1\right)\sin^2\chi\,\sin^2\vartheta$$

$$R^2{}_{332} = -R^2{}_{323}$$

$$R^3{}_{003} = \frac{\frac{d^2}{d\,t^2}\,a}{a}$$

$$R^3{}_{030} = -R^3{}_{003}$$

$$R^3{}_{113} = -\left(\left(\frac{d}{d\,t}\,a\right)^2 + 1\right)$$

$$R^3{}_{131} = -R^3{}_{113}$$

$$R^3{}_{223} = -\left(\left(\frac{d}{d\,t}\,a\right)^2 + 1\right)\sin^2\chi$$

$$R^3{}_{232} = -R^3{}_{223}$$

Ricci Tensor

$$\mathrm{Ric}_{00} = -\frac{3\left(\frac{d^2}{d\,t^2}\,a\right)}{a}$$

$$\mathrm{Ric}_{11} = a\left(\frac{d^2}{d\,t^2}\,a\right) + 2\left(\frac{d}{d\,t}\,a\right)^2 + 2$$

$$\mathrm{Ric}_{22} = \left(a\left(\frac{d^2}{d\,t^2}\,a\right) + 2\left(\frac{d}{d\,t}\,a\right)^2 + 2\right)\sin^2\chi$$

$$\mathrm{Ric}_{33} = \left(a\left(\frac{d^2}{d\,t^2}\,a\right) + 2\left(\frac{d}{d\,t}\,a\right)^2 + 2\right)\sin^2\chi\,\sin^2\vartheta$$

Ricci Scalar

$$R_{sc} = -\frac{6\left(a\left(\frac{d^2}{d\,t^2}\,a\right) + \left(\frac{d}{d\,t}\,a\right)^2 + 1\right)}{a^2}$$

Bianchi identity (Ricci cyclic equation $R^{\kappa}{}_{[\mu\nu\sigma]} = 0$)

——— o.k.

Einstein Tensor

$$G_{00} = \frac{3\left(\left(\frac{d}{d\,t}\,a\right)^2 + 1\right)}{a^2}$$

$$G_{11} = -\left(2\,a\left(\frac{d^2}{d\,t^2}\,a\right) + \left(\frac{d}{d\,t}\,a\right)^2 + 1\right)$$

$$G_{22} = -\left(2\,a\left(\frac{d^2}{d\,t^2}\,a\right) + \left(\frac{d}{d\,t}\,a\right)^2 + 1\right)\sin^2\chi$$

$$G_{33} = -\left(2\,a\left(\frac{d^2}{d\,t^2}\,a\right) + \left(\frac{d}{d\,t}\,a\right)^2 + 1\right)\sin^2\chi\,\sin^2\vartheta$$

Hodge Dual of Bianchi Identity

——— (see charge and current densities)

Scalar Charge Density (-$R^{0}{}_{i}{}^{i0}$)

$$\rho = -\frac{3\left(\frac{d^2}{d\,t^2}\,a\right)}{a}$$

Current Density Class 1 (-$R^{i}{}_{\mu}{}^{\mu j}$)

$$J_1 = -\frac{a\left(\frac{d^2}{dt^2}a\right) + 2\left(\frac{d}{dt}a\right)^2 + 2}{a^4}$$

$$J_2 = -\frac{a\left(\frac{d^2}{dt^2}a\right) + 2\left(\frac{d}{dt}a\right)^2 + 2}{a^4\sin^2\chi}$$

$$J_3 = -\frac{a\left(\frac{d^2}{dt^2}a\right) + 2\left(\frac{d}{dt}a\right)^2 + 2}{a^4\sin^2\chi\,\sin^2\vartheta}$$

Current Density Class 2 (-$R^{i}{}_{\mu}{}^{\mu j}$)

$$J_1 = 0$$

$$J_2 = 0$$

$$J_3 = 0$$

Current Density Class 3 (-$R^{i}{}_{\mu}{}^{\mu j}$)

$$J_1 = 0$$

$$J_2 = 0$$

$$J_3 = 0$$

4.4.19 Friedmann Dust metric

Metric of the Friedmann Dust universe. a is a parameter.

Coordinates

$$\mathbf{x} = \begin{pmatrix} t \\ x_1 \\ x_2 \\ x_3 \end{pmatrix}$$

Metric

$$g_{\mu\nu} = \begin{pmatrix} -1 & 0 & 0 & 0 \\ 0 & \left(\cosh\left(\frac{3t}{a}-1\right)\right)^{\frac{2}{3}} & 0 & 0 \\ 0 & 0 & \left(\cosh\left(\frac{3t}{a}-1\right)\right)^{\frac{2}{3}} & 0 \\ 0 & 0 & 0 & \left(\cosh\left(\frac{3t}{a}-1\right)\right)^{\frac{2}{3}} \end{pmatrix}$$

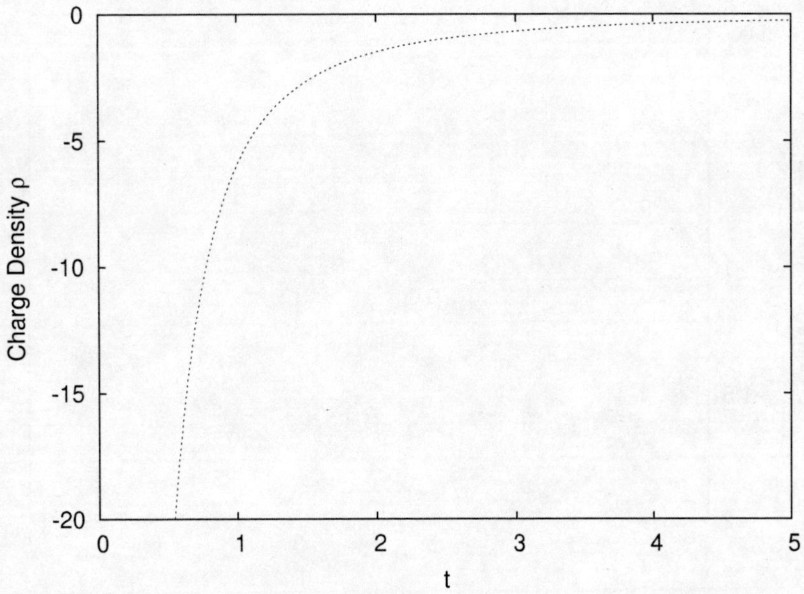

Fig. 4.47: Closed FLRW metric, charge density ρ for $a = t^2$.

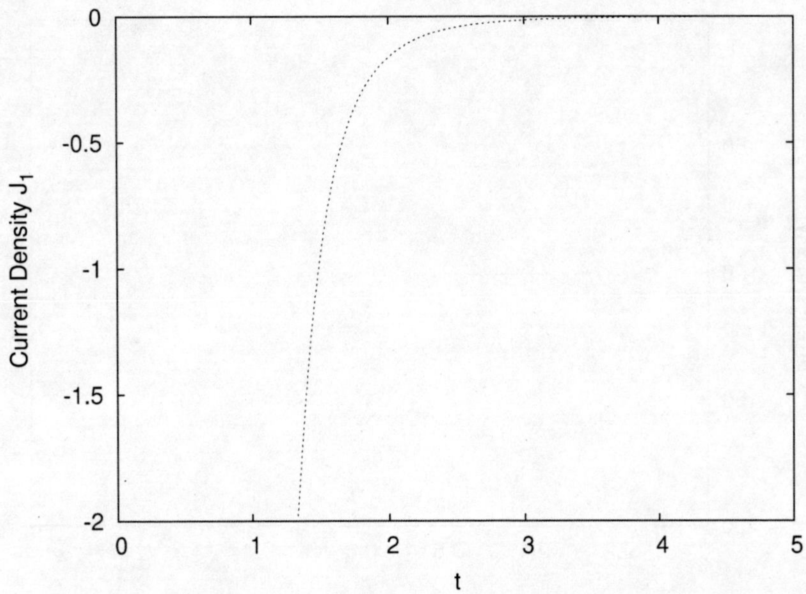

Fig. 4.48: Closed FLRW metric, current density J_χ for $a = t^2$.

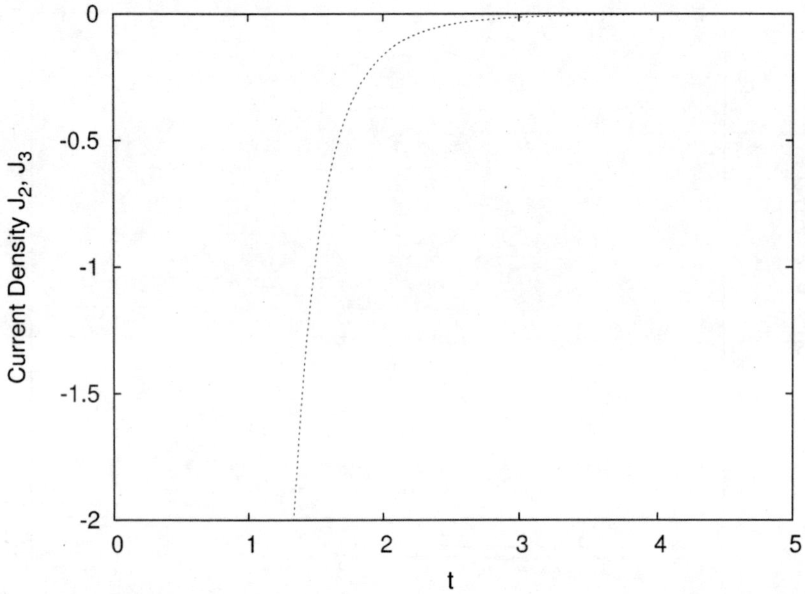

Fig. 4.49: Closed FLRW metric, current density J_ϑ, J_φ for $a = t^2, \chi = \pi/2$.

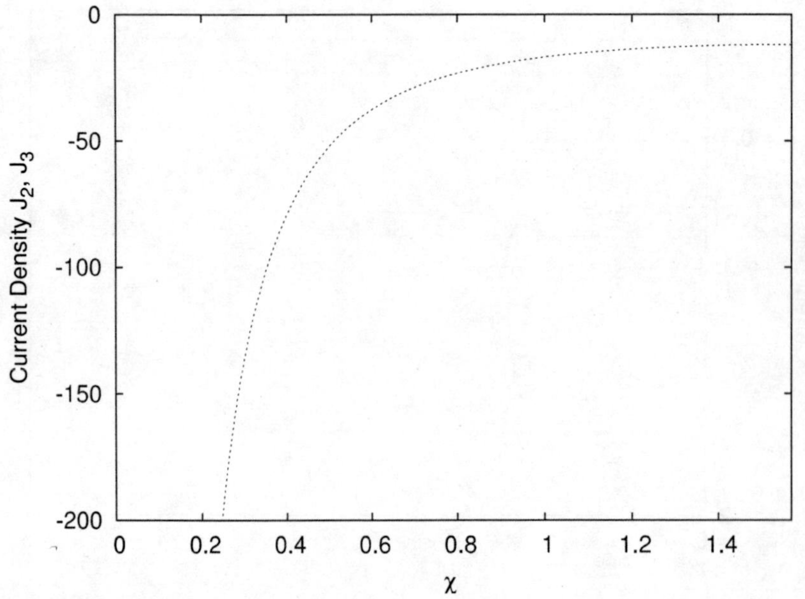

Fig. 4.50: Closed FLRW metric, current density J_ϑ, J_φ, χ dependence for $a = t^2,\ t = 1$.

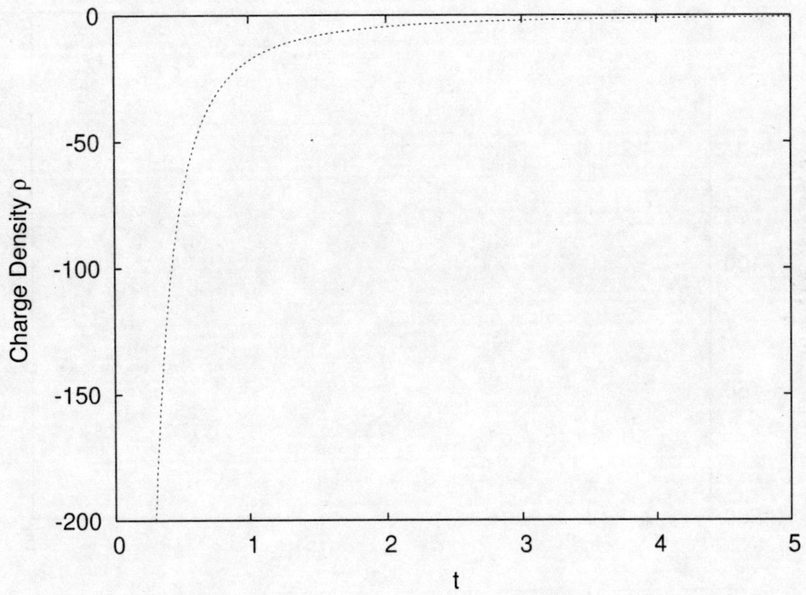

Fig. 4.51: Closed FLRW metric, charge density ρ for $a = t^{-2}$.

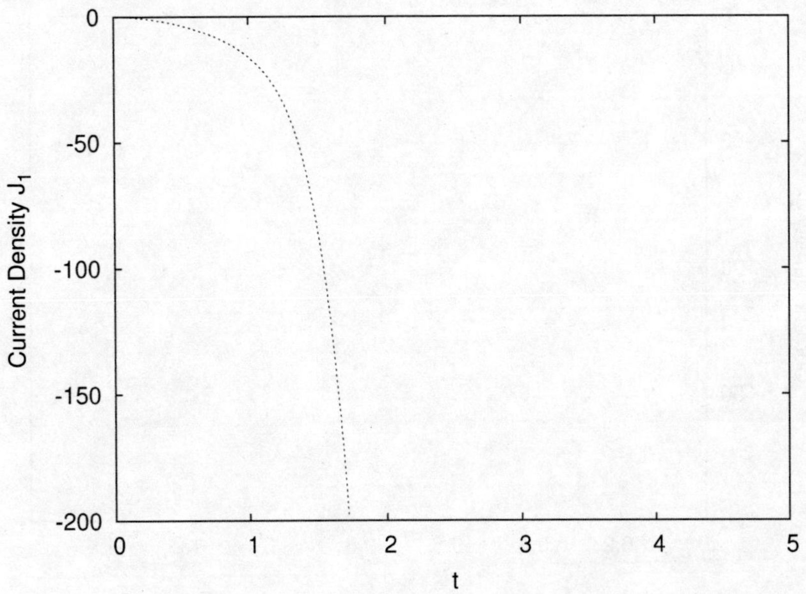

Fig. 4.52: Closed FLRW metric, current density J_χ for $a = t^{-2}$.

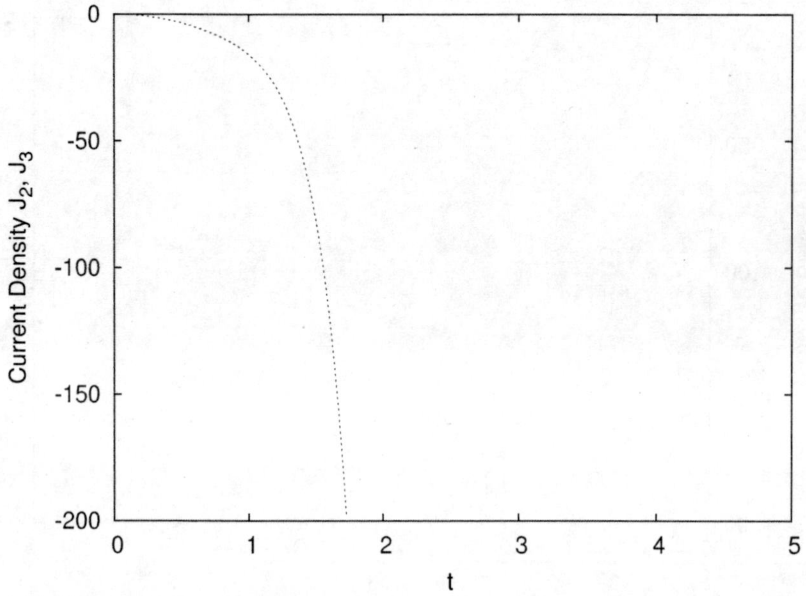

Fig. 4.53: Closed FLRW metric, current density J_ϑ, J_φ for $a = t^{-2}, \chi = \pi/2$.

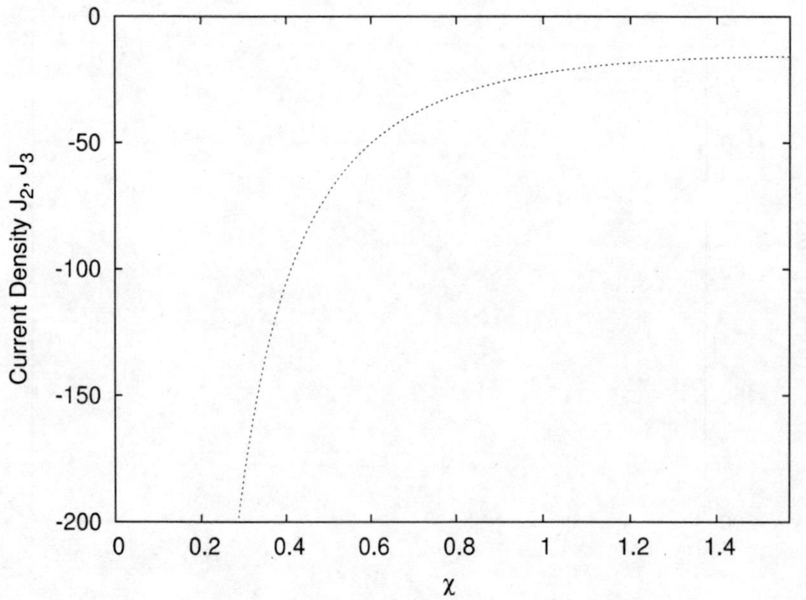

Fig. 4.54: Closed FLRW metric, current density J_ϑ, J_φ, χ dependence for $a = t^{-2}$, $t = 1$.

Contravariant Metric

$$g^{\mu\nu} = \begin{pmatrix} -1 & 0 & 0 & 0 \\ 0 & \frac{1}{\left(\cosh\left(\frac{3\,t-a}{a}\right)\right)^{\frac{2}{3}}} & 0 & 0 \\ 0 & 0 & \frac{1}{\left(\cosh\left(\frac{3\,t-a}{a}\right)\right)^{\frac{2}{3}}} & 0 \\ 0 & 0 & 0 & \frac{1}{\left(\cosh\left(\frac{3\,t-a}{a}\right)\right)^{\frac{2}{3}}} \end{pmatrix}$$

Christoffel Connection

$$\Gamma^0{}_{11} = \frac{\sinh\left(\frac{3\,t-a}{a}\right)}{a\left(\cosh\left(\frac{3\,t-a}{a}\right)\right)^{\frac{1}{3}}}$$

$$\Gamma^0{}_{22} = \frac{\sinh\left(\frac{3\,t-a}{a}\right)}{a\left(\cosh\left(\frac{3\,t-a}{a}\right)\right)^{\frac{1}{3}}}$$

$$\Gamma^0{}_{33} = \frac{\sinh\left(\frac{3\,t-a}{a}\right)}{a\left(\cosh\left(\frac{3\,t-a}{a}\right)\right)^{\frac{1}{3}}}$$

$$\Gamma^1{}_{01} = \frac{\sinh\left(\frac{3\,t-a}{a}\right)}{a\cosh\left(\frac{3\,t-a}{a}\right)}$$

$$\Gamma^1{}_{10} = \Gamma^1{}_{01}$$

$$\Gamma^2{}_{02} = \frac{\sinh\left(\frac{3\,t-a}{a}\right)}{a\cosh\left(\frac{3\,t-a}{a}\right)}$$

$$\Gamma^2{}_{20} = \Gamma^2{}_{02}$$

$$\Gamma^3{}_{03} = \frac{\sinh\left(\frac{3\,t-a}{a}\right)}{a\cosh\left(\frac{3\,t-a}{a}\right)}$$

$$\Gamma^3{}_{30} = \Gamma^3{}_{03}$$

Metric Compatibility

——— o.k.

Riemann Tensor

$$R^{0}{}_{101} = \frac{\cosh^2\left(\frac{3\,t-a}{a}\right)+2}{a^2\left(\cosh\left(\frac{3\,t-a}{a}\right)\right)^{\frac{4}{3}}}$$

$$R^{0}{}_{110} = -R^{0}{}_{101}$$

$$R^{0}{}_{202} = \frac{\cosh^2\left(\frac{3\,t-a}{a}\right)+2}{a^2\left(\cosh\left(\frac{3\,t-a}{a}\right)\right)^{\frac{4}{3}}}$$

$$R^{0}{}_{220} = -R^{0}{}_{202}$$

$$R^{0}{}_{303} = \frac{\cosh^2\left(\frac{3\,t-a}{a}\right)+2}{a^2\left(\cosh\left(\frac{3\,t-a}{a}\right)\right)^{\frac{4}{3}}}$$

$$R^{0}{}_{330} = -R^{0}{}_{303}$$

$$R^{1}{}_{001} = \frac{\cosh^2\left(\frac{3\,t-a}{a}\right)+2}{a^2\cosh^2\left(\frac{3\,t-a}{a}\right)}$$

$$R^{1}{}_{010} = -R^{1}{}_{001}$$

$$R^{1}{}_{212} = \frac{\sinh^2\left(\frac{3\,t-a}{a}\right)}{a^2\left(\cosh\left(\frac{3\,t-a}{a}\right)\right)^{\frac{4}{3}}}$$

$$R^{1}{}_{221} = -R^{1}{}_{212}$$

$$R^{1}{}_{313} = \frac{\sinh^2\left(\frac{3\,t-a}{a}\right)}{a^2\left(\cosh\left(\frac{3\,t-a}{a}\right)\right)^{\frac{4}{3}}}$$

$$R^{1}{}_{331} = -R^{1}{}_{313}$$

$$R^{2}{}_{002} = \frac{\cosh^2\left(\frac{3\,t-a}{a}\right)+2}{a^2\cosh^2\left(\frac{3\,t-a}{a}\right)}$$

$$R^{2}{}_{020} = -R^{2}{}_{002}$$

$$R^{2}{}_{112} = -\frac{\sinh^2\left(\frac{3\,t-a}{a}\right)}{a^2\left(\cosh\left(\frac{3\,t-a}{a}\right)\right)^{\frac{4}{3}}}$$

$$R^{2}{}_{121} = -R^{2}{}_{112}$$

$$R^2{}_{323} = \frac{\sinh^2\left(\frac{3\,t-a}{a}\right)}{a^2\left(\cosh\left(\frac{3\,t-a}{a}\right)\right)^{\frac{4}{3}}}$$

$$R^2{}_{332} = -R^2{}_{323}$$

$$R^3{}_{003} = \frac{\cosh^2\left(\frac{3\,t-a}{a}\right)+2}{a^2\cosh^2\left(\frac{3\,t-a}{a}\right)}$$

$$R^3{}_{030} = -R^3{}_{003}$$

$$R^3{}_{113} = -\frac{\sinh^2\left(\frac{3\,t-a}{a}\right)}{a^2\left(\cosh\left(\frac{3\,t-a}{a}\right)\right)^{\frac{4}{3}}}$$

$$R^3{}_{131} = -R^3{}_{113}$$

$$R^3{}_{223} = -\frac{\sinh^2\left(\frac{3\,t-a}{a}\right)}{a^2\left(\cosh\left(\frac{3\,t-a}{a}\right)\right)^{\frac{4}{3}}}$$

$$R^3{}_{232} = -R^3{}_{223}$$

Ricci Tensor

$$\mathrm{Ric}_{00} = -\frac{3\left(\cosh^2\left(\frac{3\,t-a}{a}\right)+2\right)}{a^2\cosh^2\left(\frac{3\,t-a}{a}\right)}$$

$$\mathrm{Ric}_{11} = \frac{3\left(\cosh\left(\frac{3\,t-a}{a}\right)\right)^{\frac{2}{3}}}{a^2}$$

$$\mathrm{Ric}_{22} = \frac{3\left(\cosh\left(\frac{3\,t-a}{a}\right)\right)^{\frac{2}{3}}}{a^2}$$

$$\mathrm{Ric}_{33} = \frac{3\left(\cosh\left(\frac{3\,t-a}{a}\right)\right)^{\frac{2}{3}}}{a^2}$$

Ricci Scalar

$$R_{sc} = \frac{6\left(2\cosh^2\left(\frac{3\,t-a}{a}\right)+1\right)}{a^2\cosh^2\left(\frac{3\,t-a}{a}\right)}$$

Bianchi identity (Ricci cyclic equation $R^{\kappa}{}_{[\mu\nu\sigma]} = 0$)

——— o.k.

Einstein Tensor

$$G_{00} = \frac{3\sinh^2\left(\frac{3t-a}{a}\right)}{a^2\cosh^2\left(\frac{3t-a}{a}\right)}$$

$$G_{11} = -\frac{3\left(\cosh^2\left(\frac{3t-a}{a}\right)+1\right)}{a^2\left(\cosh\left(\frac{3t-a}{a}\right)\right)^{\frac{4}{3}}}$$

$$G_{22} = -\frac{3\left(\cosh^2\left(\frac{3t-a}{a}\right)+1\right)}{a^2\left(\cosh\left(\frac{3t-a}{a}\right)\right)^{\frac{4}{3}}}$$

$$G_{33} = -\frac{3\left(\cosh^2\left(\frac{3t-a}{a}\right)+1\right)}{a^2\left(\cosh\left(\frac{3t-a}{a}\right)\right)^{\frac{4}{3}}}$$

Hodge Dual of Bianchi Identity

——— (see charge and current densities)

Scalar Charge Density (-$R^{0}{}_{i}{}^{i0}$)

$$\rho = -\frac{3\left(\cosh^2\left(\frac{3t-a}{a}\right)+2\right)}{a^2\cosh^2\left(\frac{3t-a}{a}\right)}$$

Current Density Class 1 (-$R^{i}{}_{\mu}{}^{\mu j}$)

$$J_1 = -\frac{3}{a^2\left(\cosh\left(\frac{3t-a}{a}\right)\right)^{\frac{2}{3}}}$$

$$J_2 = -\frac{3}{a^2\left(\cosh\left(\frac{3t-a}{a}\right)\right)^{\frac{2}{3}}}$$

$$J_3 = -\frac{3}{a^2\left(\cosh\left(\frac{3t-a}{a}\right)\right)^{\frac{2}{3}}}$$

Current Density Class 2 (-$R^{i}{}_{\mu}{}^{\mu j}$)

$$J_1 = 0$$

$$J_2 = 0$$

$$J_3 = 0$$

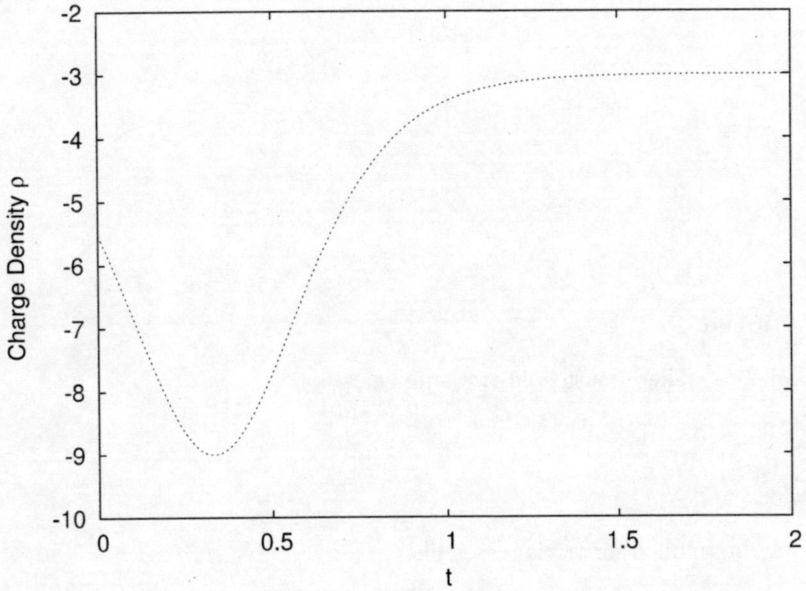

Fig. 4.55: Friedmann Dust metric, charge density ρ for $a = 1$.

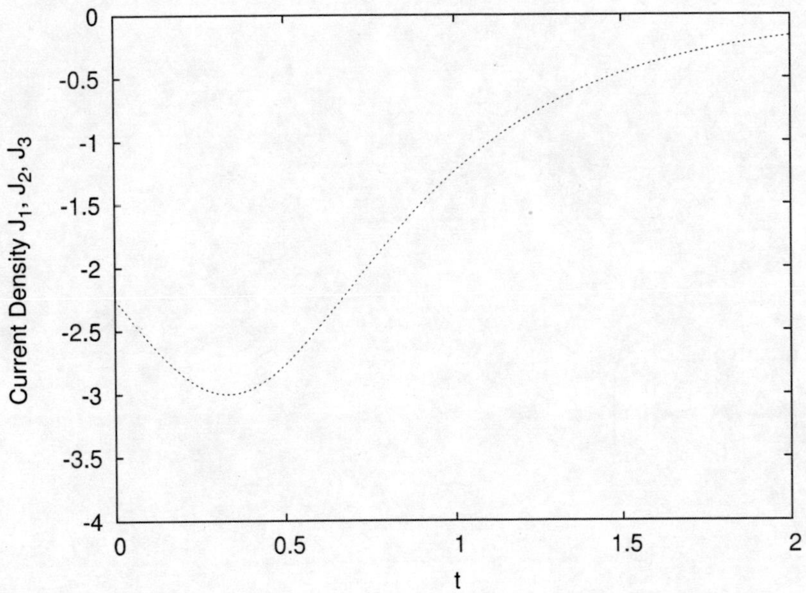

Fig. 4.56: Friedmann Dust metric, current density J_1, J_2, J_3 for $a = 1$.

Current Density Class 3 (-$R^{i}{}_{\mu}{}^{\mu j}$)

$$J_1 = 0$$

$$J_2 = 0$$

$$J_3 = 0$$

4.4.20 Kasner metric

p_1, p_2 and p_3 are parameters. There must hold two sum rules:

$$\sum_{j=1}^{3} p_j = 1, \ \sum_{j=1}^{3} p_j^2 = 1.$$

These probably can only be fulfilled for trivial cases like

$$p_1 = 1, \ p_2 = 0, \ p_3 = 0$$

and permutations. In these cases the charge and current densities are indeed zero. However, the Kasner metric does not contain dependencies on space coordinates and is nothing else than a change of one coordinate axis in time. This is not a realistic 3D model.

The plots show a case which does not obey the sum rules. Then there is a cosmological density.

Coordinates

$$\mathbf{x} = \begin{pmatrix} t \\ x_1 \\ x_2 \\ x_3 \end{pmatrix}$$

Metric

$$g_{\mu\nu} = \begin{pmatrix} -1 & 0 & 0 & 0 \\ 0 & t^{2\,p_1} & 0 & 0 \\ 0 & 0 & t^{2\,p_2} & 0 \\ 0 & 0 & 0 & t^{2\,p_3} \end{pmatrix}$$

Contravariant Metric

$$g^{\mu\nu} = \begin{pmatrix} -1 & 0 & 0 & 0 \\ 0 & \frac{1}{t^{2\,p_1}} & 0 & 0 \\ 0 & 0 & \frac{1}{t^{2\,p_2}} & 0 \\ 0 & 0 & 0 & \frac{1}{t^{2\,p_3}} \end{pmatrix}$$

Christoffel Connection

$$\Gamma^0{}_{11} = p_1\, t^{2\,p_1-1}$$

$$\Gamma^0{}_{22} = p_2\, t^{2\,p_2-1}$$

$$\Gamma^0{}_{33} = p_3\, t^{2\,p_3-1}$$

$$\Gamma^1{}_{01} = \frac{p_1}{t}$$

$$\Gamma^1{}_{10} = \Gamma^1{}_{01}$$

$$\Gamma^2{}_{02} = \frac{p_2}{t}$$

$$\Gamma^2{}_{20} = \Gamma^2{}_{02}$$

$$\Gamma^3{}_{03} = \frac{p_3}{t}$$

$$\Gamma^3{}_{30} = \Gamma^3{}_{03}$$

Metric Compatibility

——— o.k.

Riemann Tensor

$$R^0{}_{101} = (p_1 - 1)\; p_1\, t^{2\,p_1-2}$$

$$R^0{}_{110} = -R^0{}_{101}$$

$$R^0{}_{202} = (p_2 - 1)\; p_2\, t^{2\,p_2-2}$$

$$R^0{}_{220} = -R^0{}_{202}$$

$$R^0{}_{303} = (p_3 - 1)\; p_3\, t^{2\,p_3-2}$$

$$R^0{}_{330} = -R^0{}_{303}$$

$$R^1{}_{001} = \frac{(p_1 - 1)\; p_1}{t^2}$$

$$R^1{}_{010} = -R^1{}_{001}$$

$$R^1{}_{212} = p_1\; p_2\, t^{2\,p_2-2}$$

$$R^1{}_{221} = -R^1{}_{212}$$

$$R^1{}_{313} = p_1\, p_3\, t^{2\,p_3-2}$$

$$R^1{}_{331} = -R^1{}_{313}$$

$$R^2{}_{002} = \frac{(p_2-1)\, p_2}{t^2}$$

$$R^2{}_{020} = -R^2{}_{002}$$

$$R^2{}_{112} = -p_1\, p_2\, t^{2\,p_1-2}$$

$$R^2{}_{121} = -R^2{}_{112}$$

$$R^2{}_{323} = p_2\, p_3\, t^{2\,p_3-2}$$

$$R^2{}_{332} = -R^2{}_{323}$$

$$R^3{}_{003} = \frac{(p_3-1)\, p_3}{t^2}$$

$$R^3{}_{030} = -R^3{}_{003}$$

$$R^3{}_{113} = -p_1\, p_3\, t^{2\,p_1-2}$$

$$R^3{}_{131} = -R^3{}_{113}$$

$$R^3{}_{223} = -p_2\, p_3\, t^{2\,p_2-2}$$

$$R^3{}_{232} = -R^3{}_{223}$$

Ricci Tensor

$$\mathrm{Ric}_{00} = -\frac{{p_3}^2 - p_3 + {p_2}^2 - p_2 + {p_1}^2 - p_1}{t^2}$$

$$\mathrm{Ric}_{11} = p_1\,(p_3 + p_2 + p_1 - 1)\; t^{2\,p_1-2}$$

$$\mathrm{Ric}_{22} = p_2\,(p_3 + p_2 + p_1 - 1)\; t^{2\,p_2-2}$$

$$\mathrm{Ric}_{33} = p_3\,(p_3 + p_2 + p_1 - 1)\; t^{2\,p_3-2}$$

Ricci Scalar

$$R_{sc} = \frac{2\,\left({p_3}^2 + p_2\, p_3 + p_1\, p_3 - p_3 + {p_2}^2 + p_1\, p_2 - p_2 + {p_1}^2 - p_1\right)}{t^2}$$

Bianchi identity (Ricci cyclic equation $R^{\kappa}{}_{[\mu\nu\sigma]} = 0$)

——— o.k.

Einstein Tensor

$$G_{00} = \frac{p_2\, p_3 + p_1\, p_3 + p_1\, p_2}{t^2}$$

$$G_{11} = -\left(p_3{}^2 + p_2\, p_3 - p_3 + p_2{}^2 - p_2\right) t^{2\,p_1-2}$$

$$G_{22} = -\left(p_3{}^2 + p_1\, p_3 - p_3 + p_1{}^2 - p_1\right) t^{2\,p_2-2}$$

$$G_{33} = -\left(p_2{}^2 + p_1\, p_2 - p_2 + p_1{}^2 - p_1\right) t^{2\,p_3-2}$$

Hodge Dual of Bianchi Identity

——— (see charge and current densities)

Scalar Charge Density (-$R^0{}_i{}^{i0}$)

$$\rho = -\frac{p_3{}^2 - p_3 + p_2{}^2 - p_2 + p_1{}^2 - p_1}{t^2}$$

Current Density Class 1 (-$R^i{}_\mu{}^{\mu j}$)

$$J_1 = -p_1\, (p_3 + p_2 + p_1 - 1)\, t^{-2\,p_1-2}$$

$$J_2 = -p_2\, (p_3 + p_2 + p_1 - 1)\, t^{-2\,p_2-2}$$

$$J_3 = -p_3\, (p_3 + p_2 + p_1 - 1)\, t^{-2\,p_3-2}$$

Current Density Class 2 (-$R^i{}_\mu{}^{\mu j}$)

$$J_1 = 0$$

$$J_2 = 0$$

$$J_3 = 0$$

Current Density Class 3 (-$R^i{}_\mu{}^{\mu j}$)

$$J_1 = 0$$

$$J_2 = 0$$

$$J_3 = 0$$

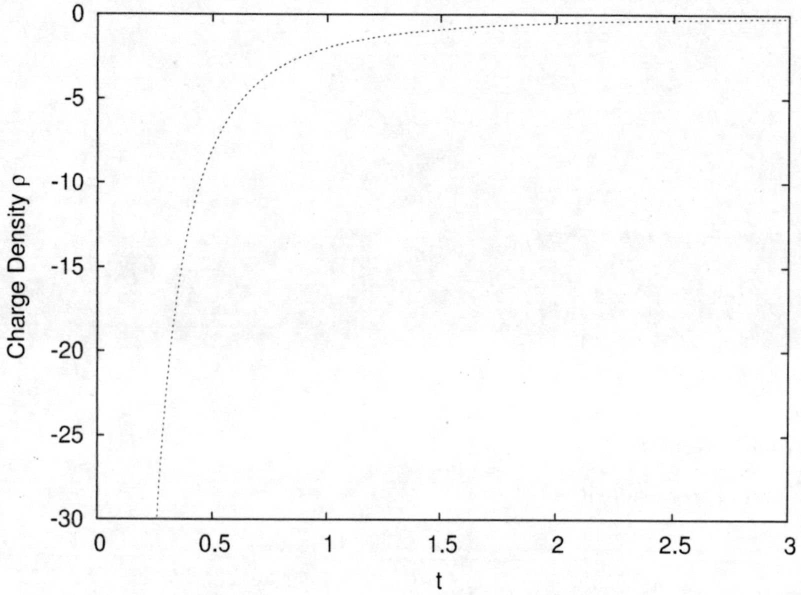

Fig. 4.57: Kasner metric, charge density ρ for $p_1 = 1, p_2 = -1, p_3 = 0$.

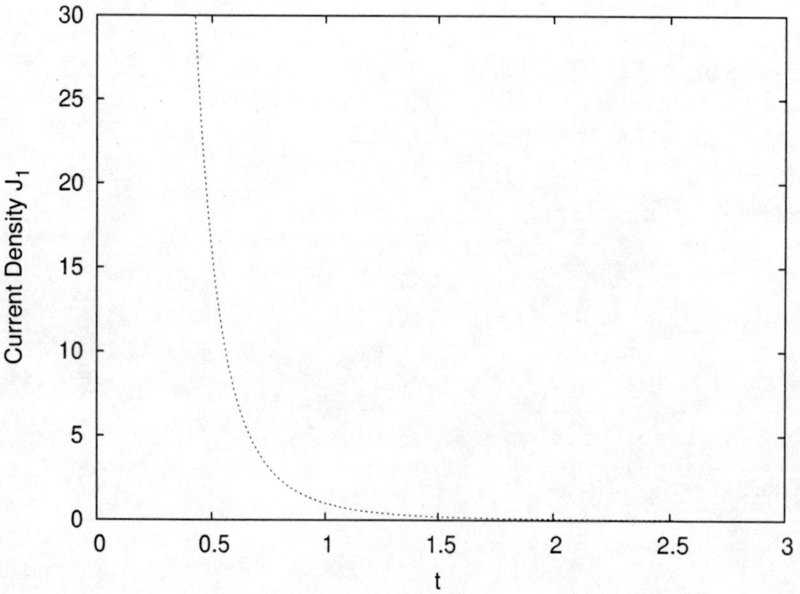

Fig. 4.58: Kasner metric, current density J_1 for $p_1 = 1, p_2 = -1, p_3 = 0$.

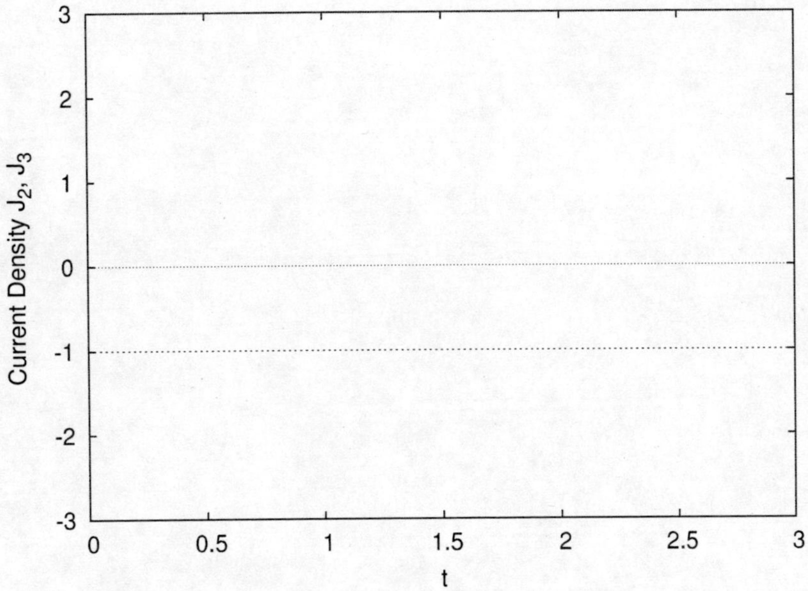

Fig. 4.59: Kasner metric, current density J_2, J_3 for $p_1 = 1, p_2 = -1, p_3 = 0$.

4.4.21 Generalized FLRW metric

The generalized form of the FLRW metric by Portugese authors. m and n are parameters.

Coordinates

$$\mathbf{x} = \begin{pmatrix} t \\ x \\ y \\ z \end{pmatrix}$$

Metric

$$g_{\mu\nu} = \begin{pmatrix} 1 & 0 & 0 & 0 \\ 0 & -\frac{t^2}{(m-n)^2} & 0 & 0 \\ 0 & 0 & -\frac{e^{-2x}}{t^{2(n+m)}} & 0 \\ 0 & 0 & 0 & -\frac{e^{2x}}{t^{2(n+m)}} \end{pmatrix}$$

Contravariant Metric

$$g^{\mu\nu} = \begin{pmatrix} 1 & 0 & 0 & 0 \\ 0 & -\frac{(n-m)^2}{t^2} & 0 & 0 \\ 0 & 0 & -t^{2n+2m}\, e^{2x} & 0 \\ 0 & 0 & 0 & -t^{2n+2m}\, e^{-2x} \end{pmatrix}$$

Christoffel Connection

$$\Gamma^0{}_{11} = \frac{t}{(n-m)^2}$$

$$\Gamma^0{}_{22} = -(n+m)\, t^{-2n-2m-1}\, e^{-2x}$$

$$\Gamma^0{}_{33} = -(n+m)\, t^{-2n-2m-1}\, e^{2x}$$

$$\Gamma^1{}_{01} = \frac{1}{t}$$

$$\Gamma^1{}_{10} = \Gamma^1{}_{01}$$

$$\Gamma^1{}_{22} = (n-m)^2\, t^{-2n-2m-2}\, e^{-2x}$$

$$\Gamma^1{}_{33} = -(n-m)^2\, t^{-2n-2m-2}\, e^{2x}$$

$$\Gamma^2{}_{02} = -\frac{n+m}{t}$$

$$\Gamma^2{}_{12} = -1$$

$$\Gamma^2{}_{20} = \Gamma^2{}_{02}$$

$$\Gamma^2{}_{21} = \Gamma^2{}_{12}$$

$$\Gamma^3{}_{03} = -\frac{n+m}{t}$$

$$\Gamma^3{}_{13} = 1$$

$$\Gamma^3{}_{30} = \Gamma^3{}_{03}$$

$$\Gamma^3{}_{31} = \Gamma^3{}_{13}$$

Metric Compatibility

——— o.k.

Riemann Tensor

$$R^0{}_{202} = (n+m)\,(n+m+1)\,t^{-2\,n-2\,m-2}\,e^{-2\,x}$$

$$R^0{}_{212} = (n+m+1)\,t^{-2\,n-2\,m-1}\,e^{-2\,x}$$

$$R^0{}_{220} = -R^0{}_{202}$$

$$R^0{}_{221} = -R^0{}_{212}$$

$$R^0{}_{303} = (n+m)\,(n+m+1)\,t^{-2\,n-2\,m-2}\,e^{2\,x}$$

$$R^0{}_{313} = -\,(n+m+1)\,t^{-2\,n-2\,m-1}\,e^{2\,x}$$

$$R^0{}_{330} = -R^0{}_{303}$$

$$R^0{}_{331} = -R^0{}_{313}$$

$$R^1{}_{202} = -\,(n-m)^2\,(n+m+1)\,t^{-2\,n-2\,m-3}\,e^{-2\,x}$$

$$R^1{}_{212} = -\left(n^2-2\,m\,n+n+m^2+m\right)\,t^{-2\,n-2\,m-2}\,e^{-2\,x}$$

$$R^1{}_{220} = -R^1{}_{202}$$

$$R^1{}_{221} = -R^1{}_{212}$$

$$R^1{}_{303} = (n-m)^2\,(n+m+1)\,t^{-2\,n-2\,m-3}\,e^{2\,x}$$

$$R^1{}_{313} = -\left(n^2-2\,m\,n+n+m^2+m\right)\,t^{-2\,n-2\,m-2}\,e^{2\,x}$$

$$R^1{}_{330} = -R^1{}_{303}$$

$$R^1{}_{331} = -R^1{}_{313}$$

$$R^2{}_{002} = \frac{(n+m)\,(n+m+1)}{t^2}$$

$$R^2{}_{012} = \frac{n+m+1}{t}$$

$$R^2{}_{020} = -R^2{}_{002}$$

$$R^2{}_{021} = -R^2{}_{012}$$

$$R^2{}_{102} = \frac{n+m+1}{t}$$

$$R^2{}_{112} = \frac{n^2 - 2\,m\,n + n + m^2 + m}{(n-m)^2}$$

$$R^2{}_{120} = -R^2{}_{102}$$

$$R^2{}_{121} = -R^2{}_{112}$$

$$R^2{}_{323} = 2\,\left(n^2 + m^2\right)\,t^{-2\,n-2\,m-2}\,e^{2\,x}$$

$$R^2{}_{332} = -R^2{}_{323}$$

$$R^3{}_{003} = \frac{(n+m)\,(n+m+1)}{t^2}$$

$$R^3{}_{013} = -\frac{n+m+1}{t}$$

$$R^3{}_{030} = -R^3{}_{003}$$

$$R^3{}_{031} = -R^3{}_{013}$$

$$R^3{}_{103} = -\frac{n+m+1}{t}$$

$$R^3{}_{113} = \frac{n^2 - 2\,m\,n + n + m^2 + m}{(n-m)^2}$$

$$R^3{}_{130} = -R^3{}_{103}$$

$$R^3{}_{131} = -R^3{}_{113}$$

$$R^3{}_{223} = -2\,\left(n^2 + m^2\right)\,t^{-2\,n-2\,m-2}\,e^{-2\,x}$$

$$R^3{}_{232} = -R^3{}_{223}$$

Ricci Tensor

$$\mathrm{Ric}_{00} = -\frac{2\,(n+m)\,(n+m+1)}{t^2}$$

$$\mathrm{Ric}_{11} = -\frac{2\,\left(n^2 - 2\,m\,n + n + m^2 + m\right)}{(n-m)^2}$$

$$\mathrm{Ric}_{22} = 2\,(n+m)^2\,t^{-2\,n-2\,m-2}\,e^{-2\,x}$$

$$\mathrm{Ric}_{33} = 2\,(n+m)^2\,t^{-2\,n-2\,m-2}\,e^{2\,x}$$

Ricci Scalar

$$R_{sc} = -\frac{4\,(n^2 + 4\,m\,n + m^2)}{t^2}$$

Bianchi identity (Ricci cyclic equation $R^{\kappa}{}_{[\mu\nu\sigma]} = 0$)

——— o.k.

Einstein Tensor

$$G_{00} = \frac{2\,(2\,m\,n - n - m)}{t^2}$$

$$G_{11} = -\frac{2\,(2\,n^2 + 2\,m\,n + n + 2\,m^2 + m)}{(n-m)^2}$$

$$G_{22} = -4\,m\,n\,t^{-2\,n-2\,m-2}\,e^{-2\,x}$$

$$G_{33} = -4\,m\,n\,t^{-2\,n-2\,m-2}\,e^{2\,x}$$

Hodge Dual of Bianchi Identity

——— (see charge and current densities)

Scalar Charge Density (-$R^{0}{}_{i}{}^{i0}$)

$$\rho = -\frac{2\,(n+m)\,(n+m+1)}{t^2}$$

Current Density Class 1 (-$R^{i}{}_{\mu}{}^{\mu j}$)

$$J_1 = \frac{2\,(n-m)^2\,(n^2 - 2\,m\,n + n + m^2 + m)}{t^4}$$

$$J_2 = -2\,(n+m)^2\,t^{2\,n+2\,m-2}\,e^{2\,x}$$

$$J_3 = -2\,(n+m)^2\,t^{2\,n+2\,m-2}\,e^{-2\,x}$$

Current Density Class 2 (-$R^{i}{}_{\mu}{}^{\mu j}$)

$$J_1 = 0$$

$$J_2 = 0$$

$$J_3 = 0$$

Current Density Class 3 ($-R^{i}{}_{\mu}{}^{\mu j}$)

$$J_1 = 0$$

$$J_2 = 0$$

$$J_3 = 0$$

4.4.22 Eddington-Finkelstein metric for black holes

This metric has non-diagonal terms and a zero on the main diagonal. G and M are the usual parameters of the spherical metric.

Coordinates

$$\mathbf{x} = \begin{pmatrix} u \\ r \\ \vartheta \\ \varphi \end{pmatrix}$$

Metric

$$g_{\mu\nu} = \begin{pmatrix} \frac{2GM}{r} - 1 & 1 & 0 & 0 \\ 1 & 0 & 0 & 0 \\ 0 & 0 & r^2 & 0 \\ 0 & 0 & 0 & r^2 \sin^2\vartheta \end{pmatrix}$$

Contravariant Metric

$$g^{\mu\nu} = \begin{pmatrix} 0 & 1 & 0 & 0 \\ 1 & -\frac{2GM-r}{r} & 0 & 0 \\ 0 & 0 & \frac{1}{r^2} & 0 \\ 0 & 0 & 0 & \frac{1}{r^2 \sin^2\vartheta} \end{pmatrix}$$

Christoffel Connection

$$\Gamma^{0}{}_{00} = \frac{GM}{r^2}$$

$$\Gamma^{0}{}_{22} = -r$$

$$\Gamma^{0}{}_{33} = -r \sin^2\vartheta$$

$$\Gamma^{1}{}_{00} = -\frac{GM\,(2GM - r)}{r^3}$$

$$\Gamma^{1}{}_{01} = -\frac{GM}{r^2}$$

$$\Gamma^1{}_{10} = \Gamma^1{}_{01}$$

$$\Gamma^1{}_{22} = 2\,G\,M - r$$

$$\Gamma^1{}_{33} = \sin^2\vartheta\,(2\,G\,M - r)$$

$$\Gamma^2{}_{12} = \frac{1}{r}$$

$$\Gamma^2{}_{21} = \Gamma^2{}_{12}$$

$$\Gamma^2{}_{33} = -\cos\vartheta\,\sin\vartheta$$

$$\Gamma^3{}_{13} = \frac{1}{r}$$

$$\Gamma^3{}_{23} = \frac{\cos\vartheta}{\sin\vartheta}$$

$$\Gamma^3{}_{31} = \Gamma^3{}_{13}$$

$$\Gamma^3{}_{32} = \Gamma^3{}_{23}$$

Metric Compatibility

——— o.k.

Riemann Tensor

$$R^0{}_{001} = \frac{2\,G\,M}{r^3}$$

$$R^0{}_{010} = -R^0{}_{001}$$

$$R^0{}_{202} = -\frac{G\,M}{r}$$

$$R^0{}_{220} = -R^0{}_{202}$$

$$R^0{}_{303} = -\frac{\sin^2\vartheta\,G\,M}{r}$$

$$R^0{}_{330} = -R^0{}_{303}$$

$$R^1{}_{001} = -\frac{2\,G\,M\,(2\,G\,M - r)}{r^4}$$

$$R^1{}_{010} = -R^1{}_{001}$$

$$R^1{}_{101} = -\frac{2\,G\,M}{r^3}$$

$$R^1{}_{110} = -R^1{}_{101}$$

$$R^1{}_{212} = -\frac{G\,M}{r}$$

$$R^1{}_{221} = -R^1{}_{212}$$

$$R^1{}_{313} = -\frac{\sin^2\vartheta\,G\,M}{r}$$

$$R^1{}_{331} = -R^1{}_{313}$$

$$R^2{}_{002} = \frac{G\,M\,(2\,G\,M - r)}{r^4}$$

$$R^2{}_{012} = \frac{G\,M}{r^3}$$

$$R^2{}_{020} = -R^2{}_{002}$$

$$R^2{}_{021} = -R^2{}_{012}$$

$$R^2{}_{102} = \frac{G\,M}{r^3}$$

$$R^2{}_{120} = -R^2{}_{102}$$

$$R^2{}_{323} = \frac{2\,\sin^2\vartheta\,G\,M}{r}$$

$$R^2{}_{332} = -R^2{}_{323}$$

$$R^3{}_{003} = \frac{G\,M\,(2\,G\,M - r)}{r^4}$$

$$R^3{}_{013} = \frac{G\,M}{r^3}$$

$$R^3{}_{030} = -R^3{}_{003}$$

$$R^3{}_{031} = -R^3{}_{013}$$

$$R^3{}_{103} = \frac{G\,M}{r^3}$$

$$R^3{}_{130} = -R^3{}_{103}$$

$$R^3{}_{223} = -\frac{2\,G\,M}{r}$$

$$R^3{}_{232} = -R^3{}_{223}$$

Ricci Tensor

——— all elements zero

Ricci Scalar

$$R_{sc} = 0$$

Bianchi identity (Ricci cyclic equation $R^{\kappa}{}_{[\mu\nu\sigma]} = 0$)

——— o.k.

Einstein Tensor

——— all elements zero

Hodge Dual of Bianchi Identity

——— (see charge and current densities)

Scalar Charge Density (-$R^{0}{}_{i}{}^{i0}$)

$$\rho = 0$$

Current Density Class 1 (-$R^{i}{}_{\mu}{}^{\mu j}$)

$$J_1 = 0$$

$$J_2 = 0$$

$$J_3 = 0$$

Current Density Class 2 (-$R^{i}{}_{\mu}{}^{\mu j}$)

$$J_1 = 0$$

$$J_2 = 0$$

$$J_3 = 0$$

Current Density Class 3 (-$R^{i}{}_{\mu}{}^{\mu j}$)

$$J_1 = 0$$

$$J_2 = 0$$

$$J_3 = 0$$

4.4.23 Kruskal coordinates metric of black hole

r is a function of the coordinates u and v.

Coordinates

$$\mathbf{x} = \begin{pmatrix} v \\ u \\ \vartheta \\ \varphi \end{pmatrix}$$

Metric

$$g_{\mu\nu} = \begin{pmatrix} -\frac{32\,G^3\,M^3\,e^{-\frac{r}{2\,G\,M}}}{r} & 0 & 0 & 0 \\ 0 & \frac{32\,G^3\,M^3\,e^{-\frac{r}{2\,G\,M}}}{r} & 0 & 0 \\ 0 & 0 & r^2 & 0 \\ 0 & 0 & 0 & r^2\,\sin^2\vartheta \end{pmatrix}$$

Contravariant Metric

$$g^{\mu\nu} = \begin{pmatrix} -\frac{r\,e^{\frac{r}{2\,G\,M}}}{32\,G^3\,M^3} & 0 & 0 & 0 \\ 0 & \frac{r\,e^{\frac{r}{2\,G\,M}}}{32\,G^3\,M^3} & 0 & 0 \\ 0 & 0 & \frac{1}{r^2} & 0 \\ 0 & 0 & 0 & \frac{1}{r^2\,\sin^2\vartheta} \end{pmatrix}$$

Christoffel Connection

$$\Gamma^0{}_{00} = -\frac{\frac{d}{d\,v}\,r\,(2\,G\,M + r)}{4\,r\,G\,M}$$

$$\Gamma^0{}_{01} = -\frac{\frac{d}{d\,u}\,r\,(2\,G\,M + r)}{4\,r\,G\,M}$$

$$\Gamma^0{}_{10} = \Gamma^0{}_{01}$$

$$\Gamma^0{}_{11} = -\frac{\frac{d}{d\,v}\,r\,(2\,G\,M + r)}{4\,r\,G\,M}$$

$$\Gamma^0{}_{22} = \frac{r^2\,\left(\frac{d}{d\,v}\,r\right)\,e^{\frac{r}{2\,G\,M}}}{32\,G^3\,M^3}$$

$$\Gamma^0{}_{33} = \frac{r^2\,\left(\frac{d}{d\,v}\,r\right)\,\sin^2\vartheta\,e^{\frac{r}{2\,G\,M}}}{32\,G^3\,M^3}$$

$$\Gamma^1{}_{00} = -\frac{\frac{d}{d\,u}\,r\,(2\,G\,M + r)}{4\,r\,G\,M}$$

$$\Gamma^1{}_{01} = -\frac{\frac{d}{d\,v}\,r\,(2\,G\,M + r)}{4\,r\,G\,M}$$

$$\Gamma^1{}_{10} = \Gamma^1{}_{01}$$

$$\Gamma^1{}_{11} = -\frac{\frac{d}{d\,u}\,r\,(2\,G\,M + r)}{4\,r\,G\,M}$$

$$\Gamma^1{}_{22} = -\frac{r^2\,\left(\frac{d}{d\,u}\,r\right)\,e^{\frac{r}{2\,G\,M}}}{32\,G^3\,M^3}$$

$$\Gamma^1{}_{33} = -\frac{r^2\,\left(\frac{d}{d\,u}\,r\right)\,\sin^2\vartheta\,e^{\frac{r}{2\,G\,M}}}{32\,G^3\,M^3}$$

$$\Gamma^2{}_{02} = \frac{\frac{d}{d\,v}\,r}{r}$$

$$\Gamma^2{}_{12} = \frac{\frac{d}{d\,u}\,r}{r}$$

$$\Gamma^2{}_{20} = \Gamma^2{}_{02}$$

$$\Gamma^2{}_{21} = \Gamma^2{}_{12}$$

$$\Gamma^2{}_{33} = -\cos\vartheta\,\sin\vartheta$$

$$\Gamma^3{}_{03} = \frac{\frac{d}{d\,v}\,r}{r}$$

$$\Gamma^3{}_{13} = \frac{\frac{d}{d\,u}\,r}{r}$$

$$\Gamma^3{}_{23} = \frac{\cos\vartheta}{\sin\vartheta}$$

$$\Gamma^3{}_{30} = \Gamma^3{}_{03}$$

$$\Gamma^3{}_{31} = \Gamma^3{}_{13}$$

$$\Gamma^3{}_{32} = \Gamma^3{}_{23}$$

Metric Compatibility

——— o.k.

Riemann Tensor

$$R^0{}_{101} = -\frac{2\,r\,\left(\frac{d^2}{d\,v^2}\,r\right)\,G\,M - 2\,\left(\frac{d}{d\,v}\,r\right)^2\,G\,M - 2\,r\,\left(\frac{d^2}{d\,u^2}\,r\right)\,G\,M + 2\,\left(\frac{d}{d\,u}\,r\right)^2\,G\,M + r^2\,\left(\frac{d^2}{d\,v^2}\,r\right) - r^2\,\left(\frac{d^2}{d\,u^2}\,r\right)}{4\,r^2\,G\,M}$$

$$R^0{}_{110} = -R^0{}_{101}$$

$$R^0{}_{202} = \frac{r\,\left(4\,r\,\left(\frac{d^2}{d\,v^2}\,r\right)\,G\,M + 2\,\left(\frac{d}{d\,v}\,r\right)^2\,G\,M + 2\,\left(\frac{d}{d\,u}\,r\right)^2\,G\,M + r\,\left(\frac{d}{d\,v}\,r\right)^2 + r\,\left(\frac{d}{d\,u}\,r\right)^2\right)\,e^{\frac{r}{2\,G\,M}}}{128\,G^4\,M^4}$$

$$R^0{}_{212} = \frac{r\left(2\left(\frac{d}{d\,u}\,r\right)\left(\frac{d}{d\,v}\,r\right)G\,M+2\,r\left(\frac{d^2}{d\,u\,d\,v}\,r\right)G\,M+r\left(\frac{d}{d\,u}\,r\right)\left(\frac{d}{d\,v}\,r\right)\right)e^{\frac{r}{2\,G\,M}}}{64\,G^4\,M^4}$$

$$R^0{}_{220} = -R^0{}_{202}$$

$$R^0{}_{221} = -R^0{}_{212}$$

$$R^0{}_{303} = \frac{r\,\sin^2\vartheta\left(4\,r\left(\frac{d^2}{d\,v^2}\,r\right)G\,M+2\left(\frac{d}{d\,v}\,r\right)^2G\,M+2\left(\frac{d}{d\,u}\,r\right)^2G\,M+r\left(\frac{d}{d\,v}\,r\right)^2+r\left(\frac{d}{d\,u}\,r\right)^2\right)e^{\frac{r}{2\,G\,M}}}{128\,G^4\,M^4}$$

$$R^0{}_{313} = \frac{r\,\sin^2\vartheta\left(2\left(\frac{d}{d\,u}\,r\right)\left(\frac{d}{d\,v}\,r\right)G\,M+2\,r\left(\frac{d^2}{d\,u\,d\,v}\,r\right)G\,M+r\left(\frac{d}{d\,u}\,r\right)\left(\frac{d}{d\,v}\,r\right)\right)e^{\frac{r}{2\,G\,M}}}{64\,G^4\,M^4}$$

$$R^0{}_{330} = -R^0{}_{303}$$

$$R^0{}_{331} = -R^0{}_{313}$$

$$R^1{}_{001} = -\frac{2\,r\left(\frac{d^2}{d\,v^2}\,r\right)G\,M-2\left(\frac{d}{d\,v}\,r\right)^2G\,M-2\,r\left(\frac{d^2}{d\,u^2}\,r\right)G\,M+2\left(\frac{d}{d\,u}\,r\right)^2G\,M+r^2\left(\frac{d^2}{d\,v^2}\,r\right)-r^2\left(\frac{d^2}{d\,u^2}\,r\right)}{4\,r^2\,G\,M}$$

$$R^1{}_{010} = -R^1{}_{001}$$

$$R^1{}_{202} = -\frac{r\left(2\left(\frac{d}{d\,u}\,r\right)\left(\frac{d}{d\,v}\,r\right)G\,M+2\,r\left(\frac{d^2}{d\,u\,d\,v}\,r\right)G\,M+r\left(\frac{d}{d\,u}\,r\right)\left(\frac{d}{d\,v}\,r\right)\right)e^{\frac{r}{2\,G\,M}}}{64\,G^4\,M^4}$$

$$R^1{}_{212} = -\frac{r\left(2\left(\frac{d}{d\,v}\,r\right)^2G\,M+4\,r\left(\frac{d^2}{d\,u^2}\,r\right)G\,M+2\left(\frac{d}{d\,u}\,r\right)^2G\,M+r\left(\frac{d}{d\,v}\,r\right)^2+r\left(\frac{d}{d\,u}\,r\right)^2\right)e^{\frac{r}{2\,G\,M}}}{128\,G^4\,M^4}$$

$$R^1{}_{220} = -R^1{}_{202}$$

$$R^1{}_{221} = -R^1{}_{212}$$

$$R^1{}_{303} = -\frac{r\,\sin^2\vartheta\left(2\left(\frac{d}{d\,u}\,r\right)\left(\frac{d}{d\,v}\,r\right)G\,M+2\,r\left(\frac{d^2}{d\,u\,d\,v}\,r\right)G\,M+r\left(\frac{d}{d\,u}\,r\right)\left(\frac{d}{d\,v}\,r\right)\right)e^{\frac{r}{2\,G\,M}}}{64\,G^4\,M^4}$$

$$R^1{}_{313} = -\frac{r\,\sin^2\vartheta\left(2\left(\frac{d}{d\,v}\,r\right)^2G\,M+4\,r\left(\frac{d^2}{d\,u^2}\,r\right)G\,M+2\left(\frac{d}{d\,u}\,r\right)^2G\,M+r\left(\frac{d}{d\,v}\,r\right)^2+r\left(\frac{d}{d\,u}\,r\right)^2\right)e^{\frac{r}{2\,G\,M}}}{128\,G^4\,M^4}$$

$$R^1{}_{330} = -R^1{}_{303}$$

$$R^1{}_{331} = -R^1{}_{313}$$

$$R^2{}_{002} = \frac{4\,r\left(\frac{d^2}{d\,v^2}\,r\right)G\,M+2\left(\frac{d}{d\,v}\,r\right)^2G\,M+2\left(\frac{d}{d\,u}\,r\right)^2G\,M+r\left(\frac{d}{d\,v}\,r\right)^2+r\left(\frac{d}{d\,u}\,r\right)^2}{4\,r^2\,G\,M}$$

$$R^2{}_{012} = \frac{2\left(\frac{d}{d\,u}\,r\right)\left(\frac{d}{d\,v}\,r\right)G\,M+2\,r\left(\frac{d^2}{d\,u\,d\,v}\,r\right)G\,M+r\left(\frac{d}{d\,u}\,r\right)\left(\frac{d}{d\,v}\,r\right)}{2\,r^2\,G\,M}$$

$$R^2{}_{020} = -R^2{}_{002}$$

$$R^2{}_{021} = -R^2{}_{012}$$

$$R^2{}_{102} = \frac{2\left(\frac{d}{d\,u}\,r\right)\left(\frac{d}{d\,v}\,r\right)G\,M + 2\,r\left(\frac{d^2}{d\,u\,d\,v}\,r\right)G\,M + r\left(\frac{d}{d\,u}\,r\right)\left(\frac{d}{d\,v}\,r\right)}{2\,r^2\,G\,M}$$

$$R^2{}_{112} = \frac{2\left(\frac{d}{d\,v}\,r\right)^2 G\,M + 4\,r\left(\frac{d^2}{d\,u^2}\,r\right)G\,M + 2\left(\frac{d}{d\,u}\,r\right)^2 G\,M + r\left(\frac{d}{d\,v}\,r\right)^2 + r\left(\frac{d}{d\,u}\,r\right)^2}{4\,r^2\,G\,M}$$

$$R^2{}_{120} = -R^2{}_{102}$$

$$R^2{}_{121} = -R^2{}_{112}$$

$$R^2{}_{323} = \frac{\sin^2\vartheta\left(r\left(\frac{d}{d\,v}\,r\right)^2 e^{\frac{r}{2\,G\,M}} - r\left(\frac{d}{d\,u}\,r\right)^2 e^{\frac{r}{2\,G\,M}} + 32\,G^3\,M^3\right)}{32\,G^3\,M^3}$$

$$R^2{}_{332} = -R^2{}_{323}$$

$$R^3{}_{003} = \frac{4\,r\left(\frac{d^2}{d\,v^2}\,r\right)G\,M + 2\left(\frac{d}{d\,v}\,r\right)^2 G\,M + 2\left(\frac{d}{d\,u}\,r\right)^2 G\,M + r\left(\frac{d}{d\,v}\,r\right)^2 + r\left(\frac{d}{d\,u}\,r\right)^2}{4\,r^2\,G\,M}$$

$$R^3{}_{013} = \frac{2\left(\frac{d}{d\,u}\,r\right)\left(\frac{d}{d\,v}\,r\right)G\,M + 2\,r\left(\frac{d^2}{d\,u\,d\,v}\,r\right)G\,M + r\left(\frac{d}{d\,u}\,r\right)\left(\frac{d}{d\,v}\,r\right)}{2\,r^2\,G\,M}$$

$$R^3{}_{030} = -R^3{}_{003}$$

$$R^3{}_{031} = -R^3{}_{013}$$

$$R^3{}_{103} = \frac{2\left(\frac{d}{d\,u}\,r\right)\left(\frac{d}{d\,v}\,r\right)G\,M + 2\,r\left(\frac{d^2}{d\,u\,d\,v}\,r\right)G\,M + r\left(\frac{d}{d\,u}\,r\right)\left(\frac{d}{d\,v}\,r\right)}{2\,r^2\,G\,M}$$

$$R^3{}_{113} = \frac{2\left(\frac{d}{d\,v}\,r\right)^2 G\,M + 4\,r\left(\frac{d^2}{d\,u^2}\,r\right)G\,M + 2\left(\frac{d}{d\,u}\,r\right)^2 G\,M + r\left(\frac{d}{d\,v}\,r\right)^2 + r\left(\frac{d}{d\,u}\,r\right)^2}{4\,r^2\,G\,M}$$

$$R^3{}_{130} = -R^3{}_{103}$$

$$R^3{}_{131} = -R^3{}_{113}$$

$$R^3{}_{223} = -\frac{r\left(\frac{d}{d\,v}\,r\right)^2 e^{\frac{r}{2\,G\,M}} - r\left(\frac{d}{d\,u}\,r\right)^2 e^{\frac{r}{2\,G\,M}} + 32\,G^3\,M^3}{32\,G^3\,M^3}$$

$$R^3{}_{232} = -R^3{}_{223}$$

Ricci Tensor

$$\mathrm{Ric}_{00} = -\frac{6\,r\left(\frac{d^2}{d\,v^2}\,r\right)G\,M + 6\left(\frac{d}{d\,v}\,r\right)^2 G\,M + 2\,r\left(\frac{d^2}{d\,u^2}\,r\right)G\,M + 2\left(\frac{d}{d\,u}\,r\right)^2 G\,M - r^2\left(\frac{d^2}{d\,v^2}\,r\right) + 2\,r\left(\frac{d}{d\,v}\,r\right)^2 + r^2\left(\frac{d^2}{d\,u^2}\,r\right) + \ldots}{4\,r^2\,G\,M}$$

$$\mathrm{Ric}_{01} = -\frac{2\left(\frac{d}{d\,u}\,r\right)\left(\frac{d}{d\,v}\,r\right)G\,M + 2\,r\left(\frac{d^2}{d\,u\,d\,v}\,r\right)G\,M + r\left(\frac{d}{d\,u}\,r\right)\left(\frac{d}{d\,v}\,r\right)}{r^2\,G\,M}$$

$$\mathrm{Ric}_{10} = \mathrm{Ric}_{01}$$

$$\mathrm{Ric}_{11} = -\frac{2\,r\left(\frac{d^2}{d\,v^2}\,r\right)G\,M + 2\left(\frac{d}{d\,v}\,r\right)^2 G\,M + 6\,r\left(\frac{d^2}{d\,u^2}\,r\right)G\,M + 6\left(\frac{d}{d\,u}\,r\right)^2 G\,M + r^2\left(\frac{d^2}{d\,v^2}\,r\right) + 2\,r\left(\frac{d}{d\,v}\,r\right)^2 - r^2\left(\frac{d^2}{d\,u^2}\,r\right) + \ldots}{4\,r^2\,G\,M}$$

$$\mathrm{Ric}_{22} = \frac{r^2\left(\frac{d^2}{d\,v^2}\,r\right)e^{\frac{r}{2\,G\,M}} + r\left(\frac{d}{d\,v}\,r\right)^2 e^{\frac{r}{2\,G\,M}} - r^2\left(\frac{d^2}{d\,u^2}\,r\right)e^{\frac{r}{2\,G\,M}} - r\left(\frac{d}{d\,u}\,r\right)^2 e^{\frac{r}{2\,G\,M}} + 32\,G^3\,M^3}{32\,G^3\,M^3}$$

$$\mathrm{Ric}_{33} = \frac{\sin^2\vartheta\left(r^2\left(\frac{d^2}{d\,v^2}\,r\right)e^{\frac{r}{2\,G\,M}} + r\left(\frac{d}{d\,v}\,r\right)^2 e^{\frac{r}{2\,G\,M}} - r^2\left(\frac{d^2}{d\,u^2}\,r\right)e^{\frac{r}{2\,G\,M}} - r\left(\frac{d}{d\,u}\,r\right)^2 e^{\frac{r}{2\,G\,M}} + 32\,G^3\,M^3\right)}{32\,G^3\,M^3}$$

Ricci Scalar

$$R_{sc} = \frac{6\,r^2\left(\frac{d^2}{d\,v^2}\,r\right)G\,M\,e^{\frac{r}{2\,G\,M}} + 6\,r\left(\frac{d}{d\,v}\,r\right)^2 G\,M\,e^{\frac{r}{2\,G\,M}} - 6\,r^2\left(\frac{d^2}{d\,u^2}\,r\right)G\,M\,e^{\frac{r}{2\,G\,M}} - 6\,r\left(\frac{d}{d\,u}\,r\right)^2 G\,M\,e^{\frac{r}{2\,G\,M}} - r^3\left(\frac{d^2}{d\,v^2}\,r\right)e^{\frac{r}{2\,G\,M}} + \ldots}{64\,r^2\,G^4\,M^4}$$

Bianchi identity (Ricci cyclic equation ${R^\kappa}_{[\mu\nu\sigma]} = 0$)

———— o.k.

Einstein Tensor

$$G_{00} = -\frac{e^{-\frac{r}{2\,G\,M}}\left(4\,r^2\left(\frac{d^2}{d\,u^2}\,r\right)G\,M\,e^{\frac{r}{2\,G\,M}} + 4\,r\left(\frac{d}{d\,u}\,r\right)^2 G\,M\,e^{\frac{r}{2\,G\,M}} + r^2\left(\frac{d}{d\,v}\,r\right)^2 e^{\frac{r}{2\,G\,M}} + r^2\left(\frac{d}{d\,u}\,r\right)^2 e^{\frac{r}{2\,G\,M}} - 64\,G^4\,M^4\right)}{2\,r^3\,G\,M}$$

$$G_{01} = -\frac{2\left(\frac{d}{d\,u}\,r\right)\left(\frac{d}{d\,v}\,r\right)G\,M + 2\,r\left(\frac{d^2}{d\,u\,d\,v}\,r\right)G\,M + r\left(\frac{d}{d\,u}\,r\right)\left(\frac{d}{d\,v}\,r\right)}{r^2\,G\,M}$$

$$G_{10} = G_{01}$$

$$G_{11} = -\frac{e^{-\frac{r}{2\,G\,M}}\left(4\,r^2\left(\frac{d^2}{d\,v^2}\,r\right)G\,M\,e^{\frac{r}{2\,G\,M}} + 4\,r\left(\frac{d}{d\,v}\,r\right)^2 G\,M\,e^{\frac{r}{2\,G\,M}} + r^2\left(\frac{d}{d\,v}\,r\right)^2 e^{\frac{r}{2\,G\,M}} + r^2\left(\frac{d}{d\,u}\,r\right)^2 e^{\frac{r}{2\,G\,M}} + 64\,G^4\,M^4\right)}{2\,r^3\,G\,M}$$

$$G_{22} = -\frac{r\left(2\,r\left(\frac{d^2}{d\,v^2}\,r\right)G\,M + 2\left(\frac{d}{d\,v}\,r\right)^2 G\,M - 2\,r\left(\frac{d^2}{d\,u^2}\,r\right)G\,M - 2\left(\frac{d}{d\,u}\,r\right)^2 G\,M - r^2\left(\frac{d^2}{d\,v^2}\,r\right) + r^2\left(\frac{d^2}{d\,u^2}\,r\right)\right)e^{\frac{r}{2\,G\,M}}}{128\,G^4\,M^4}$$

$$G_{33} = -\frac{r\,\sin^2\vartheta\left(2\,r\left(\frac{d^2}{d\,v^2}\,r\right)G\,M + 2\left(\frac{d}{d\,v}\,r\right)^2 G\,M - 2\,r\left(\frac{d^2}{d\,u^2}\,r\right)G\,M - 2\left(\frac{d}{d\,u}\,r\right)^2 G\,M - r^2\left(\frac{d^2}{d\,v^2}\,r\right) + r^2\left(\frac{d^2}{d\,u^2}\,r\right)\right)e^{\frac{r}{2\,G\,M}}}{128\,G^4\,M^4}$$

Hodge Dual of Bianchi Identity

——— (see charge and current densities)

Scalar Charge Density (-$R^{0\ \ i0}_{\ \ i}$)

$$\rho = -\frac{\left(6r\left(\frac{d^2}{d\,v^2}r\right)G\,M + 6\left(\frac{d}{d\,v}r\right)^2 G\,M + 2r\left(\frac{d^2}{d\,u^2}r\right)G\,M + 2\left(\frac{d}{d\,u}r\right)^2 G\,M - r^2\left(\frac{d^2}{d\,v^2}r\right) + 2r\left(\frac{d}{d\,v}r\right)^2 + \ldots\right)e^{\frac{r}{G\,M}}}{4096\,G^7\,M^7}$$

Current Density Class 1 (-$R^{i\ \ \mu j}_{\ \mu}$)

$$J_1 = \frac{\left(2r\left(\frac{d^2}{d\,v^2}r\right)G\,M + 2\left(\frac{d}{d\,v}r\right)^2 G\,M + 6r\left(\frac{d^2}{d\,u^2}r\right)G\,M + 6\left(\frac{d}{d\,u}r\right)^2 G\,M + r^2\left(\frac{d^2}{d\,v^2}r\right) + 2r\left(\frac{d}{d\,v}r\right)^2 - r^2\left(\frac{d^2}{d\,u^2}r\right) + \ldots\right)e^{\frac{r}{G\,M}}}{4096\,G^7\,M^7}$$

$$J_2 = -\frac{r^2\left(\frac{d^2}{d\,v^2}r\right)e^{\frac{r}{2\,G\,M}} + r\left(\frac{d}{d\,v}r\right)^2 e^{\frac{r}{2\,G\,M}} - r^2\left(\frac{d^2}{d\,u^2}r\right)e^{\frac{r}{2\,G\,M}} - r\left(\frac{d}{d\,u}r\right)^2 e^{\frac{r}{2\,G\,M}} + 32\,G^3\,M^3}{32\,r^4\,G^3\,M^3}$$

$$J_3 = -\frac{r^2\left(\frac{d^2}{d\,v^2}r\right)e^{\frac{r}{2\,G\,M}} + r\left(\frac{d}{d\,v}r\right)^2 e^{\frac{r}{2\,G\,M}} - r^2\left(\frac{d^2}{d\,u^2}r\right)e^{\frac{r}{2\,G\,M}} - r\left(\frac{d}{d\,u}r\right)^2 e^{\frac{r}{2\,G\,M}} + 32\,G^3\,M^3}{32\,r^4\,\sin^2\vartheta\,G^3\,M^3}$$

Current Density Class 2 (-$R^{i\ \ \mu j}_{\ \mu}$)

$$J_1 = 0$$

$$J_2 = 0$$

$$J_3 = 0$$

Current Density Class 3 (-$R^{i\ \ \mu j}_{\ \mu}$)

$$J_1 = 0$$

$$J_2 = 0$$

$$J_3 = 0$$

4.4.24 Einstein-Rosen bridge metric, u cordinates

Metric of the Einstein-Rosen bridge with u coordinates.

Coordinates

$$\mathbf{x} = \begin{pmatrix} t \\ u \\ \vartheta \\ \varphi \end{pmatrix}$$

Metric

$$g_{\mu\nu} = \begin{pmatrix} \frac{u^2}{u^2+2\,m} & 0 & 0 & 0 \\ 0 & -4\,(u^2+2\,m) & 0 & 0 \\ 0 & 0 & -(u^2+2\,m)^2 & 0 \\ 0 & 0 & 0 & -\sin^2\vartheta\,(u^2+2\,m)^2 \end{pmatrix}$$

Contravariant Metric

$$g^{\mu\nu} = \begin{pmatrix} \frac{u^2+2\,m}{u^2} & 0 & 0 & 0 \\ 0 & -\frac{1}{4\,(u^2+2\,m)} & 0 & 0 \\ 0 & 0 & -\frac{1}{(u^2+2\,m)^2} & 0 \\ 0 & 0 & 0 & -\frac{1}{\sin^2\vartheta\,(u^2+2\,m)^2} \end{pmatrix}$$

Christoffel Connection

$$\Gamma^0{}_{01} = \frac{2\,m}{u\,(u^2+2\,m)}$$

$$\Gamma^0{}_{10} = \Gamma^0{}_{01}$$

$$\Gamma^1{}_{00} = \frac{m\,u}{2\,(u^2+2\,m)^3}$$

$$\Gamma^1{}_{11} = \frac{u}{u^2+2\,m}$$

$$\Gamma^1{}_{22} = -\frac{u}{2}$$

$$\Gamma^1{}_{33} = -\frac{\sin^2\vartheta\,u}{2}$$

$$\Gamma^2{}_{12} = \frac{2\,u}{u^2+2\,m}$$

$$\Gamma^2{}_{21} = \Gamma^2{}_{12}$$

$$\Gamma^2{}_{33} = -\cos\vartheta\,\sin\vartheta$$

$$\Gamma^3{}_{13} = \frac{2\,u}{u^2+2\,m}$$

$$\Gamma^3{}_{23} = \frac{\cos\vartheta}{\sin\vartheta}$$

$$\Gamma^3{}_{31} = \Gamma^3{}_{13}$$

$$\Gamma^3{}_{32} = \Gamma^3{}_{23}$$

Metric Compatibility

——— o.k.

Riemann Tensor

$$R^0{}_{101} = \frac{8\,m}{(u^2+2\,m)^2}$$

$$R^0{}_{110} = -R^0{}_{101}$$

$$R^0{}_{202} = -\frac{m}{u^2+2\,m}$$

$$R^0{}_{220} = -R^0{}_{202}$$

$$R^0{}_{303} = -\frac{m\,\sin^2\vartheta}{u^2+2\,m}$$

$$R^0{}_{330} = -R^0{}_{303}$$

$$R^1{}_{001} = \frac{2\,m\,u^2}{(u^2+2\,m)^4}$$

$$R^1{}_{010} = -R^1{}_{001}$$

$$R^1{}_{212} = -\frac{m}{u^2+2\,m}$$

$$R^1{}_{221} = -R^1{}_{212}$$

$$R^1{}_{313} = -\frac{m\,\sin^2\vartheta}{u^2+2\,m}$$

$$R^1{}_{331} = -R^1{}_{313}$$

$$R^2{}_{002} = -\frac{m\,u^2}{(u^2+2\,m)^4}$$

$$R^2{}_{020} = -R^2{}_{002}$$

$$R^2{}_{112} = \frac{4\,m}{(u^2+2\,m)^2}$$

$$R^2{}_{121} = -R^2{}_{112}$$

$$R^2{}_{323} = \frac{2\,m\,\sin^2\vartheta}{u^2+2\,m}$$

$$R^2{}_{332} = -R^2{}_{323}$$

$$R^3{}_{003} = -\frac{m\,u^2}{(u^2+2\,m)^4}$$

$$R^3{}_{030} = -R^3{}_{003}$$

$$R^3{}_{113} = \frac{4\,m}{(u^2+2\,m)^2}$$

$$R^3{}_{131} = -R^3{}_{113}$$

$$R^3{}_{223} = -\frac{2\,m}{u^2+2\,m}$$

$$R^3{}_{232} = -R^3{}_{223}$$

Ricci Tensor

——— all elements zero

Ricci Scalar

$$R_{sc} = 0$$

Bianchi identity (Ricci cyclic equation $R^\kappa{}_{[\mu\nu\sigma]} = 0$)

——— o.k.

Einstein Tensor

——— all elements zero

Hodge Dual of Bianchi Identity

——— (see charge and current densities)

Scalar Charge Density (-$R^0{}_i{}^{i0}$)

$$\rho = 0$$

Current Density Class 1 (-$R^{i}{}_{\mu}{}^{\mu j}$)

$$J_1 = 0$$

$$J_2 = 0$$

$$J_3 = 0$$

Current Density Class 2 (-$R^{i}{}_{\mu}{}^{\mu j}$)

$$J_1 = 0$$

$$J_2 = 0$$

$$J_3 = 0$$

Current Density Class 3 (-$R^{i}{}_{\mu}{}^{\mu j}$)

$$J_1 = 0$$

$$J_2 = 0$$

$$J_3 = 0$$

4.4.25 Einstein-Rosen bridge metric, r coordinates

Metric of the Einstein-Rosen bridge with radial coordinates. ϵ is a parameter equivalent to charge in the Reissner-Nordstrom metric.

Coordinates

$$\mathbf{x} = \begin{pmatrix} t \\ r \\ \vartheta \\ \varphi \end{pmatrix}$$

Metric

$$g_{\mu\nu} = \begin{pmatrix} -\frac{2m}{r} - \frac{\varepsilon^2}{2r^2} + 1 & 0 & 0 & 0 \\ 0 & -\frac{1}{-\frac{2m}{r} - \frac{\varepsilon^2}{2r^2} + 1} & 0 & 0 \\ 0 & 0 & -r^2 & 0 \\ 0 & 0 & 0 & -r^2 \sin^2\vartheta \end{pmatrix}$$

Contravariant Metric

$$g^{\mu\nu} = \begin{pmatrix} \frac{2r^2}{2r^2 - 4mr - \varepsilon^2} & 0 & 0 & 0 \\ 0 & -\frac{2r^2 - 4mr - \varepsilon^2}{2r^2} & 0 & 0 \\ 0 & 0 & -\frac{1}{r^2} & 0 \\ 0 & 0 & 0 & -\frac{1}{r^2 \sin^2\vartheta} \end{pmatrix}$$

Christoffel Connection

$$\Gamma^0{}_{01} = \frac{2\,m\,r + \varepsilon^2}{r\,(2\,r^2 - 4\,m\,r - \varepsilon^2)}$$

$$\Gamma^0{}_{10} = \Gamma^0{}_{01}$$

$$\Gamma^1{}_{00} = \frac{\left(2\,m\,r + \varepsilon^2\right)\,\left(2\,r^2 - 4\,m\,r - \varepsilon^2\right)}{4\,r^5}$$

$$\Gamma^1{}_{11} = -\frac{2\,m\,r + \varepsilon^2}{r\,(2\,r^2 - 4\,m\,r - \varepsilon^2)}$$

$$\Gamma^1{}_{22} = -\frac{2\,r^2 - 4\,m\,r - \varepsilon^2}{2\,r}$$

$$\Gamma^1{}_{33} = -\frac{\left(2\,r^2 - 4\,m\,r - \varepsilon^2\right)\,\sin^2\vartheta}{2\,r}$$

$$\Gamma^2{}_{12} = \frac{1}{r}$$

$$\Gamma^2{}_{21} = \Gamma^2{}_{12}$$

$$\Gamma^2{}_{33} = -\cos\vartheta\,\sin\vartheta$$

$$\Gamma^3{}_{13} = \frac{1}{r}$$

$$\Gamma^3{}_{23} = \frac{\cos\vartheta}{\sin\vartheta}$$

$$\Gamma^3{}_{31} = \Gamma^3{}_{13}$$

$$\Gamma^3{}_{32} = \Gamma^3{}_{23}$$

Metric Compatibility

——— o.k.

Riemann Tensor

$$R^0{}_{101} = \frac{4\,m\,r + 3\,\varepsilon^2}{r^2\,\left(2\,r^2 - 4\,m\,r - \varepsilon^2\right)}$$

$$R^0{}_{110} = -R^0{}_{101}$$

$$R^0{}_{202} = -\frac{2\,m\,r + \varepsilon^2}{2\,r^2}$$

$$R^0{}_{220} = -R^0{}_{202}$$

$$R^0{}_{303} = -\frac{\left(2\,m\,r + \varepsilon^2\right)\,\sin^2\vartheta}{2\,r^2}$$

$$R^0{}_{330} = -R^0{}_{303}$$

$$R^1{}_{001} = \frac{\left(4\,m\,r + 3\,\varepsilon^2\right)\,\left(2\,r^2 - 4\,m\,r - \varepsilon^2\right)}{4\,r^6}$$

$$R^1{}_{010} = -R^1{}_{001}$$

$$R^1{}_{212} = -\frac{2\,m\,r + \varepsilon^2}{2\,r^2}$$

$$R^1{}_{221} = -R^1{}_{212}$$

$$R^1{}_{313} = -\frac{\left(2\,m\,r + \varepsilon^2\right)\,\sin^2\vartheta}{2\,r^2}$$

$$R^1{}_{331} = -R^1{}_{313}$$

$$R^2{}_{002} = -\frac{\left(2\,m\,r + \varepsilon^2\right)\,\left(2\,r^2 - 4\,m\,r - \varepsilon^2\right)}{4\,r^6}$$

$$R^2{}_{020} = -R^2{}_{002}$$

$$R^2{}_{112} = \frac{2\,m\,r + \varepsilon^2}{r^2\,\left(2\,r^2 - 4\,m\,r - \varepsilon^2\right)}$$

$$R^2{}_{121} = -R^2{}_{112}$$

$$R^2{}_{323} = \frac{\left(4\,m\,r + \varepsilon^2\right)\,\sin^2\vartheta}{2\,r^2}$$

$$R^2{}_{332} = -R^2{}_{323}$$

$$R^3{}_{003} = -\frac{(2\,m\,r + \varepsilon^2)\,(2\,r^2 - 4\,m\,r - \varepsilon^2)}{4\,r^6}$$

$$R^3{}_{030} = -R^3{}_{003}$$

$$R^3{}_{113} = \frac{2\,m\,r + \varepsilon^2}{r^2\,(2\,r^2 - 4\,m\,r - \varepsilon^2)}$$

$$R^3{}_{131} = -R^3{}_{113}$$

$$R^3{}_{223} = -\frac{4\,m\,r + \varepsilon^2}{2\,r^2}$$

$$R^3{}_{232} = -R^3{}_{223}$$

Ricci Tensor

$$\mathrm{Ric}_{00} = -\frac{\varepsilon^2\,(2\,r^2 - 4\,m\,r - \varepsilon^2)}{4\,r^6}$$

$$\mathrm{Ric}_{11} = \frac{\varepsilon^2}{r^2\,(2\,r^2 - 4\,m\,r - \varepsilon^2)}$$

$$\mathrm{Ric}_{22} = -\frac{\varepsilon^2}{2\,r^2}$$

$$\mathrm{Ric}_{33} = -\frac{\varepsilon^2\,\sin^2\vartheta}{2\,r^2}$$

Ricci Scalar

$$R_{sc} = 0$$

Bianchi identity (Ricci cyclic equation $R^{\kappa}{}_{[\mu\nu\sigma]} = 0$)

——— o.k.

Einstein Tensor

$$G_{00} = -\frac{\varepsilon^2\,(2\,r^2 - 4\,m\,r - \varepsilon^2)}{4\,r^6}$$

$$G_{11} = \frac{\varepsilon^2}{r^2\,(2\,r^2 - 4\,m\,r - \varepsilon^2)}$$

$$G_{22} = -\frac{\varepsilon^2}{2\,r^2}$$

$$G_{33} = -\frac{\varepsilon^2\,\sin^2\vartheta}{2\,r^2}$$

Hodge Dual of Bianchi Identity

——— (see charge and current densities)

Scalar Charge Density ($-R^{0}{}_{i}{}^{i0}$)

$$\rho = -\frac{\varepsilon^2}{r^2\,(2\,r^2 - 4\,m\,r - \varepsilon^2)}$$

Current Density Class 1 ($-R^{i}{}_{\mu}{}^{\mu j}$)

$$J_1 = -\frac{\varepsilon^2\,(2\,r^2 - 4\,m\,r - \varepsilon^2)}{4\,r^6}$$

$$J_2 = \frac{\varepsilon^2}{2\,r^6}$$

$$J_3 = \frac{\varepsilon^2}{2\,r^6\,\sin^2\vartheta}$$

Current Density Class 2 ($-R^{i}{}_{\mu}{}^{\mu j}$)

$$J_1 = 0$$

$$J_2 = 0$$

$$J_3 = 0$$

Current Density Class 3 ($-R^{i}{}_{\mu}{}^{\mu j}$)

$$J_1 = 0$$

$$J_2 = 0$$

$$J_3 = 0$$

4.4.26 Massless Einstein-Rosen bridge metric, r coordinates

Metric of the massless Einstein-Rosen bridge (Nandi and Xu) with radial coordinates. β is a parameter.

Coordinates

$$\mathbf{x} = \begin{pmatrix} t \\ r \\ \vartheta \\ \varphi \end{pmatrix}$$

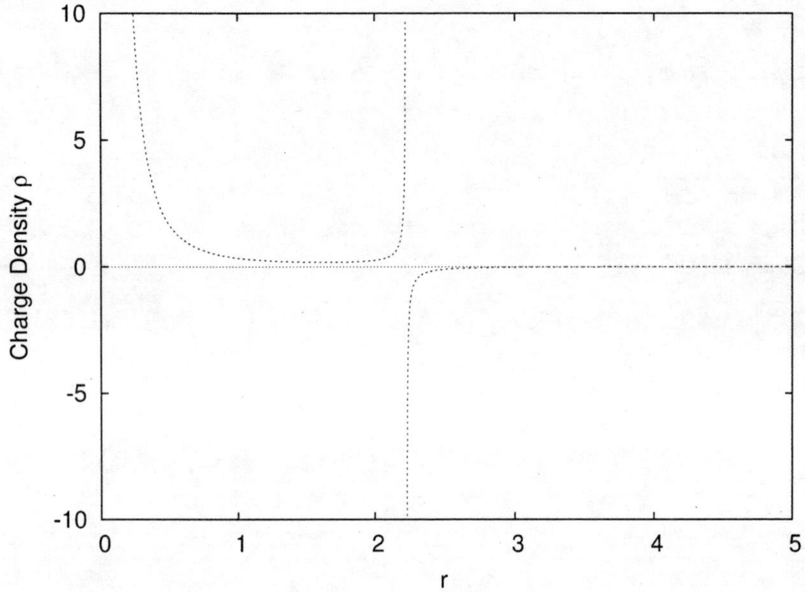

Fig. 4.60: Einstein-Rosen bridge, charge density ρ for $m = 1, \epsilon = 1$.

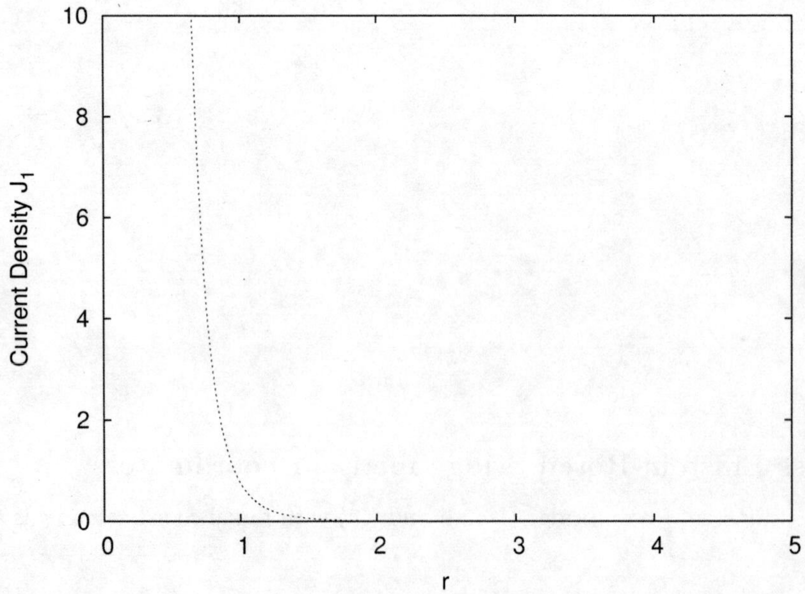

Fig. 4.61: Einstein-Rosen bridge, current density J_r for $m = 1, \epsilon = 1$.

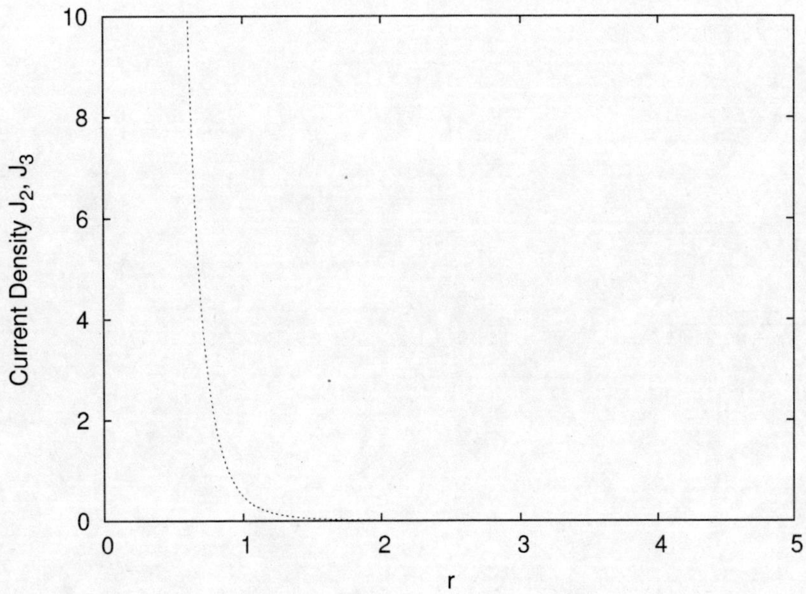

Fig. 4.62: Einstein-Rosen bridge, current density J_θ, J_φ for $m = 1, \epsilon = 1$.

Metric

$$g_{\mu\nu} = \begin{pmatrix} \frac{\left(1-\frac{m^2+\beta^2}{4\,r^2}\right)^2}{\left(\frac{m}{r}+\frac{m^2+\beta^2}{4\,r^2}+1\right)^2} & 0 & 0 & 0 \\ 0 & -\left(\frac{2\,m}{r}+\frac{m^2+\beta^2}{4\,r^2}+1\right)^2 & 0 & 0 \\ 0 & 0 & -\left(\frac{2\,m}{r}+\frac{m^2+\beta^2}{4\,r^2}+1\right)^2 r^2 & 0 \\ 0 & 0 & 0 & -\left(\frac{2\,m}{r}+\frac{m^2+\beta^2}{4\,r^2}+1\right)^2 r^2\,\sin^2\vartheta \end{pmatrix}$$

Contravariant Metric

$$g^{\mu\nu} = \begin{pmatrix} \frac{\left(4\,r^2+4\,m\,r+m^2+\beta^2\right)^2}{\left(4\,r^2-m^2-\beta^2\right)^2} & 0 & 0 & 0 \\ 0 & -\frac{16\,r^4}{\left(4\,r^2+8\,m\,r+m^2+\beta^2\right)^2} & 0 & 0 \\ 0 & 0 & -\frac{16\,r^2}{\left(4\,r^2+8\,m\,r+m^2+\beta^2\right)^2} & 0 \\ 0 & 0 & 0 & -\frac{16\,r^2}{\left(4\,r^2+8\,m\,r+m^2+\beta^2\right)^2\,\sin^2\vartheta} \end{pmatrix}$$

Christoffel Connection

$$\Gamma^0{}_{01} = \frac{4\,\left(4\,m\,r^2+4\,m^2\,r+4\,\beta^2\,r+m^3+\beta^2\,m\right)}{\left(4\,r^2-m^2-\beta^2\right)\,\left(4\,r^2+4\,m\,r+m^2+\beta^2\right)}$$

$$\Gamma^0{}_{10} = \Gamma^0{}_{01}$$

$$\Gamma^1{}_{00} = \frac{64\,r^4\,\left(4\,r^2-m^2-\beta^2\right)\,\left(4\,m\,r^2+4\,m^2\,r+4\,\beta^2\,r+m^3+\beta^2\,m\right)}{\left(4\,r^2+4\,m\,r+m^2+\beta^2\right)^3\,\left(4\,r^2+8\,m\,r+m^2+\beta^2\right)^2}$$

$$\Gamma^1{}_{11} = -\frac{2\,\left(4\,m\,r+m^2+\beta^2\right)}{r\,\left(4\,r^2+8\,m\,r+m^2+\beta^2\right)}$$

$$\Gamma^1{}_{22} = -\frac{r\,\left(4\,r^2-m^2-\beta^2\right)}{4\,r^2+8\,m\,r+m^2+\beta^2}$$

$$\Gamma^1{}_{33} = -\frac{r\,(4\,r^2 - m^2 - \beta^2)\,\sin^2\vartheta}{4\,r^2 + 8\,m\,r + m^2 + \beta^2}$$

$$\Gamma^2{}_{12} = \frac{4\,r^2 - m^2 - \beta^2}{r\,(4\,r^2 + 8\,m\,r + m^2 + \beta^2)}$$

$$\Gamma^2{}_{21} = \Gamma^2{}_{12}$$

$$\Gamma^2{}_{33} = -\cos\vartheta\,\sin\vartheta$$

$$\Gamma^3{}_{13} = \frac{4\,r^2 - m^2 - \beta^2}{r\,(4\,r^2 + 8\,m\,r + m^2 + \beta^2)}$$

$$\Gamma^3{}_{23} = \frac{\cos\vartheta}{\sin\vartheta}$$

$$\Gamma^3{}_{31} = \Gamma^3{}_{13}$$

$$\Gamma^3{}_{32} = \Gamma^3{}_{23}$$

Metric Compatibility

——— o.k.

Riemann Tensor

$$R^0{}_{101} = \frac{8\,(16\,m\,r^4 + 40\,m^2\,r^3 + 24\,\beta^2\,r^3 + 32\,m^3\,r^2 + 48\,\beta^2\,m\,r^2 + 10\,m^4\,r + 16\,\beta^2\,m^2\,r + 6\,\beta^4\,r + m^5 + 2\,\beta^2\,m^3 + \beta^4\,m)}{r\,(4\,r^2 + 4\,m\,r + m^2 + \beta^2)^2\,(4\,r^2 + 8\,m\,r + m^2 + \beta^2)}$$

$$R^0{}_{110} = -R^0{}_{101}$$

$$R^0{}_{202} = -\frac{4\,r\,(4\,m\,r^2 + 4\,m^2\,r + 4\,\beta^2\,r + m^3 + \beta^2\,m)}{(4\,r^2 + 4\,m\,r + m^2 + \beta^2)\,(4\,r^2 + 8\,m\,r + m^2 + \beta^2)}$$

$$R^0{}_{220} = -R^0{}_{202}$$

$$R^0{}_{303} = -\frac{4\,r\,(4\,m\,r^2 + 4\,m^2\,r + 4\,\beta^2\,r + m^3 + \beta^2\,m)\,\sin^2\vartheta}{(4\,r^2 + 4\,m\,r + m^2 + \beta^2)\,(4\,r^2 + 8\,m\,r + m^2 + \beta^2)}$$

$$R^0{}_{330} = -R^0{}_{303}$$

$$R^1{}_{001} = \frac{128\,r^3\,\left(4\,r^2 - m^2 - \beta^2\right)^2\,\left(16\,m\,r^4 + 40\,m^2\,r^3 + 24\,\beta^2\,r^3 + 32\,m^3\,r^2 + 48\,\beta^2\,m\,r^2 + 10\,m^4\,r + 16\,\beta^2\,m^2\,r + 6\,\beta^4\,r + m^5 + 2\,\beta^2\,m^3 + \ldots\right)}{\left(4\,r^2 + 4\,m\,r + m^2 + \beta^2\right)^4\,\left(4\,r^2 + 8\,m\,r + m^2 + \beta^2\right)^3}$$

$$R^1{}_{010} = -R^1{}_{001}$$

$$R^1{}_{212} = -\frac{8\,r\,(4\,m\,r^2 + 2\,m^2\,r + 2\,\beta^2\,r + m^3 + \beta^2\,m)}{(4\,r^2 + 8\,m\,r + m^2 + \beta^2)^2}$$

$$R^1{}_{221} = -R^1{}_{212}$$

$$R^1{}_{313} = -\frac{8\,r\,(4\,m\,r^2 + 2\,m^2\,r + 2\,\beta^2\,r + m^3 + \beta^2\,m)\,\sin^2\vartheta}{(4\,r^2 + 8\,m\,r + m^2 + \beta^2)^2}$$

$$R^1{}_{331} = -R^1{}_{313}$$

$$R^2{}_{002} = -\frac{64\,r^3\,\left(4\,r^2 - m^2 - \beta^2\right)^2\,\left(4\,m\,r^2 + 4\,m^2\,r + 4\,\beta^2\,r + m^3 + \beta^2\,m\right)}{(4\,r^2 + 4\,m\,r + m^2 + \beta^2)^3\,(4\,r^2 + 8\,m\,r + m^2 + \beta^2)^3}$$

$$R^2{}_{020} = -R^2{}_{002}$$

$$R^2{}_{112} = \frac{8\,\left(4\,m\,r^2 + 2\,m^2\,r + 2\,\beta^2\,r + m^3 + \beta^2\,m\right)}{r\,(4\,r^2 + 8\,m\,r + m^2 + \beta^2)^2}$$

$$R^2{}_{121} = -R^2{}_{112}$$

$$R^2{}_{323} = \frac{16\,r\,(r+m)\,\left(4\,m\,r + m^2 + \beta^2\right)\,\sin^2\vartheta}{(4\,r^2 + 8\,m\,r + m^2 + \beta^2)^2}$$

$$R^2{}_{332} = -R^2{}_{323}$$

$$R^3{}_{003} = -\frac{64\,r^3\,\left(4\,r^2 - m^2 - \beta^2\right)^2\,\left(4\,m\,r^2 + 4\,m^2\,r + 4\,\beta^2\,r + m^3 + \beta^2\,m\right)}{(4\,r^2 + 4\,m\,r + m^2 + \beta^2)^3\,(4\,r^2 + 8\,m\,r + m^2 + \beta^2)^3}$$

$$R^3{}_{030} = -R^3{}_{003}$$

$$R^3{}_{113} = \frac{8\,\left(4\,m\,r^2 + 2\,m^2\,r + 2\,\beta^2\,r + m^3 + \beta^2\,m\right)}{r\,(4\,r^2 + 8\,m\,r + m^2 + \beta^2)^2}$$

$$R^3{}_{131} = -R^3{}_{113}$$

$$R^3{}_{223} = -\frac{16\,r\,(r+m)\,\left(4\,m\,r + m^2 + \beta^2\right)}{(4\,r^2 + 8\,m\,r + m^2 + \beta^2)^2}$$

$$R^3{}_{232} = -R^3{}_{223}$$

Ricci Tensor

$$\mathrm{Ric}_{00} = -\frac{256\,r^4\,\left(4\,r^2 - m^2 - \beta^2\right)^2\,\left(4\,m^2\,r^2 + 4\,\beta^2\,r^2 + 4\,m^3\,r + 12\,\beta^2\,m\,r + m^4 + 2\,\beta^2\,m^2 + \beta^4\right)}{(4\,r^2 + 4\,m\,r + m^2 + \beta^2)^4\,(4\,r^2 + 8\,m\,r + m^2 + \beta^2)^3}$$

$$\mathrm{Ric}_{11} = -\frac{8\,\left(64\,m\,r^6 + 32\,m^2\,r^5 - 32\,\beta^2\,r^5 - 112\,m^3\,r^4 - 176\,\beta^2\,m\,r^4 - 112\,m^4\,r^3 - 256\,\beta^2\,m^2\,r^3 - 16\,\beta^4\,r^3 - 28\,m^5\,r^2 - 72\,\beta^2\,m^3\,r^2 - \ldots\right)}{r\,(4\,r^2 + 4\,m\,r + m^2 + \beta^2)^2\,(4\,r^2 + 8\,m\,r + m^2 + \beta^2)^2}$$

$$\mathrm{Ric}_{22} = \frac{4\,r\,\left(4\,m\,r^2 + 4\,m^2\,r - 4\,\beta^2\,r + m^3 + \beta^2\,m\right)}{(4\,r^2 + 4\,m\,r + m^2 + \beta^2)\,(4\,r^2 + 8\,m\,r + m^2 + \beta^2)}$$

$$\mathrm{Ric}_{33} = \frac{4\,r\,\left(4\,m\,r^2 + 4\,m^2\,r - 4\,\beta^2\,r + m^3 + \beta^2\,m\right)\,\sin^2\vartheta}{(4\,r^2 + 4\,m\,r + m^2 + \beta^2)\,(4\,r^2 + 8\,m\,r + m^2 + \beta^2)}$$

Ricci Scalar

$$R_{sc} = -\frac{2048\,m\,r^4\,\left(16\,m\,r^4 + 36\,m^2\,r^3 + 12\,\beta^2\,r^3 + 28\,m^3\,r^2 + 28\,\beta^2\,m\,r^2 + 9\,m^4\,r + 12\,\beta^2\,m^2\,r + 3\,\beta^4\,r + m^5 + 2\,\beta^2\,m^3 + \beta^4\,m\right)}{(4\,r^2 + 4\,m\,r + m^2 + \beta^2)^2\,(4\,r^2 + 8\,m\,r + m^2 + \beta^2)^4}$$

Bianchi identity (Ricci cyclic equation $R^{\kappa}{}_{[\mu\nu\sigma]} = 0$)

——— o.k.

Einstein Tensor

$$G_{00} = \frac{256\,(3\,m^2 - \beta^2)\,r^4\,(4\,r^2 - m^2 - \beta^2)^2}{(4\,r^2 + 4\,m\,r + m^2 + \beta^2)^2\,(4\,r^2 + 8\,m\,r + m^2 + \beta^2)^4}$$

$$G_{11} = -\frac{8\,(16\,m\,r^4 + 24\,m^2\,r^3 - 8\,\beta^2\,r^3 + 16\,m^3\,r^2 - 16\,\beta^2\,m\,r^2 + 6\,m^4\,r + 4\,\beta^2\,m^2\,r - 2\,\beta^4\,r + m^5 + 2\,\beta^2\,m^3 + \beta^4\,m)}{r\,(4\,r^2 + 4\,m\,r + m^2 + \beta^2)\,(4\,r^2 + 8\,m\,r + m^2 + \beta^2)^2}$$

$$G_{22} = \frac{4\,r\,(64\,m\,r^6 - 64\,\beta^2\,r^5 - 208\,m^3\,r^4 - 336\,\beta^2\,m\,r^4 - 192\,m^4\,r^3 - 480\,\beta^2\,m^2\,r^3 - 32\,\beta^4\,r^3 - 52\,m^5\,r^2 - 136\,\beta^2\,m^3\,r^2 - \ldots)}{(4\,r^2 + 4\,m\,r + m^2 + \beta^2)^2\,(4\,r^2 + 8\,m\,r + m^2 + \beta^2)^2}$$

$$G_{33} = \frac{4\,r\,(64\,m\,r^6 - 64\,\beta^2\,r^5 - 208\,m^3\,r^4 - 336\,\beta^2\,m\,r^4 - 192\,m^4\,r^3 - 480\,\beta^2\,m^2\,r^3 - 32\,\beta^4\,r^3 - 52\,m^5\,r^2 - 136\,\beta^2\,m^3\,r^2 - \ldots)\,\sin^2\vartheta}{(4\,r^2 + 4\,m\,r + m^2 + \beta^2)^2\,(4\,r^2 + 8\,m\,r + m^2 + \beta^2)^2}$$

Hodge Dual of Bianchi Identity

——— (see charge and current densities)

Scalar Charge Density ($-R^{0}{}_{i}{}^{i0}$)

$$\rho = -\frac{256\,r^4\,(4\,m^2\,r^2 + 4\,\beta^2\,r^2 + 4\,m^3\,r + 12\,\beta^2\,m\,r + m^4 + 2\,\beta^2\,m^2 + \beta^4)}{(4\,r^2 - m^2 - \beta^2)^2\,(4\,r^2 + 8\,m\,r + m^2 + \beta^2)^3}$$

Current Density Class 1 ($-R^{i}{}_{\mu}{}^{\mu j}$)

$$J_1 = \frac{2048\,r^7\,(64\,m\,r^6 + 32\,m^2\,r^5 - 32\,\beta^2\,r^5 - 112\,m^3\,r^4 - 176\,\beta^2\,m\,r^4 - 112\,m^4\,r^3 - 256\,\beta^2\,m^2\,r^3 - 16\,\beta^4\,r^3 - 28\,m^5\,r^2 - 72\,\beta^2\,m^3\,r^2 - \ldots)}{(4\,r^2 + 4\,m\,r + m^2 + \beta^2)^2\,(4\,r^2 + 8\,m\,r + m^2 + \beta^2)^6}$$

$$J_2 = -\frac{1024\,r^5\,(4\,m\,r^2 + 4\,m^2\,r - 4\,\beta^2\,r + m^3 + \beta^2\,m)}{(4\,r^2 + 4\,m\,r + m^2 + \beta^2)\,(4\,r^2 + 8\,m\,r + m^2 + \beta^2)^5}$$

$$J_3 = -\frac{1024\,r^5\,(4\,m\,r^2 + 4\,m^2\,r - 4\,\beta^2\,r + m^3 + \beta^2\,m)}{(4\,r^2 + 4\,m\,r + m^2 + \beta^2)\,(4\,r^2 + 8\,m\,r + m^2 + \beta^2)^5\,\sin^2\vartheta}$$

Current Density Class 2 ($-R^{i}{}_{\mu}{}^{\mu j}$)

$$J_1 = 0$$

$$J_2 = 0$$

$$J_3 = 0$$

Current Density Class 3 ($-R^{i}{}_{\mu}{}^{\mu j}$)

$$J_1 = 0$$

$$J_2 = 0$$

$$J_3 = 0$$

4.4.27 General Morris-Thorne wormhole metric

Φ and b are functions of the radial coordinate parameter R.

Coordinates

$$\mathbf{x} = \begin{pmatrix} t \\ R \\ \vartheta \\ \varphi \end{pmatrix}$$

Metric

$$g_{\mu\nu} = \begin{pmatrix} e^{2\,\Phi} & 0 & 0 & 0 \\ 0 & -\frac{1}{1-\frac{b}{R}} & 0 & 0 \\ 0 & 0 & -R^2 & 0 \\ 0 & 0 & 0 & -\sin^2\vartheta\, R^2 \end{pmatrix}$$

Contravariant Metric

$$g^{\mu\nu} = \begin{pmatrix} e^{-2\,\Phi} & 0 & 0 & 0 \\ 0 & -\frac{R-b}{R} & 0 & 0 \\ 0 & 0 & -\frac{1}{R^2} & 0 \\ 0 & 0 & 0 & -\frac{1}{\sin^2\vartheta\, R^2} \end{pmatrix}$$

Christoffel Connection

$$\Gamma^0{}_{01} = \frac{d}{d\,R}\,\Phi$$

$$\Gamma^0{}_{10} = \Gamma^0{}_{01}$$

$$\Gamma^1{}_{00} = \frac{e^{2\,\Phi}\,\left(\frac{d}{d\,R}\,\Phi\right)\,(R-b)}{R}$$

$$\Gamma^1{}_{11} = \frac{\frac{d}{d\,R}\,b\,R - b}{2\,R\,(R-b)}$$

$$\Gamma^1{}_{22} = -\,(R-b)$$

$$\Gamma^1{}_{33} = -\sin^2\vartheta\,(R-b)$$

$$\Gamma^2{}_{12} = \frac{1}{R}$$

$$\Gamma^2{}_{21} = \Gamma^2{}_{12}$$

$$\Gamma^2{}_{33} = -\cos\vartheta\,\sin\vartheta$$

$$\Gamma^3{}_{13} = \frac{1}{R}$$

$$\Gamma^3{}_{23} = \frac{\cos\vartheta}{\sin\vartheta}$$

$$\Gamma^3{}_{31} = \Gamma^3{}_{13}$$

$$\Gamma^3{}_{32} = \Gamma^3{}_{23}$$

Metric Compatibility

——— o.k.

Riemann Tensor

$$R^{0}{}_{101} = -\frac{2\left(\frac{d^2}{dR^2}\Phi\right)R^2 + 2\left(\frac{d}{dR}\Phi\right)^2 R^2 - 2b\left(\frac{d^2}{dR^2}\Phi\right)R - 2b\left(\frac{d}{dR}\Phi\right)^2 R - \frac{d}{dR}b\left(\frac{d}{dR}\Phi\right)R + b\left(\frac{d}{dR}\Phi\right)}{2R(R-b)}$$

$$R^{0}{}_{110} = -R^{0}{}_{101}$$

$$R^{0}{}_{202} = -\frac{d}{dR}\Phi\,(R-b)$$

$$R^{0}{}_{220} = -R^{0}{}_{202}$$

$$R^{0}{}_{303} = -\frac{d}{dR}\Phi\,\sin^2\vartheta\,(R-b)$$

$$R^{0}{}_{330} = -R^{0}{}_{303}$$

$$R^{1}{}_{001} = -\frac{e^{2\Phi}\left(2\left(\frac{d^2}{dR^2}\Phi\right)R^2 + 2\left(\frac{d}{dR}\Phi\right)^2 R^2 - 2b\left(\frac{d^2}{dR^2}\Phi\right)R - 2b\left(\frac{d}{dR}\Phi\right)^2 R - \frac{d}{dR}b\left(\frac{d}{dR}\Phi\right)R + b\left(\frac{d}{dR}\Phi\right)\right)}{2R^2}$$

$$R^{1}{}_{010} = -R^{1}{}_{001}$$

$$R^{1}{}_{212} = \frac{\frac{d}{dR}b\,R - b}{2R}$$

$$R^{1}{}_{221} = -R^{1}{}_{212}$$

$$R^{1}{}_{313} = \frac{\sin^2\vartheta\left(\frac{d}{dR}b\,R - b\right)}{2R}$$

$$R^{1}{}_{331} = -R^{1}{}_{313}$$

$$R^{2}{}_{002} = -\frac{e^{2\Phi}\left(\frac{d}{dR}\Phi\right)(R-b)}{R^2}$$

$$R^{2}{}_{020} = -R^{2}{}_{002}$$

$$R^{2}{}_{112} = -\frac{\frac{d}{dR}b\,R - b}{2R^2(R-b)}$$

$$R^{2}{}_{121} = -R^{2}{}_{112}$$

$$R^{2}{}_{323} = \frac{b\sin^2\vartheta}{R}$$

$$R^{2}{}_{332} = -R^{2}{}_{323}$$

$$R^{3}{}_{003} = -\frac{e^{2\Phi}\left(\frac{d}{dR}\Phi\right)(R-b)}{R^2}$$

$$R^{3}{}_{030} = -R^{3}{}_{003}$$

$$R^{3}{}_{113} = -\frac{\frac{d}{dR}b\,R - b}{2R^2(R-b)}$$

$$R^{3}{}_{131} = -R^{3}{}_{113}$$

$$R^{3}{}_{223} = -\frac{b}{R}$$

$$R^{3}{}_{232} = -R^{3}{}_{223}$$

Ricci Tensor

$$\mathrm{Ric}_{00} = \frac{e^{2\,\Phi}\left(2\left(\frac{d^2}{d\,R^2}\,\Phi\right)R^2 + 2\left(\frac{d}{d\,R}\,\Phi\right)^2 R^2 - 2\,b\left(\frac{d^2}{d\,R^2}\,\Phi\right)R - 2\,b\left(\frac{d}{d\,R}\,\Phi\right)^2 R - \frac{d}{d\,R}\,b\left(\frac{d}{d\,R}\,\Phi\right)R + 4\left(\frac{d}{d\,R}\,\Phi\right)R - 3\,b\left(\frac{d}{d\,R}\,\Phi\right)\right)}{2\,R^2}$$

$$\mathrm{Ric}_{11} = -\frac{2\left(\frac{d^2}{d\,R^2}\,\Phi\right)R^3 + 2\left(\frac{d}{d\,R}\,\Phi\right)^2 R^3 - 2\,b\left(\frac{d^2}{d\,R^2}\,\Phi\right)R^2 - 2\,b\left(\frac{d}{d\,R}\,\Phi\right)^2 R^2 - \frac{d}{d\,R}\,b\left(\frac{d}{d\,R}\,\Phi\right)R^2 + b\left(\frac{d}{d\,R}\,\Phi\right)R - 2\left(\frac{d}{d\,R}\,b\right)R + 2\,b}{2\,R^2\,(R-b)}$$

$$\mathrm{Ric}_{22} = -\frac{2\left(\frac{d}{d\,R}\,\Phi\right)R^2 - 2\,b\left(\frac{d}{d\,R}\,\Phi\right)R - \frac{d}{d\,R}\,b\,R - b}{2\,R}$$

$$\mathrm{Ric}_{33} = -\frac{\sin^2\vartheta\,\left(2\left(\frac{d}{d\,R}\,\Phi\right)R^2 - 2\,b\left(\frac{d}{d\,R}\,\Phi\right)R - \frac{d}{d\,R}\,b\,R - b\right)}{2\,R}$$

Ricci Scalar

$$R_{sc} = \frac{2\left(\frac{d^2}{d\,R^2}\,\Phi\right)R^2 + 2\left(\frac{d}{d\,R}\,\Phi\right)^2 R^2 - 2\,b\left(\frac{d^2}{d\,R^2}\,\Phi\right)R - 2\,b\left(\frac{d}{d\,R}\,\Phi\right)^2 R - \frac{d}{d\,R}\,b\left(\frac{d}{d\,R}\,\Phi\right)R + 4\left(\frac{d}{d\,R}\,\Phi\right)R - 3\,b\left(\frac{d}{d\,R}\,\Phi\right) - 2\left(\frac{d}{d\,R}\,b\right)}{R^2}$$

Bianchi identity (Ricci cyclic equation $R^{\kappa}{}_{[\mu\nu\sigma]} = 0$)

——— o.k.

Einstein Tensor

$$G_{00} = \frac{\frac{d}{d\,R}\,b\,e^{2\,\Phi}}{R^2}$$

$$G_{11} = \frac{2\left(\frac{d}{d\,R}\,\Phi\right)R^2 - 2\,b\left(\frac{d}{d\,R}\,\Phi\right)R - b}{R^2\,(R-b)}$$

$$G_{22} = \frac{2\left(\frac{d^2}{d\,R^2}\,\Phi\right)R^3 + 2\left(\frac{d}{d\,R}\,\Phi\right)^2 R^3 - 2\,b\left(\frac{d^2}{d\,R^2}\,\Phi\right)R^2 - 2\,b\left(\frac{d}{d\,R}\,\Phi\right)^2 R^2 - \frac{d}{d\,R}\,b\left(\frac{d}{d\,R}\,\Phi\right)R^2 + 2\left(\frac{d}{d\,R}\,\Phi\right)R^2 - b\left(\frac{d}{d\,R}\,\Phi\right)R - \frac{d}{d\,R}\,b\,R + b}{2\,R}$$

$$G_{33} = \frac{\sin^2\vartheta\,\left(2\left(\frac{d^2}{d\,R^2}\,\Phi\right)R^3 + 2\left(\frac{d}{d\,R}\,\Phi\right)^2 R^3 - 2\,b\left(\frac{d^2}{d\,R^2}\,\Phi\right)R^2 - 2\,b\left(\frac{d}{d\,R}\,\Phi\right)^2 R^2 - \frac{d}{d\,R}\,b\left(\frac{d}{d\,R}\,\Phi\right)R^2 + 2\left(\frac{d}{d\,R}\,\Phi\right)R^2 - b\left(\frac{d}{d\,R}\,\Phi\right)R - \ldots\right)}{2\,R}$$

Hodge Dual of Bianchi Identity

——— (see charge and current densities)

Scalar Charge Density (-$R^{0}{}_{i}{}^{i0}$)

$$\rho = \frac{e^{-2\,\Phi}\left(2\left(\frac{d^2}{d\,R^2}\,\Phi\right)R^2 + 2\left(\frac{d}{d\,R}\,\Phi\right)^2 R^2 - 2\,b\left(\frac{d^2}{d\,R^2}\,\Phi\right)R - 2\,b\left(\frac{d}{d\,R}\,\Phi\right)^2 R - \frac{d}{d\,R}\,b\left(\frac{d}{d\,R}\,\Phi\right)R + 4\left(\frac{d}{d\,R}\,\Phi\right)R - 3\,b\left(\frac{d}{d\,R}\,\Phi\right)\right)}{2\,R^2}$$

Current Density Class 1 (-$R^{i}{}_{\mu}{}^{\mu j}$)

$$J_1 = \frac{(R-b)\left(2\left(\frac{d^2}{d\,R^2}\,\Phi\right)R^3 + 2\left(\frac{d}{d\,R}\,\Phi\right)^2 R^3 - 2\,b\left(\frac{d^2}{d\,R^2}\,\Phi\right)R^2 - 2\,b\left(\frac{d}{d\,R}\,\Phi\right)^2 R^2 - \frac{d}{d\,R}\,b\left(\frac{d}{d\,R}\,\Phi\right)R^2 + b\left(\frac{d}{d\,R}\,\Phi\right)R - 2\left(\frac{d}{d\,R}\,b\right)R + 2\,b\right)}{2\,R^4}$$

$$J_2 = \frac{2\left(\frac{d}{d\,R}\,\Phi\right)R^2 - 2\,b\left(\frac{d}{d\,R}\,\Phi\right)R - \frac{d}{d\,R}\,b\,R - b}{2\,R^5}$$

$$J_3 = \frac{2\left(\frac{d}{d\,R}\,\Phi\right)R^2 - 2\,b\left(\frac{d}{d\,R}\,\Phi\right)R - \frac{d}{d\,R}\,b\,R - b}{2\,\sin^2\vartheta\,R^5}$$

Current Density Class 2 ($-R^{i}{}_{\mu}{}^{\mu j}$)

$$J_1 = 0$$

$$J_2 = 0$$

$$J_3 = 0$$

Current Density Class 3 ($-R^{i}{}_{\mu}{}^{\mu j}$)

$$J_1 = 0$$

$$J_2 = 0$$

$$J_3 = 0$$

4.4.28 Bekenstein-Hawking radiation metric

It is assumed that the angular term dX^2 is

$$dX^2 = r^2 d\theta^2 + r^2 \sin^2\theta d\phi^2$$

with

$$r = 2M + \frac{u^2}{2M}.$$

Coordinates

$$\mathbf{x} = \begin{pmatrix} t \\ u \\ \vartheta \\ \varphi \end{pmatrix}$$

Metric

$$g_{\mu\nu} = \begin{pmatrix} -\frac{u^2}{4\,M^2} & 0 & 0 & 0 \\ 0 & 4 & 0 & 0 \\ 0 & 0 & \left(2\,M + \frac{u^2}{2\,M}\right)^2 & 0 \\ 0 & 0 & 0 & \sin^2\vartheta\left(2\,M + \frac{u^2}{2\,M}\right)^2 \end{pmatrix}$$

Contravariant Metric

$$g^{\mu\nu} = \begin{pmatrix} -\frac{4\,M^2}{u^2} & 0 & 0 & 0 \\ 0 & \frac{1}{2^2} & 0 & 0 \\ 0 & 0 & \frac{4\,M^2}{(4\,M^2+u^2)^2} & 0 \\ 0 & 0 & 0 & \frac{4\,M^2}{\sin^2\vartheta\,(4\,M^2+u^2)^2} \end{pmatrix}$$

Christoffel Connection

$$\Gamma^{0}{}_{01} = \frac{1}{u}$$

$$\Gamma^{0}{}_{10} = \Gamma^{0}{}_{01}$$

$$\Gamma^{1}{}_{00} = \frac{u}{16\,M^2}$$

$$\Gamma^{1}{}_{22} = -\frac{u\,\left(4\,M^2 + u^2\right)}{8\,M^2}$$

$$\Gamma^{1}{}_{33} = -\frac{\sin^2\vartheta\,u\,\left(4\,M^2 + u^2\right)}{8\,M^2}$$

$$\Gamma^{2}{}_{12} = \frac{2\,u}{4\,M^2 + u^2}$$

$$\Gamma^{2}{}_{21} = \Gamma^{2}{}_{12}$$

$$\Gamma^{2}{}_{33} = -\cos\vartheta\,\sin\vartheta$$

$$\Gamma^{3}{}_{13} = \frac{2\,u}{4\,M^2 + u^2}$$

$$\Gamma^{3}{}_{23} = \frac{\cos\vartheta}{\sin\vartheta}$$

$$\Gamma^{3}{}_{31} = \Gamma^{3}{}_{13}$$

$$\Gamma^{3}{}_{32} = \Gamma^{3}{}_{23}$$

Metric Compatibility

——— o.k.

Riemann Tensor

$$R^{0}{}_{202} = -\frac{4\,M^2 + u^2}{8\,M^2}$$

$$R^{0}{}_{220} = -R^{0}{}_{202}$$

$$R^{0}{}_{303} = -\frac{\sin^2\vartheta\,\left(4\,M^2 + u^2\right)}{8\,M^2}$$

$$R^{0}{}_{330} = -R^{0}{}_{303}$$

$$R^1{}_{212} = -\frac{4\,M^2 + u^2}{8\,M^2}$$

$$R^1{}_{221} = -R^1{}_{212}$$

$$R^1{}_{313} = -\frac{\sin^2\vartheta\,\left(4\,M^2 + u^2\right)}{8\,M^2}$$

$$R^1{}_{331} = -R^1{}_{313}$$

$$R^2{}_{002} = -\frac{u^2}{8\,M^2\,\left(4\,M^2 + u^2\right)}$$

$$R^2{}_{020} = -R^2{}_{002}$$

$$R^2{}_{112} = \frac{2}{4\,M^2 + u^2}$$

$$R^2{}_{121} = -R^2{}_{112}$$

$$R^2{}_{323} = \frac{\sin^2\vartheta\,(2\,M - u)\,(2\,M + u)}{4\,M^2}$$

$$R^2{}_{332} = -R^2{}_{323}$$

$$R^3{}_{003} = -\frac{u^2}{8\,M^2\,\left(4\,M^2 + u^2\right)}$$

$$R^3{}_{030} = -R^3{}_{003}$$

$$R^3{}_{113} = \frac{2}{4\,M^2 + u^2}$$

$$R^3{}_{131} = -R^3{}_{113}$$

$$R^3{}_{223} = -\frac{(2\,M - u)\,(2\,M + u)}{4\,M^2}$$

$$R^3{}_{232} = -R^3{}_{223}$$

Ricci Tensor

$$\mathrm{Ric}_{00} = \frac{u^2}{4\,M^2\,\left(4\,M^2 + u^2\right)}$$

$$\mathrm{Ric}_{11} = -\frac{4}{4\,M^2 + u^2}$$

$$\mathrm{Ric}_{22} = -\frac{u^2}{2\,M^2}$$

$$\mathrm{Ric}_{33} = -\frac{\sin^2\vartheta\,u^2}{2\,M^2}$$

Ricci Scalar

$$R_{sc} = -\frac{2\,(4\,M^2 + 3\,u^2)}{(4\,M^2 + u^2)^2}$$

Bianchi identity (Ricci cyclic equation $R^{\kappa}{}_{[\mu\nu\sigma]} = 0$)

——— o.k.

Einstein Tensor

$$G_{00} = -\frac{u^4}{2\,M^2\,(4\,M^2 + u^2)^2}$$

$$G_{11} = \frac{8\,u^2}{(4\,M^2 + u^2)^2}$$

$$G_{22} = \frac{4\,M^2 + u^2}{4\,M^2}$$

$$G_{33} = \frac{\sin^2\vartheta\,(4\,M^2 + u^2)}{4\,M^2}$$

Hodge Dual of Bianchi Identity

——— (see charge and current densities)

Scalar Charge Density (-$R^{0}{}_{i}{}^{i0}$)

$$\rho = \frac{4\,M^2}{u^2\,(4\,M^2 + u^2)}$$

Current Density Class 1 (-$R^{i}{}_{\mu}{}^{\mu j}$)

$$J_1 = \frac{1}{4\,(4\,M^2 + u^2)}$$

$$J_2 = \frac{8\,u^2\,M^2}{(4\,M^2 + u^2)^4}$$

$$J_3 = \frac{8\,u^2\,M^2}{\sin^2\vartheta\,(4\,M^2 + u^2)^4}$$

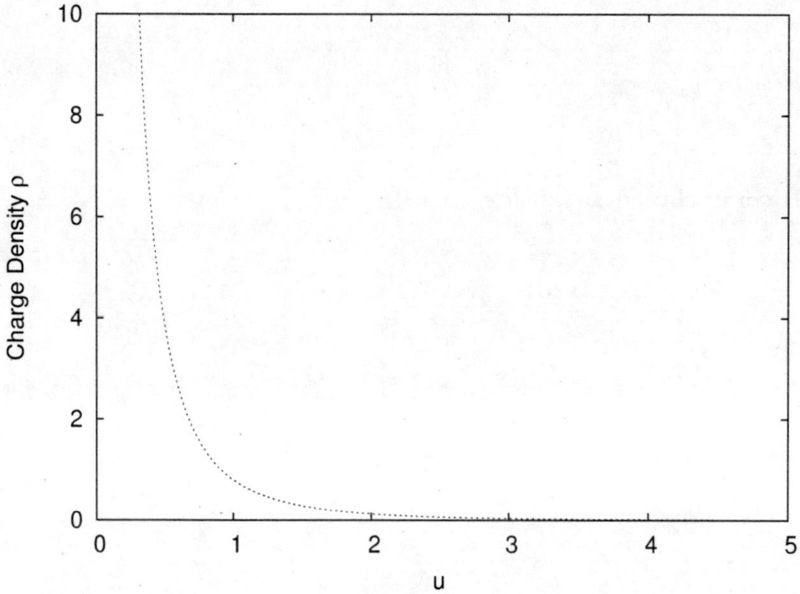

Fig. 4.63: Bekenstein-Hawking radiation, charge density ρ for $M = 1$.

Current Density Class 2 ($-R^{i}{}_{\mu}{}^{\mu j}$)

$$J_1 = 0$$

$$J_2 = 0$$

$$J_3 = 0$$

Current Density Class 3 ($-R^{i}{}_{\mu}{}^{\mu j}$)

$$J_1 = 0$$

$$J_2 = 0$$

$$J_3 = 0$$

4.4.29 Multi-cosmic string metric

Parameter $a = a_1 + ib_1$ is complex.

Coordinates

$$\mathbf{x} = \begin{pmatrix} t \\ x \\ y \\ z \end{pmatrix}$$

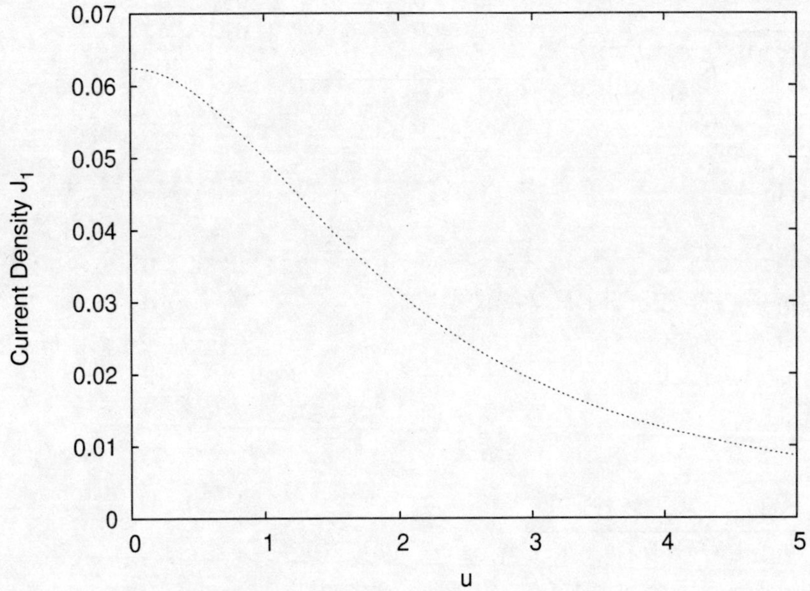

Fig. 4.64: Bekenstein-Hawking radiation, current density J_r for $M = 1$.

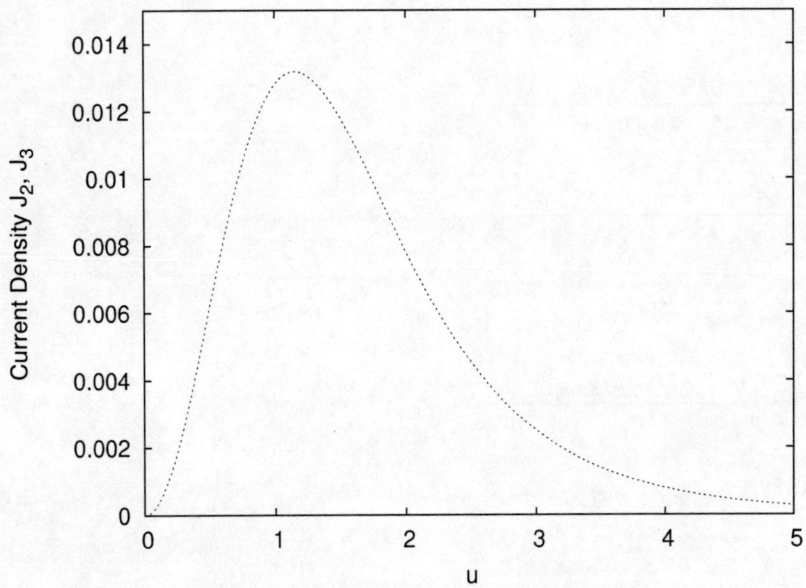

Fig. 4.65: Bekenstein-Hawking radiation, current density J_θ, J_φ for $M = 1$.

Metric

$$g_{\mu\nu} = \begin{pmatrix} 1 & 0 & 0 & 0 \\ 0 & -\frac{1}{\left((y-b_1)^2+(x-a_1)^2\right)^{4\,m_1\,G}} & 0 & 0 \\ 0 & 0 & -\frac{1}{\left((y-b_1)^2+(x-a_1)^2\right)^{4\,m_1\,G}} & 0 \\ 0 & 0 & 0 & -1 \end{pmatrix}$$

Contravariant Metric

$$g^{\mu\nu} = \begin{pmatrix} 1 & 0 & 0 & 0 \\ 0 & -\left(y^2-2\,b_1\,y+x^2-2\,a_1\,x+b_1^2+a_1^2\right)^{4\,m_1\,G} & 0 & 0 \\ 0 & 0 & -\left(y^2-2\,b_1\,y+x^2-2\,a_1\,x+b_1^2+a_1^2\right)^{4\,m_1\,G} & 0 \\ 0 & 0 & 0 & -1 \end{pmatrix}$$

Christoffel Connection

$$\Gamma^1{}_{11} = -\frac{4\,m_1\,(x-a_1)\,G}{y^2-2\,b_1\,y+x^2-2\,a_1\,x+b_1^2+a_1^2}$$

$$\Gamma^1{}_{12} = -\frac{4\,m_1\,(y-b_1)\,G}{y^2-2\,b_1\,y+x^2-2\,a_1\,x+b_1^2+a_1^2}$$

$$\Gamma^1{}_{21} = \Gamma^1{}_{12}$$

$$\Gamma^1{}_{22} = \frac{4\,m_1\,(x-a_1)\,G}{y^2-2\,b_1\,y+x^2-2\,a_1\,x+b_1^2+a_1^2}$$

$$\Gamma^2{}_{11} = \frac{4\,m_1\,(y-b_1)\,G}{y^2-2\,b_1\,y+x^2-2\,a_1\,x+b_1^2+a_1^2}$$

$$\Gamma^2{}_{12} = -\frac{4\,m_1\,(x-a_1)\,G}{y^2-2\,b_1\,y+x^2-2\,a_1\,x+b_1^2+a_1^2}$$

$$\Gamma^2{}_{21} = \Gamma^2{}_{12}$$

$$\Gamma^2{}_{22} = -\frac{4\,m_1\,(y-b_1)\,G}{y^2-2\,b_1\,y+x^2-2\,a_1\,x+b_1^2+a_1^2}$$

Metric Compatibility

——— o.k.

Riemann Tensor

——— all elements zero

Ricci Tensor

——— all elements zero

Ricci Scalar

$$R_{sc} = 0$$

Bianchi identity (Ricci cyclic equation $R^{\kappa}{}_{[\mu\nu\sigma]} = 0$)

——— o.k.

Einstein Tensor

——— all elements zero

Hodge Dual of Bianchi Identity

——— (see charge and current densities)

Scalar Charge Density (-$R^{0}{}_{i}{}^{i0}$)

$$\rho = 0$$

Current Density Class 1 (-$R^{i}{}_{\mu}{}^{\mu j}$)

$$J_1 = 0$$

$$J_2 = 0$$

$$J_3 = 0$$

Current Density Class 2 (-$R^{i}{}_{\mu}{}^{\mu j}$)

$$J_1 = 0$$

$$J_2 = 0$$

$$J_3 = 0$$

Current Density Class 3 (-$R^{i}{}_{\mu}{}^{\mu j}$)

$$J_1 = 0$$

$$J_2 = 0$$

$$J_3 = 0$$

4.4.30 Multi-cosmic string metric, bicone

Multi-cosmic string metric, describing a form of a bicone. Parameter $b = b_1 + ib_2$ is complex.

Coordinates

$$\mathbf{x} = \begin{pmatrix} t \\ x \\ y \\ z \end{pmatrix}$$

Metric

$$g_{\mu\nu} = \begin{pmatrix} 1 & 0 & 0 & 0 \\ 0 & -\frac{1}{\sqrt{\left(-y^2+x^2+b_2^2-b_1^2\right)^2+(2\,x\,y-2\,b_1\,b_2)^2}} & 0 & 0 \\ 0 & 0 & -\frac{1}{\sqrt{\left(-y^2+x^2+b_2^2-b_1^2\right)^2+(2\,x\,y-2\,b_1\,b_2)^2}} & 0 \\ 0 & 0 & 0 & -1 \end{pmatrix}$$

Contravariant Metric

$$A = y^4 + \left(2\,x^2 - 2\,b_2^2 + 2\,b_1^2\right)\,y^2 - 8\,b_1\,b_2\,x\,y + x^4 + \left(2\,b_2^2 - 2\,b_1^2\right)\,x^2 + b_2^4 + 2\,b_1^2\,b_2^2 + b_1^4$$

$$f_{11} = f_{22} = \sqrt{A}$$

$$g^{\mu\nu} = \begin{pmatrix} 1 & 0 & 0 & 0 \\ 0 & -f_{11} & 0 & 0 \\ 0 & 0 & -f_{22} & 0 \\ 0 & 0 & 0 & -1 \end{pmatrix}$$

Christoffel Connection

$${\Gamma^1}_{11} = -\frac{x\,y^2 - 2\,b_1\,b_2\,y + x^3 + b_2^2\,x - b_1^2\,x}{\left(y^2 - 2\,b_2\,y + x^2 - 2\,b_1\,x + b_2^2 + b_1^2\right)\,\left(y^2 + 2\,b_2\,y + x^2 + 2\,b_1\,x + b_2^2 + b_1^2\right)}$$

$${\Gamma^1}_{12} = -\frac{y^3 + x^2\,y - b_2^2\,y + b_1^2\,y - 2\,b_1\,b_2\,x}{\left(y^2 - 2\,b_2\,y + x^2 - 2\,b_1\,x + b_2^2 + b_1^2\right)\,\left(y^2 + 2\,b_2\,y + x^2 + 2\,b_1\,x + b_2^2 + b_1^2\right)}$$

$${\Gamma^1}_{21} = {\Gamma^1}_{12}$$

$${\Gamma^1}_{22} = \frac{x\,y^2 - 2\,b_1\,b_2\,y + x^3 + b_2^2\,x - b_1^2\,x}{\left(y^2 - 2\,b_2\,y + x^2 - 2\,b_1\,x + b_2^2 + b_1^2\right)\,\left(y^2 + 2\,b_2\,y + x^2 + 2\,b_1\,x + b_2^2 + b_1^2\right)}$$

$${\Gamma^2}_{11} = \frac{y^3 + x^2\,y - b_2^2\,y + b_1^2\,y - 2\,b_1\,b_2\,x}{\left(y^2 - 2\,b_2\,y + x^2 - 2\,b_1\,x + b_2^2 + b_1^2\right)\,\left(y^2 + 2\,b_2\,y + x^2 + 2\,b_1\,x + b_2^2 + b_1^2\right)}$$

$$\Gamma^2{}_{12} = -\frac{x\,y^2 - 2\,b_1\,b_2\,y + x^3 + b_2^2\,x - b_1^2\,x}{(y^2 - 2\,b_2\,y + x^2 - 2\,b_1\,x + b_2^2 + b_1^2)\,(y^2 + 2\,b_2\,y + x^2 + 2\,b_1\,x + b_2^2 + b_1^2)}$$

$$\Gamma^2{}_{21} = \Gamma^2{}_{12}$$

$$\Gamma^2{}_{22} = -\frac{y^3 + x^2\,y - b_2^2\,y + b_1^2\,y - 2\,b_1\,b_2\,x}{(y^2 - 2\,b_2\,y + x^2 - 2\,b_1\,x + b_2^2 + b_1^2)\,(y^2 + 2\,b_2\,y + x^2 + 2\,b_1\,x + b_2^2 + b_1^2)}$$

Metric Compatibility

——— o.k.

Riemann Tensor

——— all elements zero

Ricci Tensor

——— all elements zero

Ricci Scalar

$$R_{sc} = 0$$

Bianchi identity (Ricci cyclic equation $R^{\kappa}{}_{[\mu\nu\sigma]} = 0$)

——— o.k.

Einstein Tensor

——— all elements zero

Hodge Dual of Bianchi Identity

——— (see charge and current densities)

Scalar Charge Density (-$R^0{}_i{}^{i0}$)

$$\rho = 0$$

Current Density Class 1 (-$R^i{}_\mu{}^{\mu j}$)

$$J_1 = 0$$

$$J_2 = 0$$

$$J_3 = 0$$

Current Density Class 2 ($-R^{i}{}_{\mu}{}^{\mu j}$)

$$J_1 = 0$$

$$J_2 = 0$$

$$J_3 = 0$$

Current Density Class 3 ($-R^{i}{}_{\mu}{}^{\mu j}$)

$$J_1 = 0$$

$$J_2 = 0$$

$$J_3 = 0$$

4.4.31 Einstein-Rosen type cosmic string metric

a, b, and c are complex parameters.

Coordinates

$$\mathbf{x} = \begin{pmatrix} t \\ x \\ y \\ z \end{pmatrix}$$

Metric

$$A = \left((y - c_2)^2 + (x - c_1)^2\right)^{4\, m_1\, G} \sqrt{\left(-y^2 + 2\, a_2\, y + x^2 - 2\, a_1\, x + b_2^2 - a_2^2 - b_1^2 + a_1^2\right)^2 + \left(2\, x\, y - 2\, a_1\, y - 2\, a_2\, x - 2\, b_1\, b_2 + 2\, a_1\, a_2\right)^2}$$

$$f_{11} = -\frac{1}{A}$$

$$f_{22} = -\frac{1}{A}$$

$$g_{\mu\nu} = \begin{pmatrix} 1 & 0 & 0 & 0 \\ 0 & f_{11} & 0 & 0 \\ 0 & 0 & f_{22} & 0 \\ 0 & 0 & 0 & -1 \end{pmatrix}$$

Contravariant Metric

$$\begin{aligned}A &= y^4 - 4\,a_2\,y^3 + \left(2\,x^2 - 4\,a_1\,x - 2\,b_2^2 + 6\,a_2^2 + 2\,b_1^2 + 2\,a_1^2\right)\,y^2 \\ &+ \left(-4\,a_2\,x^2 + (8\,a_1\,a_2 - 8\,b_1\,b_2)\,x + 4\,a_2\,b_2^2 + 8\,a_1\,b_1\,b_2 - 4\,a_2^3 + \left(-4\,b_1^2 - 4\,a_1^2\right)\,a_2\right)\,y \\ &+ x^4 - 4\,a_1\,x^3 + \left(2\,b_2^2 + 2\,a_2^2 - 2\,b_1^2 + 6\,a_1^2\right)\,x^2 + \left(-4\,a_1\,b_2^2 + 8\,b_1\,a_2\,b_2 - 4\,a_1\,a_2^2 + 4\,a_1\,b_1^2 - 4\,a_1^3\right)\,x \\ &+ b_2^4 + \left(-2\,a_2^2 + 2\,b_1^2 + 2\,a_1^2\right)\,b_2^2 - 8\,a_1\,b_1\,a_2\,b_2 \\ &+ a_2^4 + \left(2\,b_1^2 + 2\,a_1^2\right)\,a_2^2 + b_1^4 - 2\,a_1^2\,b_1^2 + a_1^4\end{aligned}$$

$$\begin{aligned}B &= y^4 - 4\,a_2\,y^3 + \left(2\,x^2 - 4\,a_1\,x - 2\,b_2^2 + 6\,a_2^2 + 2\,b_1^2 + 2\,a_1^2\right)\,y^2 \\ &+ \left(-4\,a_2\,x^2 + (8\,a_1\,a_2 - 8\,b_1\,b_2)\,x + 4\,a_2\,b_2^2 + 8\,a_1\,b_1\,b_2 - 4\,a_2^3 + \left(-4\,b_1^2 - 4\,a_1^2\right)\,a_2\right)\,y \\ &+ x^4 - 4\,a_1\,x^3 + \left(2\,b_2^2 + 2\,a_2^2 - 2\,b_1^2 + 6\,a_1^2\right)\,x^2 + \left(-4\,a_1\,b_2^2 + 8\,b_1\,a_2\,b_2 - 4\,a_1\,a_2^2 + 4\,a_1\,b_1^2 - 4\,a_1^3\right)\,x \\ &+ b_2^4 + \left(-2\,a_2^2 + 2\,b_1^2 + 2\,a_1^2\right)\,b_2^2 - 8\,a_1\,b_1\,a_2\,b_2 \\ &+ a_2^4 + \left(2\,b_1^2 + 2\,a_1^2\right)\,a_2^2 + b_1^4 - 2\,a_1^2\,b_1^2 + a_1^4\end{aligned}$$

$$f_{11} = -\left(y^2 - 2\,c_2\,y + x^2 - 2\,c_1\,x + c_2^2 + c_1^2\right)^{4\,m_1\,G}\,\sqrt{A}$$

$$f_{22} = -\left(y^2 - 2\,c_2\,y + x^2 - 2\,c_1\,x + c_2^2 + c_1^2\right)^{4\,m_1\,G}\,\sqrt{B}$$

$$g^{\mu\nu} = \begin{pmatrix} 1 & 0 & 0 & 0 \\ 0 & f_{11} & 0 & 0 \\ 0 & 0 & f_{22} & 0 \\ 0 & 0 & 0 & -1 \end{pmatrix}$$

Christoffel Connection

$$\Gamma^1{}_{11} = -\frac{4\,m_1\,x\,y^4\,G - 4\,c_1\,m_1\,y^4\,G - 16\,m_1\,a_2\,x\,y^3\,G + 16\,c_1\,m_1\,a_2\,y^3\,G + 8\,m_1\,x^3\,y^2\,G - 8\,c_1\,m_1\,x^2\,y^2\,G - 16\,a_1\,m_1\,x^2\,y^2\,G - 8\,m_1\,b_2^2\,x\,y^2\,G + \ldots}{\left(y^2 - 2\,b_2\,y - 2\,a_2\,y + \ldots\right)\left(y^2 + 2\,b_2\,y - 2\,a_2\,y + x^2 + \ldots\right)\left(y^2 - 2\,c_2\,y + x^2 - 2\,c_1\,x + \ldots\right)}$$

$$\Gamma^1{}_{12} = -\frac{4\,m_1\,y^5\,G - 4\,m_1\,c_2\,y^4\,G - 16\,m_1\,a_2\,y^4\,G + 8\,m_1\,x^2\,y^3\,G - 16\,a_1\,m_1\,x\,y^3\,G + 16\,m_1\,a_2\,c_2\,y^3\,G - 8\,m_1\,b_2^2\,y^3\,G + 24\,m_1\,a_2^2\,y^3\,G + \ldots}{\left(y^2 - 2\,b_2\,y - 2\,a_2\,y + \ldots\right)\left(y^2 + 2\,b_2\,y - 2\,a_2\,y + \ldots\right)\left(y^2 - 2\,c_2\,y + x^2 - \ldots\right)}$$

$$\Gamma^1{}_{21} = \Gamma^1{}_{12}$$

$$\Gamma^1{}_{22} = \frac{4\,m_1\,x\,y^4\,G - 4\,c_1\,m_1\,y^4\,G - 16\,m_1\,a_2\,x\,y^3\,G + 16\,c_1\,m_1\,a_2\,y^3\,G + 8\,m_1\,x^3\,y^2\,G - 8\,c_1\,m_1\,x^2\,y^2\,G - 16\,a_1\,m_1\,x^2\,y^2\,G - 8\,m_1\,b_2^2\,x\,y^2\,G + \ldots}{\left(y^2 - 2\,b_2\,y - 2\,a_2\,y + \ldots\right)\left(y^2 + 2\,b_2\,y - 2\,a_2\,y + \ldots\right)\left(y^2 - 2\,c_2\,y + x^2 - \ldots\right)}$$

$$\Gamma^2{}_{11} = \frac{4\,m_1\,y^5\,G - 4\,m_1\,c_2\,y^4\,G - 16\,m_1\,a_2\,y^4\,G + 8\,m_1\,x^2\,y^3\,G - 16\,a_1\,m_1\,x\,y^3\,G + 16\,m_1\,a_2\,c_2\,y^3\,G - 8\,m_1\,b_2^2\,y^3\,G + 24\,m_1\,a_2^2\,y^3\,G + 8\,b_1^2\,m_1\,y^3\,G\ldots}{\left(y^2 - 2\,b_2\,y - 2\,a_2\,y + \ldots\right)\left(y^2 + 2\,b_2\,y - 2\,a_2\,y + \ldots\right)\left(y^2 - 2\,c_2\,y + x^2\ldots\right)}$$

$$\Gamma^2{}_{12} = -\frac{4\,m_1\,x\,y^4\,G - 4\,c_1\,m_1\,y^4\,G - 16\,m_1\,a_2\,x\,y^3\,G + 16\,c_1\,m_1\,a_2\,y^3\,G + 8\,m_1\,x^3\,y^2\,G - 8\,c_1\,m_1\,x^2\,y^2\,G - 16\,a_1\,m_1\,x^2\,y^2\,G - 8\,m_1\,b_2^2\,x\,y^2\,G + \ldots}{\left(y^2 - 2\,b_2\,y - 2\,a_2\,y + \ldots\right)\left(y^2 + 2\,b_2\,y - 2\,a_2\,y + \ldots\right)\left(y^2 - 2\,c_2\,y + x^2\ldots\right)}$$

$$\Gamma^2{}_{21} = \Gamma^2{}_{12}$$

$$\Gamma^2{}_{22} = -\frac{4\,m_1\,y^5\,G - 4\,m_1\,c_2\,y^4\,G - 16\,m_1\,a_2\,y^4\,G + 8\,m_1\,x^2\,y^3\,G - 16\,a_1\,m_1\,x\,y^3\,G + 16\,m_1\,a_2\,c_2\,y^3\,G - 8\,m_1\,b_2^2\,y^3\,G + 24\,m_1\,a_2^2\,y^3\,G + \ldots}{\left(y^2 - 2\,b_2\,y - 2\,a_2\,y + \ldots\right)\left(y^2 + 2\,b_2\,y - 2\,a_2\,y + \ldots\right)\left(y^2 - 2\,c_2\,y + x^2 - \ldots\right)}$$

Metric Compatibility

——— o.k.

Riemann Tensor

——— all elements zero

Ricci Tensor

——— all elements zero

Ricci Scalar

$$R_{sc} = 0$$

Bianchi identity (Ricci cyclic equation $R^{\kappa}{}_{[\mu\nu\sigma]} = 0$)

——— o.k.

Einstein Tensor

——— all elements zero

Hodge Dual of Bianchi Identity

——— (see charge and current densities)

Scalar Charge Density ($-R^{0}{}_{i}{}^{i0}$)

$$\rho = 0$$

Current Density Class 1 ($-R^{i}{}_{\mu}{}^{\mu j}$)

$$J_1 = 0$$

$$J_2 = 0$$

$$J_3 = 0$$

Current Density Class 2 ($-R^{i}{}_{\mu}{}^{\mu j}$)

$$J_1 = 0$$

$$J_2 = 0$$

$$J_3 = 0$$

Current Density Class 3 ($-R^{i}{}_{\mu}{}^{\mu j}$)

$$J_1 = 0$$

$$J_2 = 0$$

$$J_3 = 0$$

4.4.32 Wheeler-Misner wormhole by 2 cosmic strings

For simplicity, a, b, and c are assumed to be real (non-complex) parameters.

Coordinates

$$\mathbf{x} = \begin{pmatrix} t \\ x \\ y \\ z \end{pmatrix}$$

Metric

$$A = \frac{(-y^2+x^2-c^2)^2+4\,x^2\,y^2}{\sqrt{(y^4-6\,x^2\,y^2+2\,a^2\,y^2+x^4-2\,a^2\,x^2-b^4+a^4)^2+(-4\,x\,y^3+4\,x^3\,y-4\,a^2\,x\,y)^2}}$$

$$f_{11} = f_{22} = A$$

$$g_{\mu\nu} = \begin{pmatrix} 1 & 0 & 0 & 0 \\ 0 & -f_{11} & 0 & 0 \\ 0 & 0 & -f_{22} & 0 \\ 0 & 0 & 0 & -1 \end{pmatrix}$$

Contravariant Metric

$$A = y^8+\left(4\,x^2+4\,a^2\right)\,y^6+\left(6\,x^4+4\,a^2\,x^2-2\,b^4+6\,a^4\right)\,y^4+$$

$$+\left(4\,x^6-4\,a^2\,x^4+\left(12\,b^4-4\,a^4\right)\,x^2-4\,a^2\,b^4+4\,a^6\right)\,y^2+$$

$$+x^8-4\,a^2\,x^6+\left(6\,a^4-2\,b^4\right)\,x^4+\left(4\,a^2\,b^4-4\,a^6\right)\,x^2+b^8-2\,a^4\,b^4+a^8$$

$$f_{11} = \frac{\sqrt{A}}{(y^2+x^2-2\,c\,x+c^2)\,(y^2+x^2+2\,c\,x+c^2)}$$

$$B = y^8+\left(4\,x^2+4\,a^2\right)\,y^6+\left(6\,x^4+4\,a^2\,x^2-2\,b^4+6\,a^4\right)\,y^4+$$

$$+\left(4\,x^6-4\,a^2\,x^4+\left(12\,b^4-4\,a^4\right)\,x^2-4\,a^2\,b^4+4\,a^6\right)\,y^2+x^8-4\,a^2\,x^6+$$

$$+\left(6\,a^4-2\,b^4\right)\,x^4+\left(4\,a^2\,b^4-4\,a^6\right)\,x^2+b^8-2\,a^4\,b^4+a^8$$

$$f_{22} = \frac{\sqrt{B}}{(y^2+x^2-2\,c\,x+c^2)\,(y^2+x^2+2\,c\,x+c^2)}$$

$$g^{\mu\nu} = \begin{pmatrix} 1 & 0 & 0 & 0 \\ 0 & -f_{11} & 0 & 0 \\ 0 & 0 & -f_{22} & 0 \\ 0 & 0 & 0 & -1 \end{pmatrix}$$

Christoffel Connection

$$\Gamma^1{}_{11} = -\frac{2\,x\,\left(3\,c^2\,y^8 - 3\,a^2\,y^8 + 8\,c^2\,x^2\,y^6 - 8\,a^2\,x^2\,y^6 + c^4\,y^6 + 6\,a^2\,c^2\,y^6 + 5\,b^4\,y^6 - 7\,a^4\,y^6 + 6\,c^2\,x^4\,y^4 - 6\,a^2\,x^4\,y^4 + 3\,c^4\,x^2\,y^4 - 2\,a^2\,c^2\,x^2\,y^4 - \ldots\right)}{\left(y^2 + x^2 - 2\,c\,x + c^2\right)\left(y^2 + x^2 + 2\,c\,x + c^2\right)\ldots}$$

$$\Gamma^1{}_{12} = -\frac{2\,y\,\left(c^2\,y^8 - a^2\,y^8 + c^4\,y^6 + 2\,a^2\,c^2\,y^6 + b^4\,y^6 - 3\,a^4\,y^6 - 6\,c^2\,x^4\,y^4 + 6\,a^2\,x^4\,y^4 + 3\,c^4\,x^2\,y^4 - 6\,a^2\,c^2\,x^2\,y^4 - 9\,b^4\,x^2\,y^4 + 3\,a^4\,x^2\,y^4 + \ldots\right)}{\left(y^2 + x^2 - 2\,c\,x + c^2\right)\left(y^2 + x^2 + 2\,c\,x + c^2\right)\ldots}$$

$$\Gamma^1{}_{21} = \Gamma^1{}_{12}$$

$$\Gamma^1{}_{22} = \frac{2\,x\,\left(3\,c^2\,y^8 - 3\,a^2\,y^8 + 8\,c^2\,x^2\,y^6 - 8\,a^2\,x^2\,y^6 + c^4\,y^6 + 6\,a^2\,c^2\,y^6 + 5\,b^4\,y^6 - 7\,a^4\,y^6 + 6\,c^2\,x^4\,y^4 - 6\,a^2\,x^4\,y^4 + 3\,c^4\,x^2\,y^4 - 2\,a^2\,c^2\,x^2\,y^4 - \ldots\right)}{\left(y^2 + x^2 - 2\,c\,x + c^2\right)\left(y^2 + x^2 + 2\,c\,x + c^2\right)\ldots}$$

$$\Gamma^2{}_{11} = \frac{2\,y\,\left(c^2\,y^8 - a^2\,y^8 + c^4\,y^6 + 2\,a^2\,c^2\,y^6 + b^4\,y^6 - 3\,a^4\,y^6 - 6\,c^2\,x^4\,y^4 + 6\,a^2\,x^4\,y^4 + 3\,c^4\,x^2\,y^4 - 6\,a^2\,c^2\,x^2\,y^4 - 9\,b^4\,x^2\,y^4 + \ldots\right)}{\left(y^2 + x^2 - 2\,c\,x + c^2\right)\left(y^2 + x^2 + 2\,c\,x + c^2\right)\ldots}$$

$$\Gamma^2{}_{12} = -\frac{2\,x\,\left(3\,c^2\,y^8 - 3\,a^2\,y^8 + 8\,c^2\,x^2\,y^6 - 8\,a^2\,x^2\,y^6 + c^4\,y^6 + 6\,a^2\,c^2\,y^6 + 5\,b^4\,y^6 - 7\,a^4\,y^6 + 6\,c^2\,x^4\,y^4 - 6\,a^2\,x^4\,y^4 + 3\,c^4\,x^2\,y^4 - 2\,a^2\,c^2\,x^2\,y^4 - \ldots\right)}{\left(y^2 + x^2 - 2\,c\,x + c^2\right)\left(y^2 + x^2 + 2\,c\,x + c^2\right)\ldots}$$

$$\Gamma^2{}_{21} = \Gamma^2{}_{12}$$

$$\Gamma^2{}_{22} = -\frac{2\,y\,\left(c^2\,y^8 - a^2\,y^8 + c^4\,y^6 + 2\,a^2\,c^2\,y^6 + b^4\,y^6 - 3\,a^4\,y^6 - 6\,c^2\,x^4\,y^4 + 6\,a^2\,x^4\,y^4 + 3\,c^4\,x^2\,y^4 - 6\,a^2\,c^2\,x^2\,y^4 - 9\,b^4\,x^2\,y^4 + 3\,a^4\,x^2\,y^4 + \ldots\right)}{\left(y^2 + x^2 - 2\,c\,x + c^2\right)\left(y^2 + x^2 + 2\,c\,x + c^2\right)\ldots}$$

Metric Compatibility

——— o.k.

Riemann Tensor

——— all elements zero

Ricci Tensor

——— all elements zero

Ricci Scalar

$$R_{sc} = 0$$

Bianchi identity (Ricci cyclic equation $R^{\kappa}{}_{[\mu\nu\sigma]} = 0$)

——— o.k.

Einstein Tensor

——— all elements zero

Hodge Dual of Bianchi Identity

——— (see charge and current densities)

Scalar Charge Density (-$R^{0\ \ i0}_{\ i}$)

$\rho = 0$

Current Density Class 1 (-$R^{i\ \ \mu j}_{\ \mu}$)

$J_1 = 0$

$J_2 = 0$

$J_3 = 0$

Current Density Class 2 (-$R^{i\ \ \mu j}_{\ \mu}$)

$J_1 = 0$

$J_2 = 0$

$J_3 = 0$

Current Density Class 3 (-$R^{i\ \ \mu j}_{\ \mu}$)

$J_1 = 0$

$J_2 = 0$

$J_3 = 0$

4.4.33 Hayward-Kim-Lee wormhole type 1

Angular parts have been assumed as identical to the spherically symmetric line element. λ is a parameter. Charge and current density contain diverging terms for large r.

Coordinates

$$\mathbf{x} = \begin{pmatrix} t \\ r \\ \vartheta \\ \varphi \end{pmatrix}$$

Metric

$$g_{\mu\nu} = \begin{pmatrix} \frac{2m}{r} - 1 & 0 & 0 & 0 \\ 0 & \frac{1}{4\left(1-\frac{2m}{r}\right) r^2 \lambda^2} & 0 & 0 \\ 0 & 0 & r^2 & 0 \\ 0 & 0 & 0 & r^2 \sin^2\vartheta \end{pmatrix}$$

Contravariant Metric

$$g^{\mu\nu} = \begin{pmatrix} -\frac{r}{r-2m} & 0 & 0 & 0 \\ 0 & 4r(r-2m)\lambda^2 & 0 & 0 \\ 0 & 0 & \frac{1}{r^2} & 0 \\ 0 & 0 & 0 & \frac{1}{r^2 \sin^2\vartheta} \end{pmatrix}$$

Christoffel Connection

$$\Gamma^0{}_{01} = \frac{m}{r\,(r-2\,m)}$$

$$\Gamma^0{}_{10} = \Gamma^0{}_{01}$$

$$\Gamma^1{}_{00} = \frac{4\,m\,(r-2\,m)\,\lambda^2}{r}$$

$$\Gamma^1{}_{11} = -\frac{r-m}{r\,(r-2\,m)}$$

$$\Gamma^1{}_{22} = -4\,r^2\,(r-2\,m)\,\lambda^2$$

$$\Gamma^1{}_{33} = -4\,r^2\,(r-2\,m)\,\sin^2\vartheta\,\lambda^2$$

$$\Gamma^2{}_{12} = \frac{1}{r}$$

$$\Gamma^2{}_{21} = \Gamma^2{}_{12}$$

$$\Gamma^2{}_{33} = -\cos\vartheta\,\sin\vartheta$$

$$\Gamma^3{}_{13} = \frac{1}{r}$$

$$\Gamma^3{}_{23} = \frac{\cos\vartheta}{\sin\vartheta}$$

$$\Gamma^3{}_{31} = \Gamma^3{}_{13}$$

$$\Gamma^3{}_{32} = \Gamma^3{}_{23}$$

Metric Compatibility

——— o.k.

Riemann Tensor

$$R^0{}_{101} = \frac{m}{r^2\,(r-2\,m)}$$

$$R^0{}_{110} = -R^0{}_{101}$$

$$R^0{}_{202} = -4\,m\,r\,\lambda^2$$

$$R^0{}_{220} = -R^0{}_{202}$$

$$R^0{}_{303} = -4\,m\,r\,\sin^2\vartheta\,\lambda^2$$

$$R^0{}_{330} = -R^0{}_{303}$$

$$R^1{}_{001} = \frac{4\,m\,(r-2\,m)\,\lambda^2}{r^2}$$

$$R^1{}_{010} = -R^1{}_{001}$$

$$R^1{}_{212} = -4\,r\,(r-m)\,\lambda^2$$

$$R^1{}_{221} = -R^1{}_{212}$$

$$R^1{}_{313} = -4\,r\,(r-m)\,\sin^2\vartheta\,\lambda^2$$

$$R^1{}_{331} = -R^1{}_{313}$$

$$R^2{}_{002} = -\frac{4\,m\,(r-2\,m)\,\lambda^2}{r^2}$$

$$R^2{}_{020} = -R^2{}_{002}$$

$$R^2{}_{112} = \frac{r-m}{r^2\,(r-2\,m)}$$

$$R^2{}_{121} = -R^2{}_{112}$$

$$R^2{}_{323} = -\sin^2\vartheta\,\left(4\,r^2\,\lambda^2 - 8\,m\,r\,\lambda^2 - 1\right)$$

$$R^2{}_{332} = -R^2{}_{323}$$

$$R^3{}_{003} = -\frac{4\,m\,(r-2\,m)\,\lambda^2}{r^2}$$

$$R^3{}_{030} = -R^3{}_{003}$$

$$R^3{}_{113} = \frac{r-m}{r^2\,(r-2\,m)}$$

$$R^3{}_{131} = -R^3{}_{113}$$

$$R^3{}_{223} = 4\,r^2\,\lambda^2 - 8\,m\,r\,\lambda^2 - 1$$

$$R^3{}_{232} = -R^3{}_{223}$$

Ricci Tensor

$$\mathrm{Ric}_{00} = \frac{4\,m\,(r-2\,m)\,\lambda^2}{r^2}$$

$$\mathrm{Ric}_{11} = -\frac{2\,r-3\,m}{r^2\,(r-2\,m)}$$

$$\mathrm{Ric}_{22} = -\left(8\,r^2\,\lambda^2 - 8\,m\,r\,\lambda^2 - 1\right)$$

$$\mathrm{Ric}_{33} = -\sin^2\vartheta\,\left(8\,r^2\,\lambda^2 - 8\,m\,r\,\lambda^2 - 1\right)$$

Ricci Scalar

$$R_{sc} = -\frac{2\,\left(12\,r^2\,\lambda^2 - 12\,m\,r\,\lambda^2 - 1\right)}{r^2}$$

Bianchi identity (Ricci cyclic equation $R^{\kappa}{}_{[\mu\nu\sigma]} = 0$)

———— o.k.

Einstein Tensor

$$G_{00} = -\frac{(r-2\,m)\,\left(12\,r^2\,\lambda^2 - 16\,m\,r\,\lambda^2 - 1\right)}{r^3}$$

$$G_{11} = \frac{(2\,r\,\lambda-1)\,(2\,r\,\lambda+1)}{4\,r^3\,(r-2\,m)\,\lambda^2}$$

$$G_{22} = 4\,r\,(r-m)\,\lambda^2$$

$$G_{33} = 4\,r\,(r-m)\,\sin^2\vartheta\,\lambda^2$$

Hodge Dual of Bianchi Identity

———— (see charge and current densities)

Scalar Charge Density ($-R^{0}{}_{i}{}^{i0}$)

$$\rho = \frac{4\,m\,\lambda^2}{r-2\,m}$$

Current Density Class 1 ($-R^{i}{}_{\mu}{}^{\mu j}$)

$$J_1 = 16\,(r-2\,m)\,(2\,r-3\,m)\,\lambda^4$$

$$J_2 = \frac{8\,r^2\,\lambda^2 - 8\,m\,r\,\lambda^2 - 1}{r^4}$$

$$J_3 = \frac{8\,r^2\,\lambda^2 - 8\,m\,r\,\lambda^2 - 1}{r^4\,\sin^2\vartheta}$$

Current Density Class 2 ($-R^{i}{}_{\mu}{}^{\mu j}$)

$$J_1 = 0$$

$$J_2 = 0$$

$$J_3 = 0$$

Current Density Class 3 ($-R^{i}{}_{\mu}{}^{\mu j}$)

$$J_1 = 0$$

$$J_2 = 0$$

$$J_3 = 0$$

4.4.34 Hayward-Kim-Lee wormhole type 2

Angular parts have been assumed as identical to the spherically symmetric line element. λ and a are parameters. Charge and current density contain diverging terms for large r. Results are similar to the type 1 wormhole.

Coordinates

$$\mathbf{x} = \begin{pmatrix} t \\ r \\ \vartheta \\ \varphi \end{pmatrix}$$

Metric

$$g_{\mu\nu} = \begin{pmatrix} -\frac{2}{r} & 0 & 0 & 0 \\ 0 & \frac{1}{8\left(1-\frac{a}{r}\right) r^2 \lambda^2} & 0 & 0 \\ 0 & 0 & r^2 & 0 \\ 0 & 0 & 0 & r^2 \sin^2 \vartheta \end{pmatrix}$$

Contravariant Metric

$$g^{\mu\nu} = \begin{pmatrix} -\frac{r}{2} & 0 & 0 & 0 \\ 0 & 8 r (r-a) \lambda^2 & 0 & 0 \\ 0 & 0 & \frac{1}{r^2} & 0 \\ 0 & 0 & 0 & \frac{1}{r^2 \sin^2 \vartheta} \end{pmatrix}$$

Christoffel Connection

$$\Gamma^0{}_{01} = -\frac{1}{2\,r}$$

$$\Gamma^0{}_{10} = \Gamma^0{}_{01}$$

$$\Gamma^1{}_{00} = -\frac{8\,(r-a)\,\lambda^2}{r}$$

$$\Gamma^1{}_{11} = -\frac{2\,r-a}{2\,r\,(r-a)}$$

$$\Gamma^1{}_{22} = -8\,r^2\,(r-a)\,\lambda^2$$

$$\Gamma^1{}_{33} = -8\,r^2\,(r-a)\,\sin^2\vartheta\,\lambda^2$$

$$\Gamma^2{}_{12} = \frac{1}{r}$$

$$\Gamma^2{}_{21} = \Gamma^2{}_{12}$$

$$\Gamma^2{}_{33} = -\cos\vartheta\,\sin\vartheta$$

$$\Gamma^3{}_{13} = \frac{1}{r}$$

$$\Gamma^3{}_{23} = \frac{\cos\vartheta}{\sin\vartheta}$$

$$\Gamma^3{}_{31} = \Gamma^3{}_{13}$$

$$\Gamma^3{}_{32} = \Gamma^3{}_{23}$$

Metric Compatibility

——— o.k.

Riemann Tensor

$$R^0{}_{101} = -\frac{r-2\,a}{4\,r^2\,(r-a)}$$

$$R^0{}_{110} = -R^0{}_{101}$$

$$R^0{}_{202} = 4\,r\,(r-a)\,\lambda^2$$

$$R^0{}_{220} = -R^0{}_{202}$$

$$R^0{}_{303} = 4\,r\,(r-a)\,\sin^2\vartheta\,\lambda^2$$

$$R^0{}_{330} = -R^0{}_{303}$$

$$R^1{}_{001} = -\frac{4\,(r-2\,a)\,\lambda^2}{r^2}$$

$$R^1{}_{010} = -R^1{}_{001}$$

$$R^1{}_{212} = -4\,r\,(2\,r-a)\,\lambda^2$$

$$R^1{}_{221} = -R^1{}_{212}$$

$$R^1{}_{313} = -4\,r\,(2\,r-a)\,\sin^2\vartheta\,\lambda^2$$

$$R^1{}_{331} = -R^1{}_{313}$$

$$R^2{}_{002} = \frac{8\,(r-a)\,\lambda^2}{r^2}$$

$$R^2{}_{020} = -R^2{}_{002}$$

$$R^2{}_{112} = \frac{2\,r-a}{2\,r^2\,(r-a)}$$

$$R^2{}_{121} = -R^2{}_{112}$$

$$R^2{}_{323} = -\sin^2\vartheta\,\left(8\,r^2\,\lambda^2 - 8\,a\,r\,\lambda^2 - 1\right)$$

$$R^2{}_{332} = -R^2{}_{323}$$

$$R^3{}_{003} = \frac{8\,(r-a)\,\lambda^2}{r^2}$$

$$R^3{}_{030} = -R^3{}_{003}$$

$$R^3{}_{113} = \frac{2\,r-a}{2\,r^2\,(r-a)}$$

$$R^3{}_{131} = -R^3{}_{113}$$

$$R^3{}_{223} = 8\,r^2\,\lambda^2 - 8\,a\,r\,\lambda^2 - 1$$

$$R^3{}_{232} = -R^3{}_{223}$$

Ricci Tensor

$$\mathrm{Ric}_{00} = -\frac{4\,(3\,r - 2\,a)\,\lambda^2}{r^2}$$

$$\mathrm{Ric}_{11} = -\frac{3\,(3\,r - 2\,a)}{4\,r^2\,(r - a)}$$

$$\mathrm{Ric}_{22} = -\left(12\,r^2\,\lambda^2 - 8\,a\,r\,\lambda^2 - 1\right)$$

$$\mathrm{Ric}_{33} = -\sin^2\vartheta\,\left(12\,r^2\,\lambda^2 - 8\,a\,r\,\lambda^2 - 1\right)$$

Ricci Scalar

$$R_{sc} = -\frac{2\,\left(18\,r^2\,\lambda^2 - 12\,a\,r\,\lambda^2 - 1\right)}{r^2}$$

Bianchi identity (Ricci cyclic equation $R^{\kappa}{}_{[\mu\nu\sigma]} = 0$)

——— o.k.

Einstein Tensor

$$G_{00} = -\frac{2\,\left(24\,r^2\,\lambda^2 - 16\,a\,r\,\lambda^2 - 1\right)}{r^3}$$

$$G_{11} = -\frac{1}{8\,r^3\,(r - a)\,\lambda^2}$$

$$G_{22} = 2\,r\,(3\,r - 2\,a)\,\lambda^2$$

$$G_{33} = 2\,r\,(3\,r - 2\,a)\,\sin^2\vartheta\,\lambda^2$$

Hodge Dual of Bianchi Identity

——— (see charge and current densities)

Scalar Charge Density (-$R^{0}{}_{i}{}^{i0}$)

$$\rho = -(3\,r - 2\,a)\,\lambda^2$$

Current Density Class 1 (-$R^{i}{}_{\mu}{}^{\mu j}$)

$$J_1 = 48\,(r - a)\,(3\,r - 2\,a)\,\lambda^4$$

$$J_2 = \frac{12\,r^2\,\lambda^2 - 8\,a\,r\,\lambda^2 - 1}{r^4}$$

$$J_3 = \frac{12\,r^2\,\lambda^2 - 8\,a\,r\,\lambda^2 - 1}{r^4\,\sin^2\vartheta}$$

Current Density Class 2 (-$R^{i}{}_{\mu}{}^{\mu j}$)

$$J_1 = 0$$

$$J_2 = 0$$

$$J_3 = 0$$

Current Density Class 3 (-$R^{i}{}_{\mu}{}^{\mu j}$)

$$J_1 = 0$$

$$J_2 = 0$$

$$J_3 = 0$$

4.4.35 Simple wormhole metric

Metric of a simple form of a wormhole. k is a paremeter.

Coordinates

$$\mathbf{x} = \begin{pmatrix} t \\ l \\ \vartheta \\ \varphi \end{pmatrix}$$

Metric

$$g_{\mu\nu} = \begin{pmatrix} -c^2 & 0 & 0 & 0 \\ 0 & 1 & 0 & 0 \\ 0 & 0 & l^2 + k^2 & 0 \\ 0 & 0 & 0 & \left(l^2 + k^2\right) \sin^2 \vartheta \end{pmatrix}$$

Contravariant Metric

$$g^{\mu\nu} = \begin{pmatrix} -\frac{1}{c^2} & 0 & 0 & 0 \\ 0 & 1 & 0 & 0 \\ 0 & 0 & \frac{1}{l^2+k^2} & 0 \\ 0 & 0 & 0 & \frac{1}{(l^2+k^2)\sin^2 \vartheta} \end{pmatrix}$$

Christoffel Connection

$$\Gamma^1{}_{22} = -l$$

$$\Gamma^1{}_{33} = -l \sin^2 \vartheta$$

$$\Gamma^2{}_{12} = \frac{l}{l^2 + k^2}$$

$$\Gamma^2{}_{21} = \Gamma^2{}_{12}$$

$$\Gamma^2{}_{33} = -\cos \vartheta \sin \vartheta$$

$$\Gamma^3{}_{13} = \frac{l}{l^2 + k^2}$$

$$\Gamma^3{}_{23} = \frac{\cos \vartheta}{\sin \vartheta}$$

$$\Gamma^3{}_{31} = \Gamma^3{}_{13}$$

$$\Gamma^3{}_{32} = \Gamma^3{}_{23}$$

Metric Compatibility

———— o.k.

Riemann Tensor

$$R^1{}_{212} = -\frac{k^2}{l^2 + k^2}$$

$$R^1{}_{221} = -R^1{}_{212}$$

$$R^1{}_{313} = -\frac{k^2 \sin^2 \vartheta}{l^2 + k^2}$$

$$R^1{}_{331} = -R^1{}_{313}$$

$$R^2{}_{112} = \frac{k^2}{(l^2 + k^2)^2}$$

$$R^2{}_{121} = -R^2{}_{112}$$

$$R^2{}_{323} = \frac{k^2 \sin^2 \vartheta}{l^2 + k^2}$$

$$R^2{}_{332} = -R^2{}_{323}$$

$$R^3{}_{113} = \frac{k^2}{(l^2+k^2)^2}$$

$$R^3{}_{131} = -R^3{}_{113}$$

$$R^3{}_{223} = -\frac{k^2}{l^2+k^2}$$

$$R^3{}_{232} = -R^3{}_{223}$$

Ricci Tensor

$$\mathrm{Ric}_{11} = -\frac{2\,k^2}{(l^2+k^2)^2}$$

Ricci Scalar

$$R_{sc} = -\frac{2\,k^2}{(l^2+k^2)^2}$$

Bianchi identity (Ricci cyclic equation $R^{\kappa}{}_{[\mu\nu\sigma]} = 0$)

——— o.k.

Einstein Tensor

$$G_{00} = -\frac{c^2\,k^2}{(l^2+k^2)^2}$$

$$G_{11} = -\frac{k^2}{(l^2+k^2)^2}$$

$$G_{22} = \frac{k^2}{l^2+k^2}$$

$$G_{33} = \frac{k^2\,\sin^2\vartheta}{l^2+k^2}$$

Hodge Dual of Bianchi Identity

——— (see charge and current densities)

Scalar Charge Density (-$R^{0}{}_{i}{}^{i0}$)

$$\rho = 0$$

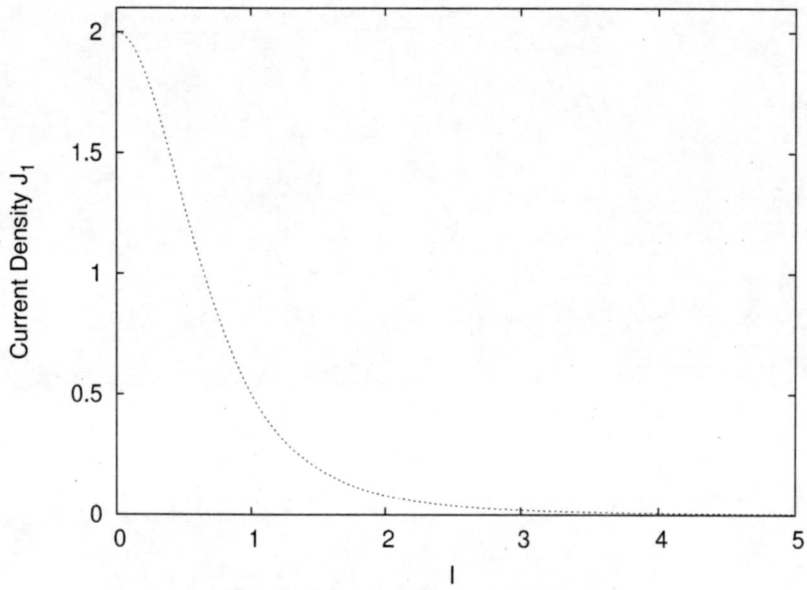

Fig. 4.66: Simple Wormhole metric, current density J_l for $k = 1$.

Current Density Class 1 ($-R^{i}{}_{\mu}{}^{\mu j}$)

$$J_1 = \frac{2\,k^2}{\left(l^2 + k^2\right)^2}$$

$$J_2 = 0$$

$$J_3 = 0$$

Current Density Class 2 ($-R^{i}{}_{\mu}{}^{\mu j}$)

$$J_1 = 0$$

$$J_2 = 0$$

$$J_3 = 0$$

Current Density Class 3 ($-R^{i}{}_{\mu}{}^{\mu j}$)

$$J_1 = 0$$

$$J_2 = 0$$

$$J_3 = 0$$

4.4.36 Simple wormhole metric with varying cosmological constant

Metric of a simple form of a wormhole with a varying cosmological constant. Here $\Gamma(r)/2$ is the redshift function and $b(r)$ is the shape function determining the shape of the wormhole. The cosmological constant is contained in the form of b(r).

Coordinates

$$\mathbf{x} = \begin{pmatrix} t \\ r \\ \vartheta \\ \varphi \end{pmatrix}$$

Metric

$$g_{\mu\nu} = \begin{pmatrix} -e^{\Gamma(r)} & 0 & 0 & 0 \\ 0 & \frac{1}{1-\frac{b(r)}{r}} & 0 & 0 \\ 0 & 0 & r^2 & 0 \\ 0 & 0 & 0 & r^2 \sin^2\vartheta \end{pmatrix}$$

Contravariant Metric

$$g^{\mu\nu} = \begin{pmatrix} -e^{-\Gamma(r)} & 0 & 0 & 0 \\ 0 & -\frac{b(r)-r}{r} & 0 & 0 \\ 0 & 0 & \frac{1}{r^2} & 0 \\ 0 & 0 & 0 & \frac{1}{r^2 \sin^2\vartheta} \end{pmatrix}$$

Christoffel Connection

$$\Gamma^0{}_{01} = \frac{\frac{d}{dr}\Gamma(r)}{2}$$

$$\Gamma^0{}_{10} = \Gamma^0{}_{01}$$

$$\Gamma^1{}_{00} = -\frac{(b(r)-r)\, e^{\Gamma(r)} \left(\frac{d}{dr}\Gamma(r)\right)}{2\,r}$$

$$\Gamma^1{}_{11} = -\frac{r\left(\frac{d}{dr}b(r)\right) - b(r)}{2\,r\,(b(r)-r)}$$

$$\Gamma^1{}_{22} = b(r) - r$$

$$\Gamma^1{}_{33} = (b(r)-r)\sin^2\vartheta$$

$$\Gamma^2{}_{12} = \frac{1}{r}$$

$$\Gamma^2{}_{21} = \Gamma^2{}_{12}$$

$$\Gamma^2{}_{33} = -\cos\vartheta\,\sin\vartheta$$

$$\Gamma^3{}_{13} = \frac{1}{r}$$

$$\Gamma^3{}_{23} = \frac{\cos\vartheta}{\sin\vartheta}$$

$$\Gamma^3{}_{31} = \Gamma^3{}_{13}$$

$$\Gamma^3{}_{32} = \Gamma^3{}_{23}$$

Metric Compatibility

——— o.k.

Riemann Tensor

$$R^{0}{}_{101} = -\frac{2\,r\,b(r)\left(\frac{d^2}{d\,r^2}\Gamma(r)\right) - 2\,r^2\left(\frac{d^2}{d\,r^2}\Gamma(r)\right) + r\,b(r)\left(\frac{d}{d\,r}\Gamma(r)\right)^2 - r^2\left(\frac{d}{d\,r}\Gamma(r)\right)^2 + r\left(\frac{d}{d\,r}b(r)\right)\left(\frac{d}{d\,r}\Gamma(r)\right) - b(r)\left(\frac{d}{d\,r}\Gamma(r)\right)}{4\,r\,(b(r)-r)}$$

$$R^{0}{}_{110} = -R^{0}{}_{101}$$

$$R^{0}{}_{202} = \frac{(b(r)-r)\left(\frac{d}{d\,r}\Gamma(r)\right)}{2}$$

$$R^{0}{}_{220} = -R^{0}{}_{202}$$

$$R^{0}{}_{303} = \frac{(b(r)-r)\left(\frac{d}{d\,r}\Gamma(r)\right)\sin^2\vartheta}{2}$$

$$R^{0}{}_{330} = -R^{0}{}_{303}$$

$$R^{1}{}_{001} = \frac{e^{\Gamma(r)}\left(2\,r\,b(r)\left(\frac{d^2}{d\,r^2}\Gamma(r)\right) - 2\,r^2\left(\frac{d^2}{d\,r^2}\Gamma(r)\right) + r\,b(r)\left(\frac{d}{d\,r}\Gamma(r)\right)^2 - r^2\left(\frac{d}{d\,r}\Gamma(r)\right)^2 + r\left(\frac{d}{d\,r}b(r)\right)\left(\frac{d}{d\,r}\Gamma(r)\right) - b(r)\left(\frac{d}{d\,r}\Gamma(r)\right)\right)}{4\,r^2}$$

$$R^{1}{}_{010} = -R^{1}{}_{001}$$

$$R^{1}{}_{212} = \frac{r\left(\frac{d}{d\,r}b(r)\right) - b(r)}{2\,r}$$

$$R^{1}{}_{221} = -R^{1}{}_{212}$$

$$R^{1}{}_{313} = \frac{\left(r\left(\frac{d}{d\,r}b(r)\right) - b(r)\right)\sin^2\vartheta}{2\,r}$$

$$R^{1}{}_{331} = -R^{1}{}_{313}$$

$$R^{2}{}_{002} = \frac{(b(r)-r)\,e^{\Gamma(r)}\left(\frac{d}{d\,r}\Gamma(r)\right)}{2\,r^2}$$

$$R^{2}{}_{020} = -R^{2}{}_{002}$$

$$R^{2}{}_{112} = \frac{r\left(\frac{d}{d\,r}b(r)\right) - b(r)}{2\,r^2\,(b(r)-r)}$$

$$R^2{}_{121} = -R^2{}_{112}$$

$$R^2{}_{323} = \frac{b(r)\,\sin^2\vartheta}{r}$$

$$R^2{}_{332} = -R^2{}_{323}$$

$$R^3{}_{003} = \frac{(b(r) - r)\, e^{\Gamma(r)} \left(\frac{d}{dr}\Gamma(r)\right)}{2\,r^2}$$

$$R^3{}_{030} = -R^3{}_{003}$$

$$R^3{}_{113} = \frac{r\left(\frac{d}{dr}b(r)\right) - b(r)}{2\,r^2\,(b(r) - r)}$$

$$R^3{}_{131} = -R^3{}_{113}$$

$$R^3{}_{223} = -\frac{b(r)}{r}$$

$$R^3{}_{232} = -R^3{}_{223}$$

Ricci Tensor

$$\mathrm{Ric}_{00} = -\frac{e^{\Gamma(r)}\left(2\,r\,b(r)\left(\frac{d^2}{dr^2}\Gamma(r)\right) - 2\,r^2\left(\frac{d^2}{dr^2}\Gamma(r)\right) + r\,b(r)\left(\frac{d}{dr}\Gamma(r)\right)^2 - r^2\left(\frac{d}{dr}\Gamma(r)\right)^2 + r\left(\frac{d}{dr}b(r)\right)\left(\frac{d}{dr}\Gamma(r)\right) + \ldots\right)}{4\,r^2}$$

$$\mathrm{Ric}_{11} = -\frac{2\,r^2\,b(r)\left(\frac{d^2}{dr^2}\Gamma(r)\right) - 2\,r^3\left(\frac{d^2}{dr^2}\Gamma(r)\right) + r^2\,b(r)\left(\frac{d}{dr}\Gamma(r)\right)^2 - r^3\left(\frac{d}{dr}\Gamma(r)\right)^2 + r^2\left(\frac{d}{dr}b(r)\right)\left(\frac{d}{dr}\Gamma(r)\right) - r\,b(r)\,(\ldots) + \ldots}{4\,r^2\,(b(r) - r)}$$

$$\mathrm{Ric}_{22} = \frac{r\,b(r)\left(\frac{d}{dr}\Gamma(r)\right) - r^2\left(\frac{d}{dr}\Gamma(r)\right) + r\left(\frac{d}{dr}b(r)\right) + b(r)}{2\,r}$$

$$\mathrm{Ric}_{33} = \frac{\left(r\,b(r)\left(\frac{d}{dr}\Gamma(r)\right) - r^2\left(\frac{d}{dr}\Gamma(r)\right) + r\left(\frac{d}{dr}b(r)\right) + b(r)\right)\sin^2\vartheta}{2\,r}$$

Ricci Scalar

$$R_{sc} = \frac{2\,r\,b(r)\left(\frac{d^2}{dr^2}\Gamma(r)\right) - 2\,r^2\left(\frac{d^2}{dr^2}\Gamma(r)\right) + r\,b(r)\left(\frac{d}{dr}\Gamma(r)\right)^2 - r^2\left(\frac{d}{dr}\Gamma(r)\right)^2 + r\left(\frac{d}{dr}b(r)\right)\left(\frac{d}{dr}\Gamma(r)\right) + 3\,b(r)\left(\frac{d}{dr}\Gamma(r)\right) - \ldots}{2\,r^2}$$

Bianchi identity (Ricci cyclic equation $R^\kappa{}_{[\mu\nu\sigma]} = 0$)

——— o.k.

Einstein Tensor

$$G_{00} = \frac{e^{\Gamma(r)} \left(\frac{d}{d\,r} b\left(r\right)\right)}{r^2}$$

$$G_{11} = \frac{r\, b\left(r\right) \left(\frac{d}{d\,r} \Gamma\left(r\right)\right) - r^2 \left(\frac{d}{d\,r} \Gamma\left(r\right)\right) + b\left(r\right)}{r^2 \left(b\left(r\right) - r\right)}$$

$$G_{22} = -\frac{2\,r^2\, b\left(r\right) \left(\frac{d^2}{d\,r^2} \Gamma\left(r\right)\right) - 2\,r^3 \left(\frac{d^2}{d\,r^2} \Gamma\left(r\right)\right) + r^2\, b\left(r\right) \left(\frac{d}{d\,r} \Gamma\left(r\right)\right)^2 - r^3 \left(\frac{d}{d\,r} \Gamma\left(r\right)\right)^2 + r^2 \left(\frac{d}{d\,r} b\left(r\right)\right) \left(\frac{d}{d\,r} \Gamma\left(r\right)\right) + r\, b\left(r\right) \left(...\right) - ...}{4\,r}$$

$$G_{33} = -\frac{\left(2\,r^2\, b\left(r\right) \left(\frac{d^2}{d\,r^2} \Gamma\left(r\right)\right) - 2\,r^3 \left(\frac{d^2}{d\,r^2} \Gamma\left(r\right)\right) + r^2\, b\left(r\right) \left(\frac{d}{d\,r} \Gamma\left(r\right)\right)^2 - r^3 \left(\frac{d}{d\,r} \Gamma\left(r\right)\right)^2 + r^2 \left(\frac{d}{d\,r} b\left(r\right)\right) \left(\frac{d}{d\,r} \Gamma\left(r\right)\right) + ...\right) \sin^2 \vartheta}{4\,r}$$

Hodge Dual of Bianchi Identity

——— (see charge and current densities)

Scalar Charge Density ($-R^{0\ \ i0}_{\ i}$)

$$\rho = -\frac{e^{-\Gamma(r)} \left(2\,r\, b\left(r\right) \left(\frac{d^2}{d\,r^2} \Gamma\left(r\right)\right) - 2\,r^2 \left(\frac{d^2}{d\,r^2} \Gamma\left(r\right)\right) + r\, b\left(r\right) \left(\frac{d}{d\,r} \Gamma\left(r\right)\right)^2 - r^2 \left(\frac{d}{d\,r} \Gamma\left(r\right)\right)^2 + r \left(\frac{d}{d\,r} b\left(r\right)\right) \left(\frac{d}{d\,r} \Gamma\left(r\right)\right) + ...\right)}{4\,r^2}$$

Current Density Class 1 ($-R^{i\ \ \mu j}_{\ \mu}$)

$$J_1 = \frac{\left(b\left(r\right) - r\right) \left(2\,r^2\, b\left(r\right) \left(\frac{d^2}{d\,r^2} \Gamma\left(r\right)\right) - 2\,r^3 \left(\frac{d^2}{d\,r^2} \Gamma\left(r\right)\right) + r^2\, b\left(r\right) \left(\frac{d}{d\,r} \Gamma\left(r\right)\right)^2 - r^3 \left(\frac{d}{d\,r} \Gamma\left(r\right)\right)^2 + r^2 \left(\frac{d}{d\,r} b\left(r\right)\right) \left(\frac{d}{d\,r} \Gamma\left(r\right)\right) - ...\right)}{4\,r^4}$$

$$J_2 = -\frac{r\, b\left(r\right) \left(\frac{d}{d\,r} \Gamma\left(r\right)\right) - r^2 \left(\frac{d}{d\,r} \Gamma\left(r\right)\right) + r \left(\frac{d}{d\,r} b\left(r\right)\right) + b\left(r\right)}{2\,r^5}$$

$$J_3 = -\frac{r\, b\left(r\right) \left(\frac{d}{d\,r} \Gamma\left(r\right)\right) - r^2 \left(\frac{d}{d\,r} \Gamma\left(r\right)\right) + r \left(\frac{d}{d\,r} b\left(r\right)\right) + b\left(r\right)}{2\,r^5\, \sin^2 \vartheta}$$

Current Density Class 2 ($-R^{i\ \ \mu j}_{\ \mu}$)

$$J_1 = 0$$

$$J_2 = 0$$

$$J_3 = 0$$

Current Density Class 3 ($-R^{i\ \ \mu j}_{\ \mu}$)

$$J_1 = 0$$

$$J_2 = 0$$

$$J_3 = 0$$

4.4.37 Evans metric

The general spherical metric contains four functions $A(t,r), B(r), C(r), D(r)$ by which all so-called vacuum metrics can be described. The terms for the cosmological charge and current densities can be used to determine the functions in a way so that these densities disappear, leading to the most general condition to describe a true vacuum in the EH theory:

$$\rho(A,B,C,D) = 0$$
$$J_1(A,B,C,D) = 0$$
$$J_2(A,B,C,D) = 0$$
$$J_3(A,B,C,D) = 0$$

However, the Einstein tensor, which describes the energy momentum density, has also to vanish in order to describe a true vacuum.

Coordinates

$$\mathbf{x} = \begin{pmatrix} t \\ r \\ \vartheta \\ \varphi \end{pmatrix}$$

Metric

$$g_{\mu\nu} = \begin{pmatrix} A & 0 & 0 & 0 \\ 0 & B & 0 & 0 \\ 0 & 0 & C & 0 \\ 0 & 0 & 0 & \sin^2\vartheta\, D \end{pmatrix}$$

Contravariant Metric

$$g^{\mu\nu} = \begin{pmatrix} \frac{1}{A} & 0 & 0 & 0 \\ 0 & \frac{1}{B} & 0 & 0 \\ 0 & 0 & \frac{1}{C} & 0 \\ 0 & 0 & 0 & \frac{1}{\sin^2\vartheta\, D} \end{pmatrix}$$

Christoffel Connection

$$\Gamma^0{}_{00} = \frac{\frac{d}{dt} A}{2\,A}$$

$$\Gamma^0{}_{01} = \frac{\frac{d}{dr} A}{2\,A}$$

$$\Gamma^0{}_{10} = \Gamma^0{}_{01}$$

$$\Gamma^1{}_{00} = -\frac{\frac{d}{dr} A}{2\,B}$$

$$\Gamma^1{}_{11} = \frac{\frac{d}{dr} B}{2\,B}$$

$$\Gamma^1{}_{22} = -\frac{\frac{d}{dr} C}{2\,B}$$

$$\Gamma^1{}_{33} = -\frac{\sin^2\vartheta\,\left(\frac{d}{d\,r}\,D\right)}{2\,B}$$

$$\Gamma^2{}_{12} = \frac{\frac{d}{d\,r}\,C}{2\,C}$$

$$\Gamma^2{}_{21} = \Gamma^2{}_{12}$$

$$\Gamma^2{}_{33} = -\frac{\cos\vartheta\,\sin\vartheta\,D}{C}$$

$$\Gamma^3{}_{13} = \frac{\frac{d}{d\,r}\,D}{2\,D}$$

$$\Gamma^3{}_{23} = \frac{\cos\vartheta}{\sin\vartheta}$$

$$\Gamma^3{}_{31} = \Gamma^3{}_{13}$$

$$\Gamma^3{}_{32} = \Gamma^3{}_{23}$$

Metric Compatibility

——— o.k.

Riemann Tensor

$$R^0{}_{101} = \frac{A\,\left(\frac{d}{d\,r}\,A\right)\,\left(\frac{d}{d\,r}\,B\right) - 2\,A\,\left(\frac{d^2}{d\,r^2}\,A\right)\,B + \left(\frac{d}{d\,r}\,A\right)^2\,B}{4\,A^2\,B}$$

$$R^0{}_{110} = -R^0{}_{101}$$

$$R^0{}_{202} = -\frac{\frac{d}{d\,r}\,A\,\left(\frac{d}{d\,r}\,C\right)}{4\,A\,B}$$

$$R^0{}_{220} = -R^0{}_{202}$$

$$R^0{}_{303} = -\frac{\sin^2\vartheta\,\left(\frac{d}{d\,r}\,A\right)\,\left(\frac{d}{d\,r}\,D\right)}{4\,A\,B}$$

$$R^0{}_{330} = -R^0{}_{303}$$

$$R^1{}_{001} = -\frac{A\,\left(\frac{d}{d\,r}\,A\right)\,\left(\frac{d}{d\,r}\,B\right) - 2\,A\,\left(\frac{d^2}{d\,r^2}\,A\right)\,B + \left(\frac{d}{d\,r}\,A\right)^2\,B}{4\,A\,B^2}$$

$$R^1{}_{010} = -R^1{}_{001}$$

$$R^1{}_{212} = -\frac{2\,B\,C\,\left(\frac{d^2}{d\,r^2}\,C\right) - B\,\left(\frac{d}{d\,r}\,C\right)^2 - \frac{d}{d\,r}\,B\,C\,\left(\frac{d}{d\,r}\,C\right)}{4\,B^2\,C}$$

$$R^1{}_{221} = -R^1{}_{212}$$

$$R^1{}_{313} = -\frac{\sin^2\vartheta \left(2\,B\,D\left(\frac{d^2}{d\,r^2}\,D\right) - B\left(\frac{d}{d\,r}\,D\right)^2 - \frac{d}{d\,r}\,B\,D\left(\frac{d}{d\,r}\,D\right)\right)}{4\,B^2\,D}$$

$$R^1{}_{323} = -\frac{\cos\vartheta\,\sin\vartheta\left(C\left(\frac{d}{d\,r}\,D\right) - \frac{d}{d\,r}\,C\,D\right)}{2\,B\,C}$$

$$R^1{}_{331} = -R^1{}_{313}$$

$$R^1{}_{332} = -R^1{}_{323}$$

$$R^2{}_{002} = \frac{\frac{d}{d\,r}\,A\left(\frac{d}{d\,r}\,C\right)}{4\,B\,C}$$

$$R^2{}_{020} = -R^2{}_{002}$$

$$R^2{}_{112} = \frac{2\,B\,C\left(\frac{d^2}{d\,r^2}\,C\right) - B\left(\frac{d}{d\,r}\,C\right)^2 - \frac{d}{d\,r}\,B\,C\left(\frac{d}{d\,r}\,C\right)}{4\,B\,C^2}$$

$$R^2{}_{121} = -R^2{}_{112}$$

$$R^2{}_{313} = -\frac{\cos\vartheta\,\sin\vartheta\left(C\left(\frac{d}{d\,r}\,D\right) - \frac{d}{d\,r}\,C\,D\right)}{2\,C^2}$$

$$R^2{}_{323} = -\frac{\sin^2\vartheta\left(\frac{d}{d\,r}\,C\left(\frac{d}{d\,r}\,D\right) - 4\,B\,D\right)}{4\,B\,C}$$

$$R^2{}_{331} = -R^2{}_{313}$$

$$R^2{}_{332} = -R^2{}_{323}$$

$$R^3{}_{003} = \frac{\frac{d}{d\,r}\,A\left(\frac{d}{d\,r}\,D\right)}{4\,B\,D}$$

$$R^3{}_{030} = -R^3{}_{003}$$

$$R^3{}_{113} = \frac{2\,B\,D\left(\frac{d^2}{d\,r^2}\,D\right) - B\left(\frac{d}{d\,r}\,D\right)^2 - \frac{d}{d\,r}\,B\,D\left(\frac{d}{d\,r}\,D\right)}{4\,B\,D^2}$$

$$R^3{}_{123} = \frac{\cos\vartheta\left(C\left(\frac{d}{d\,r}\,D\right) - \frac{d}{d\,r}\,C\,D\right)}{2\,\sin\vartheta\,C\,D}$$

$$R^3{}_{131} = -R^3{}_{113}$$

$$R^3{}_{132} = -R^3{}_{123}$$

$$R^3{}_{213} = \frac{\cos\vartheta\left(C\left(\frac{d}{d\,r}\,D\right) - \frac{d}{d\,r}\,C\,D\right)}{2\,\sin\vartheta\,C\,D}$$

$$R^3{}_{223} = \frac{\frac{d}{d\,r}\,C\left(\frac{d}{d\,r}\,D\right) - 4\,B\,D}{4\,B\,D}$$

$$R^3{}_{231} = -R^3{}_{213}$$

$$R^3{}_{232} = -R^3{}_{223}$$

Ricci Tensor

$$\mathrm{Ric}_{00} = -\frac{A\left(\frac{d}{d\,r}A\right)B\,C\left(\frac{d}{d\,r}D\right) + A\left(\frac{d}{d\,r}A\right)B\left(\frac{d}{d\,r}C\right)D - A\left(\frac{d}{d\,r}A\right)\left(\frac{d}{d\,r}B\right)C\,D + 2\,A\left(\frac{d^2}{d\,r^2}A\right)B\,C\,D - \left(\frac{d}{d\,r}A\right)^2 B\,C\,D}{4\,A\,B^2\,C\,D}$$

$$\mathrm{Ric}_{11} = -\frac{2\,A^2\,B\,C^2\,D\left(\frac{d^2}{d\,r^2}D\right) - A^2\,B\,C^2\left(\frac{d}{d\,r}D\right)^2 - A^2\left(\frac{d}{d\,r}B\right)C^2\,D\left(\frac{d}{d\,r}D\right) + 2\,A^2\,B\,C\left(\frac{d^2}{d\,r^2}C\right)D^2 - A^2\,B\left(\frac{d}{d\,r}C\right)^2 D^2 - \dots}{4\,A^2\,B\,C^2\,D^2}$$

$$\mathrm{Ric}_{12} = -\frac{\cos\vartheta\left(C\left(\frac{d}{d\,r}D\right) - \frac{d}{d\,r}C\,D\right)}{2\,\sin\vartheta\,C\,D}$$

$$\mathrm{Ric}_{21} = \mathrm{Ric}_{12}$$

$$\mathrm{Ric}_{22} = -\frac{A\,B\,C\left(\frac{d}{d\,r}C\right)\left(\frac{d}{d\,r}D\right) + 2\,A\,B\,C\left(\frac{d^2}{d\,r^2}C\right)D - A\,B\left(\frac{d}{d\,r}C\right)^2 D - A\left(\frac{d}{d\,r}B\right)C\left(\frac{d}{d\,r}C\right)D + \frac{d}{d\,r}A\,B\,C\left(\frac{d}{d\,r}C\right)D - \dots}{4\,A\,B^2\,C\,D}$$

$$\mathrm{Ric}_{33} = -\frac{\sin^2\vartheta\left(2\,A\,B\,C\,D\left(\frac{d^2}{d\,r^2}D\right) - A\,B\,C\left(\frac{d}{d\,r}D\right)^2 + A\,B\left(\frac{d}{d\,r}C\right)D\left(\frac{d}{d\,r}D\right) - A\left(\frac{d}{d\,r}B\right)C\,D\left(\frac{d}{d\,r}D\right) + \dots\right)}{4\,A\,B^2\,C\,D}$$

Ricci Scalar

$$\begin{aligned}R_{sc} = &-\frac{2\,A^2\,B\,C^2\,D\left(\frac{d^2}{d\,r^2}D\right)}{2\,A^2\,B^2\,C^2\,D^2} - \frac{A^2\,B\,C^2\left(\frac{d}{d\,r}D\right)^2}{2\,A^2\,B^2\,C^2\,D^2} + \frac{A^2\,B\,C\left(\frac{d}{d\,r}C\right)D\left(\frac{d}{d\,r}D\right)}{2\,A^2\,B^2\,C^2\,D^2}\\
&-\frac{A^2\left(\frac{d}{d\,r}B\right)C^2\,D\left(\frac{d}{d\,r}D\right)}{2\,A^2\,B^2\,C^2\,D^2} + \frac{A\left(\frac{d}{d\,r}A\right)B\,C^2\,D\left(\frac{d}{d\,r}D\right)}{2\,A^2\,B^2\,C^2\,D^2} + \frac{2\,A^2\,B\,C\left(\frac{d^2}{d\,r^2}C\right)D^2}{2\,A^2\,B^2\,C^2\,D^2}\\
&-\frac{A^2\,B\left(\frac{d}{d\,r}C\right)^2 D^2 - A^2\left(\frac{d}{d\,r}B\right)C\left(\frac{d}{d\,r}C\right)D^2}{2\,A^2\,B^2\,C^2\,D^2} + \frac{A\left(\frac{d}{d\,r}A\right)B\,C\left(\frac{d}{d\,r}C\right)D^2}{2\,A^2\,B^2\,C^2\,D^2}\\
&-\frac{A\left(\frac{d}{d\,r}A\right)\left(\frac{d}{d\,r}B\right)C^2\,D^2}{2\,A^2\,B^2\,C^2\,D^2} + \frac{2\,A\left(\frac{d^2}{d\,r^2}A\right)B\,C^2\,D^2}{2\,A^2\,B^2\,C^2\,D^2} - \frac{\left(\frac{d}{d\,r}A\right)^2 B\,C^2\,D^2 - 4\,A^2\,B^2\,C\,D^2}{2\,A^2\,B^2\,C^2\,D^2}\end{aligned}$$

Bianchi identity (Ricci cyclic equation $R^{\kappa}{}_{[\mu\nu\sigma]} = 0$)

——— o.k.

Einstein Tensor

$$G_{00} = \frac{A\left(2\,B\,C^2\,D\left(\frac{d^2}{d\,r^2}D\right) - B\,C^2\left(\frac{d}{d\,r}D\right)^2 + B\,C\left(\frac{d}{d\,r}C\right)D\left(\frac{d}{d\,r}D\right) - \frac{d}{d\,r}B\,C^2\,D\left(\frac{d}{d\,r}D\right) + 2\,B\,C\left(\frac{d^2}{d\,r^2}C\right)D^2 - \dots\right)}{4\,B^2\,C^2\,D^2}$$

$$G_{11} = \frac{A\left(\frac{d}{d\,r}C\right)\left(\frac{d}{d\,r}D\right) + \frac{d}{d\,r}A\,C\left(\frac{d}{d\,r}D\right) + \frac{d}{d\,r}A\left(\frac{d}{d\,r}C\right)D - 4\,A\,B\,D}{4\,A\,C\,D}$$

$$G_{12} = -\frac{\cos\vartheta\left(C\left(\frac{d}{d\,r}D\right) - \frac{d}{d\,r}C\,D\right)}{2\,\sin\vartheta\,C\,D}$$

$$G_{21} = G_{12}$$

$$G_{22} = \frac{C\left(2\,A^2\,B\,D\left(\frac{d^2}{d\,r^2}\,D\right) - A^2\,B\left(\frac{d}{d\,r}\,D\right)^2 - A^2\left(\frac{d}{d\,r}\,B\right)D\left(\frac{d}{d\,r}\,D\right) + A\left(\frac{d}{d\,r}\,A\right)B\,D\left(\frac{d}{d\,r}\,D\right) - A\left(\frac{d}{d\,r}\,A\right)\left(\frac{d}{d\,r}\,B\right)D^2 + \ldots\right)}{4\,A^2\,B^2\,D^2}$$

$$G_{33} = \frac{\sin^2\vartheta\left(2\,A^2\,B\,C\left(\frac{d^2}{d\,r^2}\,C\right) - A^2\,B\left(\frac{d}{d\,r}\,C\right)^2 - A^2\left(\frac{d}{d\,r}\,B\right)C\left(\frac{d}{d\,r}\,C\right) + A\left(\frac{d}{d\,r}\,A\right)B\,C\left(\frac{d}{d\,r}\,C\right) - A\left(\frac{d}{d\,r}\,A\right)\left(\frac{d}{d\,r}\,B\right)C^2 + \ldots\right)D}{4\,A^2\,B^2\,C^2}$$

Hodge Dual of Bianchi Identity

——— (see charge and current densities)

Scalar Charge Density ($-R^{0\ i0}_{\ i}$)

$$\rho = -\frac{A\left(\frac{d}{d\,r}\,A\right)B\,C\left(\frac{d}{d\,r}\,D\right) + A\left(\frac{d}{d\,r}\,A\right)B\left(\frac{d}{d\,r}\,C\right)D - A\left(\frac{d}{d\,r}\,A\right)\left(\frac{d}{d\,r}\,B\right)C\,D + 2\,A\left(\frac{d^2}{d\,r^2}\,A\right)B\,C\,D - \left(\frac{d}{d\,r}\,A\right)^2 B\,C\,D}{4\,A^3\,B^2\,C\,D}$$

Current Density Class 1 ($-R^{i\ \mu j}_{\ \mu}$)

$$J_1 = \frac{2\,A^2\,B\,C^2\,D\left(\frac{d^2}{d\,r^2}\,D\right) - A^2\,B\,C^2\left(\frac{d}{d\,r}\,D\right)^2 - A^2\left(\frac{d}{d\,r}\,B\right)C^2\,D\left(\frac{d}{d\,r}\,D\right) + 2\,A^2\,B\,C\left(\frac{d^2}{d\,r^2}\,C\right)D^2 - A^2\,B\left(\frac{d}{d\,r}\,C\right)^2 D^2 - \ldots}{4\,A^2\,B^3\,C^2\,D^2}$$

$$J_2 = \frac{A\,B\,C\left(\frac{d}{d\,r}\,C\right)\left(\frac{d}{d\,r}\,D\right) + 2\,A\,B\,C\left(\frac{d^2}{d\,r^2}\,C\right)D - A\,B\left(\frac{d}{d\,r}\,C\right)^2 D - A\left(\frac{d}{d\,r}\,B\right)C\left(\frac{d}{d\,r}\,C\right)D + \frac{d}{d\,r}\,A\,B\,C\left(\frac{d}{d\,r}\,C\right)D - 4\,A\,B^2\,C\,D}{4\,A\,B^2\,C^3\,D}$$

$$J_3 = \frac{2\,A\,B\,C\,D\left(\frac{d^2}{d\,r^2}\,D\right) - A\,B\,C\left(\frac{d}{d\,r}\,D\right)^2 + A\,B\left(\frac{d}{d\,r}\,C\right)D\left(\frac{d}{d\,r}\,D\right) - A\left(\frac{d}{d\,r}\,B\right)C\,D\left(\frac{d}{d\,r}\,D\right) + \frac{d}{d\,r}\,A\,B\,C\,D\left(\frac{d}{d\,r}\,D\right) - 4\,A\,B^2\,D^2}{4\,\sin^2\vartheta\,A\,B^2\,C\,D^3}$$

Current Density Class 2 ($-R^{i\ \mu j}_{\ \mu}$)

$$J_1 = 0$$

$$J_2 = \frac{\cos\vartheta\left(C\left(\frac{d}{d\,r}\,D\right) - \frac{d}{d\,r}\,C\,D\right)}{2\,\sin\vartheta\,B\,C^2\,D}$$

$$J_3 = 0$$

Current Density Class 3 ($-R^{i\ \mu j}_{\ \mu}$)

$$J_1 = \frac{\cos\vartheta\left(C\left(\frac{d}{d\,r}\,D\right) - \frac{d}{d\,r}\,C\,D\right)}{2\,\sin\vartheta\,B\,C^2\,D}$$

$$J_2 = 0$$

$$J_3 = 0$$

4.4.38 Perfect spherical fluid metric

There are similar versions of this metric called homogeneous perfect fluid. a and b are parameters.

Coordinates

$$\mathbf{x} = \begin{pmatrix} t \\ r \\ \vartheta \\ \varphi \end{pmatrix}$$

Metric

$$g_{\mu\nu} = \begin{pmatrix} -a\,r^2 - 1 & 0 & 0 & 0 \\ 0 & \frac{(1-3\,a\,r^2)^{\frac{2}{3}}}{(3\,a\,r^2+1)^{\frac{2}{3}} - b\,r^2} & 0 & 0 \\ 0 & 0 & r^2 & 0 \\ 0 & 0 & 0 & r^2\,\sin^2\vartheta \end{pmatrix}$$

Contravariant Metric

$$g^{\mu\nu} = \begin{pmatrix} -\frac{1}{a\,r^2+1} & 0 & 0 & 0 \\ 0 & \frac{(3\,a\,r^2+1)^{\frac{2}{3}} - b\,r^2}{(1-3\,a\,r^2)^{\frac{2}{3}}} & 0 & 0 \\ 0 & 0 & \frac{1}{r^2} & 0 \\ 0 & 0 & 0 & \frac{1}{r^2\,\sin^2\vartheta} \end{pmatrix}$$

Christoffel Connection

$$\Gamma^0{}_{01} = \frac{a\,r}{a\,r^2+1}$$

$$\Gamma^0{}_{10} = \Gamma^0{}_{01}$$

$$\Gamma^1{}_{00} = \frac{a\,r\left((3\,a\,r^2+1)^{\frac{2}{3}} - b\,r^2\right)}{(1-3\,a\,r^2)^{\frac{2}{3}}}$$

$$\Gamma^1{}_{11} = -\frac{r\left(3\,a^2\,b\,r^4\,(3\,a\,r^2+1)^{\frac{2}{3}} - 10\,a\,b\,r^2\,(3\,a\,r^2+1)^{\frac{2}{3}} - b\,(3\,a\,r^2+1)^{\frac{2}{3}} + a\,b^3\,r^6\,(3\,a\,r^2+1)^{\frac{1}{3}} - b^3\,r^4\,(3\,a\,r^2+1)^{\frac{1}{3}} + \ldots\right)}{(3\,a\,r^2-1)\left(9\,a\,b\,r^4\,(3\,a\,r^2+1)^{\frac{2}{3}} + 3\,b\,r^2\,(3\,a\,r^2+1)^{\frac{2}{3}} + b^3\,r^6\,(3\,a\,r^2+1)^{\frac{1}{3}} - 9\,a^2\,r^4\,(3\,a\,r^2+1)^{\frac{1}{3}} - \ldots\right)}$$

$$\Gamma^1{}_{22} = -\frac{r\left((3\,a\,r^2+1)^{\frac{2}{3}} - b\,r^2\right)}{(1-3\,a\,r^2)^{\frac{2}{3}}}$$

$$\Gamma^1{}_{33} = -\frac{r\left((3\,a\,r^2+1)^{\frac{2}{3}} - b\,r^2\right)\,\sin^2\vartheta}{(1-3\,a\,r^2)^{\frac{2}{3}}}$$

$$\Gamma^2{}_{12} = \frac{1}{r}$$

$$\Gamma^2{}_{21} = \Gamma^2{}_{12}$$

$$\Gamma^2{}_{33} = -\cos\vartheta\,\sin\vartheta$$

$$\Gamma^3{}_{13} = \frac{1}{r}$$

$$\Gamma^3{}_{23} = \frac{\cos\vartheta}{\sin\vartheta}$$

$$\Gamma^3{}_{31} = \Gamma^3{}_{13}$$

$$\Gamma^3{}_{32} = \Gamma^3{}_{23}$$

Metric Compatibility

——— o.k.

Riemann Tensor

$$R^0{}_{101} = -\frac{a\left(3a^3 b r^8 \left(3ar^2+1\right)^{\frac{2}{3}} + 20a^2 b r^6 \left(3ar^2+1\right)^{\frac{2}{3}} - 11 a b r^4 \left(3ar^2+1\right)^{\frac{2}{3}} - 4 b r^2 \left(3ar^2+1\right)^{\frac{2}{3}} + a^2 b^3 r^{10} \left(3ar^2+1\right)^{\frac{1}{3}} + \ldots\right)}{\left(ar^2+1\right)^2 \left(3ar^2-1\right)\left(9abr^4 \left(3ar^2+1\right)^{\frac{2}{3}} + 3br^2 \left(3ar^2+1\right)^{\frac{2}{3}} + b^3 r^6 \left(3ar^2+1\right)^{\frac{1}{3}} - 9a^2 r^4 \left(3ar^2+1\right)^{\frac{1}{3}} - \ldots\right)}$$

$$R^0{}_{110} = -R^0{}_{101}$$

$$R^0{}_{202} = -\frac{a r^2 \left(\left(3ar^2+1\right)^{\frac{2}{3}} - b r^2\right)}{\left(1-3ar^2\right)^{\frac{2}{3}} \left(ar^2+1\right)}$$

$$R^0{}_{220} = -R^0{}_{202}$$

$$R^0{}_{303} = -\frac{a r^2 \left(\left(3ar^2+1\right)^{\frac{2}{3}} - b r^2\right)\sin^2\vartheta}{\left(1-3ar^2\right)^{\frac{2}{3}} \left(ar^2+1\right)}$$

$$R^0{}_{330} = -R^0{}_{303}$$

$$R^1{}_{001} = \frac{a\left(a^2 b^4 r^{12} \left(3ar^2+1\right)^{\frac{2}{3}} + 3ab^4 r^{10} \left(3ar^2+1\right)^{\frac{2}{3}} - 9a^4 b r^{10} \left(3ar^2+1\right)^{\frac{2}{3}} - 2b^4 r^8 \left(3ar^2+1\right)^{\frac{2}{3}} - 78 a^3 b r^8 \left(3ar^2+1\right)^{\frac{2}{3}} + \ldots\right)}{\left(1-3ar^2\right)^{\frac{2}{3}} \left(ar^2+1\right)\left(3ar^2-1\right)\left(b^3 r^6 \left(3ar^2+1\right)^{\frac{2}{3}} - 9a^2 r^4 \left(3ar^2+1\right)^{\frac{2}{3}} - 6ar^2 \left(3ar^2+1\right)^{\frac{2}{3}} - \ldots\right)}$$

$$R^1{}_{010} = -R^1{}_{001}$$

$$R^1{}_{212} = \frac{r^2\left(ab^4 r^8 \left(3ar^2+1\right)^{\frac{2}{3}} - b^4 r^6 \left(3ar^2+1\right)^{\frac{2}{3}} - 9a^3 b r^6 \left(3ar^2+1\right)^{\frac{2}{3}} + 39 a^2 b r^4 \left(3ar^2+1\right)^{\frac{2}{3}} + 17 a b r^2 \left(3ar^2+1\right)^{\frac{2}{3}} + \ldots\right)}{\left(1-3ar^2\right)^{\frac{2}{3}} \left(3ar^2-1\right)\left(b^3 r^6 \left(3ar^2+1\right)^{\frac{2}{3}} - 9a^2 r^4 \left(3ar^2+1\right)^{\frac{2}{3}} - 6ar^2 \left(3ar^2+1\right)^{\frac{2}{3}} - \ldots\right)}$$

$$R^1{}_{221} = -R^1{}_{212}$$

$$R^1{}_{313} = \frac{r^2 \left(a\,b^4\,r^8\,(3\,a\,r^2+1)^{\frac{2}{3}} - b^4\,r^6\,(3\,a\,r^2+1)^{\frac{2}{3}} - 9\,a^3\,b\,r^6\,(3\,a\,r^2+1)^{\frac{2}{3}} + 39\,a^2\,b\,r^4\,(3\,a\,r^2+1)^{\frac{2}{3}} + ...\right)\sin^2\vartheta}{(1-3\,a\,r^2)^{\frac{2}{3}}\,(3\,a\,r^2-1)\left(b^3\,r^6\,(3\,a\,r^2+1)^{\frac{2}{3}} - 9\,a^2\,r^4\,(3\,a\,r^2+1)^{\frac{2}{3}} - 6\,a\,r^2\,(3\,a\,r^2+1)^{\frac{2}{3}} - ...\right)}$$

$$R^1{}_{331} = -R^1{}_{313}$$

$$R^2{}_{002} = -\frac{a\left((3\,a\,r^2+1)^{\frac{2}{3}} - b\,r^2\right)}{(1-3\,a\,r^2)^{\frac{2}{3}}}$$

$$R^2{}_{020} = -R^2{}_{002}$$

$$R^2{}_{112} = \frac{3\,a^2\,b\,r^4\,(3\,a\,r^2+1)^{\frac{2}{3}} - 10\,a\,b\,r^2\,(3\,a\,r^2+1)^{\frac{2}{3}} - b\,(3\,a\,r^2+1)^{\frac{2}{3}} + a\,b^3\,r^6\,(3\,a\,r^2+1)^{\frac{1}{3}} - b^3\,r^4\,(3\,a\,r^2+1)^{\frac{1}{3}} + ...}{(3\,a\,r^2-1)\left(9\,a\,b\,r^4\,(3\,a\,r^2+1)^{\frac{2}{3}} + 3\,b\,r^2\,(3\,a\,r^2+1)^{\frac{2}{3}} + b^3\,r^6\,(3\,a\,r^2+1)^{\frac{1}{3}} - 9\,a^2\,r^4\,(3\,a\,r^2+1)^{\frac{1}{3}} - ...\right)}$$

$$R^2{}_{121} = -R^2{}_{112}$$

$$R^2{}_{323} = -\frac{\left((3\,a\,r^2+1)^{\frac{2}{3}} - (1-3\,a\,r^2)^{\frac{2}{3}} - b\,r^2\right)\sin^2\vartheta}{(1-3\,a\,r^2)^{\frac{2}{3}}}$$

$$R^2{}_{332} = -R^2{}_{323}$$

$$R^3{}_{003} = -\frac{a\left((3\,a\,r^2+1)^{\frac{2}{3}} - b\,r^2\right)}{(1-3\,a\,r^2)^{\frac{2}{3}}}$$

$$R^3{}_{030} = -R^3{}_{003}$$

$$R^3{}_{113} = \frac{3\,a^2\,b\,r^4\,(3\,a\,r^2+1)^{\frac{2}{3}} - 10\,a\,b\,r^2\,(3\,a\,r^2+1)^{\frac{2}{3}} - b\,(3\,a\,r^2+1)^{\frac{2}{3}} + a\,b^3\,r^6\,(3\,a\,r^2+1)^{\frac{1}{3}} - b^3\,r^4\,(3\,a\,r^2+1)^{\frac{1}{3}} + ...}{(3\,a\,r^2-1)\left(9\,a\,b\,r^4\,(3\,a\,r^2+1)^{\frac{2}{3}} + 3\,b\,r^2\,(3\,a\,r^2+1)^{\frac{2}{3}} + b^3\,r^6\,(3\,a\,r^2+1)^{\frac{1}{3}} - 9\,a^2\,r^4\,(3\,a\,r^2+1)^{\frac{1}{3}} - ...\right)}$$

$$R^3{}_{131} = -R^3{}_{113}$$

$$R^3{}_{223} = \frac{(3\,a\,r^2+1)^{\frac{2}{3}} - (1-3\,a\,r^2)^{\frac{2}{3}} - b\,r^2}{(1-3\,a\,r^2)^{\frac{2}{3}}}$$

$$R^3{}_{232} = -R^3{}_{223}$$

Ricci Tensor

$$\mathrm{Ric}_{00} = \frac{a\left(1-3\,a\,r^2\right)^{\frac{1}{3}}\left(7\,a^2\,b^4\,r^{12}\left(3\,a\,r^2+1\right)^{\frac{2}{3}}+7\,a\,b^4\,r^{10}\left(3\,a\,r^2+1\right)^{\frac{2}{3}}-225\,a^4\,b\,r^{10}\left(3\,a\,r^2+1\right)^{\frac{2}{3}}-4\,b^4\,r^8\left(3\,a\,r^2+1\right)^{\frac{2}{3}}-\ldots\right)}{\left(a\,r^2+1\right)\left(3\,a\,r^2-1\right)^2\left(b^3\,r^6\left(3\,a\,r^2+1\right)^{\frac{2}{3}}-9\,a^2\,r^4\left(3\,a\,r^2+1\right)^{\frac{2}{3}}-6\,a\,r^2\left(3\,a\,r^2+1\right)^{\frac{2}{3}}-\ldots\right)}$$

$$\mathrm{Ric}_{11} = -\frac{3\,a^3\,b^6\,r^{16}\left(3\,a\,r^2+1\right)^{\frac{2}{3}}+5\,a^2\,b^6\,r^{14}\left(3\,a\,r^2+1\right)^{\frac{2}{3}}-270\,a^5\,b^3\,r^{14}\left(3\,a\,r^2+1\right)^{\frac{2}{3}}-4\,a\,b^6\,r^{12}\left(3\,a\,r^2+1\right)^{\frac{2}{3}}-\ldots}{\left(a\,r^2+1\right)^2\left(3\,a\,r^2-1\right)\left(b^6\,r^{12}\left(3\,a\,r^2+1\right)^{\frac{2}{3}}-180\,a^2\,b^3\,r^{10}\left(3\,a\,r^2+1\right)^{\frac{2}{3}}-120\,a\,b^3\,r^8\left(3\,a\,r^2+1\right)^{\frac{2}{3}}+81\,a^4\,r^8\left(3\,a\,r^2+1\right)^{\frac{2}{3}}-\ldots\right)}$$

$$\mathrm{Ric}_{22} = \frac{3\,a^2\,b^3\,r^{10}\left(1-3\,a\,r^2\right)^{\frac{2}{3}}\left(3\,a\,r^2+1\right)^{\frac{2}{3}}+2\,a\,b^3\,r^8\left(1-3\,a\,r^2\right)^{\frac{2}{3}}\left(3\,a\,r^2+1\right)^{\frac{2}{3}}-27\,a^4\,r^8\left(1-3\,a\,r^2\right)^{\frac{2}{3}}\left(3\,a\,r^2+1\right)^{\frac{2}{3}}-\ldots}{\left(1-3\,a\,r^2\right)^{\frac{2}{3}}\left(a\,r^2+1\right)\left(3\,a\,r^2-1\right)\left(b^3\,r^6\left(3\,a\,r^2+1\right)^{\frac{2}{3}}-9\,a^2\,r^4\left(3\,a\,r^2+1\right)^{\frac{2}{3}}-6\,a\,r^2\left(3\,a\,r^2+1\right)^{\frac{2}{3}}-\ldots\right)}$$

$$\mathrm{Ric}_{33} = \frac{\left(3\,a^2\,b^3\,r^{10}\left(1-3\,a\,r^2\right)^{\frac{2}{3}}\left(3\,a\,r^2+1\right)^{\frac{2}{3}}+2\,a\,b^3\,r^8\left(1-3\,a\,r^2\right)^{\frac{2}{3}}\left(3\,a\,r^2+1\right)^{\frac{2}{3}}-27\,a^4\,r^8\left(1-3\,a\,r^2\right)^{\frac{2}{3}}\left(3\,a\,r^2+1\right)^{\frac{2}{3}}-\ldots\right)\sin^2\vartheta}{\left(1-3\,a\,r^2\right)^{\frac{2}{3}}\left(a\,r^2+1\right)\left(3\,a\,r^2-1\right)\left(b^3\,r^6\left(3\,a\,r^2+1\right)^{\frac{2}{3}}-9\,a^2\,r^4\left(3\,a\,r^2+1\right)^{\frac{2}{3}}-6\,a\,r^2\left(3\,a\,r^2+1\right)^{\frac{2}{3}}-\left(3\,a\,r^2+1\right)^{\frac{2}{3}}-\ldots\right)}$$

Ricci Scalar

$$R_{sc} = -\frac{2\left(459\,a^4\,b^{12}\,r^{32}\left(1-3\,a\,r^2\right)^{\frac{1}{3}}\left(3\,a\,r^2+1\right)^{\frac{2}{3}}+681\,a^3\,b^{12}\,r^{30}\left(1-3\,a\,r^2\right)^{\frac{1}{3}}\left(3\,a\,r^2+1\right)^{\frac{2}{3}}-\ldots\right)}{r^2\left(a\,r^2+1\right)^2\left(3\,a\,r^2-1\right)^2\left(3\,a\,r^2+1\right)\left(12\,b^{11}\,r^{22}\left(3\,a\,r^2+1\right)^{\frac{2}{3}}-4455\,a^2\,b^8\,r^{20}\left(3\,a\,r^2+1\right)^{\frac{2}{3}}-2970\,a\,b^8\,r^{18}\left(3\,a\,r^2+1\right)^{\frac{2}{3}}+\ldots\right)}$$

Bianchi identity (Ricci cyclic equation $R^{\kappa}{}_{[\mu\nu\sigma]} = 0$)

——— o.k.

Einstein Tensor

$$G_{00} = -\frac{\left(a\,r^2+1\right)\left(5\,a\,b^{16}\,r^{34}\left(1-3\,a\,r^2\right)^{\frac{1}{3}}\left(3\,a\,r^2+1\right)^{\frac{2}{3}}-3\,b^{16}\,r^{32}\left(1-3\,a\,r^2\right)^{\frac{1}{3}}\left(3\,a\,r^2+1\right)^{\frac{2}{3}}-\ldots\right)}{r^2\left(3\,a\,r^2-1\right)^2\left(b^{15}\,r^{30}\left(3\,a\,r^2+1\right)^{\frac{2}{3}}-4095\,a^2\,b^{12}\,r^{28}\left(3\,a\,r^2+1\right)^{\frac{2}{3}}-2730\,a\,b^{12}\,r^{26}\left(3\,a\,r^2+1\right)^{\frac{2}{3}}+\ldots\right)}$$

$$G_{11} = \frac{a\,b^{18}\,r^{38}\left(1-3\,a\,r^2\right)^{\frac{2}{3}}\left(3\,a\,r^2+1\right)^{\frac{2}{3}}+b^{18}\,r^{36}\left(1-3\,a\,r^2\right)^{\frac{2}{3}}\left(3\,a\,r^2+1\right)^{\frac{2}{3}}-7344\,a^3\,b^{15}\,r^{36}\left(1-3\,a\,r^2\right)^{\frac{2}{3}}\left(3\,a\,r^2+1\right)^{\frac{2}{3}}-\ldots}{r^2\left(a\,r^2+1\right)\left(b^{19}\,r^{38}\left(3\,a\,r^2+1\right)^{\frac{2}{3}}-8721\,a^2\,b^{16}\,r^{36}\left(3\,a\,r^2+1\right)^{\frac{2}{3}}-5814\,a\,b^{16}\,r^{34}\left(3\,a\,r^2+1\right)^{\frac{2}{3}}+\ldots\right)}$$

$$G_{22} = -\frac{r^2\left(5\,a^3\,b^{16}\,r^{36}\left(3\,a\,r^2+1\right)^{\frac{2}{3}}+6\,a^2\,b^{16}\,r^{34}\left(3\,a\,r^2+1\right)^{\frac{2}{3}}-23310\,a^5\,b^{13}\,r^{34}\left(3\,a\,r^2+1\right)^{\frac{2}{3}}-4\,a\,b^{16}\,r^{32}\left(3\,a\,r^2+1\right)^{\frac{2}{3}}-\ldots\right)}{\left(1-3\,a\,r^2\right)^{\frac{2}{3}}\left(a\,r^2+1\right)^2\left(3\,a\,r^2-1\right)\left(b^{15}\,r^{30}\left(3\,a\,r^2+1\right)^{\frac{2}{3}}-4095\,a^2\,b^{12}\,r^{28}\left(3\,a\,r^2+1\right)^{\frac{2}{3}}-2730\,a\,b^{12}\,r^{26}\left(3\,a\,r^2+1\right)^{\frac{2}{3}}+\ldots\right)}$$

$$G_{33} = -\frac{r^2\left(5\,a^3\,b^{16}\,r^{36}\left(3\,a\,r^2+1\right)^{\frac{2}{3}}+6\,a^2\,b^{16}\,r^{34}\left(3\,a\,r^2+1\right)^{\frac{2}{3}}-23310\,a^5\,b^{13}\,r^{34}\left(3\,a\,r^2+1\right)^{\frac{2}{3}}-4\,a\,b^{16}\,r^{32}\left(3\,a\,r^2+1\right)^{\frac{2}{3}}-\ldots\right)\sin^2\vartheta}{\left(1-3\,a\,r^2\right)^{\frac{2}{3}}\left(a\,r^2+1\right)^2\left(3\,a\,r^2-1\right)\left(b^{15}\,r^{30}\left(3\,a\,r^2+1\right)^{\frac{2}{3}}-4095\,a^2\,b^{12}\,r^{28}\left(3\,a\,r^2+1\right)^{\frac{2}{3}}-2730\,a\,b^{12}\,r^{26}\left(3\,a\,r^2+1\right)^{\frac{2}{3}}+\ldots\right)}$$

Hodge Dual of Bianchi Identity

——— (see charge and current densities)

Scalar Charge Density ($-R^{0\ \ i0}_{\ i}$)

$$\rho = \frac{a\,(1-3\,a\,r^2)^{\frac{1}{3}}\,\left(117\,a^3\,b^2\,r^{10}\,(3\,a\,r^2+1)^{\frac{2}{3}}+153\,a^2\,b^2\,r^8\,(3\,a\,r^2+1)^{\frac{2}{3}}-33\,a\,b^2\,r^6\,(3\,a\,r^2+1)^{\frac{2}{3}}-21\,b^2\,r^4\,(3\,a\,r^2+1)^{\frac{2}{3}}+...\right)}{(a\,r^2+1)^3\,(3\,a\,r^2-1)^2\,\left(9\,a\,b\,r^4\,(3\,a\,r^2+1)^{\frac{2}{3}}+3\,b\,r^2\,(3\,a\,r^2+1)^{\frac{2}{3}}+b^3\,r^6\,(3\,a\,r^2+1)^{\frac{1}{3}}-...\right)}$$

Current Density Class 1 ($-R^{i\ \ \mu j}_{\ \mu}$)

$$J_1 = -\frac{189\,a^4\,b^6\,r^{18}\,(3\,a\,r^2+1)^{\frac{2}{3}}+357\,a^3\,b^6\,r^{16}\,(3\,a\,r^2+1)^{\frac{2}{3}}-1701\,a^6\,b^3\,r^{16}\,(3\,a\,r^2+1)^{\frac{2}{3}}-287\,a^2\,b^6\,r^{14}\,(3\,a\,r^2+1)^{\frac{2}{3}}-...}{(1-3\,a\,r^2)^{\frac{1}{3}}\,(a\,r^2+1)^2\,(3\,a\,r^2-1)^2\,\left(45\,a\,b^4\,r^{10}\,(3\,a\,r^2+1)^{\frac{2}{3}}+15\,b^4\,r^8\,(3\,a\,r^2+1)^{\frac{2}{3}}-162\,a^3\,b\,r^8\,(3\,a\,r^2+1)^{\frac{2}{3}}-...\right)}$$

$$J_2 = -\frac{27\,a^3\,b\,r^8\,(1-3\,a\,r^2)^{\frac{2}{3}}\,(3\,a\,r^2+1)^{\frac{2}{3}}+27\,a^2\,b\,r^6\,(1-3\,a\,r^2)^{\frac{2}{3}}\,(3\,a\,r^2+1)^{\frac{2}{3}}-3\,a\,b\,r^4\,(1-3\,a\,r^2)^{\frac{2}{3}}\,(3\,a\,r^2+1)^{\frac{2}{3}}-...}{r^4\,(1-3\,a\,r^2)^{\frac{2}{3}}\,(a\,r^2+1)\,(3\,a\,r^2-1)\,\left(9\,a\,b\,r^4\,(3\,a\,r^2+1)^{\frac{2}{3}}+3\,b\,r^2\,(3\,a\,r^2+1)^{\frac{2}{3}}+b^3\,r^6\,(3\,a\,r^2+1)^{\frac{1}{3}}-...\right)}$$

$$J_3 = -\frac{27\,a^3\,b\,r^8\,(1-3\,a\,r^2)^{\frac{2}{3}}\,(3\,a\,r^2+1)^{\frac{2}{3}}+27\,a^2\,b\,r^6\,(1-3\,a\,r^2)^{\frac{2}{3}}\,(3\,a\,r^2+1)^{\frac{2}{3}}-3\,a\,b\,r^4\,(1-3\,a\,r^2)^{\frac{2}{3}}\,(3\,a\,r^2+1)^{\frac{2}{3}}-...}{r^4\,(1-3\,a\,r^2)^{\frac{2}{3}}\,(a\,r^2+1)\,(3\,a\,r^2-1)\,\left(9\,a\,b\,r^4\,(3\,a\,r^2+1)^{\frac{2}{3}}+3\,b\,r^2\,(3\,a\,r^2+1)^{\frac{2}{3}}+b^3\,r^6\,(3\,a\,r^2+1)^{\frac{1}{3}}-...\right)\sin^2\vartheta}$$

Current Density Class 2 ($-R^{i\ \ \mu j}_{\ \mu}$)

$$J_1 = 0$$

$$J_2 = 0$$

$$J_3 = 0$$

Current Density Class 3 ($-R^{i\ \ \mu j}_{\ \mu}$)

$$J_1 = 0$$

$$J_2 = 0$$

$$J_3 = 0$$

4.4.39 Carmeli metric for spiral galaxies

R_1 and τ are parameters.

Coordinates

$$\mathbf{x} = \begin{pmatrix} \eta \\ \chi \\ \vartheta \\ \varphi \end{pmatrix}$$

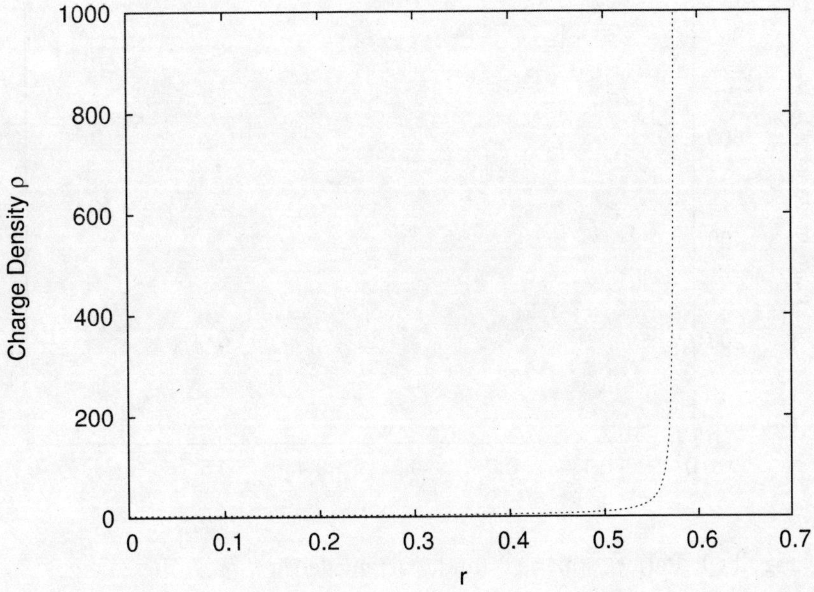

Fig. 4.67: Perfect spherical fluid, charge density ρ for $a = b = 1$.

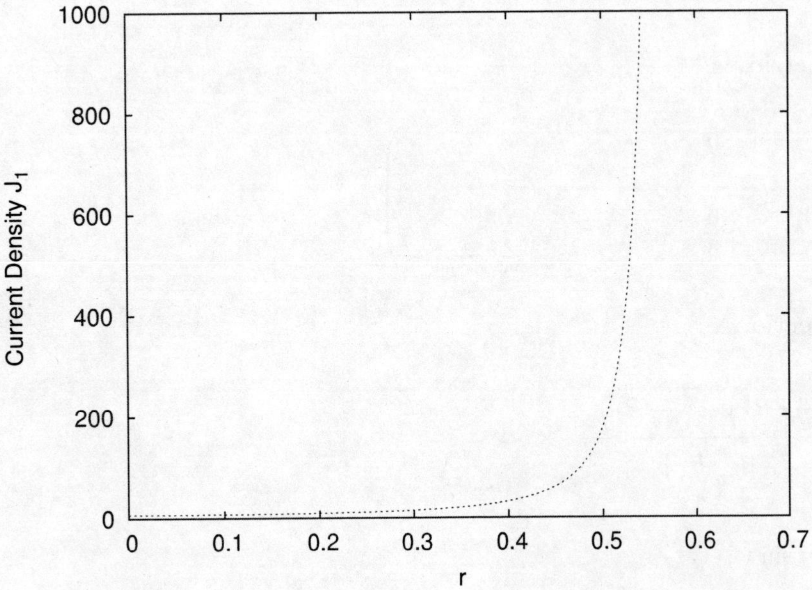

Fig. 4.68: Perfect spherical fluid, current density J_r for $a = b = 1$.

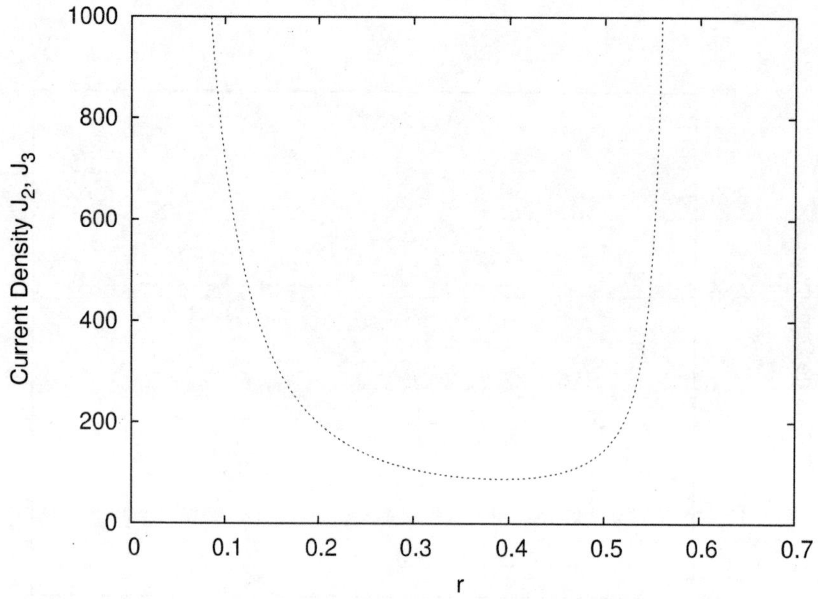

Fig. 4.69: Perfect spherical fluid, current density J_θ, J_φ for $a = b = 1$.

Metric

$$g_{\mu\nu} = \begin{pmatrix} \frac{\tau\left(\frac{d}{d\eta} R_1\right)^2}{R_1{}^2} & 0 & 0 & 0 \\ 0 & -\tau & 0 & 0 \\ 0 & 0 & -\sin^2\chi\,\tau & 0 \\ 0 & 0 & 0 & -\sin^2\chi\,\tau\,\sin^2\vartheta \end{pmatrix}$$

Contravariant Metric

$$g^{\mu\nu} = \begin{pmatrix} \frac{R_1{}^2}{\tau\left(\frac{d}{d\eta} R_1\right)^2} & 0 & 0 & 0 \\ 0 & -\frac{1}{\tau} & 0 & 0 \\ 0 & 0 & -\frac{1}{\sin^2\chi\,\tau} & 0 \\ 0 & 0 & 0 & -\frac{1}{\sin^2\chi\,\tau\,\sin^2\vartheta} \end{pmatrix}$$

Christoffel Connection

$$\Gamma^0{}_{00} = \frac{R_1\left(\frac{d^2}{d\eta^2} R_1\right) - \left(\frac{d}{d\eta} R_1\right)^2}{R_1\left(\frac{d}{d\eta} R_1\right)}$$

$$\Gamma^1{}_{22} = -\cos\chi\,\sin\chi$$

$$\Gamma^1{}_{33} = -\cos\chi\,\sin\chi\,\sin^2\vartheta$$

$$\Gamma^2{}_{12} = \frac{\cos\chi}{\sin\chi}$$

$$\Gamma^2{}_{21} = \Gamma^2{}_{12}$$

$$\Gamma^2{}_{33} = -\cos\vartheta\,\sin\vartheta$$

$$\Gamma^3{}_{13} = \frac{\cos\chi}{\sin\chi}$$

$$\Gamma^3{}_{23} = \frac{\cos\vartheta}{\sin\vartheta}$$

$$\Gamma^3{}_{31} = \Gamma^3{}_{13}$$

$$\Gamma^3{}_{32} = \Gamma^3{}_{23}$$

Metric Compatibility

——— o.k.

Riemann Tensor

$$R^1{}_{212} = \sin^2\chi$$

$$R^1{}_{221} = -R^1{}_{212}$$

$$R^1{}_{313} = \sin^2\chi\,\sin^2\vartheta$$

$$R^1{}_{331} = -R^1{}_{313}$$

$$R^2{}_{112} = -1$$

$$R^2{}_{121} = -R^2{}_{112}$$

$$R^2{}_{323} = \sin^2\chi\,\sin^2\vartheta$$

$$R^2{}_{332} = -R^2{}_{323}$$

$$R^3{}_{113} = -1$$

$$R^3{}_{131} = -R^3{}_{113}$$

$$R^3{}_{223} = (\cos\chi - 1)\,(\cos\chi + 1)$$

$$R^3{}_{232} = -R^3{}_{223}$$

Ricci Tensor

$$\mathrm{Ric}_{11} = 2$$

$$\mathrm{Ric}_{22} = 2\,\sin^2\chi$$

$$\mathrm{Ric}_{33} = 2\,\sin^2\chi\,\sin^2\vartheta$$

Ricci Scalar

$$R_{sc} = -\frac{6}{\tau}$$

Bianchi identity (Ricci cyclic equation $R^{\kappa}{}_{[\mu\nu\sigma]} = 0$)

——— o.k.

Einstein Tensor

$$G_{00} = \frac{3\left(\frac{d}{d\eta}\,R_1\right)^2}{{R_1}^2}$$

$$G_{11} = -1$$

$$G_{22} = -\sin^2\chi$$

$$G_{33} = -\sin^2\chi\,\sin^2\vartheta$$

Hodge Dual of Bianchi Identity

——— (see charge and current densities)

Scalar Charge Density (-$R^{0}{}_{i}{}^{i0}$)

$$\rho = 0$$

Current Density Class 1 (-$R^{i}{}_{\mu}{}^{\mu j}$)

$$J_1 = -\frac{2}{\tau^2}$$

$$J_2 = -\frac{2}{\sin^2\chi\,\tau^2}$$

$$J_3 = -\frac{2}{\sin^2\chi\,\tau^2\,\sin^2\vartheta}$$

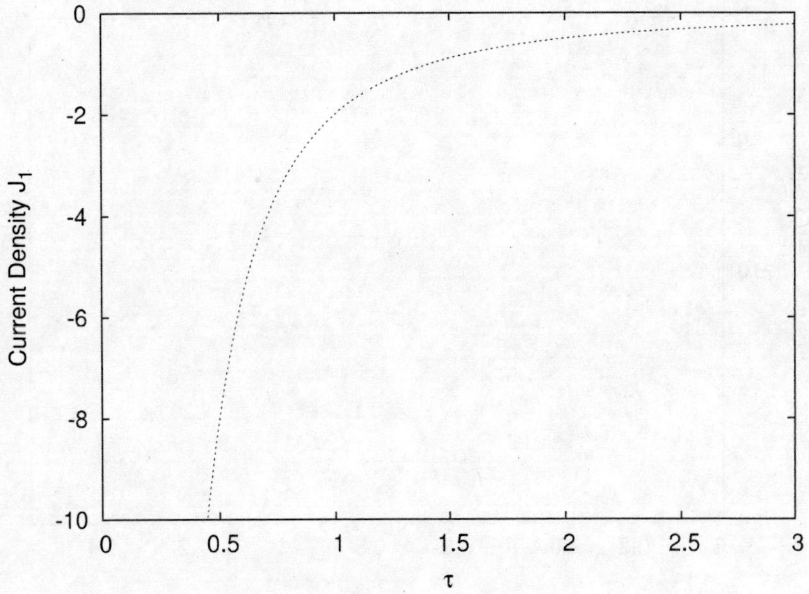

Fig. 4.70: Carmeli metric, current density J_χ, τ-dependence.

Current Density Class 2 ($-R^{i\ \ \mu j}_{\ \mu}$)

$$J_1 = 0$$

$$J_2 = 0$$

$$J_3 = 0$$

Current Density Class 3 ($-R^{i\ \ \mu j}_{\ \mu}$)

$$J_1 = 0$$

$$J_2 = 0$$

$$J_3 = 0$$

4.4.40 Dirac metric

This metric is reported by J. Dunning-Davies and assumed to go back to Dirac. m, μ and τ are parameters. m and μ have been set to unity in the plots.

Coordinates

$$\mathbf{x} = \begin{pmatrix} \tau \\ \rho \\ \vartheta \\ \varphi \end{pmatrix}$$

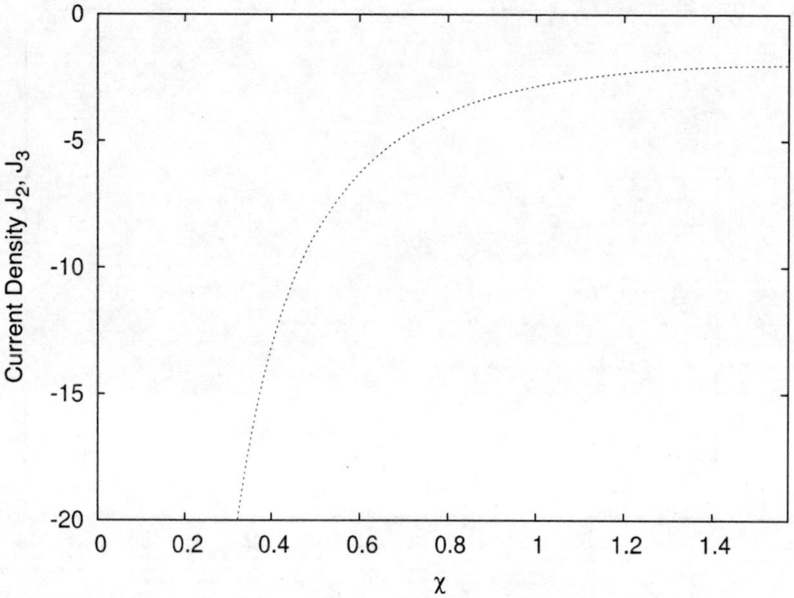

Fig. 4.71: Carmeli metric, current density J_ϑ, J_φ, χ-dependence with $\tau = 1$.

Metric

$$g_{\mu\nu} = \begin{pmatrix} 1 & 0 & 0 & 0 \\ 0 & -\frac{2\,m}{\mu\,(\rho-\tau)^{\frac{2}{3}}} & 0 & 0 \\ 0 & 0 & -\mu^2\,(\rho-\tau)^{\frac{4}{3}} & 0 \\ 0 & 0 & 0 & -\mu^2\,(\rho-\tau)^{\frac{4}{3}}\,\sin^2\vartheta \end{pmatrix}$$

Contravariant Metric

$$g^{\mu\nu} = \begin{pmatrix} 1 & 0 & 0 & 0 \\ 0 & -\frac{\mu\,(\rho-\tau)^{\frac{2}{3}}}{2\,m} & 0 & 0 \\ 0 & 0 & \frac{1}{\mu^2\,(\rho-\tau)^{\frac{1}{3}}\,(\tau-\rho)} & 0 \\ 0 & 0 & 0 & \frac{1}{\mu^2\,(\rho-\tau)^{\frac{1}{3}}\,(\tau-\rho)\,\sin^2\vartheta} \end{pmatrix}$$

Christoffel Connection

$$\Gamma^0{}_{11} = -\frac{2\,m}{3\,\mu\,(\rho-\tau)^{\frac{2}{3}}\,(\tau-\rho)}$$

$$\Gamma^0{}_{22} = -\frac{2\,\mu^2\,(\rho-\tau)^{\frac{1}{3}}}{3}$$

$$\Gamma^0{}_{33} = -\frac{2\,\mu^2\,(\rho-\tau)^{\frac{1}{3}}\,\sin^2\vartheta}{3}$$

$$\Gamma^1{}_{01} = -\frac{1}{3\,(\tau-\rho)}$$

$$\Gamma^1{}_{10} = \Gamma^1{}_{01}$$

$$\Gamma^1{}_{11} = \frac{1}{3\,(\tau-\rho)}$$

$$\Gamma^1{}_{22} = \frac{\mu^3\,(\tau-\rho)}{3\,m}$$

$$\Gamma^1{}_{33} = \frac{\mu^3\,(\tau-\rho)\,\sin^2\vartheta}{3\,m}$$

$$\Gamma^2{}_{02} = \frac{2}{3\,(\tau-\rho)}$$

$$\Gamma^2{}_{12} = -\frac{2}{3\,(\tau-\rho)}$$

$$\Gamma^2{}_{20} = \Gamma^2{}_{02}$$

$$\Gamma^2{}_{21} = \Gamma^2{}_{12}$$

$$\Gamma^2{}_{33} = -\cos\vartheta\,\sin\vartheta$$

$$\Gamma^3{}_{03} = \frac{2}{3\,(\tau-\rho)}$$

$$\Gamma^3{}_{13} = -\frac{2}{3\,(\tau-\rho)}$$

$$\Gamma^3{}_{23} = \frac{\cos\vartheta}{\sin\vartheta}$$

$$\Gamma^3{}_{30} = \Gamma^3{}_{03}$$

$$\Gamma^3{}_{31} = \Gamma^3{}_{13}$$

$$\Gamma^3{}_{32} = \Gamma^3{}_{23}$$

Metric Compatibility

——— o.k.

Riemann Tensor

$$R^0{}_{101} = -\frac{8\,m\,(\rho-\tau)^{\frac{1}{3}}}{9\,\mu\,(\tau-\rho)^3}$$

$$R^0{}_{110} = -R^0{}_{101}$$

$$R^0{}_{202} = -\frac{2\,\mu^2}{9\,(\rho-\tau)^{\frac{2}{3}}}$$

$$R^0{}_{220} = -R^0{}_{202}$$

$$R^0{}_{303} = -\frac{2\,\mu^2\,\sin^2\vartheta}{9\,(\rho-\tau)^{\frac{2}{3}}}$$

$$R^0{}_{330} = -R^0{}_{303}$$

$$R^1{}_{001} = \frac{4}{9\,(\tau-\rho)^2}$$

$$R^1{}_{010} = -R^1{}_{001}$$

$$R^1{}_{212} = \frac{2\,\mu^2\,(\rho-\tau)^{\frac{1}{3}}}{9\,(\tau-\rho)}$$

$$R^1{}_{221} = -R^1{}_{212}$$

$$R^1{}_{313} = \frac{2\,\mu^2\,(\rho-\tau)^{\frac{1}{3}}\,\sin^2\vartheta}{9\,(\tau-\rho)}$$

$$R^1{}_{331} = -R^1{}_{313}$$

$$R^2{}_{002} = -\frac{2}{9\,(\tau-\rho)^2}$$

$$R^2{}_{020} = -R^2{}_{002}$$

$$R^2{}_{112} = \frac{4\,m}{9\,\mu\,(\rho-\tau)^{\frac{2}{3}}\,(\tau-\rho)^2}$$

$$R^2{}_{121} = -R^2{}_{112}$$

$$R^2{}_{323} = -\frac{\left(2\,\mu^3\,\tau - 9\,m\,\tau + 4\,m\,\mu^2\,(\rho-\tau)^{\frac{1}{3}} - 2\,\mu^3\,\rho + 9\,m\,\rho\right)\,\sin^2\vartheta}{9\,m\,(\tau-\rho)}$$

$$R^2{}_{332} = -R^2{}_{323}$$

$$R^3{}_{003} = -\frac{2}{9\,(\tau-\rho)^2}$$

$$R^3{}_{030} = -R^3{}_{003}$$

$$R^3{}_{113} = \frac{4\,m}{9\,\mu\,(\rho-\tau)^{\frac{2}{3}}\,(\tau-\rho)^2}$$

$$R^3{}_{131} = -R^3{}_{113}$$

$$R^3{}_{223} = \frac{2\,\mu^3\,\tau - 9\,m\,\tau + 4\,m\,\mu^2\,(\rho-\tau)^{\frac{1}{3}} - 2\,\mu^3\,\rho + 9\,m\,\rho}{9\,m\,(\tau-\rho)}$$

$$R^3{}_{232} = -R^3{}_{223}$$

Ricci Tensor

$$\mathrm{Ric}_{22} = -\frac{2\,\mu^3 - 9\,m}{9\,m}$$

$$\mathrm{Ric}_{33} = -\frac{(2\,\mu^3 - 9\,m)\,\sin^2\vartheta}{9\,m}$$

Ricci Scalar

$$R_{sc} = -\frac{2\,(2\,\mu^3 - 9\,m)}{9\,m\,\mu^2\,(\rho-\tau)^{\frac{1}{3}}\,(\tau-\rho)}$$

Bianchi identity (Ricci cyclic equation $R^{\kappa}{}_{[\mu\nu\sigma]} = 0$)

——— o.k.

Einstein Tensor

$$G_{00} = \frac{2\,\mu^3 - 9\,m}{9\,m\,\mu^2\,(\rho-\tau)^{\frac{1}{3}}\,(\tau-\rho)}$$

$$G_{11} = \frac{2\,(2\,\mu^3 - 9\,m)}{9\,\mu^3\,(\tau-\rho)^2}$$

Hodge Dual of Bianchi Identity

——— (see charge and current densities)

Scalar Charge Density (-$R^{0}{}_{i}{}^{i0}$)

$$\rho = 0$$

Current Density Class 1 (-$R^{i}{}_{\mu}{}^{\mu j}$)

$$J_1 = 0$$

$$J_2 = -\frac{\left(2\,\mu^3 - 9\,m\right)\,\left(\rho - \tau\right)^{\frac{1}{3}}}{9\,m\,\mu^4\,\left(\tau - \rho\right)^3}$$

$$J_3 = -\frac{\left(2\,\mu^3 - 9\,m\right)\,\left(\rho - \tau\right)^{\frac{1}{3}}}{9\,m\,\mu^4\,\left(\tau - \rho\right)^3\,\sin^2\vartheta}$$

Current Density Class 2 (-$R^{i}{}_{\mu}{}^{\mu j}$)

$$J_1 = 0$$

$$J_2 = 0$$

$$J_3 = 0$$

Current Density Class 3 (-$R^{i}{}_{\mu}{}^{\mu j}$)

$$J_1 = 0$$

$$J_2 = 0$$

$$J_3 = 0$$

4.4.41 Alcubierre metric

This metric has been synthesized to model a "spaceship drive" by space-time. The parameters are

$$v_s = \frac{\partial x_s(t)}{\partial t},$$

$$r_s = \sqrt{\left(x_1 - x_s(t)\right)^2 - x_2^2 - x_3^2}.$$

For the figures we used

$$f(r_s) = \frac{\tanh(\sigma(r_s + R)) - \tanh(\sigma(r_s - R))}{2\tanh(\sigma R)}.$$

Interestingly, the charge and current densities are a counterpart of the intended space-time curvature.

Coordinates

$$\mathbf{x} = \begin{pmatrix} t \\ x_1 \\ x_2 \\ x_3 \end{pmatrix}$$

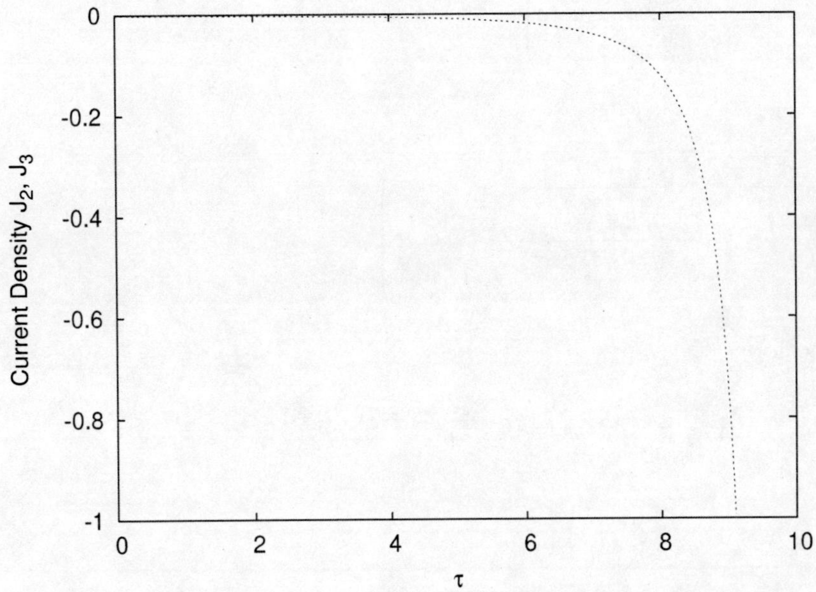

Fig. 4.72: Dirac metric, current density J_θ, J_φ, τ-dependence for $\rho = 10$.

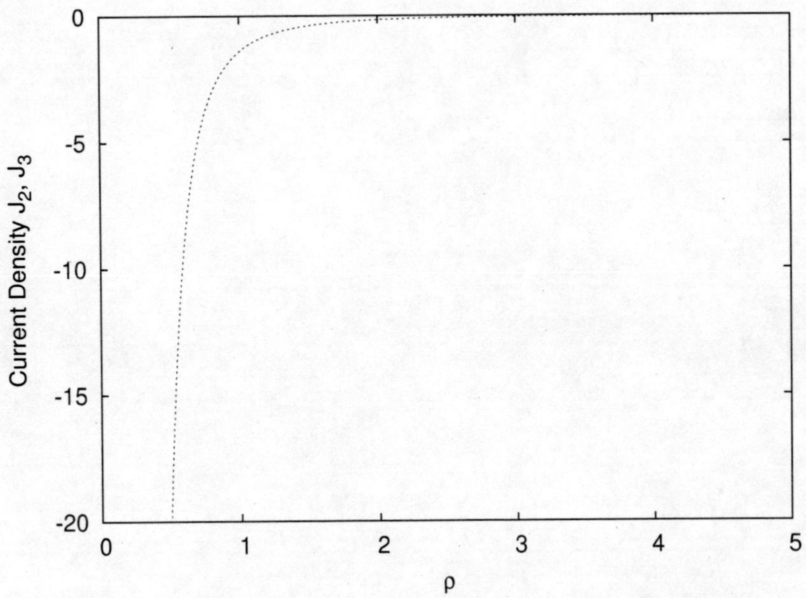

Fig. 4.73: Dirac metric, current density J_θ, J_φ, ρ-dependence for $\tau = 0.2$.

Metric

$$g_{\mu\nu} = \begin{pmatrix} f^2\left(\frac{d}{d\,t}\,xs\right)^2 - 1 & -2\,f\,\left(\frac{d}{d\,t}\,xs\right) & 0 & 0 \\ -2\,f\,\left(\frac{d}{d\,t}\,xs\right) & 1 & 0 & 0 \\ 0 & 0 & 1 & 0 \\ 0 & 0 & 0 & 1 \end{pmatrix}$$

Contravariant Metric

$$g^{\mu\nu} = \begin{pmatrix} -\frac{1}{3\,f^2\left(\frac{d}{d\,t}\,xs\right)^2+1} & -\frac{2\,f\,\left(\frac{d}{d\,t}\,xs\right)}{3\,f^2\left(\frac{d}{d\,t}\,xs\right)^2+1} & 0 & 0 \\ -\frac{2\,f\,\left(\frac{d}{d\,t}\,xs\right)}{3\,f^2\left(\frac{d}{d\,t}\,xs\right)^2+1} & -\frac{\left(f\,\left(\frac{d}{d\,t}\,xs\right)-1\right)\left(f\,\left(\frac{d}{d\,t}\,xs\right)+1\right)}{3\,f^2\left(\frac{d}{d\,t}\,xs\right)^2+1} & 0 & 0 \\ 0 & 0 & 1 & 0 \\ 0 & 0 & 0 & 1 \end{pmatrix}$$

Christoffel Connection

$$\Gamma^0{}_{00} = \frac{f\,\left(\frac{d}{d\,t}\,xs\right)\left(3\,f\,\left(\frac{d^2}{d\,t^2}\,xs\right) + 2\,f\,\left(\frac{d}{d\,x_1}\,f\right)\left(\frac{d}{d\,t}\,xs\right)^2 + 3\,\left(\frac{d}{d\,t}\,f\right)\left(\frac{d}{d\,t}\,xs\right)\right)}{3\,f^2\,\left(\frac{d}{d\,t}\,xs\right)^2 + 1}$$

$$\Gamma^0{}_{01} = -\frac{f\,\left(\frac{d}{d\,x_1}\,f\right)\left(\frac{d}{d\,t}\,xs\right)^2}{3\,f^2\,\left(\frac{d}{d\,t}\,xs\right)^2 + 1}$$

$$\Gamma^0{}_{02} = \frac{f\,\left(\frac{d}{d\,x_2}\,f\right)\left(\frac{d}{d\,t}\,xs\right)^2}{3\,f^2\,\left(\frac{d}{d\,t}\,xs\right)^2 + 1}$$

$$\Gamma^0{}_{03} = \frac{f\,\left(\frac{d}{d\,x_3}\,f\right)\left(\frac{d}{d\,t}\,xs\right)^2}{3\,f^2\,\left(\frac{d}{d\,t}\,xs\right)^2 + 1}$$

$$\Gamma^0{}_{10} = \Gamma^0{}_{01}$$

$$\Gamma^0{}_{11} = \frac{2\,\left(\frac{d}{d\,x_1}\,f\right)\left(\frac{d}{d\,t}\,xs\right)}{3\,f^2\,\left(\frac{d}{d\,t}\,xs\right)^2 + 1}$$

$$\Gamma^0{}_{12} = \frac{\frac{d}{d\,x_2}\,f\,\left(\frac{d}{d\,t}\,xs\right)}{3\,f^2\,\left(\frac{d}{d\,t}\,xs\right)^2 + 1}$$

$$\Gamma^0{}_{13} = \frac{\frac{d}{d\,x_3}\,f\,\left(\frac{d}{d\,t}\,xs\right)}{3\,f^2\,\left(\frac{d}{d\,t}\,xs\right)^2 + 1}$$

$$\Gamma^0{}_{20} = \Gamma^0{}_{02}$$

$$\Gamma^0{}_{21} = \Gamma^0{}_{12}$$

$$\Gamma^0{}_{30} = \Gamma^0{}_{03}$$

$$\Gamma^0{}_{31} = \Gamma^0{}_{13}$$

$$\Gamma^1{}_{00} = -\frac{2\,f\,\left(\frac{d^2}{d\,t^2}\,xs\right) - f^3\,\left(\frac{d}{d\,x_1}\,f\right)\left(\frac{d}{d\,t}\,xs\right)^4 + f\,\left(\frac{d}{d\,x_1}\,f\right)\left(\frac{d}{d\,t}\,xs\right)^2 + 2\,\left(\frac{d}{d\,t}\,f\right)\left(\frac{d}{d\,t}\,xs\right)}{3\,f^2\,\left(\frac{d}{d\,t}\,xs\right)^2 + 1}$$

$$\Gamma^{1}{}_{01} = -\frac{2\, f^2 \left(\frac{d}{d\, x_1} f\right) \left(\frac{d}{d\, t}\, xs\right)^3}{3\, f^2 \left(\frac{d}{d\, t}\, xs\right)^2 + 1}$$

$$\Gamma^{1}{}_{02} = -\frac{\frac{d}{d\, x_2} f \left(\frac{d}{d\, t}\, xs\right) \left(f^2 \left(\frac{d}{d\, t}\, xs\right)^2 + 1\right)}{3\, f^2 \left(\frac{d}{d\, t}\, xs\right)^2 + 1}$$

$$\Gamma^{1}{}_{03} = -\frac{\frac{d}{d\, x_3} f \left(\frac{d}{d\, t}\, xs\right) \left(f^2 \left(\frac{d}{d\, t}\, xs\right)^2 + 1\right)}{3\, f^2 \left(\frac{d}{d\, t}\, xs\right)^2 + 1}$$

$$\Gamma^{1}{}_{10} = \Gamma^{1}{}_{01}$$

$$\Gamma^{1}{}_{11} = \frac{4\, f \left(\frac{d}{d\, x_1} f\right) \left(\frac{d}{d\, t}\, xs\right)^2}{3\, f^2 \left(\frac{d}{d\, t}\, xs\right)^2 + 1}$$

$$\Gamma^{1}{}_{12} = \frac{2\, f \left(\frac{d}{d\, x_2} f\right) \left(\frac{d}{d\, t}\, xs\right)^2}{3\, f^2 \left(\frac{d}{d\, t}\, xs\right)^2 + 1}$$

$$\Gamma^{1}{}_{13} = \frac{2\, f \left(\frac{d}{d\, x_3} f\right) \left(\frac{d}{d\, t}\, xs\right)^2}{3\, f^2 \left(\frac{d}{d\, t}\, xs\right)^2 + 1}$$

$$\Gamma^{1}{}_{20} = \Gamma^{1}{}_{02}$$

$$\Gamma^{1}{}_{21} = \Gamma^{1}{}_{12}$$

$$\Gamma^{1}{}_{30} = \Gamma^{1}{}_{03}$$

$$\Gamma^{1}{}_{31} = \Gamma^{1}{}_{13}$$

$$\Gamma^{2}{}_{00} = -f \left(\frac{d}{d\, x_2} f\right) \left(\frac{d}{d\, t}\, xs\right)^2$$

$$\Gamma^{2}{}_{01} = \frac{d}{d\, x_2} f \left(\frac{d}{d\, t}\, xs\right)$$

$$\Gamma^{2}{}_{10} = \Gamma^{2}{}_{01}$$

$$\Gamma^{3}{}_{00} = -f \left(\frac{d}{d\, x_3} f\right) \left(\frac{d}{d\, t}\, xs\right)^2$$

$$\Gamma^{3}{}_{01} = \frac{d}{d\, x_3} f \left(\frac{d}{d\, t}\, xs\right)$$

$$\Gamma^{3}{}_{10} = \Gamma^{3}{}_{01}$$

Metric Compatibility

——— o.k.

Riemann Tensor

$$R^{0}{}_{001} = -\frac{2\,f\,\left(\frac{d}{d\,t}\,xs\right)\left(2\left(\frac{d}{d\,x_1}\,f\right)\left(\frac{d^2}{d\,t^2}\,xs\right) - 3\,f^2\left(\frac{d}{d\,x_3}\,f\right)^2\left(\frac{d}{d\,t}\,xs\right)^4 - 3\,f^2\left(\frac{d}{d\,x_2}\,f\right)^2\left(\frac{d}{d\,t}\,xs\right)^4 + 3\,f^3\left(\frac{d^2}{d\,x_1{}^2}\,f\right)\left(\frac{d}{d\,t}\,xs\right)^4 - \ldots\right)}{\left(3\,f^2\left(\frac{d}{d\,t}\,xs\right)^2 + 1\right)^2}$$

$$R^{0}{}_{002} = -\frac{2\,f\,\left(\frac{d}{d\,t}\,xs\right)\left(\frac{d}{d\,x_2}\,f\left(\frac{d^2}{d\,t^2}\,xs\right) + 3\,f^3\left(\frac{d^2}{d\,x_1\,d\,x_2}\,f\right)\left(\frac{d}{d\,t}\,xs\right)^4 - 3\,f\left(\frac{d}{d\,t}\,f\right)\left(\frac{d}{d\,x_2}\,f\right)\left(\frac{d}{d\,t}\,xs\right)^3 + 3\,f^2\left(\frac{d^2}{d\,t\,d\,x_2}\,f\right)\left(\frac{d}{d\,t}\,xs\right)^3 + \ldots\right)}{\left(3\,f^2\left(\frac{d}{d\,t}\,xs\right)^2 + 1\right)^2}$$

$$R^{0}{}_{003} = -\frac{2\,f\,\left(\frac{d}{d\,t}\,xs\right)\left(\frac{d}{d\,x_3}\,f\left(\frac{d^2}{d\,t^2}\,xs\right) + 3\,f^3\left(\frac{d^2}{d\,x_1\,d\,x_3}\,f\right)\left(\frac{d}{d\,t}\,xs\right)^4 - 3\,f\left(\frac{d}{d\,t}\,f\right)\left(\frac{d}{d\,x_3}\,f\right)\left(\frac{d}{d\,t}\,xs\right)^3 + 3\,f^2\left(\frac{d^2}{d\,t\,d\,x_3}\,f\right)\left(\frac{d}{d\,t}\,xs\right)^3 + \ldots\right)}{\left(3\,f^2\left(\frac{d}{d\,t}\,xs\right)^2 + 1\right)^2}$$

$$R^{0}{}_{010} = -R^{0}{}_{001}$$

$$R^{0}{}_{012} = -\frac{2\,f\,\left(\frac{d}{d\,t}\,xs\right)^2\left(3\,f\left(\frac{d}{d\,x_1}\,f\right)\left(\frac{d}{d\,x_2}\,f\right)\left(\frac{d}{d\,t}\,xs\right)^2 - 3\,f^2\left(\frac{d^2}{d\,x_1\,d\,x_2}\,f\right)\left(\frac{d}{d\,t}\,xs\right)^2 - \frac{d^2}{d\,x_1\,d\,x_2}\,f\right)}{\left(3\,f^2\left(\frac{d}{d\,t}\,xs\right)^2 + 1\right)^2}$$

$$R^{0}{}_{013} = -\frac{2\,f\,\left(\frac{d}{d\,t}\,xs\right)^2\left(3\,f\left(\frac{d}{d\,x_1}\,f\right)\left(\frac{d}{d\,x_3}\,f\right)\left(\frac{d}{d\,t}\,xs\right)^2 - 3\,f^2\left(\frac{d^2}{d\,x_1\,d\,x_3}\,f\right)\left(\frac{d}{d\,t}\,xs\right)^2 - \frac{d^2}{d\,x_1\,d\,x_3}\,f\right)}{\left(3\,f^2\left(\frac{d}{d\,t}\,xs\right)^2 + 1\right)^2}$$

$$R^{0}{}_{020} = -R^{0}{}_{002}$$

$$R^{0}{}_{021} = -R^{0}{}_{012}$$

$$R^{0}{}_{030} = -R^{0}{}_{003}$$

$$R^{0}{}_{031} = -R^{0}{}_{013}$$

$$R^{0}{}_{101} = \frac{2\left(\frac{d}{d\,x_1}\,f\right)\left(\frac{d^2}{d\,t^2}\,xs\right) - 3\,f^2\left(\frac{d}{d\,x_3}\,f\right)^2\left(\frac{d}{d\,t}\,xs\right)^4 - 3\,f^2\left(\frac{d}{d\,x_2}\,f\right)^2\left(\frac{d}{d\,t}\,xs\right)^4 + 3\,f^3\left(\frac{d^2}{d\,x_1{}^2}\,f\right)\left(\frac{d}{d\,t}\,xs\right)^4 - 6\,f\left(\frac{d}{d\,t}\,f\right)\left(\frac{d}{d\,x_1}\,f\right)\left(\frac{d}{d\,t}\,xs\right)^3 + \ldots}{\left(3\,f^2\left(\frac{d}{d\,t}\,xs\right)^2 + 1\right)^2}$$

$$R^{0}{}_{102} = \frac{\frac{d}{d\,x_2}\,f\left(\frac{d^2}{d\,t^2}\,xs\right) + 3\,f^3\left(\frac{d^2}{d\,x_1\,d\,x_2}\,f\right)\left(\frac{d}{d\,t}\,xs\right)^4 - 3\,f\left(\frac{d}{d\,t}\,f\right)\left(\frac{d}{d\,x_2}\,f\right)\left(\frac{d}{d\,t}\,xs\right)^3 + 3\,f^2\left(\frac{d^2}{d\,t\,d\,x_2}\,f\right)\left(\frac{d}{d\,t}\,xs\right)^3 + \frac{d}{d\,x_1}\,f\left(\frac{d}{d\,x_2}\,f\right)\left(\frac{d}{d\,t}\,xs\right)^2 + \ldots}{\left(3\,f^2\left(\frac{d}{d\,t}\,xs\right)^2 + 1\right)^2}$$

$$R^{0}{}_{103} = \frac{\frac{d}{d\,x_3}\,f\left(\frac{d^2}{d\,t^2}\,xs\right) + 3\,f^3\left(\frac{d^2}{d\,x_1\,d\,x_3}\,f\right)\left(\frac{d}{d\,t}\,xs\right)^4 - 3\,f\left(\frac{d}{d\,t}\,f\right)\left(\frac{d}{d\,x_3}\,f\right)\left(\frac{d}{d\,t}\,xs\right)^3 + 3\,f^2\left(\frac{d^2}{d\,t\,d\,x_3}\,f\right)\left(\frac{d}{d\,t}\,xs\right)^3 + \frac{d}{d\,x_1}\,f\left(\frac{d}{d\,x_3}\,f\right)\left(\frac{d}{d\,t}\,xs\right)^2 + \ldots}{\left(3\,f^2\left(\frac{d}{d\,t}\,xs\right)^2 + 1\right)^2}$$

$$R^{0}{}_{110} = -R^{0}{}_{101}$$

$$R^{0}{}_{112} = \frac{\frac{d}{dt}\,xs\,\left(3\,f\,\left(\frac{d}{d\,x_1}\,f\right)\,\left(\frac{d}{d\,x_2}\,f\right)\,\left(\frac{d}{dt}\,xs\right)^2 - 3\,f^2\,\left(\frac{d^2}{d\,x_1\,d\,x_2}\,f\right)\,\left(\frac{d}{dt}\,xs\right)^2 - \frac{d^2}{d\,x_1\,d\,x_2}\,f\right)}{\left(3\,f^2\,\left(\frac{d}{dt}\,xs\right)^2+1\right)^2}$$

$$R^{0}{}_{113} = \frac{\frac{d}{dt}\,xs\,\left(3\,f\,\left(\frac{d}{d\,x_1}\,f\right)\,\left(\frac{d}{d\,x_3}\,f\right)\,\left(\frac{d}{dt}\,xs\right)^2 - 3\,f^2\,\left(\frac{d^2}{d\,x_1\,d\,x_3}\,f\right)\,\left(\frac{d}{dt}\,xs\right)^2 - \frac{d^2}{d\,x_1\,d\,x_3}\,f\right)}{\left(3\,f^2\,\left(\frac{d}{dt}\,xs\right)^2+1\right)^2}$$

$$R^{0}{}_{120} = -R^{0}{}_{102}$$

$$R^{0}{}_{121} = -R^{0}{}_{112}$$

$$R^{0}{}_{130} = -R^{0}{}_{103}$$

$$R^{0}{}_{131} = -R^{0}{}_{113}$$

$$R^{0}{}_{201} = \frac{\frac{d}{d\,x_2}\,f\,\left(\frac{d^2}{d\,t^2}\,xs\right) + 6\,f^2\,\left(\frac{d}{d\,x_1}\,f\right)\,\left(\frac{d}{d\,x_2}\,f\right)\,\left(\frac{d}{dt}\,xs\right)^4 - 3\,f^3\,\left(\frac{d^2}{d\,x_1\,d\,x_2}\,f\right)\,\left(\frac{d}{dt}\,xs\right)^4 - 3\,f\,\left(\frac{d}{dt}\,f\right)\,\left(\frac{d}{d\,x_2}\,f\right)\,\left(\frac{d}{dt}\,xs\right)^3 + \ldots}{\left(3\,f^2\,\left(\frac{d}{dt}\,xs\right)^2+1\right)^2}$$

$$R^{0}{}_{202} = -\frac{f\,\left(\frac{d}{dt}\,xs\right)^2\,\left(3\,f^2\,\left(\frac{d^2}{d\,x_2{}^2}\,f\right)\,\left(\frac{d}{dt}\,xs\right)^2 - 3\,f\,\left(\frac{d}{d\,x_2}\,f\right)^2\,\left(\frac{d}{dt}\,xs\right)^2 + \frac{d^2}{d\,x_2{}^2}\,f\right)}{\left(3\,f^2\,\left(\frac{d}{dt}\,xs\right)^2+1\right)^2}$$

$$R^{0}{}_{203} = \frac{f\,\left(\frac{d}{dt}\,xs\right)^2\,\left(3\,f\,\left(\frac{d}{d\,x_2}\,f\right)\,\left(\frac{d}{d\,x_3}\,f\right)\,\left(\frac{d}{dt}\,xs\right)^2 - 3\,f^2\,\left(\frac{d^2}{d\,x_2\,d\,x_3}\,f\right)\,\left(\frac{d}{dt}\,xs\right)^2 - \frac{d^2}{d\,x_2\,d\,x_3}\,f\right)}{\left(3\,f^2\,\left(\frac{d}{dt}\,xs\right)^2+1\right)^2}$$

$$R^{0}{}_{210} = -R^{0}{}_{201}$$

$$R^{0}{}_{212} = -\frac{\frac{d}{dt}\,xs\,\left(3\,f^2\,\left(\frac{d^2}{d\,x_2{}^2}\,f\right)\,\left(\frac{d}{dt}\,xs\right)^2 - 3\,f\,\left(\frac{d}{d\,x_2}\,f\right)^2\,\left(\frac{d}{dt}\,xs\right)^2 + \frac{d^2}{d\,x_2{}^2}\,f\right)}{\left(3\,f^2\,\left(\frac{d}{dt}\,xs\right)^2+1\right)^2}$$

$$R^{0}{}_{213} = \frac{\frac{d}{dt}\,xs\,\left(3\,f\,\left(\frac{d}{d\,x_2}\,f\right)\,\left(\frac{d}{d\,x_3}\,f\right)\,\left(\frac{d}{dt}\,xs\right)^2 - 3\,f^2\,\left(\frac{d^2}{d\,x_2\,d\,x_3}\,f\right)\,\left(\frac{d}{dt}\,xs\right)^2 - \frac{d^2}{d\,x_2\,d\,x_3}\,f\right)}{\left(3\,f^2\,\left(\frac{d}{dt}\,xs\right)^2+1\right)^2}$$

$$R^{0}{}_{220} = -R^{0}{}_{202}$$

$$R^{0}{}_{221} = -R^{0}{}_{212}$$

$$R^{0}{}_{230} = -R^{0}{}_{203}$$

$$R^{0}{}_{231} = -R^{0}{}_{213}$$

$$R^{0}{}_{301} = \frac{\frac{d}{d\,x_3}\,f\,\left(\frac{d^2}{d\,t^2}\,xs\right) + 6\,f^2\,\left(\frac{d}{d\,x_1}\,f\right)\left(\frac{d}{d\,x_3}\,f\right)\left(\frac{d}{d\,t}\,xs\right)^4 - 3\,f^3\,\left(\frac{d^2}{d\,x_1\,d\,x_3}\,f\right)\left(\frac{d}{d\,t}\,xs\right)^4 - 3\,f\,\left(\frac{d}{d\,t}\,f\right)\left(\frac{d}{d\,x_3}\,f\right)\left(\frac{d}{d\,t}\,xs\right)^3 + ...}{\left(3\,f^2\,\left(\frac{d}{d\,t}\,xs\right)^2 + 1\right)^2}$$

$$R^{0}{}_{302} = \frac{f\,\left(\frac{d}{d\,t}\,xs\right)^2\,\left(3\,f\,\left(\frac{d}{d\,x_2}\,f\right)\left(\frac{d}{d\,x_3}\,f\right)\left(\frac{d}{d\,t}\,xs\right)^2 - 3\,f^2\,\left(\frac{d^2}{d\,x_2\,d\,x_3}\,f\right)\left(\frac{d}{d\,t}\,xs\right)^2 - \frac{d^2}{d\,x_2\,d\,x_3}\,f\right)}{\left(3\,f^2\,\left(\frac{d}{d\,t}\,xs\right)^2 + 1\right)^2}$$

$$R^{0}{}_{303} = -\frac{f\,\left(\frac{d}{d\,t}\,xs\right)^2\,\left(3\,f^2\,\left(\frac{d^2}{d\,x_3{}^2}\,f\right)\left(\frac{d}{d\,t}\,xs\right)^2 - 3\,f\,\left(\frac{d}{d\,x_3}\,f\right)^2\left(\frac{d}{d\,t}\,xs\right)^2 + \frac{d^2}{d\,x_3{}^2}\,f\right)}{\left(3\,f^2\,\left(\frac{d}{d\,t}\,xs\right)^2 + 1\right)^2}$$

$$R^{0}{}_{310} = -R^{0}{}_{301}$$

$$R^{0}{}_{312} = \frac{\frac{d}{d\,t}\,xs\,\left(3\,f\,\left(\frac{d}{d\,x_2}\,f\right)\left(\frac{d}{d\,x_3}\,f\right)\left(\frac{d}{d\,t}\,xs\right)^2 - 3\,f^2\,\left(\frac{d^2}{d\,x_2\,d\,x_3}\,f\right)\left(\frac{d}{d\,t}\,xs\right)^2 - \frac{d^2}{d\,x_2\,d\,x_3}\,f\right)}{\left(3\,f^2\,\left(\frac{d}{d\,t}\,xs\right)^2 + 1\right)^2}$$

$$R^{0}{}_{313} = -\frac{\frac{d}{d\,t}\,xs\,\left(3\,f^2\,\left(\frac{d^2}{d\,x_3{}^2}\,f\right)\left(\frac{d}{d\,t}\,xs\right)^2 - 3\,f\,\left(\frac{d}{d\,x_3}\,f\right)^2\left(\frac{d}{d\,t}\,xs\right)^2 + \frac{d^2}{d\,x_3{}^2}\,f\right)}{\left(3\,f^2\,\left(\frac{d}{d\,t}\,xs\right)^2 + 1\right)^2}$$

$$R^{0}{}_{320} = -R^{0}{}_{302}$$

$$R^{0}{}_{321} = -R^{0}{}_{312}$$

$$R^{0}{}_{330} = -R^{0}{}_{303}$$

$$R^{0}{}_{331} = -R^{0}{}_{313}$$

$$R^{1}{}_{001} = -\frac{\left(f\,\left(\frac{d}{d\,t}\,xs\right) - 1\right)\left(f\,\left(\frac{d}{d\,t}\,xs\right) + 1\right)\left(2\,\left(\frac{d}{d\,x_1}\,f\right)\left(\frac{d^2}{d\,t^2}\,xs\right) - 3\,f^2\,\left(\frac{d}{d\,x_3}\,f\right)^2\left(\frac{d}{d\,t}\,xs\right)^4 - 3\,f^2\,\left(\frac{d}{d\,x_2}\,f\right)^2\left(\frac{d}{d\,t}\,xs\right)^4 + ...\right)}{\left(3\,f^2\,\left(\frac{d}{d\,t}\,xs\right)^2 + 1\right)^2}$$

$$R^{1}{}_{002} = -\frac{\left(f\,\left(\frac{d}{d\,t}\,xs\right) - 1\right)\left(f\,\left(\frac{d}{d\,t}\,xs\right) + 1\right)\left(\frac{d}{d\,x_2}\,f\,\left(\frac{d^2}{d\,t^2}\,xs\right) + 3\,f^3\,\left(\frac{d^2}{d\,x_1\,d\,x_2}\,f\right)\left(\frac{d}{d\,t}\,xs\right)^4 - 3\,f\,\left(\frac{d}{d\,t}\,f\right)\left(\frac{d}{d\,x_2}\,f\right)\left(\frac{d}{d\,t}\,xs\right)^3 + ...\right)}{\left(3\,f^2\,\left(\frac{d}{d\,t}\,xs\right)^2 + 1\right)^2}$$

$$R^{1}{}_{003} = -\frac{\left(f\,\left(\frac{d}{d\,t}\,xs\right) - 1\right)\left(f\,\left(\frac{d}{d\,t}\,xs\right) + 1\right)\left(\frac{d}{d\,x_3}\,f\,\left(\frac{d^2}{d\,t^2}\,xs\right) + 3\,f^3\,\left(\frac{d^2}{d\,x_1\,d\,x_3}\,f\right)\left(\frac{d}{d\,t}\,xs\right)^4 - ...\right)}{\left(3\,f^2\,\left(\frac{d}{d\,t}\,xs\right)^2 + 1\right)^2}$$

$$R^{1}{}_{010} = -R^{1}{}_{001}$$

$$R^{1}{}_{012} = -\frac{\frac{d}{dt}xs\,\left(f\,\left(\frac{d}{dt}xs\right)-1\right)\left(f\,\left(\frac{d}{dt}xs\right)+1\right)\left(3\,f\,\left(\frac{d}{dx_1}f\right)\left(\frac{d}{dx_2}f\right)\left(\frac{d}{dt}xs\right)^2-3\,f^2\,\left(\frac{d^2}{dx_1\,dx_2}f\right)\left(\frac{d}{dt}xs\right)^2-\frac{d^2}{dx_1\,dx_2}f\right)}{\left(3\,f^2\,\left(\frac{d}{dt}xs\right)^2+1\right)^2}$$

$$R^{1}{}_{013} = -\frac{\frac{d}{dt}xs\,\left(f\,\left(\frac{d}{dt}xs\right)-1\right)\left(f\,\left(\frac{d}{dt}xs\right)+1\right)\left(3\,f\,\left(\frac{d}{dx_1}f\right)\left(\frac{d}{dx_3}f\right)\left(\frac{d}{dt}xs\right)^2-3\,f^2\,\left(\frac{d^2}{dx_1\,dx_3}f\right)\left(\frac{d}{dt}xs\right)^2-\frac{d^2}{dx_1\,dx_3}f\right)}{\left(3\,f^2\,\left(\frac{d}{dt}xs\right)^2+1\right)^2}$$

$$R^{1}{}_{020} = -R^{1}{}_{002}$$

$$R^{1}{}_{021} = -R^{1}{}_{012}$$

$$R^{1}{}_{030} = -R^{1}{}_{003}$$

$$R^{1}{}_{031} = -R^{1}{}_{013}$$

$$R^{1}{}_{101} = \frac{2\,f\,\left(\frac{d}{dt}xs\right)\left(2\,\left(\frac{d}{dx_1}f\right)\left(\frac{d^2}{dt^2}xs\right)-3\,f^2\,\left(\frac{d}{dx_3}f\right)^2\left(\frac{d}{dt}xs\right)^4-3\,f^2\,\left(\frac{d}{dx_2}f\right)^2\left(\frac{d}{dt}xs\right)^4+3\,f^3\,\left(\frac{d^2}{d{x_1}^2}f\right)\left(\frac{d}{dt}xs\right)^4-\ldots\right)}{\left(3\,f^2\,\left(\frac{d}{dt}xs\right)^2+1\right)^2}$$

$$R^{1}{}_{102} = \frac{2\,f\,\left(\frac{d}{dt}xs\right)\left(\frac{d}{dx_2}f\,\left(\frac{d^2}{dt^2}xs\right)+3\,f^3\,\left(\frac{d^2}{dx_1\,dx_2}f\right)\left(\frac{d}{dt}xs\right)^4-3\,f\,\left(\frac{d}{dt}f\right)\left(\frac{d}{dx_2}f\right)\left(\frac{d}{dt}xs\right)^3+3\,f^2\,\left(\frac{d^2}{dt\,dx_2}f\right)\left(\frac{d}{dt}xs\right)^3+\ldots\right)}{\left(3\,f^2\,\left(\frac{d}{dt}xs\right)^2+1\right)^2}$$

$$R^{1}{}_{103} = \frac{2\,f\,\left(\frac{d}{dt}xs\right)\left(\frac{d}{dx_3}f\,\left(\frac{d^2}{dt^2}xs\right)+3\,f^3\,\left(\frac{d^2}{dx_1\,dx_3}f\right)\left(\frac{d}{dt}xs\right)^4-3\,f\,\left(\frac{d}{dt}f\right)\left(\frac{d}{dx_3}f\right)\left(\frac{d}{dt}xs\right)^3+3\,f^2\,\left(\frac{d^2}{dt\,dx_3}f\right)\left(\frac{d}{dt}xs\right)^3+\ldots\right)}{\left(3\,f^2\,\left(\frac{d}{dt}xs\right)^2+1\right)^2}$$

$$R^{1}{}_{110} = -R^{1}{}_{101}$$

$$R^{1}{}_{112} = \frac{2\,f\,\left(\frac{d}{dt}xs\right)^2\left(3\,f\,\left(\frac{d}{dx_1}f\right)\left(\frac{d}{dx_2}f\right)\left(\frac{d}{dt}xs\right)^2-3\,f^2\,\left(\frac{d^2}{dx_1\,dx_2}f\right)\left(\frac{d}{dt}xs\right)^2-\frac{d^2}{dx_1\,dx_2}f\right)}{\left(3\,f^2\,\left(\frac{d}{dt}xs\right)^2+1\right)^2}$$

$$R^{1}{}_{113} = \frac{2\,f\,\left(\frac{d}{dt}xs\right)^2\left(3\,f\,\left(\frac{d}{dx_1}f\right)\left(\frac{d}{dx_3}f\right)\left(\frac{d}{dt}xs\right)^2-3\,f^2\,\left(\frac{d^2}{dx_1\,dx_3}f\right)\left(\frac{d}{dt}xs\right)^2-\frac{d^2}{dx_1\,dx_3}f\right)}{\left(3\,f^2\,\left(\frac{d}{dt}xs\right)^2+1\right)^2}$$

$$R^{1}{}_{120} = -R^{1}{}_{102}$$

$$R^{1}{}_{121} = -R^{1}{}_{112}$$

$$R^1{}_{130} = -R^1{}_{103}$$

$$R^1{}_{131} = -R^1{}_{113}$$

$$R^1{}_{201} = \frac{\frac{d}{d\,t}\,xs\,\left(2\,f\,\left(\frac{d}{d\,x_2}\,f\right)\left(\frac{d^2}{d\,t^2}\,xs\right)+3\,f^3\,\left(\frac{d}{d\,x_1}\,f\right)\left(\frac{d}{d\,x_2}.f\right)\left(\frac{d}{d\,t}\,xs\right)^4+3\,f^4\,\left(\frac{d^2}{d\,x_1\,d\,x_2}\,f\right)\left(\frac{d}{d\,t}\,xs\right)^4-6\,f^2\,\left(\frac{d}{d\,t}\,f\right)\left(\frac{d}{d\,x_2}\,f\right)\left(\frac{d}{d\,t}\,xs\right)^3+\ldots\right)}{\left(3\,f^2\,\left(\frac{d}{d\,t}\,xs\right)^2+1\right)^2}$$

$$R^1{}_{202} = \frac{\frac{d}{d\,t}\,xs\,\left(3\,f^4\,\left(\frac{d^2}{d\,x_2{}^2}\,f\right)\left(\frac{d}{d\,t}\,xs\right)^4+3\,f^3\,\left(\frac{d}{d\,x_2}\,f\right)^2\left(\frac{d}{d\,t}\,xs\right)^4+4\,f^2\,\left(\frac{d^2}{d\,x_2{}^2}\,f\right)\left(\frac{d}{d\,t}\,xs\right)^2-f\,\left(\frac{d}{d\,x_2}\,f\right)^2\left(\frac{d}{d\,t}\,xs\right)^2+\frac{d^2}{d\,x_2{}^2}\,f\right)}{\left(3\,f^2\,\left(\frac{d}{d\,t}\,xs\right)^2+1\right)^2}$$

$$R^1{}_{203} = \frac{\frac{d}{d\,t}\,xs\,\left(3\,f^3\,\left(\frac{d}{d\,x_2}\,f\right)\left(\frac{d}{d\,x_3}\,f\right)\left(\frac{d}{d\,t}\,xs\right)^4+3\,f^4\,\left(\frac{d^2}{d\,x_2\,d\,x_3}\,f\right)\left(\frac{d}{d\,t}\,xs\right)^4-f\,\left(\frac{d}{d\,x_2}\,f\right)\left(\frac{d}{d\,x_3}\,f\right)\left(\frac{d}{d\,t}\,xs\right)^2+4\,f^2\,\left(\frac{d^2}{d\,x_2\,d\,x_3}\,f\right)\left(\frac{d}{d\,t}\,xs\right)^2+\ldots\right)}{\left(3\,f^2\,\left(\frac{d}{d\,t}\,xs\right)^2+1\right)^2}$$

$$R^1{}_{210} = -R^1{}_{201}$$

$$R^1{}_{212} = -\frac{\left(\frac{d}{d\,t}\,xs\right)^2\left(6\,f^3\,\left(\frac{d^2}{d\,x_2{}^2}\,f\right)\left(\frac{d}{d\,t}\,xs\right)^2-3\,f^2\,\left(\frac{d}{d\,x_2}\,f\right)^2\left(\frac{d}{d\,t}\,xs\right)^2+2\,f\,\left(\frac{d^2}{d\,x_2{}^2}\,f\right)+\left(\frac{d}{d\,x_2}\,f\right)^2\right)}{\left(3\,f^2\,\left(\frac{d}{d\,t}\,xs\right)^2+1\right)^2}$$

$$R^1{}_{213} = \frac{\left(\frac{d}{d\,t}\,xs\right)^2\left(3\,f^2\,\left(\frac{d}{d\,x_2}\,f\right)\left(\frac{d}{d\,x_3}\,f\right)\left(\frac{d}{d\,t}\,xs\right)^2-6\,f^3\,\left(\frac{d^2}{d\,x_2\,d\,x_3}\,f\right)\left(\frac{d}{d\,t}\,xs\right)^2-\frac{d}{d\,x_2}\,f\,\left(\frac{d}{d\,x_3}\,f\right)-2\,f\,\left(\frac{d^2}{d\,x_2\,d\,x_3}\,f\right)\right)}{\left(3\,f^2\,\left(\frac{d}{d\,t}\,xs\right)^2+1\right)^2}$$

$$R^1{}_{220} = -R^1{}_{202}$$

$$R^1{}_{221} = -R^1{}_{212}$$

$$R^1{}_{230} = -R^1{}_{203}$$

$$R^1{}_{231} = -R^1{}_{213}$$

$$R^1{}_{301} = \frac{\frac{d}{d\,t}\,xs\,\left(2\,f\,\left(\frac{d}{d\,x_3}\,f\right)\left(\frac{d^2}{d\,t^2}\,xs\right)+3\,f^3\,\left(\frac{d}{d\,x_1}\,f\right)\left(\frac{d}{d\,x_3}\,f\right)\left(\frac{d}{d\,t}\,xs\right)^4+3\,f^4\,\left(\frac{d^2}{d\,x_1\,d\,x_3}\,f\right)\left(\frac{d}{d\,t}\,xs\right)^4-6\,f^2\,\left(\frac{d}{d\,t}\,f\right)\left(\frac{d}{d\,x_3}\,f\right)\left(\frac{d}{d\,t}\,xs\right)^3+\ldots\right)}{\left(3\,f^2\,\left(\frac{d}{d\,t}\,xs\right)^2+1\right)^2}$$

$$R^1{}_{302} = \frac{\frac{d}{d\,t}\,xs\,\left(3\,f^3\,\left(\frac{d}{d\,x_2}\,f\right)\left(\frac{d}{d\,x_3}\,f\right)\left(\frac{d}{d\,t}\,xs\right)^4+3\,f^4\,\left(\frac{d^2}{d\,x_2\,d\,x_3}\,f\right)\left(\frac{d}{d\,t}\,xs\right)^4-f\,\left(\frac{d}{d\,x_2}\,f\right)\left(\frac{d}{d\,x_3}\,f\right)\left(\frac{d}{d\,t}\,xs\right)^2+4\,f^2\,\left(\frac{d^2}{d\,x_2\,d\,x_3}\,f\right)\left(\frac{d}{d\,t}\,xs\right)^2+\ldots\right)}{\left(3\,f^2\,\left(\frac{d}{d\,t}\,xs\right)^2+1\right)^2}$$

$$R^{1}{}_{303} = \frac{\frac{d}{d\,t}\,xs\,\left(3\,f^4\,\left(\frac{d^2}{d\,x_3{}^2}\,f\right)\,\left(\frac{d}{d\,t}\,xs\right)^4 + 3\,f^3\,\left(\frac{d}{d\,x_3}\,f\right)^2\,\left(\frac{d}{d\,t}\,xs\right)^4 + 4\,f^2\,\left(\frac{d^2}{d\,x_3{}^2}\,f\right)\,\left(\frac{d}{d\,t}\,xs\right)^2 - f\,\left(\frac{d}{d\,x_3}\,f\right)^2\,\left(\frac{d}{d\,t}\,xs\right)^2 + \frac{d^2}{d\,x_3{}^2}\,f\right)}{\left(3\,f^2\,\left(\frac{d}{d\,t}\,xs\right)^2 + 1\right)^2}$$

$$R^{1}{}_{310} = -R^{1}{}_{301}$$

$$R^{1}{}_{312} = \frac{\left(\frac{d}{d\,t}\,xs\right)^2\,\left(3\,f^2\,\left(\frac{d}{d\,x_2}\,f\right)\,\left(\frac{d}{d\,x_3}\,f\right)\,\left(\frac{d}{d\,t}\,xs\right)^2 - 6\,f^3\,\left(\frac{d^2}{d\,x_2\,d\,x_3}\,f\right)\,\left(\frac{d}{d\,t}\,xs\right)^2 - \frac{d}{d\,x_2}\,f\,\left(\frac{d}{d\,x_3}\,f\right) - 2\,f\,\left(\frac{d^2}{d\,x_2\,d\,x_3}\,f\right)\right)}{\left(3\,f^2\,\left(\frac{d}{d\,t}\,xs\right)^2 + 1\right)^2}$$

$$R^{1}{}_{313} = -\frac{\left(\frac{d}{d\,t}\,xs\right)^2\,\left(6\,f^3\,\left(\frac{d^2}{d\,x_3{}^2}\,f\right)\,\left(\frac{d}{d\,t}\,xs\right)^2 - 3\,f^2\,\left(\frac{d}{d\,x_3}\,f\right)^2\,\left(\frac{d}{d\,t}\,xs\right)^2 + 2\,f\,\left(\frac{d^2}{d\,x_3{}^2}\,f\right) + \left(\frac{d}{d\,x_3}\,f\right)^2\right)}{\left(3\,f^2\,\left(\frac{d}{d\,t}\,xs\right)^2 + 1\right)^2}$$

$$R^{1}{}_{320} = -R^{1}{}_{302}$$

$$R^{1}{}_{321} = -R^{1}{}_{312}$$

$$R^{1}{}_{330} = -R^{1}{}_{303}$$

$$R^{1}{}_{331} = -R^{1}{}_{313}$$

$$R^{2}{}_{001} = \frac{\frac{d}{d\,x_2}\,f\,\left(\frac{d^2}{d\,t^2}\,xs\right) + 3\,f^3\,\left(\frac{d^2}{d\,x_1\,d\,x_2}\,f\right)\,\left(\frac{d}{d\,t}\,xs\right)^4 - 3\,f\,\left(\frac{d}{d\,t}\,f\right)\,\left(\frac{d}{d\,x_2}\,f\right)\,\left(\frac{d}{d\,t}\,xs\right)^3 + 3\,f^2\,\left(\frac{d^2}{d\,t\,d\,x_2}\,f\right)\,\left(\frac{d}{d\,t}\,xs\right)^3 + \frac{d}{d\,x_1}\,f\,\left(\frac{d}{d\,x_2}\,f\right)\,\left(\frac{d}{d\,t}\,xs\right)^2 + \ldots}{3\,f^2\,\left(\frac{d}{d\,t}\,xs\right)^2 + 1}$$

$$R^{2}{}_{002} = \frac{f\,\left(\frac{d}{d\,t}\,xs\right)^2\,\left(3\,f^2\,\left(\frac{d^2}{d\,x_2{}^2}\,f\right)\,\left(\frac{d}{d\,t}\,xs\right)^2 + f\,\left(\frac{d}{d\,x_2}\,f\right)^2\,\left(\frac{d}{d\,t}\,xs\right)^2 + \frac{d^2}{d\,x_2{}^2}\,f\right)}{3\,f^2\,\left(\frac{d}{d\,t}\,xs\right)^2 + 1}$$

$$R^{2}{}_{003} = \frac{f\,\left(\frac{d}{d\,t}\,xs\right)^2\,\left(f\,\left(\frac{d}{d\,x_2}\,f\right)\,\left(\frac{d}{d\,x_3}\,f\right)\,\left(\frac{d}{d\,t}\,xs\right)^2 + 3\,f^2\,\left(\frac{d^2}{d\,x_2\,d\,x_3}\,f\right)\,\left(\frac{d}{d\,t}\,xs\right)^2 + \frac{d^2}{d\,x_2\,d\,x_3}\,f\right)}{3\,f^2\,\left(\frac{d}{d\,t}\,xs\right)^2 + 1}$$

$$R^{2}{}_{010} = -R^{2}{}_{001}$$

$$R^{2}{}_{012} = -\frac{\frac{d}{d\,t}\,xs\,\left(3\,f^2\,\left(\frac{d^2}{d\,x_2{}^2}\,f\right)\,\left(\frac{d}{d\,t}\,xs\right)^2 - f\,\left(\frac{d}{d\,x_2}\,f\right)^2\,\left(\frac{d}{d\,t}\,xs\right)^2 + \frac{d^2}{d\,x_2{}^2}\,f\right)}{3\,f^2\,\left(\frac{d}{d\,t}\,xs\right)^2 + 1}$$

$$R^{2}{}_{013} = \frac{\frac{d}{d\,t}\,xs\,\left(f\,\left(\frac{d}{d\,x_2}\,f\right)\,\left(\frac{d}{d\,x_3}\,f\right)\,\left(\frac{d}{d\,t}\,xs\right)^2 - 3\,f^2\,\left(\frac{d^2}{d\,x_2\,d\,x_3}\,f\right)\,\left(\frac{d}{d\,t}\,xs\right)^2 - \frac{d^2}{d\,x_2\,d\,x_3}\,f\right)}{3\,f^2\,\left(\frac{d}{d\,t}\,xs\right)^2 + 1}$$

$$R^{2}{}_{020} = -R^{2}{}_{002}$$

$$R^2{}_{021} = -R^2{}_{012}$$

$$R^2{}_{030} = -R^2{}_{003}$$

$$R^2{}_{031} = -R^2{}_{013}$$

$$R^2{}_{101} = \frac{\frac{d}{dt}\,xs\left(3\,f\left(\frac{d}{d\,x_1}\,f\right)\left(\frac{d}{d\,x_2}\,f\right)\left(\frac{d}{dt}\,xs\right)^2 - 3\,f^2\left(\frac{d^2}{d\,x_1\,d\,x_2}\,f\right)\left(\frac{d}{dt}\,xs\right)^2 - \frac{d^2}{d\,x_1\,d\,x_2}\,f\right)}{3\,f^2\left(\frac{d}{dt}\,xs\right)^2+1}$$

$$R^2{}_{102} = -\frac{\frac{d}{dt}\,xs\left(3\,f^2\left(\frac{d^2}{d\,{x_2}^2}\,f\right)\left(\frac{d}{dt}\,xs\right)^2 - f\left(\frac{d}{d\,x_2}\,f\right)^2\left(\frac{d}{dt}\,xs\right)^2 + \frac{d^2}{d\,{x_2}^2}\,f\right)}{3\,f^2\left(\frac{d}{dt}\,xs\right)^2+1}$$

$$R^2{}_{103} = \frac{\frac{d}{dt}\,xs\left(f\left(\frac{d}{d\,x_2}\,f\right)\left(\frac{d}{d\,x_3}\,f\right)\left(\frac{d}{dt}\,xs\right)^2 - 3\,f^2\left(\frac{d^2}{d\,x_2\,d\,x_3}\,f\right)\left(\frac{d}{dt}\,xs\right)^2 - \frac{d^2}{d\,x_2\,d\,x_3}\,f\right)}{3\,f^2\left(\frac{d}{dt}\,xs\right)^2+1}$$

$$R^2{}_{110} = -R^2{}_{101}$$

$$R^2{}_{112} = \frac{\left(\frac{d}{d\,x_2}\,f\right)^2\left(\frac{d}{dt}\,xs\right)^2}{3\,f^2\left(\frac{d}{dt}\,xs\right)^2+1}$$

$$R^2{}_{113} = \frac{\frac{d}{d\,x_2}\,f\left(\frac{d}{d\,x_3}\,f\right)\left(\frac{d}{dt}\,xs\right)^2}{3\,f^2\left(\frac{d}{dt}\,xs\right)^2+1}$$

$$R^2{}_{120} = -R^2{}_{102}$$

$$R^2{}_{121} = -R^2{}_{112}$$

$$R^2{}_{130} = -R^2{}_{103}$$

$$R^2{}_{131} = -R^2{}_{113}$$

$$R^3{}_{001} = \frac{\frac{d}{d\,x_3}\,f\left(\frac{d^2}{d\,t^2}\,xs\right) + 3\,f^3\left(\frac{d^2}{d\,x_1\,d\,x_3}\,f\right)\left(\frac{d}{dt}\,xs\right)^4 - 3\,f\left(\frac{d}{dt}\,f\right)\left(\frac{d}{d\,x_3}\,f\right)\left(\frac{d}{dt}\,xs\right)^3 + 3\,f^2\left(\frac{d^2}{d\,t\,d\,x_3}\,f\right)\left(\frac{d}{dt}\,xs\right)^3 + \ldots}{3\,f^2\left(\frac{d}{dt}\,xs\right)^2+1}$$

$$R^3{}_{002} = \frac{f\left(\frac{d}{dt}\,xs\right)^2\left(f\left(\frac{d}{d\,x_2}\,f\right)\left(\frac{d}{d\,x_3}\,f\right)\left(\frac{d}{dt}\,xs\right)^2 + 3\,f^2\left(\frac{d^2}{d\,x_2\,d\,x_3}\,f\right)\left(\frac{d}{dt}\,xs\right)^2 + \frac{d^2}{d\,x_2\,d\,x_3}\,f\right)}{3\,f^2\left(\frac{d}{dt}\,xs\right)^2+1}$$

$$R^3{}_{003} = \frac{f\,\left(\frac{d}{d\,t}\,xs\right)^2\,\left(3\,f^2\,\left(\frac{d^2}{d\,x_3{}^2}\,f\right)\,\left(\frac{d}{d\,t}\,xs\right)^2 + f\,\left(\frac{d}{d\,x_3}\,f\right)^2\,\left(\frac{d}{d\,t}\,xs\right)^2 + \frac{d^2}{d\,x_3{}^2}\,f\right)}{3\,f^2\,\left(\frac{d}{d\,t}\,xs\right)^2 + 1}$$

$$R^3{}_{010} = -R^3{}_{001}$$

$$R^3{}_{012} = \frac{\frac{d}{d\,t}\,xs\,\left(f\,\left(\frac{d}{d\,x_2}\,f\right)\,\left(\frac{d}{d\,x_3}\,f\right)\,\left(\frac{d}{d\,t}\,xs\right)^2 - 3\,f^2\,\left(\frac{d^2}{d\,x_2\,d\,x_3}\,f\right)\,\left(\frac{d}{d\,t}\,xs\right)^2 - \frac{d^2}{d\,x_2\,d\,x_3}\,f\right)}{3\,f^2\,\left(\frac{d}{d\,t}\,xs\right)^2 + 1}$$

$$R^3{}_{013} = -\frac{\frac{d}{d\,t}\,xs\,\left(3\,f^2\,\left(\frac{d^2}{d\,x_3{}^2}\,f\right)\,\left(\frac{d}{d\,t}\,xs\right)^2 - f\,\left(\frac{d}{d\,x_3}\,f\right)^2\,\left(\frac{d}{d\,t}\,xs\right)^2 + \frac{d^2}{d\,x_3{}^2}\,f\right)}{3\,f^2\,\left(\frac{d}{d\,t}\,xs\right)^2 + 1}$$

$$R^3{}_{020} = -R^3{}_{002}$$

$$R^3{}_{021} = -R^3{}_{012}$$

$$R^3{}_{030} = -R^3{}_{003}$$

$$R^3{}_{031} = -R^3{}_{013}$$

$$R^3{}_{101} = \frac{\frac{d}{d\,t}\,xs\,\left(3\,f\,\left(\frac{d}{d\,x_1}\,f\right)\,\left(\frac{d}{d\,x_3}\,f\right)\,\left(\frac{d}{d\,t}\,xs\right)^2 - 3\,f^2\,\left(\frac{d^2}{d\,x_1\,d\,x_3}\,f\right)\,\left(\frac{d}{d\,t}\,xs\right)^2 - \frac{d^2}{d\,x_1\,d\,x_3}\,f\right)}{3\,f^2\,\left(\frac{d}{d\,t}\,xs\right)^2 + 1}$$

$$R^3{}_{102} = \frac{\frac{d}{d\,t}\,xs\,\left(f\,\left(\frac{d}{d\,x_2}\,f\right)\,\left(\frac{d}{d\,x_3}\,f\right)\,\left(\frac{d}{d\,t}\,xs\right)^2 - 3\,f^2\,\left(\frac{d^2}{d\,x_2\,d\,x_3}\,f\right)\,\left(\frac{d}{d\,t}\,xs\right)^2 - \frac{d^2}{d\,x_2\,d\,x_3}\,f\right)}{3\,f^2\,\left(\frac{d}{d\,t}\,xs\right)^2 + 1}$$

$$R^3{}_{103} = -\frac{\frac{d}{d\,t}\,xs\,\left(3\,f^2\,\left(\frac{d^2}{d\,x_3{}^2}\,f\right)\,\left(\frac{d}{d\,t}\,xs\right)^2 - f\,\left(\frac{d}{d\,x_3}\,f\right)^2\,\left(\frac{d}{d\,t}\,xs\right)^2 + \frac{d^2}{d\,x_3{}^2}\,f\right)}{3\,f^2\,\left(\frac{d}{d\,t}\,xs\right)^2 + 1}$$

$$R^3{}_{110} = -R^3{}_{101}$$

$$R^3{}_{112} = \frac{\frac{d}{d\,x_2}\,f\,\left(\frac{d}{d\,x_3}\,f\right)\,\left(\frac{d}{d\,t}\,xs\right)^2}{3\,f^2\,\left(\frac{d}{d\,t}\,xs\right)^2 + 1}$$

$$R^3{}_{113} = \frac{\left(\frac{d}{d\,x_3}\,f\right)^2\,\left(\frac{d}{d\,t}\,xs\right)^2}{3\,f^2\,\left(\frac{d}{d\,t}\,xs\right)^2 + 1}$$

$$R^3{}_{120} = -R^3{}_{102}$$

$$R^3{}_{121} = -R^3{}_{112}$$

$$R^3{}_{130} = -R^3{}_{103}$$

$$R^3{}_{131} = -R^3{}_{113}$$

Ricci Tensor

$$\mathrm{Ric}_{00} = \frac{2 f^2 \left(\frac{d}{d x_1} f\right) \left(\frac{d}{d t} xs\right)^2 \left(\frac{d^2}{d t^2} xs\right) - 2 \left(\frac{d}{d x_1} f\right) \left(\frac{d^2}{d t^2} xs\right) - 9 f^5 \left(\frac{d^2}{d {x_3}^2} f\right) \left(\frac{d}{d t} xs\right)^6 - 6 f^4 \left(\frac{d}{d x_3} f\right)^2 \left(\frac{d}{d t} xs\right)^6 - \ldots}{\left(3 f^2 \left(\frac{d}{d t} xs\right)^2 + 1\right)^2}$$

$$\mathrm{Ric}_{01} = -\frac{\frac{d}{d t} xs \left(4 f \left(\frac{d}{d x_1} f\right) \left(\frac{d^2}{d t^2} xs\right) - 9 f^4 \left(\frac{d^2}{d {x_3}^2} f\right) \left(\frac{d}{d t} xs\right)^4 - 3 f^3 \left(\frac{d}{d x_3} f\right)^2 \left(\frac{d}{d t} xs\right)^4 - 9 f^4 \left(\frac{d^2}{d {x_2}^2} f\right) \left(\frac{d}{d t} xs\right)^4 - \ldots\right)}{\left(3 f^2 \left(\frac{d}{d t} xs\right)^2 + 1\right)^2}$$

$$\mathrm{Ric}_{02} = -\frac{\frac{d}{d t} xs \left(2 f \left(\frac{d}{d x_2} f\right) \left(\frac{d^2}{d t^2} xs\right) + 3 f^3 \left(\frac{d}{d x_1} f\right) \left(\frac{d}{d x_2} f\right) \left(\frac{d}{d t} xs\right)^4 + 3 f^4 \left(\frac{d^2}{d x_1\, d x_2} f\right) \left(\frac{d}{d t} xs\right)^4 - 6 f^2 \left(\frac{d}{d t} f\right) \left(\frac{d}{d x_2} f\right) \left(\frac{d}{d t} xs\right)^3 + \ldots\right)}{\left(3 f^2 \left(\frac{d}{d t} xs\right)^2 + 1\right)^2}$$

$$\mathrm{Ric}_{03} = -\frac{\frac{d}{d t} xs \left(2 f \left(\frac{d}{d x_3} f\right) \left(\frac{d^2}{d t^2} xs\right) + 3 f^3 \left(\frac{d}{d x_1} f\right) \left(\frac{d}{d x_3} f\right) \left(\frac{d}{d t} xs\right)^4 + 3 f^4 \left(\frac{d^2}{d x_1\, d x_3} f\right) \left(\frac{d}{d t} xs\right)^4 - 6 f^2 \left(\frac{d}{d t} f\right) \left(\frac{d}{d x_3} f\right) \left(\frac{d}{d t} xs\right)^3 + \ldots\right)}{\left(3 f^2 \left(\frac{d}{d t} xs\right)^2 + 1\right)^2}$$

$$\mathrm{Ric}_{10} = \mathrm{Ric}_{01}$$

$$\mathrm{Ric}_{11} = \frac{2 \left(\frac{d}{d x_1} f\right) \left(\frac{d^2}{d t^2} xs\right) - 6 f^2 \left(\frac{d}{d x_3} f\right)^2 \left(\frac{d}{d t} xs\right)^4 - 6 f^2 \left(\frac{d}{d x_2} f\right)^2 \left(\frac{d}{d t} xs\right)^4 + 3 f^3 \left(\frac{d^2}{d {x_1}^2} f\right) \left(\frac{d}{d t} xs\right)^4 - 6 f \left(\frac{d}{d t} f\right) \left(\frac{d}{d x_1} f\right) \left(\frac{d}{d t} xs\right)^3 + \ldots}{\left(3 f^2 \left(\frac{d}{d t} xs\right)^2 + 1\right)^2}$$

$$\mathrm{Ric}_{12} = \frac{\frac{d}{d x_2} f \left(\frac{d^2}{d t^2} xs\right) + 6 f^2 \left(\frac{d}{d x_1} f\right) \left(\frac{d}{d x_2} f\right) \left(\frac{d}{d t} xs\right)^4 - 3 f^3 \left(\frac{d^2}{d x_1\, d x_2} f\right) \left(\frac{d}{d t} xs\right)^4 - 3 f \left(\frac{d}{d t} f\right) \left(\frac{d}{d x_2} f\right) \left(\frac{d}{d t} xs\right)^3 + \ldots}{\left(3 f^2 \left(\frac{d}{d t} xs\right)^2 + 1\right)^2}$$

$$\mathrm{Ric}_{13} = \frac{\frac{d}{d x_3} f \left(\frac{d^2}{d t^2} xs\right) + 6 f^2 \left(\frac{d}{d x_1} f\right) \left(\frac{d}{d x_3} f\right) \left(\frac{d}{d t} xs\right)^4 - 3 f^3 \left(\frac{d^2}{d x_1\, d x_3} f\right) \left(\frac{d}{d t} xs\right)^4 - 3 f \left(\frac{d}{d t} f\right) \left(\frac{d}{d x_3} f\right) \left(\frac{d}{d t} xs\right)^3 + \ldots}{\left(3 f^2 \left(\frac{d}{d t} xs\right)^2 + 1\right)^2}$$

$$\mathrm{Ric}_{20} = \mathrm{Ric}_{02}$$

$$\mathrm{Ric}_{21} = \mathrm{Ric}_{12}$$

$$\mathrm{Ric}_{22} = -\frac{\left(\frac{d}{d t} xs\right)^2 \left(9 f^3 \left(\frac{d^2}{d {x_2}^2} f\right) \left(\frac{d}{d t} xs\right)^2 - 6 f^2 \left(\frac{d}{d x_2} f\right)^2 \left(\frac{d}{d t} xs\right)^2 + 3 f \left(\frac{d^2}{d {x_2}^2} f\right) + \left(\frac{d}{d x_2} f\right)^2\right)}{\left(3 f^2 \left(\frac{d}{d t} xs\right)^2 + 1\right)^2}$$

$$\mathrm{Ric}_{23} = \frac{\left(\frac{d}{d t} xs\right)^2 \left(6 f^2 \left(\frac{d}{d x_2} f\right) \left(\frac{d}{d x_3} f\right) \left(\frac{d}{d t} xs\right)^2 - 9 f^3 \left(\frac{d^2}{d x_2\, d x_3} f\right) \left(\frac{d}{d t} xs\right)^2 - \frac{d}{d x_2} f \left(\frac{d}{d x_3} f\right) - 3 f \left(\frac{d^2}{d x_2\, d x_3} f\right)\right)}{\left(3 f^2 \left(\frac{d}{d t} xs\right)^2 + 1\right)^2}$$

$$\mathrm{Ric}_{30} = \mathrm{Ric}_{03}$$

$$\mathrm{Ric}_{31} = \mathrm{Ric}_{13}$$

$$\mathrm{Ric}_{32} = \mathrm{Ric}_{23}$$

$$\mathrm{Ric}_{33} = -\frac{\left(\frac{d}{d t} xs\right)^2 \left(9 f^3 \left(\frac{d^2}{d {x_3}^2} f\right) \left(\frac{d}{d t} xs\right)^2 - 6 f^2 \left(\frac{d}{d x_3} f\right)^2 \left(\frac{d}{d t} xs\right)^2 + 3 f \left(\frac{d^2}{d {x_3}^2} f\right) + \left(\frac{d}{d x_3} f\right)^2\right)}{\left(3 f^2 \left(\frac{d}{d t} xs\right)^2 + 1\right)^2}$$

Ricci Scalar

$$R_{sc} = \frac{2\left(2\left(\frac{d}{d x_1} f\right)\left(\frac{d^2}{d t^2} xs\right) - 9 f^3 \left(\frac{d^2}{d x_3{}^2} f\right)\left(\frac{d}{d t} xs\right)^4 + 3 f^2 \left(\frac{d}{d x_3} f\right)^2 \left(\frac{d}{d t} xs\right)^4 - 9 f^3 \left(\frac{d^2}{d x_2{}^2} f\right)\left(\frac{d}{d t} xs\right)^4 + 3 f^2 \left(\frac{d}{d x_2} f\right)^2 \left(\frac{d}{d t} xs\right)^4 + \ldots\right)}{\left(3 f^2 \left(\frac{d}{d t} xs\right)^2 + 1\right)^2}$$

Bianchi identity (Ricci cyclic equation $R^{\kappa}{}_{[\mu\nu\sigma]} = 0$)

——— o.k.

Einstein Tensor

$$G_{00} = -\frac{\left(\frac{d}{d t} xs\right)^2 \left(9 f^4 \left(\frac{d}{d x_3} f\right)^2 \left(\frac{d}{d t} xs\right)^4 + 9 f^4 \left(\frac{d}{d x_2} f\right)^2 \left(\frac{d}{d t} xs\right)^4 + 12 f^3 \left(\frac{d^2}{d x_3{}^2} f\right)\left(\frac{d}{d t} xs\right)^2 - 6 f^2 \left(\frac{d}{d x_3} f\right)^2 \left(\frac{d}{d t} xs\right)^2 + \ldots\right)}{\left(3 f^2 \left(\frac{d}{d t} xs\right)^2 + 1\right)^2}$$

$$G_{01} = -\frac{\frac{d}{d t} xs \left(3 f^2 \left(\frac{d}{d t} xs\right)^2 - 1\right)\left(3 f^2 \left(\frac{d^2}{d x_3{}^2} f\right)\left(\frac{d}{d t} xs\right)^2 - 3 f \left(\frac{d}{d x_3} f\right)^2 \left(\frac{d}{d t} xs\right)^2 + 3 f^2 \left(\frac{d^2}{d x_2{}^2} f\right)\left(\frac{d}{d t} xs\right)^2 - 3 f \left(\frac{d}{d x_2} f\right)^2 \left(\frac{d}{d t} xs\right)^2 + \ldots\right)}{\left(3 f^2 \left(\frac{d}{d t} xs\right)^2 + 1\right)^2}$$

$$G_{02} = -\frac{\frac{d}{d t} xs \left(2 f \left(\frac{d}{d x_2} f\right)\left(\frac{d^2}{d t^2} xs\right) + 3 f^3 \left(\frac{d}{d x_1} f\right)\left(\frac{d}{d x_2} f\right)\left(\frac{d}{d t} xs\right)^4 + 3 f^4 \left(\frac{d^2}{d x_1\, d x_2} f\right)\left(\frac{d}{d t} xs\right)^4 - 6 f^2 \left(\frac{d}{d t} f\right)\left(\frac{d}{d x_2} f\right)\left(\frac{d}{d t} xs\right)^3 + \ldots\right)}{\left(3 f^2 \left(\frac{d}{d t} xs\right)^2 + 1\right)^2}$$

$$G_{03} = -\frac{\frac{d}{d t} xs \left(2 f \left(\frac{d}{d x_3} f\right)\left(\frac{d^2}{d t^2} xs\right) + 3 f^3 \left(\frac{d}{d x_1} f\right)\left(\frac{d}{d x_3} f\right)\left(\frac{d}{d t} xs\right)^4 + 3 f^4 \left(\frac{d^2}{d x_1\, d x_3} f\right)\left(\frac{d}{d t} xs\right)^4 - 6 f^2 \left(\frac{d}{d t} f\right)\left(\frac{d}{d x_3} f\right)\left(\frac{d}{d t} xs\right)^3 + \ldots\right)}{\left(3 f^2 \left(\frac{d}{d t} xs\right)^2 + 1\right)^2}$$

$$G_{10} = G_{01}$$

$$G_{11} = \frac{3 f \left(\frac{d}{d t} xs\right)^2 \left(3 f^2 \left(\frac{d^2}{d x_3{}^2} f\right)\left(\frac{d}{d t} xs\right)^2 - 3 f \left(\frac{d}{d x_3} f\right)^2 \left(\frac{d}{d t} xs\right)^2 + 3 f^2 \left(\frac{d^2}{d x_2{}^2} f\right)\left(\frac{d}{d t} xs\right)^2 - 3 f \left(\frac{d}{d x_2} f\right)^2 \left(\frac{d}{d t} xs\right)^2 + \frac{d^2}{d x_3{}^2} f + \frac{d^2}{d x_2{}^2} f\right)}{\left(3 f^2 \left(\frac{d}{d t} xs\right)^2 + 1\right)^2}$$

$$G_{12} = \frac{\frac{d}{d x_2} f \left(\frac{d^2}{d t^2} xs\right) + 6 f^2 \left(\frac{d}{d x_1} f\right)\left(\frac{d}{d x_2} f\right)\left(\frac{d}{d t} xs\right)^4 - 3 f^3 \left(\frac{d^2}{d x_1\, d x_2} f\right)\left(\frac{d}{d t} xs\right)^4 - 3 f \left(\frac{d}{d t} f\right)\left(\frac{d}{d x_2} f\right)\left(\frac{d}{d t} xs\right)^3 + \ldots}{\left(3 f^2 \left(\frac{d}{d t} xs\right)^2 + 1\right)^2}$$

$$G_{13} = \frac{\frac{d}{d x_3} f \left(\frac{d^2}{d t^2} xs\right) + 6 f^2 \left(\frac{d}{d x_1} f\right)\left(\frac{d}{d x_3} f\right)\left(\frac{d}{d t} xs\right)^4 - 3 f^3 \left(\frac{d^2}{d x_1\, d x_3} f\right)\left(\frac{d}{d t} xs\right)^4 - 3 f \left(\frac{d}{d t} f\right)\left(\frac{d}{d x_3} f\right)\left(\frac{d}{d t} xs\right)^3 + \ldots}{\left(3 f^2 \left(\frac{d}{d t} xs\right)^2 + 1\right)^2}$$

$$G_{20} = G_{02}$$

$$G_{21} = G_{12}$$

$$G_{22} = -\frac{2\left(\frac{d}{d x_1} f\right)\left(\frac{d^2}{d t^2} xs\right) - 9 f^3 \left(\frac{d^2}{d x_3{}^2} f\right)\left(\frac{d}{d t} xs\right)^4 + 3 f^2 \left(\frac{d}{d x_3} f\right)^2 \left(\frac{d}{d t} xs\right)^4 - 3 f^2 \left(\frac{d}{d x_2} f\right)^2 \left(\frac{d}{d t} xs\right)^4 + 3 f^3 \left(\frac{d^2}{d x_1{}^2} f\right)\left(\frac{d}{d t} xs\right)^4 - \ldots}{\left(3 f^2 \left(\frac{d}{d t} xs\right)^2 + 1\right)^2}$$

$$G_{23} = \frac{\left(\frac{d}{dt}\,xs\right)^2 \left(6\,f^2 \left(\frac{d}{d\,x_2}\,f\right)\left(\frac{d}{d\,x_3}\,f\right)\left(\frac{d}{dt}\,xs\right)^2 - 9\,f^3 \left(\frac{d^2}{d\,x_2\,d\,x_3}\,f\right)\left(\frac{d}{dt}\,xs\right)^2 - \frac{d}{d\,x_2}\,f\left(\frac{d}{d\,x_3}\,f\right) - 3\,f\left(\frac{d^2}{d\,x_2\,d\,x_3}\,f\right)\right)}{\left(3\,f^2\left(\frac{d}{dt}\,xs\right)^2+1\right)^2}$$

$$G_{30} = G_{03}$$

$$G_{31} = G_{13}$$

$$G_{32} = G_{23}$$

$$G_{33} = -\frac{2\left(\frac{d}{d\,x_1}\,f\right)\left(\frac{d^2}{d\,t^2}\,xs\right) - 3\,f^2\left(\frac{d}{d\,x_3}\,f\right)^2\left(\frac{d}{dt}\,xs\right)^4 - 9\,f^3\left(\frac{d^2}{d\,x_2{}^2}\,f\right)\left(\frac{d}{dt}\,xs\right)^4 + 3\,f^2\left(\frac{d}{d\,x_2}\,f\right)^2\left(\frac{d}{dt}\,xs\right)^4 + 3\,f^3\left(\frac{d^2}{d\,x_1{}^2}\,f\right)\left(\frac{d}{dt}\,xs\right)^4 - \ldots}{\left(3\,f^2\left(\frac{d}{dt}\,xs\right)^2+1\right)^2}$$

Hodge Dual of Bianchi Identity

——— (see charge and current densities)

Scalar Charge Density ($-R^{0}{}_{i}{}^{i0}$)

$$\rho = -\frac{2\left(\frac{d}{d\,x_1}\,f\right)\left(\frac{d^2}{d\,t^2}\,xs\right) - 9\,f^3\left(\frac{d^2}{d\,x_3{}^2}\,f\right)\left(\frac{d}{dt}\,xs\right)^4 + 6\,f^2\left(\frac{d}{d\,x_3}\,f\right)^2\left(\frac{d}{dt}\,xs\right)^4 - 9\,f^3\left(\frac{d^2}{d\,x_2{}^2}\,f\right)\left(\frac{d}{dt}\,xs\right)^4 + 6\,f^2\left(\frac{d}{d\,x_2}\,f\right)^2\left(\frac{d}{dt}\,xs\right)^4 + \ldots}{\left(3\,f^2\left(\frac{d}{dt}\,xs\right)^2+1\right)^3}$$

Current Density Class 1 ($-R^{i}{}_{\mu}{}^{\mu j}$)

$$J_1 = \frac{2\,f^2\left(\frac{d}{d\,x_1}\,f\right)\left(\frac{d}{dt}\,xs\right)^2\left(\frac{d^2}{d\,t^2}\,xs\right) - 2\left(\frac{d}{d\,x_1}\,f\right)\left(\frac{d^2}{d\,t^2}\,xs\right) + 6\,f^4\left(\frac{d}{d\,x_3}\,f\right)^2\left(\frac{d}{dt}\,xs\right)^6 + 6\,f^4\left(\frac{d}{d\,x_2}\,f\right)^2\left(\frac{d}{dt}\,xs\right)^6 + 3\,f^5\left(\frac{d^2}{d\,x_1{}^2}\,f\right)\left(\frac{d}{dt}\,xs\right)^6 - \ldots}{\left(3\,f^2\left(\frac{d}{dt}\,xs\right)^2+1\right)^3}$$

$$J_2 = \frac{\left(\frac{d}{dt}\,xs\right)^2\left(9\,f^3\left(\frac{d^2}{d\,x_2{}^2}\,f\right)\left(\frac{d}{dt}\,xs\right)^2 - 6\,f^2\left(\frac{d}{d\,x_2}\,f\right)^2\left(\frac{d}{dt}\,xs\right)^2 + 3\,f\left(\frac{d^2}{d\,x_2{}^2}\,f\right) + \left(\frac{d}{d\,x_2}\,f\right)^2\right)}{\left(3\,f^2\left(\frac{d}{dt}\,xs\right)^2+1\right)^2}$$

$$J_3 = \frac{\left(\frac{d}{dt}\,xs\right)^2\left(9\,f^3\left(\frac{d^2}{d\,x_3{}^2}\,f\right)\left(\frac{d}{dt}\,xs\right)^2 - 6\,f^2\left(\frac{d}{d\,x_3}\,f\right)^2\left(\frac{d}{dt}\,xs\right)^2 + 3\,f\left(\frac{d^2}{d\,x_3{}^2}\,f\right) + \left(\frac{d}{d\,x_3}\,f\right)^2\right)}{\left(3\,f^2\left(\frac{d}{dt}\,xs\right)^2+1\right)^2}$$

Current Density Class 2 ($-R^{i}{}_{\mu}{}^{\mu j}$)

$$J_1 = -\frac{\frac{d}{d\,x_3}\,f\left(\frac{d^2}{d\,t^2}\,xs\right) + 3\,f^3\left(\frac{d^2}{d\,x_1\,d\,x_3}\,f\right)\left(\frac{d}{dt}\,xs\right)^4 - 3\,f\left(\frac{d}{dt}\,f\right)\left(\frac{d}{d\,x_3}\,f\right)\left(\frac{d}{dt}\,xs\right)^3 + 3\,f^2\left(\frac{d^2}{d\,t\,d\,x_3}\,f\right)\left(\frac{d}{dt}\,xs\right)^3 + \frac{d}{d\,x_1}\,f\left(\frac{d}{d\,x_3}\,f\right)\left(\frac{d}{dt}\,xs\right)^2 + \ldots}{\left(3\,f^2\left(\frac{d}{dt}\,xs\right)^2+1\right)^2}$$

$$J_2 = -\frac{\frac{d}{d\,x_2}\,f\left(\frac{d^2}{d\,t^2}\,xs\right) + 3\,f^3\left(\frac{d^2}{d\,x_1\,d\,x_2}\,f\right)\left(\frac{d}{dt}\,xs\right)^4 - 3\,f\left(\frac{d}{dt}\,f\right)\left(\frac{d}{d\,x_2}\,f\right)\left(\frac{d}{dt}\,xs\right)^3 + 3\,f^2\left(\frac{d^2}{d\,t\,d\,x_2}\,f\right)\left(\frac{d}{dt}\,xs\right)^3 + \frac{d}{d\,x_1}\,f\left(\frac{d}{d\,x_2}\,f\right)\left(\frac{d}{dt}\,xs\right)^2 + \ldots}{\left(3\,f^2\left(\frac{d}{dt}\,xs\right)^2+1\right)^2}$$

$$J_3 = -\frac{\left(\frac{d}{dt}\,xs\right)^2 \left(6\,f^2 \left(\frac{d}{d\,x_2}\,f\right)\left(\frac{d}{d\,x_3}\,f\right)\left(\frac{d}{dt}\,xs\right)^2 - 9\,f^3 \left(\frac{d^2}{d\,x_2\,d\,x_3}\,f\right)\left(\frac{d}{dt}\,xs\right)^2 - \frac{d}{d\,x_2}\,f\left(\frac{d}{d\,x_3}\,f\right) - 3\,f\left(\frac{d^2}{d\,x_2\,d\,x_3}\,f\right)\right)}{\left(3\,f^2\left(\frac{d}{dt}\,xs\right)^2+1\right)^2}$$

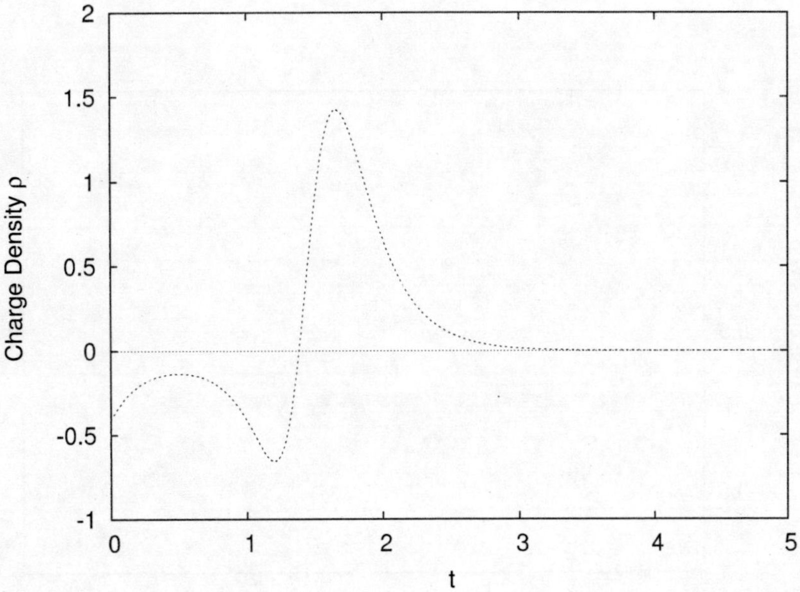

Fig. 4.74: Alcubierre metric, charge density ρ, time dependence for $x_1 = 1, x_2 = 0, x_3 = 0$.

Current Density Class 3 (-$R^{i}{}_{\mu}{}^{\mu j}$)

$$J_1 = -\frac{\frac{d}{d\,x_2} f \left(\frac{d^2}{d\,t^2}\,xs\right) + 3\,f^3 \left(\frac{d^2}{d\,x_1\,d\,x_2}\,f\right) \left(\frac{d}{d\,t}\,xs\right)^4 - 3\,f \left(\frac{d}{d\,t}\,f\right) \left(\frac{d}{d\,x_2}\,f\right) \left(\frac{d}{d\,t}\,xs\right)^3 + 3\,f^2 \left(\frac{d^2}{d\,t\,d\,x_2}\,f\right) \left(\frac{d}{d\,t}\,xs\right)^3 + \frac{d}{d\,x_1}\,f \left(\frac{d}{d\,x_2}\,f\right) \left(\frac{d}{d\,t}\,xs\right)^2 + \ldots}{\left(3\,f^2 \left(\frac{d}{d\,t}\,xs\right)^2 + 1\right)^2}$$

$$J_2 = -\frac{\left(\frac{d}{d\,t}\,xs\right)^2 \left(6\,f^2 \left(\frac{d}{d\,x_2}\,f\right) \left(\frac{d}{d\,x_3}\,f\right) \left(\frac{d}{d\,t}\,xs\right)^2 - 9\,f^3 \left(\frac{d^2}{d\,x_2\,d\,x_3}\,f\right) \left(\frac{d}{d\,t}\,xs\right)^2 - \frac{d}{d\,x_2}\,f \left(\frac{d}{d\,x_3}\,f\right) - 3\,f \left(\frac{d^2}{d\,x_2\,d\,x_3}\,f\right)\right)}{\left(3\,f^2 \left(\frac{d}{d\,t}\,xs\right)^2 + 1\right)^2}$$

$$J_3 = -\frac{\frac{d}{d\,x_3} f \left(\frac{d^2}{d\,t^2}\,xs\right) + 3\,f^3 \left(\frac{d^2}{d\,x_1\,d\,x_3}\,f\right) \left(\frac{d}{d\,t}\,xs\right)^4 - 3\,f \left(\frac{d}{d\,t}\,f\right) \left(\frac{d}{d\,x_3}\,f\right) \left(\frac{d}{d\,t}\,xs\right)^3 + 3\,f^2 \left(\frac{d^2}{d\,t\,d\,x_3}\,f\right) \left(\frac{d}{d\,t}\,xs\right)^3 + \frac{d}{d\,x_1}\,f \left(\frac{d}{d\,x_3}\,f\right) \left(\frac{d}{d\,t}\,xs\right)^2 + \ldots}{\left(3\,f^2 \left(\frac{d}{d\,t}\,xs\right)^2 + 1\right)^2}$$

4.4.42 Homogeneous space-time

This metric is a generalization of the Robertson-Walker metric. In addition to the coordinates, it contains functions $\Sigma(x,k) =: \Sigma$ and $\Sigma(z,k') =: \Theta$.

Coordinates

$$\mathbf{x} = \begin{pmatrix} t \\ x \\ y \\ z \end{pmatrix}$$

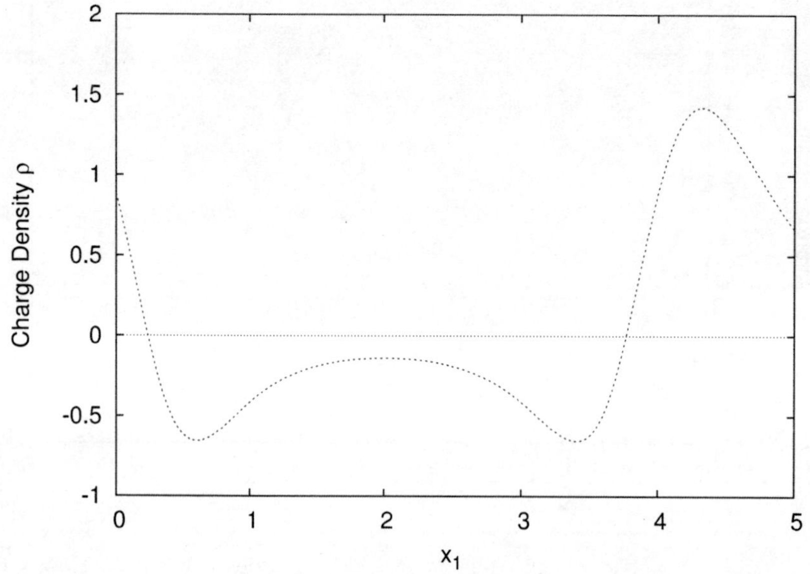

Fig. 4.75: Alcubierre metric, charge density, x_1 dependence for $t = 1, x_2 = 0, x_3 = 0$.

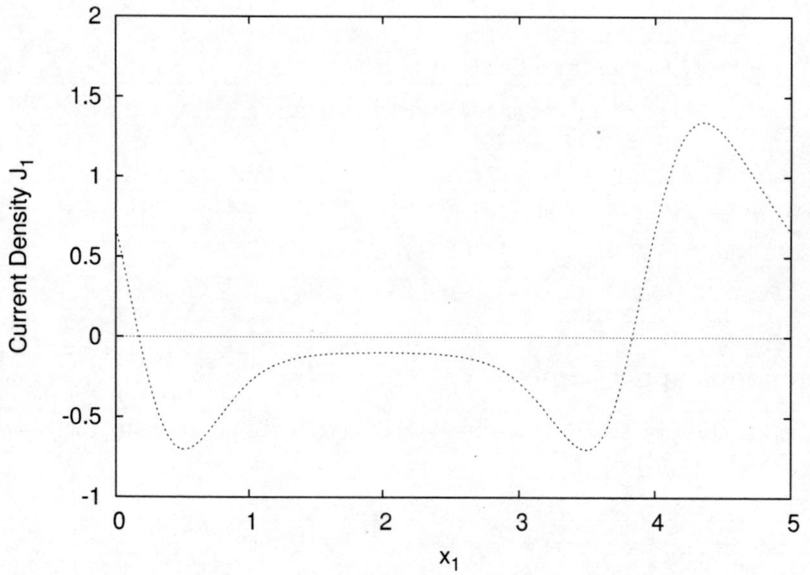

Fig. 4.76: Alcubierre metric, current density J_1, x_1 dependence for $t = 1, x_2 = 0, x_3 = 0$.

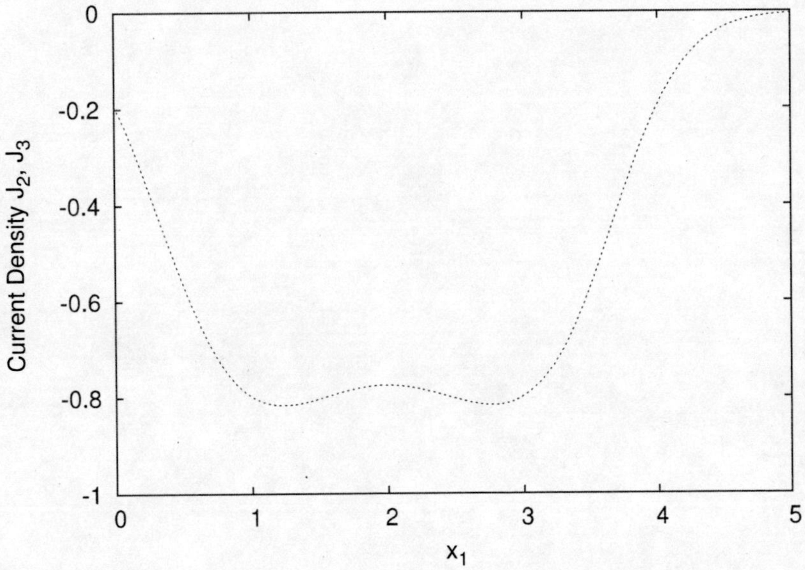

Fig. 4.77: Alcubierre metric, current density J_2, J_3, x_1 dependence for $t = 1, x_2 = 0, x_3 = 0$.

Metric

$$g_{\mu\nu} = \begin{pmatrix} -\Theta^2\, B^2 & 0 & 0 & 0 \\ 0 & A^2 & 0 & 0 \\ 0 & 0 & \Sigma^2\, A^2 & 0 \\ 0 & 0 & 0 & B^2 \end{pmatrix}$$

Contravariant Metric

$$g^{\mu\nu} = \begin{pmatrix} -\frac{1}{\Theta^2\, B^2} & 0 & 0 & 0 \\ 0 & \frac{1}{A^2} & 0 & 0 \\ 0 & 0 & \frac{1}{\Sigma^2\, A^2} & 0 \\ 0 & 0 & 0 & \frac{1}{B^2} \end{pmatrix}$$

Christoffel Connection

$$\Gamma^0{}_{03} = \frac{\frac{d}{d\,z}\,\Theta}{\Theta}$$

$$\Gamma^0{}_{30} = \Gamma^0{}_{03}$$

$$\Gamma^1{}_{22} = -\Sigma\,\left(\frac{d}{d\,x}\,\Sigma\right)$$

$$\Gamma^2{}_{12} = \frac{\frac{d}{d\,x}\,\Sigma}{\Sigma}$$

$$\Gamma^2{}_{21} = \Gamma^2{}_{12}$$

$$\Gamma^3{}_{00} = \Theta\,\left(\frac{d}{d\,z}\,\Theta\right)$$

Metric Compatibility

——— o.k.

Riemann Tensor

$$R^0{}_{303} = -\frac{\frac{d^2}{dz^2}\Theta}{\Theta}$$

$$R^0{}_{330} = -R^0{}_{303}$$

$$R^1{}_{212} = -\Sigma\left(\frac{d^2}{dx^2}\Sigma\right)$$

$$R^1{}_{221} = -R^1{}_{212}$$

$$R^2{}_{112} = \frac{\frac{d^2}{dx^2}\Sigma}{\Sigma}$$

$$R^2{}_{121} = -R^2{}_{112}$$

$$R^3{}_{003} = -\Theta\left(\frac{d^2}{dz^2}\Theta\right)$$

$$R^3{}_{030} = -R^3{}_{003}$$

Ricci Tensor

$$\mathrm{Ric}_{00} = \Theta\left(\frac{d^2}{dz^2}\Theta\right)$$

$$\mathrm{Ric}_{11} = -\frac{\frac{d^2}{dx^2}\Sigma}{\Sigma}$$

$$\mathrm{Ric}_{22} = -\Sigma\left(\frac{d^2}{dx^2}\Sigma\right)$$

$$\mathrm{Ric}_{33} = -\frac{\frac{d^2}{dz^2}\Theta}{\Theta}$$

Ricci Scalar

$$R_{sc} = -\frac{2\left(\frac{d^2}{dx^2}\Sigma\,\Theta\,B^2 + \Sigma\left(\frac{d^2}{dz^2}\Theta\right)A^2\right)}{\Sigma\,\Theta\,A^2\,B^2}$$

Bianchi identity (Ricci cyclic equation $R^{\kappa}{}_{[\mu\nu\sigma]} = 0$)

——— o.k.

Einstein Tensor

$$G_{00} = -\frac{\frac{d^2}{d\,x^2}\,\Sigma\,\Theta^2\,B^2}{\Sigma\,A^2}$$

$$G_{11} = \frac{\frac{d^2}{d\,z^2}\,\Theta\,A^2}{\Theta\,B^2}$$

$$G_{22} = \frac{\Sigma^2\left(\frac{d^2}{d\,z^2}\,\Theta\right)\,A^2}{\Theta\,B^2}$$

$$G_{33} = \frac{\frac{d^2}{d\,x^2}\,\Sigma\,B^2}{\Sigma\,A^2}$$

Hodge Dual of Bianchi Identity

——— (see charge and current densities)

Scalar Charge Density (-$R^{0}{}_{i}{}^{i0}$)

$$\rho = \frac{\frac{d^2}{d\,z^2}\,\Theta}{\Theta^3\,B^4}$$

Current Density Class 1 (-$R^{i}{}_{\mu}{}^{\mu j}$)

$$J_1 = \frac{\frac{d^2}{d\,x^2}\,\Sigma}{\Sigma\,A^4}$$

$$J_2 = \frac{\frac{d^2}{d\,x^2}\,\Sigma}{\Sigma^3\,A^4}$$

$$J_3 = \frac{\frac{d^2}{d\,z^2}\,\Theta}{\Theta\,B^4}$$

Current Density Class 2 (-$R^{i}{}_{\mu}{}^{\mu j}$)

$$J_1 = 0$$

$$J_2 = 0$$

$$J_3 = 0$$

Current Density Class 3 (-$R^{i}{}_{\mu}{}^{\mu j}$)

$$J_1 = 0$$

$$J_2 = 0$$

$$J_3 = 0$$

4.4.43 Robertson-Walker metric

$a(t)$ is a time-dependent scaing factor, $\Sigma(r,k) =: \Sigma$ depends on k which determines if the universe is expanding, static or contracting.

Coordinates

$$\mathbf{x} = \begin{pmatrix} t \\ r \\ \vartheta \\ \varphi \end{pmatrix}$$

Metric

$$g_{\mu\nu} = \begin{pmatrix} -1 & 0 & 0 & 0 \\ 0 & a^2 & 0 & 0 \\ 0 & 0 & a^2\,\Sigma^2 & 0 \\ 0 & 0 & 0 & a^2\,\Sigma^2\,\sin^2\vartheta \end{pmatrix}$$

Contravariant Metric

$$g^{\mu\nu} = \begin{pmatrix} -1 & 0 & 0 & 0 \\ 0 & \frac{1}{a^2} & 0 & 0 \\ 0 & 0 & \frac{1}{a^2\,\Sigma^2} & 0 \\ 0 & 0 & 0 & \frac{1}{a^2\,\Sigma^2\,\sin^2\vartheta} \end{pmatrix}$$

Christoffel Connection

$$\Gamma^{0}{}_{11} = a\left(\frac{d}{dt}\,a\right)$$

$$\Gamma^{0}{}_{22} = a\left(\frac{d}{dt}\,a\right)\,\Sigma^2$$

$$\Gamma^{0}{}_{33} = a\left(\frac{d}{dt}\,a\right)\,\Sigma^2\,\sin^2\vartheta$$

$$\Gamma^{1}{}_{01} = \frac{\frac{d}{dt}\,a}{a}$$

$$\Gamma^{1}{}_{10} = \Gamma^{1}{}_{01}$$

$$\Gamma^1{}_{22} = -\Sigma \left(\frac{d}{d\,r} \Sigma \right)$$

$$\Gamma^1{}_{33} = -\Sigma \left(\frac{d}{d\,r} \Sigma \right) \sin^2 \vartheta$$

$$\Gamma^2{}_{02} = \frac{\frac{d}{d\,t} a}{a}$$

$$\Gamma^2{}_{12} = \frac{\frac{d}{d\,r} \Sigma}{\Sigma}$$

$$\Gamma^2{}_{20} = \Gamma^2{}_{02}$$

$$\Gamma^2{}_{21} = \Gamma^2{}_{12}$$

$$\Gamma^2{}_{33} = -\cos \vartheta \sin \vartheta$$

$$\Gamma^3{}_{03} = \frac{\frac{d}{d\,t} a}{a}$$

$$\Gamma^3{}_{13} = \frac{\frac{d}{d\,r} \Sigma}{\Sigma}$$

$$\Gamma^3{}_{23} = \frac{\cos \vartheta}{\sin \vartheta}$$

$$\Gamma^3{}_{30} = \Gamma^3{}_{03}$$

$$\Gamma^3{}_{31} = \Gamma^3{}_{13}$$

$$\Gamma^3{}_{32} = \Gamma^3{}_{23}$$

Metric Compatibility

——— o.k.

Riemann Tensor

$$R^0{}_{101} = a \left(\frac{d^2}{d\,t^2} a \right)$$

$$R^0{}_{110} = -R^0{}_{101}$$

$$R^0{}_{202} = a \left(\frac{d^2}{d\,t^2} a \right) \Sigma^2$$

$$R^{0}{}_{220} = -R^{0}{}_{202}$$

$$R^{0}{}_{303} = a\left(\frac{d^2}{d\,t^2}\,a\right)\Sigma^2\sin^2\vartheta$$

$$R^{0}{}_{330} = -R^{0}{}_{303}$$

$$R^{1}{}_{001} = \frac{\frac{d^2}{d\,t^2}\,a}{a}$$

$$R^{1}{}_{010} = -R^{1}{}_{001}$$

$$R^{1}{}_{212} = -\Sigma\left(\frac{d^2}{d\,r^2}\,\Sigma - \left(\frac{d}{d\,t}\,a\right)^2\Sigma\right)$$

$$R^{1}{}_{221} = -R^{1}{}_{212}$$

$$R^{1}{}_{313} = -\Sigma\left(\frac{d^2}{d\,r^2}\,\Sigma - \left(\frac{d}{d\,t}\,a\right)^2\Sigma\right)\sin^2\vartheta$$

$$R^{1}{}_{331} = -R^{1}{}_{313}$$

$$R^{2}{}_{002} = \frac{\frac{d^2}{d\,t^2}\,a}{a}$$

$$R^{2}{}_{020} = -R^{2}{}_{002}$$

$$R^{2}{}_{112} = \frac{\frac{d^2}{d\,r^2}\,\Sigma - \left(\frac{d}{d\,t}\,a\right)^2\Sigma}{\Sigma}$$

$$R^{2}{}_{121} = -R^{2}{}_{112}$$

$$R^{2}{}_{323} = -\left(\left(\frac{d}{d\,r}\,\Sigma\right)^2 - \left(\frac{d}{d\,t}\,a\right)^2\Sigma^2 - 1\right)\sin^2\vartheta$$

$$R^{2}{}_{332} = -R^{2}{}_{323}$$

$$R^{3}{}_{003} = \frac{\frac{d^2}{d\,t^2}\,a}{a}$$

$$R^{3}{}_{030} = -R^{3}{}_{003}$$

$$R^{3}{}_{113} = \frac{\frac{d^2}{d\,r^2}\,\Sigma - \left(\frac{d}{d\,t}\,a\right)^2\Sigma}{\Sigma}$$

$$R^{3}{}_{131} = -R^{3}{}_{113}$$

$$R^{3}{}_{223} = \left(\frac{d}{d\,r}\,\Sigma\right)^2 - \left(\frac{d}{d\,t}\,a\right)^2\Sigma^2 - 1$$

$$R^{3}{}_{232} = -R^{3}{}_{223}$$

Ricci Tensor

$$\mathrm{Ric}_{00} = -\frac{3\left(\frac{d^2}{d\,t^2}\,a\right)}{a}$$

$$\mathrm{Ric}_{11} = -\frac{2\left(\frac{d^2}{d\,r^2}\,\Sigma\right) - a\left(\frac{d^2}{d\,t^2}\,a\right)\Sigma - 2\left(\frac{d}{d\,t}\,a\right)^2\Sigma}{\Sigma}$$

$$\mathrm{Ric}_{22} = -\left(\Sigma\left(\frac{d^2}{d\,r^2}\,\Sigma\right) + \left(\frac{d}{d\,r}\,\Sigma\right)^2 - a\left(\frac{d^2}{d\,t^2}\,a\right)\Sigma^2 - 2\left(\frac{d}{d\,t}\,a\right)^2\Sigma^2 - 1\right)$$

$$\mathrm{Ric}_{33} = -\left(\Sigma\left(\frac{d^2}{d\,r^2}\,\Sigma\right) + \left(\frac{d}{d\,r}\,\Sigma\right)^2 - a\left(\frac{d^2}{d\,t^2}\,a\right)\Sigma^2 - 2\left(\frac{d}{d\,t}\,a\right)^2\Sigma^2 - 1\right)\sin^2\vartheta$$

Ricci Scalar

$$R_{sc} = -\frac{2\left(2\,\Sigma\left(\frac{d^2}{d\,r^2}\,\Sigma\right) + \left(\frac{d}{d\,r}\,\Sigma\right)^2 - 3\,a\left(\frac{d^2}{d\,t^2}\,a\right)\Sigma^2 - 3\left(\frac{d}{d\,t}\,a\right)^2\Sigma^2 - 1\right)}{a^2\,\Sigma^2}$$

Bianchi identity (Ricci cyclic equation $R^{\kappa}{}_{[\mu\nu\sigma]} = 0$)

——— o.k.

Einstein Tensor

$$G_{00} = -\frac{2\,\Sigma\left(\frac{d^2}{d\,r^2}\,\Sigma\right) + \left(\frac{d}{d\,r}\,\Sigma\right)^2 - 3\left(\frac{d}{d\,t}\,a\right)^2\Sigma^2 - 1}{a^2\,\Sigma^2}$$

$$G_{11} = \frac{\left(\frac{d}{d\,r}\,\Sigma\right)^2 - 2\,a\left(\frac{d^2}{d\,t^2}\,a\right)\Sigma^2 - \left(\frac{d}{d\,t}\,a\right)^2\Sigma^2 - 1}{\Sigma^2}$$

$$G_{22} = \Sigma\left(\frac{d^2}{d\,r^2}\,\Sigma - 2\,a\left(\frac{d^2}{d\,t^2}\,a\right)\Sigma - \left(\frac{d}{d\,t}\,a\right)^2\Sigma\right)$$

$$G_{33} = \Sigma\left(\frac{d^2}{d\,r^2}\,\Sigma - 2\,a\left(\frac{d^2}{d\,t^2}\,a\right)\Sigma - \left(\frac{d}{d\,t}\,a\right)^2\Sigma\right)\sin^2\vartheta$$

Hodge Dual of Bianchi Identity

——— (see charge and current densities)

Scalar Charge Density ($-R^{0}{}_{i}{}^{i0}$)

$$\rho = -\frac{3\left(\frac{d^2}{d\,t^2}\,a\right)}{a}$$

Current Density Class 1 ($-R^{i}{}_{\mu}{}^{\mu j}$)

$$J_1 = \frac{2\left(\frac{d^2}{d\,r^2}\,\Sigma\right) - a\left(\frac{d^2}{d\,t^2}\,a\right)\Sigma - 2\left(\frac{d}{d\,t}\,a\right)^2\Sigma}{a^4\,\Sigma}$$

$$J_2 = \frac{\Sigma\left(\frac{d^2}{d\,r^2}\,\Sigma\right) + \left(\frac{d}{d\,r}\,\Sigma\right)^2 - a\left(\frac{d^2}{d\,t^2}\,a\right)\Sigma^2 - 2\left(\frac{d}{d\,t}\,a\right)^2\Sigma^2 - 1}{a^4\,\Sigma^4}$$

$$J_3 = \frac{\Sigma\left(\frac{d^2}{d\,r^2}\,\Sigma\right) + \left(\frac{d}{d\,r}\,\Sigma\right)^2 - a\left(\frac{d^2}{d\,t^2}\,a\right)\Sigma^2 - 2\left(\frac{d}{d\,t}\,a\right)^2\Sigma^2 - 1}{a^4\,\Sigma^4\,\sin^2\vartheta}$$

Current Density Class 2 ($-R^{i}{}_{\mu}{}^{\mu j}$)

$$J_1 = 0$$

$$J_2 = 0$$

$$J_3 = 0$$

Current Density Class 3 ($-R^{i}{}_{\mu}{}^{\mu j}$)

$$J_1 = 0$$

$$J_2 = 0$$

$$J_3 = 0$$

4.4.44 Anti-Mach metric

Anti-Mach metric of plane waves of homogeneous vacuum. This is a vacuum metric.

Coordinates

$$\mathbf{x} = \begin{pmatrix} u \\ v \\ x \\ y \end{pmatrix}$$

Metric

$$g_{\mu\nu} = \begin{pmatrix} -2\left(\cos(2\,u)\left(x^2 - y^2\right) - 2\sin(2\,u)\,x\,y\right) & -2 & 0 & 0 \\ -2 & 0 & 0 & 0 \\ 0 & 0 & 1 & 0 \\ 0 & 0 & 0 & 1 \end{pmatrix}$$

Contravariant Metric

$$g^{\mu\nu} = \begin{pmatrix} 0 & -\frac{1}{2} & 0 & 0 \\ -\frac{1}{2} & -\frac{\cos(2\,u)\,y^2+2\,\sin(2\,u)\,x\,y-\cos(2\,u)\,x^2}{2} & 0 & 0 \\ 0 & 0 & 1 & 0 \\ 0 & 0 & 0 & 1 \end{pmatrix}$$

Christoffel Connection

$$\Gamma^1{}_{00} = \sin(2\,u)\;y^2 - 2\,\cos(2\,u)\;x\,y - \sin(2\,u)\;x^2$$

$$\Gamma^1{}_{02} = -\left(\sin(2\,u)\;y - \cos(2\,u)\;x\right)$$

$$\Gamma^1{}_{03} = -\left(\cos(2\,u)\;y + \sin(2\,u)\;x\right)$$

$$\Gamma^1{}_{20} = \Gamma^1{}_{02}$$

$$\Gamma^1{}_{30} = \Gamma^1{}_{03}$$

$$\Gamma^2{}_{00} = -2\,\left(\sin(2\,u)\;y - \cos(2\,u)\;x\right)$$

$$\Gamma^3{}_{00} = -2\,\left(\cos(2\,u)\;y + \sin(2\,u)\;x\right)$$

Metric Compatibility

——— o.k.

Riemann Tensor

$$R^1{}_{202} = -\cos(2\,u)$$

$$R^1{}_{203} = \sin(2\,u)$$

$$R^1{}_{220} = -R^1{}_{202}$$

$$R^1{}_{230} = -R^1{}_{203}$$

$$R^1{}_{302} = \sin(2\,u)$$

$$R^1{}_{303} = \cos(2\,u)$$

$$R^1{}_{320} = -R^1{}_{302}$$

$$R^1{}_{330} = -R^1{}_{303}$$

$$R^2{}_{002} = -2\,\cos(2\,u)$$

$$R^2{}_{003} = 2\sin(2u)$$

$$R^2{}_{020} = -R^2{}_{002}$$

$$R^2{}_{030} = -R^2{}_{003}$$

$$R^3{}_{002} = 2\sin(2u)$$

$$R^3{}_{003} = 2\cos(2u)$$

$$R^3{}_{020} = -R^3{}_{002}$$

$$R^3{}_{030} = -R^3{}_{003}$$

Ricci Tensor

——— all elements zero

Ricci Scalar

$$R_{sc} = 0$$

Bianchi identity (Ricci cyclic equation $R^{\kappa}{}_{[\mu\nu\sigma]} = 0$)

——— o.k.

Einstein Tensor

——— all elements zero

Hodge Dual of Bianchi Identity

——— (see charge and current densities)

Scalar Charge Density (-$R^0{}_i{}^{i0}$)

$$\rho = 0$$

Current Density Class 1 (-$R^i{}_\mu{}^{\mu j}$)

$$J_1 = 0$$

$$J_2 = 0$$

$$J_3 = 0$$

Current Density Class 2 (-$R^{i}{}_{\mu}{}^{\mu j}$)

$$J_1 = 0$$

$$J_2 = 0$$

$$J_3 = 0$$

Current Density Class 3 (-$R^{i}{}_{\mu}{}^{\mu j}$)

$$J_1 = 0$$

$$J_2 = 0$$

$$J_3 = 0$$

4.4.45 Petrov metric

This metric is a special case of the Anti-Mach metric of plane waves.

Coordinates

$$\mathbf{x} = \begin{pmatrix} t \\ x \\ y \\ z \end{pmatrix}$$

Metric

$$g_{\mu\nu} = \begin{pmatrix} -e^{x}\cos\left(\sqrt{3}\,x\right) & 0 & 0 & -2\sin\left(\sqrt{3}\,x\right) \\ 0 & 1 & 0 & 0 \\ 0 & 0 & e^{-2\,x} & 0 \\ -2\sin\left(\sqrt{3}\,x\right) & 0 & 0 & e^{x}\cos\left(\sqrt{3}\,x\right) \end{pmatrix}$$

Contravariant Metric

$$g^{\mu\nu} = \begin{pmatrix} -\frac{e^{x}\cos(\sqrt{3}\,x)}{4\sin^2(\sqrt{3}\,x)+e^{2\,x}\cos^2(\sqrt{3}\,x)} & 0 & 0 & -\frac{2\sin(\sqrt{3}\,x)}{4\sin^2(\sqrt{3}\,x)+e^{2\,x}\cos^2(\sqrt{3}\,x)} \\ 0 & 1 & 0 & 0 \\ 0 & 0 & e^{2\,x} & 0 \\ -\frac{2\sin(\sqrt{3}\,x)}{4\sin^2(\sqrt{3}\,x)+e^{2\,x}\cos^2(\sqrt{3}\,x)} & 0 & 0 & \frac{e^{x}\cos(\sqrt{3}\,x)}{4\sin^2(\sqrt{3}\,x)+e^{2\,x}\cos^2(\sqrt{3}\,x)} \end{pmatrix}$$

Christoffel Connection

$$\Gamma^{0}{}_{01} = -\frac{\cos\left(\sqrt{3}\,x\right)\left(\sqrt{3}\,e^{2\,x}\sin\left(\sqrt{3}\,x\right) - 4\sqrt{3}\sin\left(\sqrt{3}\,x\right) - e^{2\,x}\cos\left(\sqrt{3}\,x\right)\right)}{2\left(e^{2\,x}\cos^2\left(\sqrt{3}\,x\right) - 4\cos^2\left(\sqrt{3}\,x\right) + 4\right)}$$

$$\Gamma^{0}{}_{10} = \Gamma^{0}{}_{01}$$

$$\Gamma^{0}{}_{13} = -\frac{e^{x}\left(\cos\left(\sqrt{3}\,x\right)\sin\left(\sqrt{3}\,x\right) - \sqrt{3}\right)}{e^{2\,x}\cos^2\left(\sqrt{3}\,x\right) - 4\cos^2\left(\sqrt{3}\,x\right) + 4}$$

$$\Gamma^0{}_{31} = \Gamma^0{}_{13}$$

$$\Gamma^1{}_{00} = -\frac{e^x \left(\sqrt{3}\sin\left(\sqrt{3}\,x\right) - \cos\left(\sqrt{3}\,x\right)\right)}{2}$$

$$\Gamma^1{}_{03} = \sqrt{3}\cos\left(\sqrt{3}\,x\right)$$

$$\Gamma^1{}_{22} = e^{-2x}$$

$$\Gamma^1{}_{30} = \Gamma^1{}_{03}$$

$$\Gamma^1{}_{33} = \frac{e^x \left(\sqrt{3}\sin\left(\sqrt{3}\,x\right) - \cos\left(\sqrt{3}\,x\right)\right)}{2}$$

$$\Gamma^2{}_{12} = -1$$

$$\Gamma^2{}_{21} = \Gamma^2{}_{12}$$

$$\Gamma^3{}_{01} = \frac{e^x \left(\cos\left(\sqrt{3}\,x\right)\sin\left(\sqrt{3}\,x\right) - \sqrt{3}\right)}{e^{2x}\cos^2\left(\sqrt{3}\,x\right) - 4\cos^2\left(\sqrt{3}\,x\right) + 4}$$

$$\Gamma^3{}_{10} = \Gamma^3{}_{01}$$

$$\Gamma^3{}_{13} = -\frac{\cos\left(\sqrt{3}\,x\right)\left(\sqrt{3}\,e^{2x}\sin\left(\sqrt{3}\,x\right) - 4\sqrt{3}\sin\left(\sqrt{3}\,x\right) - e^{2x}\cos\left(\sqrt{3}\,x\right)\right)}{2\left(e^{2x}\cos^2\left(\sqrt{3}\,x\right) - 4\cos^2\left(\sqrt{3}\,x\right) + 4\right)}$$

$$\Gamma^3{}_{31} = \Gamma^3{}_{13}$$

Metric Compatibility

——— o.k.

Riemann Tensor

$$R^0{}_{003} = \frac{\sin\left(\sqrt{3}\,x\right)\left(2e^{2x}\sin^2\left(\sqrt{3}\,x\right) - 12\sin^2\left(\sqrt{3}\,x\right) - 2\sqrt{3}\,e^{2x}\cos\left(\sqrt{3}\,x\right)\sin\left(\sqrt{3}\,x\right) + e^{2x} + 12\right)}{2\left(e^{2x}\cos^2\left(\sqrt{3}\,x\right) - 4\cos^2\left(\sqrt{3}\,x\right) + 4\right)}$$

$$R^0{}_{030} = -R^0{}_{003}$$

$$R^0{}_{101} = \frac{2\sqrt{3}\,e^{4x}\cos^3\left(\sqrt{3}\,x\right)\sin\left(\sqrt{3}\,x\right) - 8\sqrt{3}\,e^{2x}\cos^3\left(\sqrt{3}\,x\right)\sin\left(\sqrt{3}\,x\right) + 24\sqrt{3}\,e^{2x}\cos\left(\sqrt{3}\,x\right)\sin\left(\sqrt{3}\,x\right) + 2e^{4x}\cos^4\left(\sqrt{3}\,x\right) - \ldots}{4\left(e^{2x}\cos^2\left(\sqrt{3}\,x\right) - 4\cos^2\left(\sqrt{3}\,x\right) + 4\right)^2}$$

$$R^0{}_{110} = -R^0{}_{101}$$

$$R^0{}_{113} = \frac{e^x\left(4\cos^3\left(\sqrt{3}\,x\right)\sin\left(\sqrt{3}\,x\right) + 3e^{2x}\cos\left(\sqrt{3}\,x\right)\sin\left(\sqrt{3}\,x\right) - 16\cos\left(\sqrt{3}\,x\right)\sin\left(\sqrt{3}\,x\right) - \sqrt{3}\,e^{2x}\cos^4\left(\sqrt{3}\,x\right) + 4\sqrt{3}\cos^4\left(\sqrt{3}\,x\right) - \ldots\right)}{\left(e^{2x}\cos^2\left(\sqrt{3}\,x\right) - 4\cos^2\left(\sqrt{3}\,x\right) + 4\right)^2}$$

$$R^{0}{}_{131} = -R^{0}{}_{113}$$

$$R^{0}{}_{202} = -\frac{e^{-2x}\cos\left(\sqrt{3}\,x\right)\left(\sqrt{3}\,e^{2x}\sin\left(\sqrt{3}\,x\right) - 4\sqrt{3}\sin\left(\sqrt{3}\,x\right) - e^{2x}\cos\left(\sqrt{3}\,x\right)\right)}{2\left(e^{2x}\cos^2\left(\sqrt{3}\,x\right) - 4\cos^2\left(\sqrt{3}\,x\right) + 4\right)}$$

$$R^{0}{}_{220} = -R^{0}{}_{202}$$

$$R^{0}{}_{223} = \frac{e^{-x}\left(\cos\left(\sqrt{3}\,x\right)\sin\left(\sqrt{3}\,x\right) - \sqrt{3}\right)}{e^{2x}\cos^2\left(\sqrt{3}\,x\right) - 4\cos^2\left(\sqrt{3}\,x\right) + 4}$$

$$R^{0}{}_{232} = -R^{0}{}_{223}$$

$$R^{0}{}_{303} = \frac{e^{x}\cos\left(\sqrt{3}\,x\right)\left(2\sqrt{3}\,e^{2x}\cos\left(\sqrt{3}\,x\right)\sin\left(\sqrt{3}\,x\right) + 2\,e^{2x}\cos^2\left(\sqrt{3}\,x\right) - 12\cos^2\left(\sqrt{3}\,x\right) - 3\,e^{2x}\right)}{4\left(e^{2x}\cos^2\left(\sqrt{3}\,x\right) - 4\cos^2\left(\sqrt{3}\,x\right) + 4\right)}$$

$$R^{0}{}_{330} = -R^{0}{}_{303}$$

$$R^{1}{}_{001} = \frac{e^{x}\left(2\sqrt{3}\,e^{2x}\cos^2\left(\sqrt{3}\,x\right)\sin\left(\sqrt{3}\,x\right) - 8\sqrt{3}\cos^2\left(\sqrt{3}\,x\right)\sin\left(\sqrt{3}\,x\right) + 16\sqrt{3}\sin\left(\sqrt{3}\,x\right) + 2\,e^{2x}\cos^3\left(\sqrt{3}\,x\right) - 4\cos^3\left(\sqrt{3}\,x\right) + \ldots\right)}{4\left(e^{2x}\cos^2\left(\sqrt{3}\,x\right) - 4\cos^2\left(\sqrt{3}\,x\right) + 4\right)}$$

$$R^{1}{}_{010} = -R^{1}{}_{001}$$

$$R^{1}{}_{013} = \frac{2\,e^{2x}\sin^3\left(\sqrt{3}\,x\right) - 12\sin^3\left(\sqrt{3}\,x\right) + e^{2x}\sin\left(\sqrt{3}\,x\right) - 12\sin\left(\sqrt{3}\,x\right) - 2\sqrt{3}\,e^{2x}\cos\left(\sqrt{3}\,x\right)}{2\left(e^{2x}\cos^2\left(\sqrt{3}\,x\right) - 4\cos^2\left(\sqrt{3}\,x\right) + 4\right)}$$

$$R^{1}{}_{031} = -R^{1}{}_{013}$$

$$R^{1}{}_{212} = -e^{-2x}$$

$$R^{1}{}_{221} = -R^{1}{}_{212}$$

$$R^{1}{}_{301} = \frac{2\,e^{2x}\cos^2\left(\sqrt{3}\,x\right)\sin\left(\sqrt{3}\,x\right) - 12\cos^2\left(\sqrt{3}\,x\right)\sin\left(\sqrt{3}\,x\right) - 3\,e^{2x}\sin\left(\sqrt{3}\,x\right) + 24\sin\left(\sqrt{3}\,x\right) + 2\sqrt{3}\,e^{2x}\cos\left(\sqrt{3}\,x\right)}{2\left(e^{2x}\cos^2\left(\sqrt{3}\,x\right) - 4\cos^2\left(\sqrt{3}\,x\right) + 4\right)}$$

$$R^{1}{}_{310} = -R^{1}{}_{301}$$

$$R^{1}{}_{313} = \frac{e^{x}\left(2\sqrt{3}\,e^{2x}\cos^{2}\left(\sqrt{3}\,x\right)\sin\left(\sqrt{3}\,x\right) - 8\sqrt{3}\cos^{2}\left(\sqrt{3}\,x\right)\sin\left(\sqrt{3}\,x\right) + 16\sqrt{3}\sin\left(\sqrt{3}\,x\right) + 2\,e^{2x}\cos^{3}\left(\sqrt{3}\,x\right) - 4\cos^{3}\left(\sqrt{3}\,x\right) + \ldots\right)}{4\left(e^{2x}\cos^{2}\left(\sqrt{3}\,x\right) - 4\cos^{2}\left(\sqrt{3}\,x\right) + 4\right)}$$

$$R^{1}{}_{331} = -R^{1}{}_{313}$$

$$R^{2}{}_{002} = -\frac{e^{x}\left(\sqrt{3}\sin\left(\sqrt{3}\,x\right) - \cos\left(\sqrt{3}\,x\right)\right)}{2}$$

$$R^{2}{}_{020} = -R^{2}{}_{002}$$

$$R^{2}{}_{023} = -\sqrt{3}\cos\left(\sqrt{3}\,x\right)$$

$$R^{2}{}_{032} = -R^{2}{}_{023}$$

$$R^{2}{}_{112} = 1$$

$$R^{2}{}_{121} = -R^{2}{}_{112}$$

$$R^{2}{}_{302} = \sqrt{3}\cos\left(\sqrt{3}\,x\right)$$

$$R^{2}{}_{320} = -R^{2}{}_{302}$$

$$R^{2}{}_{323} = -\frac{e^{x}\left(\sqrt{3}\sin\left(\sqrt{3}\,x\right) - \cos\left(\sqrt{3}\,x\right)\right)}{2}$$

$$R^{2}{}_{332} = -R^{2}{}_{323}$$

$$R^{3}{}_{003} = \frac{e^{x}\cos\left(\sqrt{3}\,x\right)\left(2\sqrt{3}\,e^{2x}\cos\left(\sqrt{3}\,x\right)\sin\left(\sqrt{3}\,x\right) + 2\,e^{2x}\cos^{2}\left(\sqrt{3}\,x\right) - 12\cos^{2}\left(\sqrt{3}\,x\right) - 3\,e^{2x}\right)}{4\left(e^{2x}\cos^{2}\left(\sqrt{3}\,x\right) - 4\cos^{2}\left(\sqrt{3}\,x\right) + 4\right)}$$

$$R^{3}{}_{030} = -R^{3}{}_{003}$$

$$R^{3}{}_{101} = \frac{e^{x}\left(4\cos^{3}\left(\sqrt{3}\,x\right)\sin\left(\sqrt{3}\,x\right) + 3\,e^{2x}\cos\left(\sqrt{3}\,x\right)\sin\left(\sqrt{3}\,x\right) - 16\cos\left(\sqrt{3}\,x\right)\sin\left(\sqrt{3}\,x\right) - \sqrt{3}\,e^{2x}\cos^{4}\left(\sqrt{3}\,x\right) + 4\sqrt{3}\cos^{4}\left(\sqrt{3}\,x\right) - \ldots\right)}{\left(e^{2x}\cos^{2}\left(\sqrt{3}\,x\right) - 4\cos^{2}\left(\sqrt{3}\,x\right) + 4\right)^{2}}$$

$$R^{3}{}_{110} = -R^{3}{}_{101}$$

$$R^3{}_{113} = -\frac{2\sqrt{3}\,e^{4x}\cos^3\left(\sqrt{3}\,x\right)\sin\left(\sqrt{3}\,x\right) - 8\sqrt{3}\,e^{2x}\cos^3\left(\sqrt{3}\,x\right)\sin\left(\sqrt{3}\,x\right) + 24\sqrt{3}\,e^{2x}\cos\left(\sqrt{3}\,x\right)\sin\left(\sqrt{3}\,x\right) + 2\,e^{4x}\cos^4\left(\sqrt{3}\,x\right) - \ldots}{4\left(e^{2x}\cos^2\left(\sqrt{3}\,x\right) - 4\cos^2\left(\sqrt{3}\,x\right) + 4\right)^2}$$

$$R^3{}_{131} = -R^3{}_{113}$$

$$R^3{}_{202} = \frac{e^{-x}\left(\cos\left(\sqrt{3}\,x\right)\sin\left(\sqrt{3}\,x\right) - \sqrt{3}\right)}{e^{2x}\cos^2\left(\sqrt{3}\,x\right) - 4\cos^2\left(\sqrt{3}\,x\right) + 4}$$

$$R^3{}_{220} = -R^3{}_{202}$$

$$R^3{}_{223} = \frac{e^{-2x}\cos\left(\sqrt{3}\,x\right)\left(\sqrt{3}\,e^{2x}\sin\left(\sqrt{3}\,x\right) - 4\sqrt{3}\sin\left(\sqrt{3}\,x\right) - e^{2x}\cos\left(\sqrt{3}\,x\right)\right)}{2\left(e^{2x}\cos^2\left(\sqrt{3}\,x\right) - 4\cos^2\left(\sqrt{3}\,x\right) + 4\right)}$$

$$R^3{}_{232} = -R^3{}_{223}$$

$$R^3{}_{303} = -\frac{\sin\left(\sqrt{3}\,x\right)\left(2\,e^{2x}\sin^2\left(\sqrt{3}\,x\right) - 12\sin^2\left(\sqrt{3}\,x\right) - 2\sqrt{3}\,e^{2x}\cos\left(\sqrt{3}\,x\right)\sin\left(\sqrt{3}\,x\right) + e^{2x} + 12\right)}{2\left(e^{2x}\cos^2\left(\sqrt{3}\,x\right) - 4\cos^2\left(\sqrt{3}\,x\right) + 4\right)}$$

$$R^3{}_{330} = -R^3{}_{303}$$

Ricci Tensor

$$\mathrm{Ric}_{00} = -\frac{e^{x}\left(\sqrt{3}\,e^{2x}\cos^2\left(\sqrt{3}\,x\right)\sin\left(\sqrt{3}\,x\right) + 4\sqrt{3}\sin\left(\sqrt{3}\,x\right) + 3\,e^{2x}\cos^3\left(\sqrt{3}\,x\right) - 12\cos^3\left(\sqrt{3}\,x\right)\right)}{2\left(e^{2x}\cos^2\left(\sqrt{3}\,x\right) - 4\cos^2\left(\sqrt{3}\,x\right) + 4\right)}$$

$$\mathrm{Ric}_{03} = \frac{2\,e^{2x}\sin^3\left(\sqrt{3}\,x\right) - 12\sin^3\left(\sqrt{3}\,x\right) - 4\sqrt{3}\cos\left(\sqrt{3}\,x\right)\sin^2\left(\sqrt{3}\,x\right) + e^{2x}\sin\left(\sqrt{3}\,x\right) - 2\sqrt{3}\,e^{2x}\cos\left(\sqrt{3}\,x\right)}{e^{2x}\cos^2\left(\sqrt{3}\,x\right) - 4\cos^2\left(\sqrt{3}\,x\right) + 4}$$

$$\mathrm{Ric}_{11} = \frac{2\sqrt{3}\,e^{4x}\cos^3\left(\sqrt{3}\,x\right)\sin\left(\sqrt{3}\,x\right) - 8\sqrt{3}\,e^{2x}\cos^3\left(\sqrt{3}\,x\right)\sin\left(\sqrt{3}\,x\right) + 24\sqrt{3}\,e^{2x}\cos\left(\sqrt{3}\,x\right)\sin\left(\sqrt{3}\,x\right) + 4\,e^{2x}\cos^4\left(\sqrt{3}\,x\right) + \ldots}{2\left(e^{2x}\cos^2\left(\sqrt{3}\,x\right) - 4\cos^2\left(\sqrt{3}\,x\right) + 4\right)^2}$$

$$\mathrm{Ric}_{22} = -\frac{e^{-2x}\sin\left(\sqrt{3}\,x\right)\left(4\sin\left(\sqrt{3}\,x\right) + \sqrt{3}\,e^{2x}\cos\left(\sqrt{3}\,x\right) - 4\sqrt{3}\cos\left(\sqrt{3}\,x\right)\right)}{e^{2x}\cos^2\left(\sqrt{3}\,x\right) - 4\cos^2\left(\sqrt{3}\,x\right) + 4}$$

$$\mathrm{Ric}_{30} = \mathrm{Ric}_{03}$$

$$\mathrm{Ric}_{33} = \frac{e^{x}\left(\sqrt{3}\,e^{2x}\cos^2\left(\sqrt{3}\,x\right)\sin\left(\sqrt{3}\,x\right) + 4\sqrt{3}\sin\left(\sqrt{3}\,x\right) + 3\,e^{2x}\cos^3\left(\sqrt{3}\,x\right) - 12\cos^3\left(\sqrt{3}\,x\right)\right)}{2\left(e^{2x}\cos^2\left(\sqrt{3}\,x\right) - 4\cos^2\left(\sqrt{3}\,x\right) + 4\right)}$$

Ricci Scalar

$$R_{sc} = \frac{2\sqrt{3}\,e^{4x}\cos^3\left(\sqrt{3}\,x\right)\sin\left(\sqrt{3}\,x\right) + 8\sqrt{3}\,e^{2x}\cos^3\left(\sqrt{3}\,x\right)\sin\left(\sqrt{3}\,x\right) - 64\sqrt{3}\cos^3\left(\sqrt{3}\,x\right)\sin\left(\sqrt{3}\,x\right) + 40\sqrt{3}\,e^{2x}\cos\left(\sqrt{3}\,x\right)\sin\left(\sqrt{3}\,x\right) + \ldots}{2\left(e^{2x}\cos^2\left(\sqrt{3}\,x\right) - 4\cos^2\left(\sqrt{3}\,x\right) + 4\right)^2}$$

Bianchi identity (Ricci cyclic equation ${R^{\kappa}}_{[\mu\nu\sigma]} = 0$)

——— o.k.

Einstein Tensor

$$G_{00} = \frac{e^{x}\left(16\sqrt{3}\,e^{2x}\cos^4\left(\sqrt{3}\,x\right)\sin\left(\sqrt{3}\,x\right) - 64\sqrt{3}\cos^4\left(\sqrt{3}\,x\right)\sin\left(\sqrt{3}\,x\right) + 24\sqrt{3}\,e^{2x}\cos^2\left(\sqrt{3}\,x\right)\sin\left(\sqrt{3}\,x\right) + \ldots\right)}{4\left(e^{2x}\cos^2\left(\sqrt{3}\,x\right) - 4\cos^2\left(\sqrt{3}\,x\right) + 4\right)^2}$$

$$G_{03} = \frac{2\,e^{4x}\cos^4\left(\sqrt{3}\,x\right)\sin\left(\sqrt{3}\,x\right) + 12\,e^{2x}\cos^4\left(\sqrt{3}\,x\right)\sin\left(\sqrt{3}\,x\right) - 16\cos^4\left(\sqrt{3}\,x\right)\sin\left(\sqrt{3}\,x\right) + 9\,e^{4x}\cos^2\left(\sqrt{3}\,x\right)\sin\left(\sqrt{3}\,x\right) - \ldots}{2\left(e^{2x}\cos^2\left(\sqrt{3}\,x\right) - 4\cos^2\left(\sqrt{3}\,x\right) + 4\right)^2}$$

$$G_{11} = \frac{2\sqrt{3}\,e^{2x}\cos\left(\sqrt{3}\,x\right)\sin\left(\sqrt{3}\,x\right) - 16\sqrt{3}\cos\left(\sqrt{3}\,x\right)\sin\left(\sqrt{3}\,x\right) - 6\,e^{2x}\cos^2\left(\sqrt{3}\,x\right) + 12\cos^2\left(\sqrt{3}\,x\right) + 3\,e^{2x}}{4\left(e^{2x}\cos^2\left(\sqrt{3}\,x\right) - 4\cos^2\left(\sqrt{3}\,x\right) + 4\right)}$$

$$G_{22} = -\frac{e^{-2x}\left(6\sqrt{3}\,e^{4x}\cos^3\left(\sqrt{3}\,x\right)\sin\left(\sqrt{3}\,x\right) - 24\sqrt{3}\,e^{2x}\cos^3\left(\sqrt{3}\,x\right)\sin\left(\sqrt{3}\,x\right) + 56\sqrt{3}\,e^{2x}\cos\left(\sqrt{3}\,x\right)\sin\left(\sqrt{3}\,x\right) + 6\,e^{4x}\cos^4\left(\sqrt{3}\,x\right) - \ldots\right)}{4\left(e^{2x}\cos^2\left(\sqrt{3}\,x\right) - 4\cos^2\left(\sqrt{3}\,x\right) + 4\right)^2}$$

$$G_{30} = G_{03}$$

$$G_{33} = -\frac{e^{x}\left(16\sqrt{3}\,e^{2x}\cos^4\left(\sqrt{3}\,x\right)\sin\left(\sqrt{3}\,x\right) - 64\sqrt{3}\cos^4\left(\sqrt{3}\,x\right)\sin\left(\sqrt{3}\,x\right) + 24\sqrt{3}\,e^{2x}\cos^2\left(\sqrt{3}\,x\right)\sin\left(\sqrt{3}\,x\right) + \ldots\right)}{4\left(e^{2x}\cos^2\left(\sqrt{3}\,x\right) - 4\cos^2\left(\sqrt{3}\,x\right) + 4\right)^2}$$

Hodge Dual of Bianchi Identity

——— (see charge and current densities)

Scalar Charge Density ($-{R^0}_i{}^{i0}$)

$$\rho = -\frac{e^{x}\left(\sqrt{3}\,e^{4x}\cos^4\left(\sqrt{3}\,x\right)\sin\left(\sqrt{3}\,x\right) + 4\sqrt{3}\,e^{2x}\cos^4\left(\sqrt{3}\,x\right)\sin\left(\sqrt{3}\,x\right) - 32\sqrt{3}\cos^4\left(\sqrt{3}\,x\right)\sin\left(\sqrt{3}\,x\right) + \ldots\right)}{2\left(e^{2x}\cos^2\left(\sqrt{3}\,x\right) - 4\cos^2\left(\sqrt{3}\,x\right) + 4\right)^3}$$

Current Density Class 1 ($-{R^i}_\mu{}^{\mu j}$)

$$J_1 = -\frac{2\sqrt{3}\,e^{4x}\cos^3\left(\sqrt{3}\,x\right)\sin\left(\sqrt{3}\,x\right) - 8\sqrt{3}\,e^{2x}\cos^3\left(\sqrt{3}\,x\right)\sin\left(\sqrt{3}\,x\right) + 24\sqrt{3}\,e^{2x}\cos\left(\sqrt{3}\,x\right)\sin\left(\sqrt{3}\,x\right) + 4\,e^{2x}\cos^4\left(\sqrt{3}\,x\right) + \ldots}{2\left(e^{2x}\cos^2\left(\sqrt{3}\,x\right) - 4\cos^2\left(\sqrt{3}\,x\right) + 4\right)^2}$$

$$J_2 = \frac{e^{2x}\sin\left(\sqrt{3}\,x\right)\left(4\sin\left(\sqrt{3}\,x\right) + \sqrt{3}\,e^{2x}\cos\left(\sqrt{3}\,x\right) - 4\sqrt{3}\cos\left(\sqrt{3}\,x\right)\right)}{e^{2x}\cos^2\left(\sqrt{3}\,x\right) - 4\cos^2\left(\sqrt{3}\,x\right) + 4}$$

$$J_3 = -\frac{e^{x}\left(\sqrt{3}\,e^{4x}\cos^4\left(\sqrt{3}\,x\right)\sin\left(\sqrt{3}\,x\right) + 4\sqrt{3}\,e^{2x}\cos^4\left(\sqrt{3}\,x\right)\sin\left(\sqrt{3}\,x\right) - 32\sqrt{3}\cos^4\left(\sqrt{3}\,x\right)\sin\left(\sqrt{3}\,x\right) + \ldots\right)}{2\left(e^{2x}\cos^2\left(\sqrt{3}\,x\right) - 4\cos^2\left(\sqrt{3}\,x\right) + 4\right)^3}$$

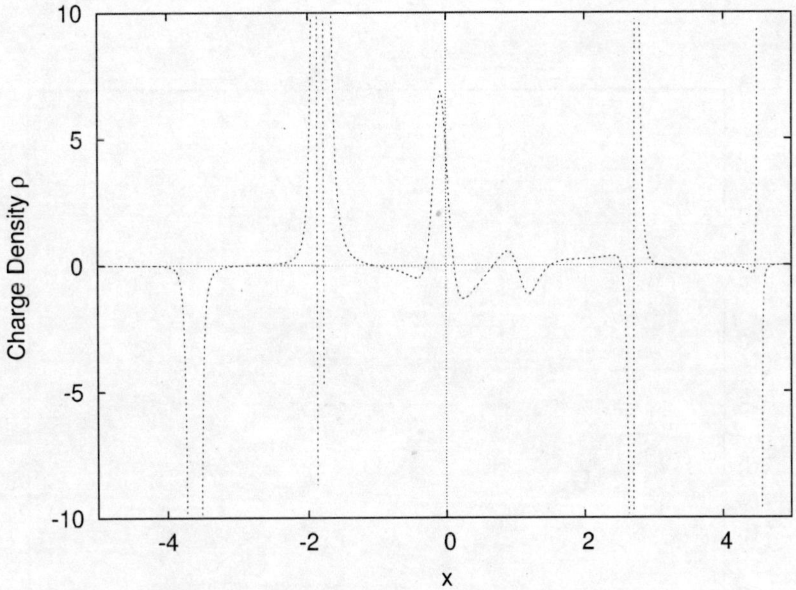

Fig. 4.78: Petrov metric, charge density ρ.

Current Density Class 2 (-$R^{i}{}_{\mu}{}^{\mu j}$)

$$J_1 = 0$$

$$J_2 = 0$$

$$J_3 = 0$$

Current Density Class 3 (-$R^{i}{}_{\mu}{}^{\mu j}$)

$$J_1 = 0$$

$$J_2 = 0$$

$$J_3 = 0$$

4.4.46 Homogeneous non-null electromagnetic fields, type 1

This metric describes Homogeneous non-null electromagnetic Fields. k is a parameter.

Coordinates

$$\mathbf{x} = \begin{pmatrix} t \\ x \\ \vartheta \\ \varphi \end{pmatrix}$$

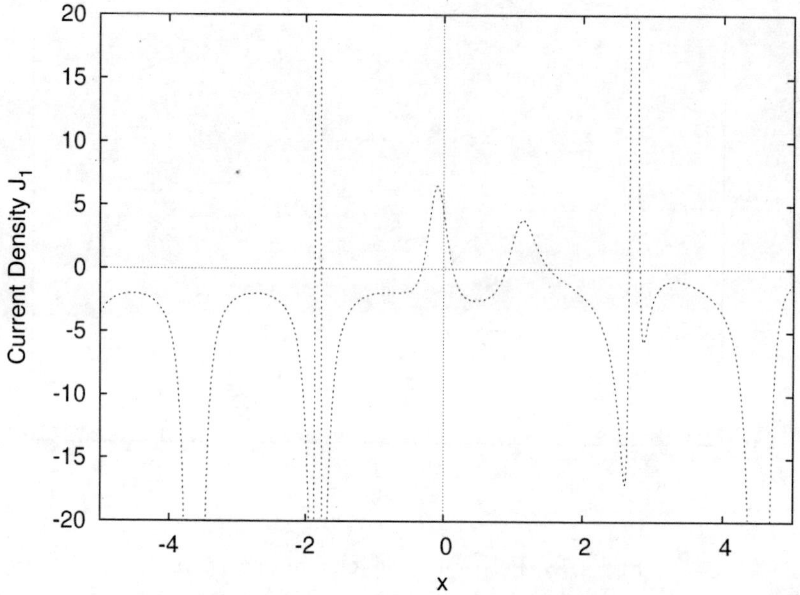

Fig. 4.79: Petrov metric, current density J_1.

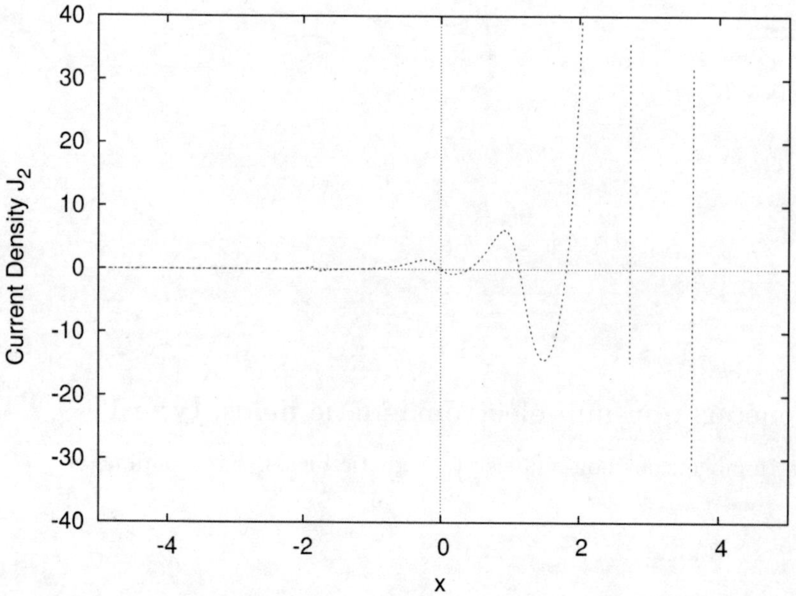

Fig. 4.80: Petrov metric, current density J_2.

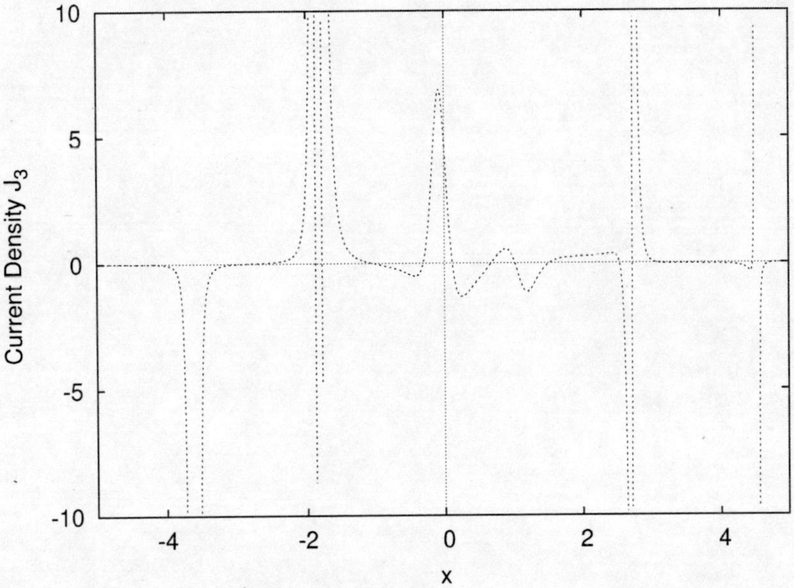

Fig. 4.81: Petrov metric, current density J_3.

Metric

$$g_{\mu\nu} = \begin{pmatrix} -k^2 \sinh^2 x & 0 & 0 & 0 \\ 0 & k^2 & 0 & 0 \\ 0 & 0 & k^2 & 0 \\ 0 & 0 & 0 & k^2 \sin^2 \vartheta \end{pmatrix}$$

Contravariant Metric

$$g^{\mu\nu} = \begin{pmatrix} -\frac{1}{k^2 \sinh^2 x} & 0 & 0 & 0 \\ 0 & \frac{1}{k^2} & 0 & 0 \\ 0 & 0 & \frac{1}{k^2} & 0 \\ 0 & 0 & 0 & \frac{1}{k^2 \sin^2 \vartheta} \end{pmatrix}$$

Christoffel Connection

$$\Gamma^0{}_{01} = \frac{\cosh x}{\sinh x}$$

$$\Gamma^0{}_{10} = \Gamma^0{}_{01}$$

$$\Gamma^1{}_{00} = \cosh x \sinh x$$

$$\Gamma^2{}_{33} = -\cos\vartheta \sin\vartheta$$

$$\Gamma^3{}_{23} = \frac{\cos\vartheta}{\sin\vartheta}$$

$$\Gamma^3{}_{32} = \Gamma^3{}_{23}$$

Metric Compatibility

——— o.k.

Riemann Tensor

$$R^0{}_{101} = -1$$

$$R^0{}_{110} = -R^0{}_{101}$$

$$R^1{}_{001} = -\sinh^2 x$$

$$R^1{}_{010} = -R^1{}_{001}$$

$$R^2{}_{323} = \sin^2 \vartheta$$

$$R^2{}_{332} = -R^2{}_{323}$$

$$R^3{}_{223} = -1$$

$$R^3{}_{232} = -R^3{}_{223}$$

Ricci Tensor

$$\mathrm{Ric}_{00} = \sinh^2 x$$

$$\mathrm{Ric}_{11} = -1$$

$$\mathrm{Ric}_{22} = 1$$

$$\mathrm{Ric}_{33} = \sin^2 \vartheta$$

Ricci Scalar

$$R_{sc} = 0$$

Bianchi identity (Ricci cyclic equation $R^{\kappa}{}_{[\mu\nu\sigma]} = 0$)

——— o.k.

Einstein Tensor

$$G_{00} = \sinh^2 x$$

$$G_{11} = -1$$

$$G_{22} = 1$$

$$G_{33} = \sin^2 \vartheta$$

Hodge Dual of Bianchi Identity

——— (see charge and current densities)

Scalar Charge Density $(-R^{0\ \ i0}_{\ \ i})$

$$\rho = \frac{1}{k^4 \sinh^2 x}$$

Current Density Class 1 $(-R^{i\ \ \mu j}_{\ \mu})$

$$J_1 = \frac{1}{k^4}$$

$$J_2 = -\frac{1}{k^4}$$

$$J_3 = -\frac{1}{k^4 \sin^2 \vartheta}$$

Current Density Class 2 $(-R^{i\ \ \mu j}_{\ \mu})$

$$J_1 = 0$$

$$J_2 = 0$$

$$J_3 = 0$$

Current Density Class 3 $(-R^{i\ \ \mu j}_{\ \mu})$

$$J_1 = 0$$

$$J_2 = 0$$

$$J_3 = 0$$

4.4.47 Homogeneous non-null electromagnetic fields, type 2

This metric describes Homogeneous non-null electromagnetic Fields. a is a parameter.

Coordinates

$$\mathbf{x} = \begin{pmatrix} t \\ x \\ y \\ \varphi \end{pmatrix}$$

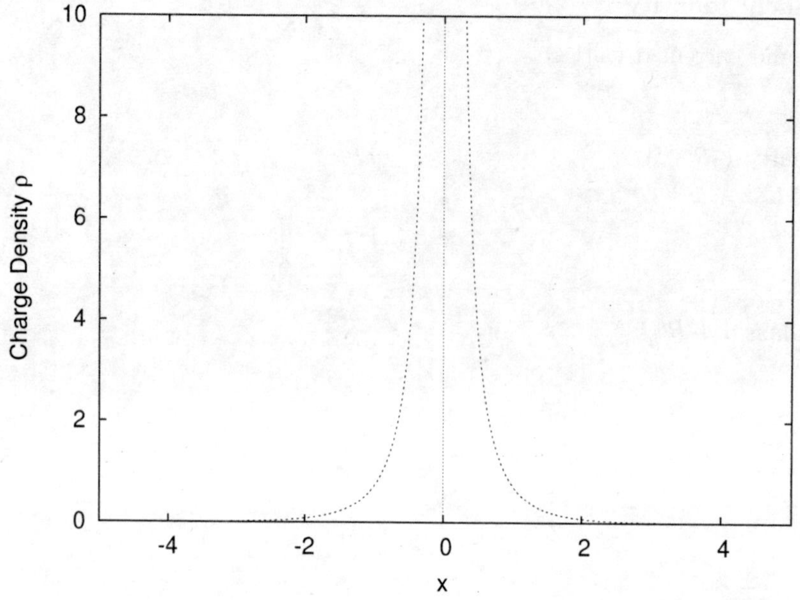

Fig. 4.82: Homogeneous non-null electromagnetic Fields, type 1, charge density ρ for $k = 1$.

Metric

$$g_{\mu\nu} = \begin{pmatrix} -1 & 0 & 0 & 4\,y \\ 0 & \frac{a^2}{x^2} & 0 & 0 \\ 0 & 0 & \frac{a^2}{x^2} & 0 \\ 4\,y & 0 & 0 & x^2 - 4\,y^2 \end{pmatrix}$$

Contravariant Metric

$$g^{\mu\nu} = \begin{pmatrix} \frac{(2\,y-x)\,(2\,y+x)}{12\,y^2+x^2} & 0 & 0 & \frac{4\,y}{12\,y^2+x^2} \\ 0 & \frac{x^2}{a^2} & 0 & 0 \\ 0 & 0 & \frac{x^2}{a^2} & 0 \\ \frac{4\,y}{12\,y^2+x^2} & 0 & 0 & \frac{1}{12\,y^2+x^2} \end{pmatrix}$$

Christoffel Connection

$$\Gamma^0{}_{02} = \frac{8\,y}{12\,y^2 + x^2}$$

$$\Gamma^0{}_{13} = \frac{4\,x\,y}{12\,y^2 + x^2}$$

$$\Gamma^0{}_{20} = \Gamma^0{}_{02}$$

$$\Gamma^0{}_{23} = -\frac{2\,\left(4\,y^2 + x^2\right)}{12\,y^2 + x^2}$$

$$\Gamma^0{}_{31} = \Gamma^0{}_{13}$$

$$\Gamma^0{}_{32} = \Gamma^0{}_{23}$$

$$\Gamma^1{}_{11} = -\frac{1}{x}$$

$$\Gamma^1{}_{22} = \frac{1}{x}$$

$$\Gamma^1{}_{33} = -\frac{x^3}{a^2}$$

$$\Gamma^2{}_{03} = -\frac{2\,x^2}{a^2}$$

$$\Gamma^2{}_{12} = -\frac{1}{x}$$

$$\Gamma^2{}_{21} = \Gamma^2{}_{12}$$

$$\Gamma^2{}_{30} = \Gamma^2{}_{03}$$

$$\Gamma^2{}_{33} = \frac{4\,x^2\,y}{a^2}$$

$$\Gamma^3{}_{02} = \frac{2}{12\,y^2 + x^2}$$

$$\Gamma^3{}_{13} = \frac{x}{12\,y^2 + x^2}$$

$$\Gamma^3{}_{20} = \Gamma^3{}_{02}$$

$$\Gamma^3{}_{23} = \frac{4\,y}{12\,y^2 + x^2}$$

$$\Gamma^3{}_{31} = \Gamma^3{}_{13}$$

$$\Gamma^3{}_{32} = \Gamma^3{}_{23}$$

Metric Compatibility

——— o.k.

Riemann Tensor

$$R^0{}_{003} = -\frac{16\,x^2\,y}{a^2\,(12\,y^2+x^2)}$$

$$R^0{}_{012} = -\frac{8\,x\,y}{(12\,y^2+x^2)^2}$$

$$R^0{}_{021} = -R^0{}_{012}$$

$$R^0{}_{030} = -R^0{}_{003}$$

$$R^0{}_{102} = -\frac{8\,y}{x\,(12\,y^2+x^2)}$$

$$R^0{}_{113} = \frac{4\,y\,\left(24\,y^2+x^2\right)}{(12\,y^2+x^2)^2}$$

$$R^0{}_{120} = -R^0{}_{102}$$

$$R^0{}_{123} = -\frac{8\,y^2\,\left(12\,y^2+7\,x^2\right)}{x\,(12\,y^2+x^2)^2}$$

$$R^0{}_{131} = -R^0{}_{113}$$

$$R^0{}_{132} = -R^0{}_{123}$$

$$R^0{}_{201} = -\frac{96\,y^3}{x\,(12\,y^2+x^2)^2}$$

$$R^0{}_{202} = \frac{4\,\left(12\,y^2-x^2\right)}{(12\,y^2+x^2)^2}$$

$$R^0{}_{210} = -R^0{}_{201}$$

$$R^0{}_{213} = -\frac{2\,\left(48\,y^4+24\,x^2\,y^2+x^4\right)}{x\,(12\,y^2+x^2)^2}$$

$$R^0{}_{220} = -R^0{}_{202}$$

$$R^0{}_{223} = -\frac{4\,y\,(6\,y-x)\,(6\,y+x)}{(12\,y^2+x^2)^2}$$

$$R^0{}_{231} = -R^0{}_{213}$$

$$R^0{}_{232} = -R^0{}_{223}$$

$$R^0{}_{303} = \frac{4\,x^2\,(2\,y-x)\,(2\,y+x)}{a^2\,(12\,y^2+x^2)}$$

$$R^0{}_{312} = \frac{2\,x\,(2\,y-x)\,(2\,y+x)}{(12\,y^2+x^2)^2}$$

$$R^0{}_{321} = -R^0{}_{312}$$

$$R^0{}_{330} = -R^0{}_{303}$$

$$R^1{}_{023} = -\frac{24\,x\,y^2}{a^2\,(12\,y^2+x^2)}$$

$$R^1{}_{032} = -R^1{}_{023}$$

$$R^1{}_{203} = \frac{2\,x^3}{a^2\,(12\,y^2+x^2)}$$

$$R^1{}_{212} = -\frac{1}{x^2}$$

$$R^1{}_{221} = -R^1{}_{212}$$

$$R^1{}_{230} = -R^1{}_{203}$$

$$R^1{}_{302} = \frac{2\,x}{a^2}$$

$$R^1{}_{313} = -\frac{x^2\,(24\,y^2+x^2)}{a^2\,(12\,y^2+x^2)}$$

$$R^1{}_{320} = -R^1{}_{302}$$

$$R^1{}_{323} = \frac{8\,x\,y\,(6\,y^2+x^2)}{a^2\,(12\,y^2+x^2)}$$

$$R^1{}_{331} = -R^1{}_{313}$$

$$R^1{}_{332} = -R^1{}_{323}$$

$$R^2{}_{002} = -\frac{4\,x^2}{a^2\,(12\,y^2+x^2)}$$

$$R^2{}_{013} = -\frac{2\,x}{a^2}$$

$$R^2{}_{020} = -R^2{}_{002}$$

$$R^2{}_{023} = \frac{8\,x^2\,y}{a^2\,(12\,y^2 + x^2)}$$

$$R^2{}_{031} = -R^2{}_{013}$$

$$R^2{}_{032} = -R^2{}_{023}$$

$$R^2{}_{103} = -\frac{2\,x^3}{a^2\,(12\,y^2 + x^2)}$$

$$R^2{}_{112} = \frac{1}{x^2}$$

$$R^2{}_{121} = -R^2{}_{112}$$

$$R^2{}_{130} = -R^2{}_{103}$$

$$R^2{}_{301} = \frac{24\,x\,y^2}{a^2\,(12\,y^2 + x^2)}$$

$$R^2{}_{302} = -\frac{8\,x^2\,y}{a^2\,(12\,y^2 + x^2)}$$

$$R^2{}_{310} = -R^2{}_{301}$$

$$R^2{}_{313} = \frac{8\,x\,y\,\left(6\,y^2 + x^2\right)}{a^2\,(12\,y^2 + x^2)}$$

$$R^2{}_{320} = -R^2{}_{302}$$

$$R^2{}_{323} = \frac{x^2\,\left(28\,y^2 + x^2\right)}{a^2\,(12\,y^2 + x^2)}$$

$$R^2{}_{331} = -R^2{}_{313}$$

$$R^2{}_{332} = -R^2{}_{323}$$

$$R^3{}_{003} = -\frac{4\,x^2}{a^2\,(12\,y^2 + x^2)}$$

$$R^3{}_{012} = -\frac{2\,x}{\left(12\,y^2 + x^2\right)^2}$$

$$R^3{}_{021} = -R^3{}_{012}$$

$$R^3{}_{030} = -R^3{}_{003}$$

$$R^3{}_{102} = -\frac{2}{x\,(12\,y^2 + x^2)}$$

$$R^3{}_{113} = \frac{24\,y^2 + x^2}{(12\,y^2 + x^2)^2}$$

$$R^3{}_{120} = -R^3{}_{102}$$

$$R^3{}_{123} = \frac{8\,y\,(6\,y^2 - x^2)}{x\,(12\,y^2 + x^2)^2}$$

$$R^3{}_{131} = -R^3{}_{113}$$

$$R^3{}_{132} = -R^3{}_{123}$$

$$R^3{}_{201} = -\frac{24\,y^2}{x\,(12\,y^2 + x^2)^2}$$

$$R^3{}_{202} = \frac{24\,y}{(12\,y^2 + x^2)^2}$$

$$R^3{}_{210} = -R^3{}_{201}$$

$$R^3{}_{213} = \frac{48\,y^3}{x\,(12\,y^2 + x^2)^2}$$

$$R^3{}_{220} = -R^3{}_{202}$$

$$R^3{}_{223} = -\frac{60\,y^2 + x^2}{(12\,y^2 + x^2)^2}$$

$$R^3{}_{231} = -R^3{}_{213}$$

$$R^3{}_{232} = -R^3{}_{223}$$

$$R^3{}_{303} = \frac{16\,x^2\,y}{a^2\,(12\,y^2 + x^2)}$$

$$R^3{}_{312} = \frac{8\,x\,y}{(12\,y^2 + x^2)^2}$$

$$R^3{}_{321} = -R^3{}_{312}$$

$$R^3{}_{330} = -R^3{}_{303}$$

Ricci Tensor

$$\mathrm{Ric}_{00} = \frac{8\,x^2}{a^2\,(12\,y^2 + x^2)}$$

$$\mathrm{Ric}_{03} = -\frac{8\,x^2\,y}{a^2\,(12\,y^2 + x^2)}$$

$$\mathrm{Ric}_{11} = -\frac{2\,(72\,y^4 + 24\,x^2\,y^2 + x^4)}{x^2\,(12\,y^2 + x^2)^2}$$

$$\mathrm{Ric}_{12} = -\frac{144\,y^3}{x\,(12\,y^2 + x^2)^2}$$

$$\mathrm{Ric}_{21} = \mathrm{Ric}_{12}$$

$$\mathrm{Ric}_{22} = -\frac{4\,(6\,y^2 - 3\,x\,y - x^2)\,(6\,y^2 + 3\,x\,y - x^2)}{x^2\,(12\,y^2 + x^2)^2}$$

$$\mathrm{Ric}_{30} = \mathrm{Ric}_{03}$$

$$\mathrm{Ric}_{33} = \frac{4\,x^2\,(5\,y^2 - x^2)}{a^2\,(12\,y^2 + x^2)}$$

Ricci Scalar

$$R_{sc} = -\frac{6\,(48\,y^4 - 4\,x^2\,y^2 + 3\,x^4)}{a^2\,(12\,y^2 + x^2)^2}$$

Bianchi identity (Ricci cyclic equation $R^{\kappa}{}_{[\mu\nu\sigma]} = 0$)

——— o.k.

Einstein Tensor

$$G_{00} = -\frac{144\,y^4 - 108\,x^2\,y^2 + x^4}{a^2\,(12\,y^2 + x^2)^2}$$

$$G_{03} = \frac{4\,y\,(144\,y^4 - 36\,x^2\,y^2 + 7\,x^4)}{a^2\,(12\,y^2 + x^2)^2}$$

$$G_{11} = -\frac{60\,y^2 - 7\,x^2}{(12\,y^2 + x^2)^2}$$

$$G_{12} = -\frac{144\,y^3}{x\,(12\,y^2 + x^2)^2}$$

$$G_{21} = G_{12}$$

$$G_{22} = \frac{72\,y^2 + 5\,x^2}{\left(12\,y^2 + x^2\right)^2}$$

$$G_{30} = G_{03}$$

$$G_{33} = -\frac{576\,y^6 - 432\,x^2\,y^4 + 76\,x^4\,y^2 - 5\,x^6}{a^2\,\left(12\,y^2 + x^2\right)^2}$$

Hodge Dual of Bianchi Identity

——— (see charge and current densities)

Scalar Charge Density ($-R^{0}{}_{i}{}^{i0}$)

$$\rho = \frac{8\,x^2\,\left(24\,y^4 - 8\,x^2\,y^2 + x^4\right)}{a^2\,\left(12\,y^2 + x^2\right)^3}$$

Current Density Class 1 ($-R^{i}{}_{\mu}{}^{\mu j}$)

$$J_1 = \frac{2\,x^2\,\left(72\,y^4 + 24\,x^2\,y^2 + x^4\right)}{a^4\,\left(12\,y^2 + x^2\right)^2}$$

$$J_2 = \frac{4\,x^2\,\left(6\,y^2 - 3\,x\,y - x^2\right)\,\left(6\,y^2 + 3\,x\,y - x^2\right)}{a^4\,\left(12\,y^2 + x^2\right)^2}$$

$$J_3 = -\frac{4\,x^2\,\left(21\,y^2 - x^2\right)}{a^2\,\left(12\,y^2 + x^2\right)^3}$$

Current Density Class 2 ($-R^{i}{}_{\mu}{}^{\mu j}$)

$$J_1 = 0$$

$$J_2 = \frac{144\,x^3\,y^3}{a^4\,\left(12\,y^2 + x^2\right)^2}$$

$$J_3 = 0$$

Current Density Class 3 ($-R^{i}{}_{\mu}{}^{\mu j}$)

$$J_1 = \frac{144\,x^3\,y^3}{a^4\,\left(12\,y^2 + x^2\right)^2}$$

$$J_2 = 0$$

$$J_3 = 0$$

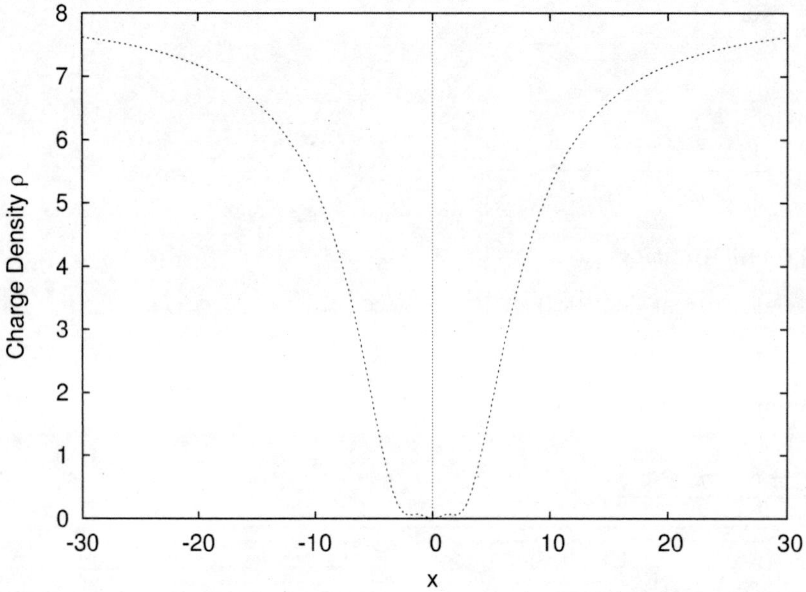

Fig. 4.83: Homogeneous non-null electromagnetic fields, type 2, charge density ρ for $a = 1, y = 1$.

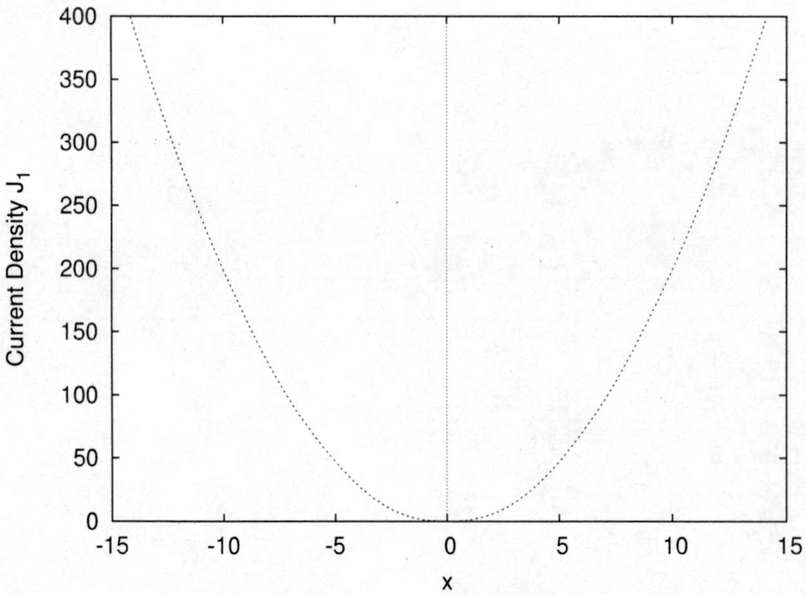

Fig. 4.84: Homogeneous non-null electromagnetic fields, type 2, current density J_x for $a = 1, y = 1$.

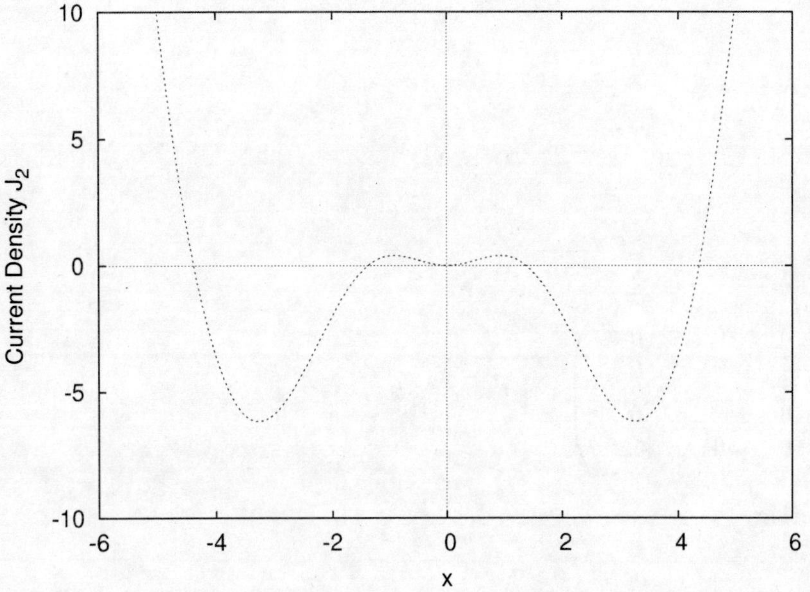

Fig. 4.85: Homogeneous non-null electromagnetic fields, type 2, current density J_y for $a = 1, y = 1$.

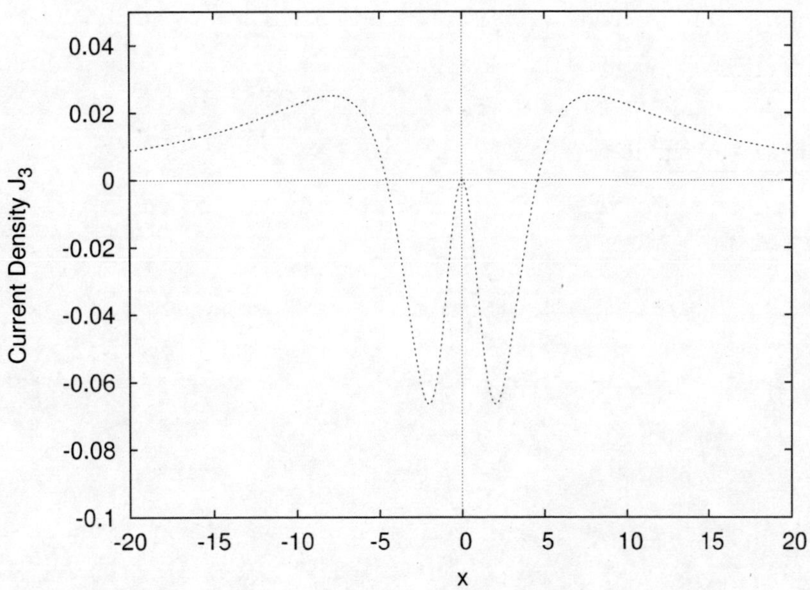

Fig. 4.86: Homogeneous non-null electromagnetic fields, type 2, current density J_z for $a = 1, y = 1$.

4.4.48 Homogeneous perfect fluid, spherical

This metric describes a homogeneous perfect fluid in spherical coordinates. a is a parameter.

Coordinates

$$\mathbf{x} = \begin{pmatrix} t \\ r \\ \vartheta \\ \varphi \end{pmatrix}$$

Metric

$$g_{\mu\nu} = \begin{pmatrix} -1 & 0 & 0 & 0 \\ 0 & \frac{1}{1-\frac{r^2}{a^2}} & 0 & 0 \\ 0 & 0 & r^2 & 0 \\ 0 & 0 & 0 & r^2 \sin^2 \vartheta \end{pmatrix}$$

Contravariant Metric

$$g^{\mu\nu} = \begin{pmatrix} -1 & 0 & 0 & 0 \\ 0 & -\frac{(r-a)(r+a)}{a^2} & 0 & 0 \\ 0 & 0 & \frac{1}{r^2} & 0 \\ 0 & 0 & 0 & \frac{1}{r^2 \sin^2 \vartheta} \end{pmatrix}$$

Christoffel Connection

$$\Gamma^1{}_{11} = -\frac{r}{(r-a)\,(r+a)}$$

$$\Gamma^1{}_{22} = \frac{r\,(r-a)\,(r+a)}{a^2}$$

$$\Gamma^1{}_{33} = \frac{r\,(r-a)\,(r+a)\sin^2\vartheta}{a^2}$$

$$\Gamma^2{}_{12} = \frac{1}{r}$$

$$\Gamma^2{}_{21} = \Gamma^2{}_{12}$$

$$\Gamma^2{}_{33} = -\cos\vartheta\,\sin\vartheta$$

$$\Gamma^3{}_{13} = \frac{1}{r}$$

$$\Gamma^3{}_{23} = \frac{\cos\vartheta}{\sin\vartheta}$$

$$\Gamma^3{}_{31} = \Gamma^3{}_{13}$$

$$\Gamma^3{}_{32} = \Gamma^3{}_{23}$$

Metric Compatibility

———— o.k.

Riemann Tensor

$$R^1{}_{212} = \frac{r^2}{a^2}$$

$$R^1{}_{221} = -R^1{}_{212}$$

$$R^1{}_{313} = \frac{r^2 \sin^2 \vartheta}{a^2}$$

$$R^1{}_{331} = -R^1{}_{313}$$

$$R^2{}_{112} = \frac{1}{(r-a)\,(r+a)}$$

$$R^2{}_{121} = -R^2{}_{112}$$

$$R^2{}_{323} = \frac{r^2 \sin^2 \vartheta}{a^2}$$

$$R^2{}_{332} = -R^2{}_{323}$$

$$R^3{}_{113} = \frac{1}{(r-a)\,(r+a)}$$

$$R^3{}_{131} = -R^3{}_{113}$$

$$R^3{}_{223} = -\frac{r^2}{a^2}$$

$$R^3{}_{232} = -R^3{}_{223}$$

Ricci Tensor

$$\mathrm{Ric}_{11} = -\frac{2}{(r-a)\,(r+a)}$$

$$\mathrm{Ric}_{22} = \frac{2\,r^2}{a^2}$$

$$\mathrm{Ric}_{33} = \frac{2\,r^2 \sin^2 \vartheta}{a^2}$$

Ricci Scalar

$$R_{sc} = \frac{6}{a^2}$$

Bianchi identity (Ricci cyclic equation $R^{\kappa}{}_{[\mu\nu\sigma]} = 0$)

——— o.k.

Einstein Tensor

$$G_{00} = \frac{3}{a^2}$$

$$G_{11} = \frac{1}{(r-a)\,(r+a)}$$

$$G_{22} = -\frac{r^2}{a^2}$$

$$G_{33} = -\frac{r^2 \sin^2 \vartheta}{a^2}$$

Hodge Dual of Bianchi Identity

——— (see charge and current densities)

Scalar Charge Density (-$R^{0}{}_{i}{}^{i0}$)

$$\rho = 0$$

Current Density Class 1 (-$R^{i}{}_{\mu}{}^{\mu j}$)

$$J_1 = \frac{2\,(r-a)\,(r+a)}{a^4}$$

$$J_2 = -\frac{2}{a^2\, r^2}$$

$$J_3 = -\frac{2}{a^2\, r^2\, \sin^2 \vartheta}$$

Current Density Class 2 (-$R^{i}{}_{\mu}{}^{\mu j}$)

$$J_1 = 0$$

$$J_2 = 0$$

$$J_3 = 0$$

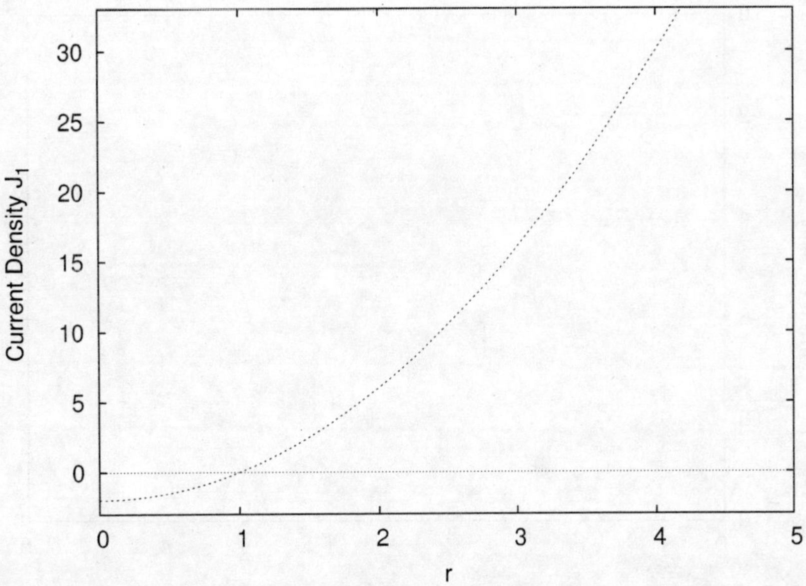

Fig. 4.87: Homogeneous perfect fluid, spherical, current density J_r for $a = 1$.

Current Density Class 3 ($-R^i{}_\mu{}^{\mu j}$)

$$J_1 = 0$$

$$J_2 = 0$$

$$J_3 = 0$$

4.4.49 Homogeneous perfect fluid, Cartesian

This metric describes a homogeneous perfect fluid in Cartesian coordinates. a is a parameter.

Coordinates

$$\mathbf{x} = \begin{pmatrix} t \\ x \\ y \\ z \end{pmatrix}$$

Metric

$$g_{\mu\nu} = \begin{pmatrix} -a^2 & 0 & 0 & -2\,e^x \\ 0 & a^2 & 0 & 0 \\ 0 & 0 & a^2 & 0 \\ -2\,e^x & 0 & 0 & -\frac{a^2\,e^{2\,x}}{2} \end{pmatrix}$$

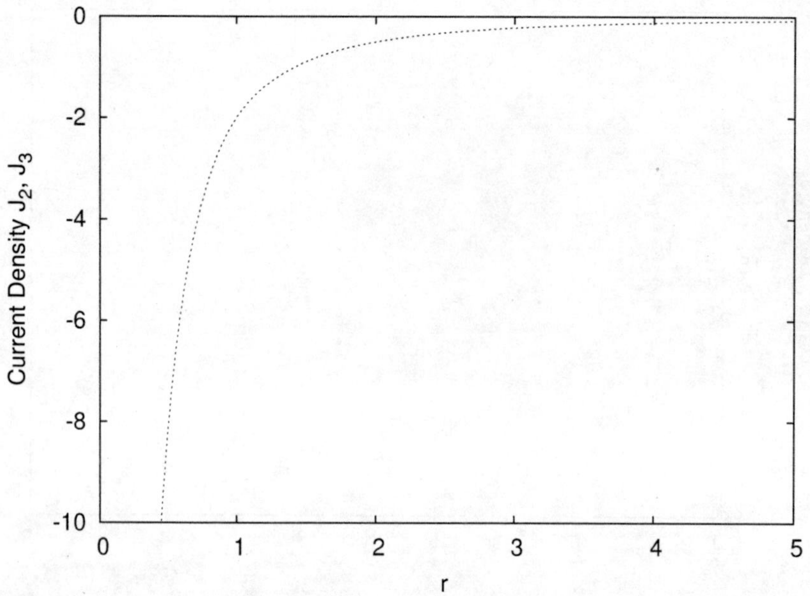

Fig. 4.88: Homogeneous perfect fluid, spherical, current density J_ϑ, J_φ for $a = 1$.

Contravariant Metric

$$g^{\mu\nu} = \begin{pmatrix} -\frac{a^2}{a^4-8} & 0 & 0 & \frac{4\,e^{-x}}{a^4-8} \\ 0 & \frac{1}{a^2} & 0 & 0 \\ 0 & 0 & \frac{1}{a^2} & 0 \\ \frac{4\,e^{-x}}{a^4-8} & 0 & 0 & -\frac{2\,a^2\,e^{-2\,x}}{a^4-8} \end{pmatrix}$$

Christoffel Connection

$$\Gamma^0{}_{01} = -\frac{4}{a^4-8}$$

$$\Gamma^0{}_{10} = \Gamma^0{}_{01}$$

$$\Gamma^0{}_{13} = -\frac{a^2\,e^x}{a^4-8}$$

$$\Gamma^0{}_{31} = \Gamma^0{}_{13}$$

$$\Gamma^1{}_{03} = \frac{e^x}{a^2}$$

$$\Gamma^1{}_{30} = \Gamma^1{}_{03}$$

$$\Gamma^1{}_{33} = \frac{e^{2\,x}}{2}$$

$$\Gamma^3{}_{01} = \frac{2\,a^2\,e^{-x}}{a^4-8}$$

$$\Gamma^3{}_{10} = \Gamma^3{}_{01}$$

$$\Gamma^3{}_{13} = \frac{\left(a^2-2\right)\,\left(a^2+2\right)}{a^4-8}$$

$$\Gamma^3{}_{31} = \Gamma^3{}_{13}$$

Metric Compatibility

——— o.k.

Riemann Tensor

$$R^0{}_{003} = -\frac{4\,e^x}{a^2\,\left(a^4-8\right)}$$

$$R^0{}_{030} = -R^0{}_{003}$$

$$R^0{}_{101} = \frac{2}{a^4-8}$$

$$R^0{}_{110} = -R^0{}_{101}$$

$$R^0{}_{113} = -\frac{2\,a^2\,e^x}{a^4-8}$$

$$R^0{}_{131} = -R^0{}_{113}$$

$$R^0{}_{303} = -\frac{e^{2\,x}}{a^4-8}$$

$$R^0{}_{330} = -R^0{}_{303}$$

$$R^1{}_{001} = \frac{2}{a^4-8}$$

$$R^1{}_{010} = -R^1{}_{001}$$

$$R^1{}_{013} = -\frac{4\,e^x}{a^2\,\left(a^4-8\right)}$$

$$R^1{}_{031} = -R^1{}_{013}$$

$$R^1{}_{301} = \frac{4\,e^x}{a^2\,\left(a^4-8\right)}$$

$$R^1{}_{310} = -R^1{}_{301}$$

$$R^1{}_{313} = \frac{\left(a^4 - 10\right) e^{2x}}{2\,(a^4 - 8)}$$

$$R^1{}_{331} = -R^1{}_{313}$$

$$R^3{}_{003} = \frac{2}{a^4 - 8}$$

$$R^3{}_{030} = -R^3{}_{003}$$

$$R^3{}_{113} = \frac{a^4 - 2}{a^4 - 8}$$

$$R^3{}_{131} = -R^3{}_{113}$$

$$R^3{}_{303} = \frac{4\,e^x}{a^2\,(a^4 - 8)}$$

$$R^3{}_{330} = -R^3{}_{303}$$

Ricci Tensor

$$\mathrm{Ric}_{00} = -\frac{4}{a^4 - 8}$$

$$\mathrm{Ric}_{03} = -\frac{8\,e^x}{a^2\,(a^4 - 8)}$$

$$\mathrm{Ric}_{11} = -\frac{\left(a^2 - 2\right)\,\left(a^2 + 2\right)}{a^4 - 8}$$

$$\mathrm{Ric}_{30} = \mathrm{Ric}_{03}$$

$$\mathrm{Ric}_{33} = \frac{\left(a^4 - 12\right) e^{2x}}{2\,(a^4 - 8)}$$

Ricci Scalar

$$R_{sc} = -\frac{2\,\left(a^4 - 6\right)}{a^2\,(a^4 - 8)}$$

Bianchi identity (Ricci cyclic equation $R^\kappa{}_{[\mu\nu\sigma]} = 0$)

——— o.k.

Einstein Tensor

$$G_{00} = -\frac{a^4 - 2}{a^4 - 8}$$

$$G_{03} = -\frac{2\left(a^4 - 2\right) e^x}{a^2 \left(a^4 - 8\right)}$$

$$G_{11} = -\frac{2}{a^4 - 8}$$

$$G_{22} = \frac{a^4 - 6}{a^4 - 8}$$

$$G_{30} = G_{03}$$

$$G_{33} = -\frac{3\,e^{2\,x}}{a^4 - 8}$$

Hodge Dual of Bianchi Identity

——— (see charge and current densities)

Scalar Charge Density ($-R^{0}{}_{i}{}^{i0}$)

$$\rho = \frac{4}{\left(a^4 - 8\right)^2}$$

Current Density Class 1 ($-R^{i}{}_{\mu}{}^{\mu j}$)

$$J_1 = \frac{\left(a^2 - 2\right)\left(a^2 + 2\right)}{a^4 \left(a^4 - 8\right)}$$

$$J_2 = 0$$

$$J_3 = -\frac{2\left(a^2 - 2\right)\left(a^2 + 2\right) e^{-2\,x}}{\left(a^4 - 8\right)^2}$$

Current Density Class 2 ($-R^{i}{}_{\mu}{}^{\mu j}$)

$$J_1 = 0$$

$$J_2 = 0$$

$$J_3 = 0$$

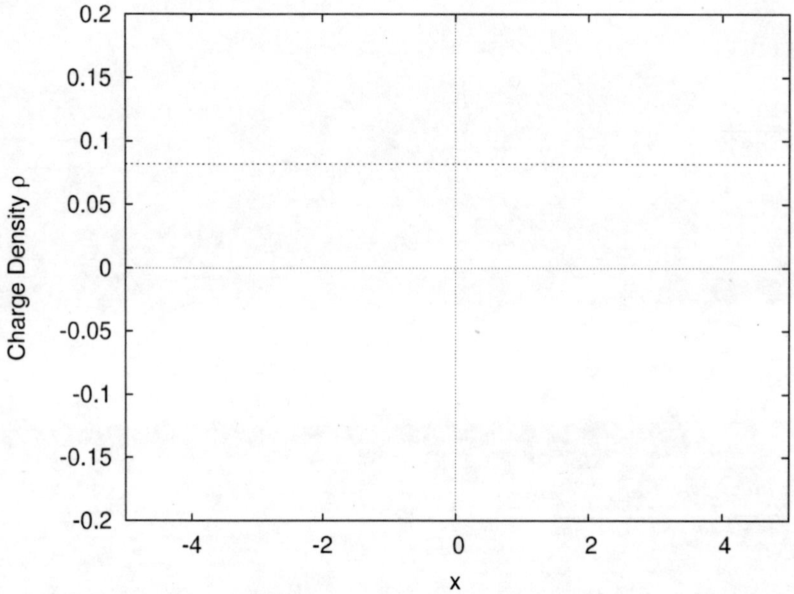

Fig. 4.89: Homogeneous perfect fluid, cartesian, charge density ρ for $a = 1$.

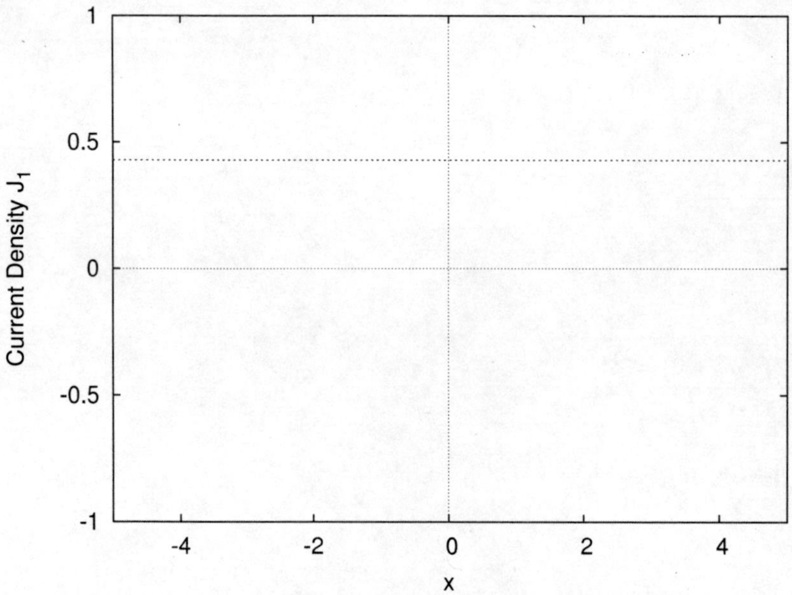

Fig. 4.90: Homogeneous perfect fluid, cartesian, current density J_r for $a = 1$.

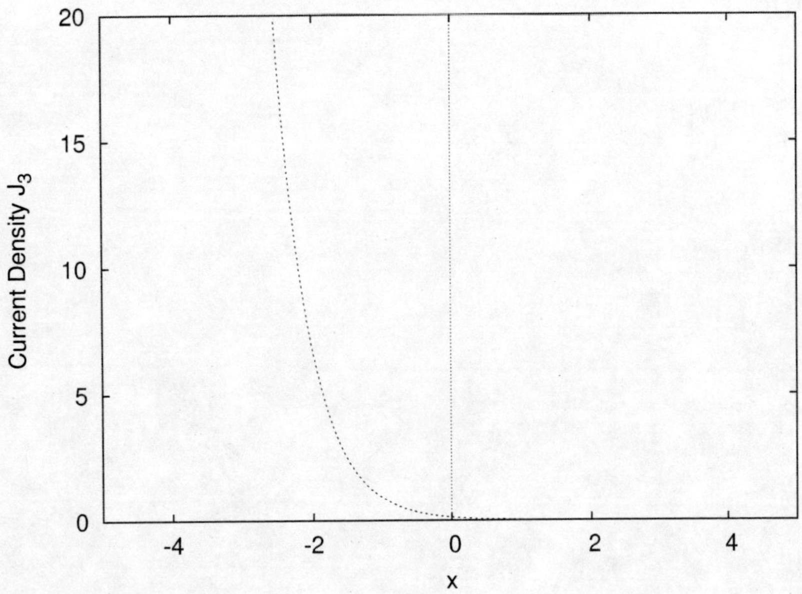

Fig. 4.91: Homogeneous perfect fluid, cartesian, current density J_φ for $a = 1$.

Current Density Class 3 (-$R^{i}{}_{\mu}{}^{\mu j}$)

$$J_1 = 0$$

$$J_2 = 0$$

$$J_3 = 0$$

4.4.50 Petrov type N metric

In this metric there is no diagonal element for v. ρ is a parameter. The radial charge density is rising exponentially.

Coordinates

$$\mathbf{x} = \begin{pmatrix} u \\ v \\ x \\ y \end{pmatrix}$$

Metric

$$g_{\mu\nu} = \begin{pmatrix} -2\,e^{2\,\rho\,x} & 2 & 0 & 0 \\ 2 & 0 & 0 & 0 \\ 0 & 0 & 1 & 0 \\ 0 & 0 & 0 & 1 \end{pmatrix}$$

Contravariant Metric

$$g^{\mu\nu} = \begin{pmatrix} 0 & \frac{1}{2} & 0 & 0 \\ \frac{1}{2} & \frac{e^{2\rho x}}{2} & 0 & 0 \\ 0 & 0 & 1 & 0 \\ 0 & 0 & 0 & 1 \end{pmatrix}$$

Christoffel Connection

$$\Gamma^{1}{}_{02} = -\rho\, e^{2\rho x}$$

$$\Gamma^{1}{}_{20} = \Gamma^{1}{}_{02}$$

$$\Gamma^{2}{}_{00} = 2\,\rho\, e^{2\rho x}$$

Metric Compatibility

———— o.k.

Riemann Tensor

$$R^{1}{}_{202} = 2\,\rho^{2}\, e^{2\rho x}$$

$$R^{1}{}_{220} = -R^{1}{}_{202}$$

$$R^{2}{}_{002} = -4\,\rho^{2}\, e^{2\rho x}$$

$$R^{2}{}_{020} = -R^{2}{}_{002}$$

Ricci Tensor

$$\mathrm{Ric}_{00} = 4\,\rho^{2}\, e^{2\rho x}$$

Ricci Scalar

$$R_{sc} = 0$$

Bianchi identity (Ricci cyclic equation $R^{\kappa}{}_{[\mu\nu\sigma]} = 0$)

———— o.k.

Einstein Tensor

$$G_{00} = 4\,\rho^{2}\, e^{2\rho x}$$

Hodge Dual of Bianchi Identity

——— (see charge and current densities)

Scalar Charge Density (-$R^{0}{}_{i}{}^{i0}$)

$$\rho = 0$$

Current Density Class 1 (-$R^{i}{}_{\mu}{}^{\mu j}$)

$$J_1 = -\rho^2 \, e^{2\,\rho\,x}$$

$$J_2 = 0$$

$$J_3 = 0$$

Current Density Class 2 (-$R^{i}{}_{\mu}{}^{\mu j}$)

$$J_1 = 0$$

$$J_2 = 0$$

$$J_3 = 0$$

Current Density Class 3 (-$R^{i}{}_{\mu}{}^{\mu j}$)

$$J_1 = 0$$

$$J_2 = 0$$

$$J_3 = 0$$

4.4.51 Space rotationally isotropic metric

$\epsilon = \pm 1$ is a parameter. $A(t)$ and $B(t)$ are time-dependent functions.

Coordinates

$$\mathbf{x} = \begin{pmatrix} t \\ x \\ y \\ z \end{pmatrix}$$

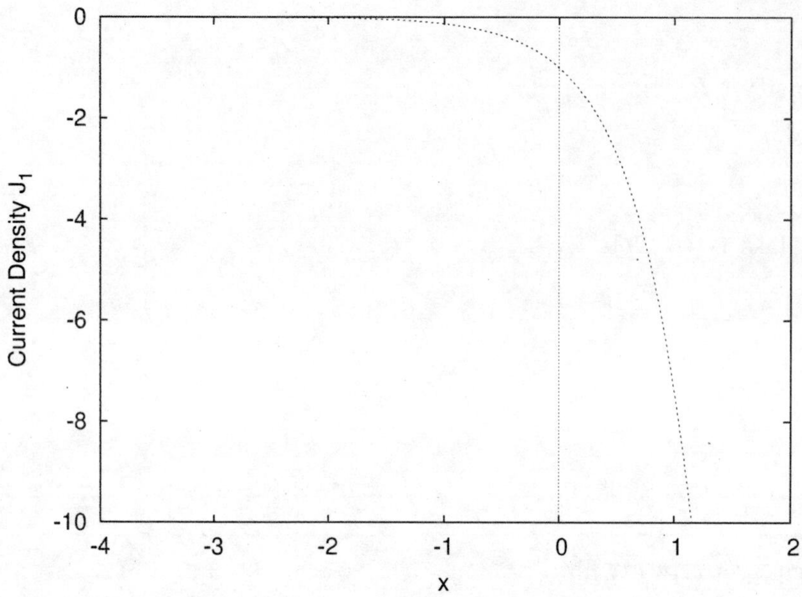

Fig. 4.92: Petrov type N metric, current density J_1 for $\rho = 1$.

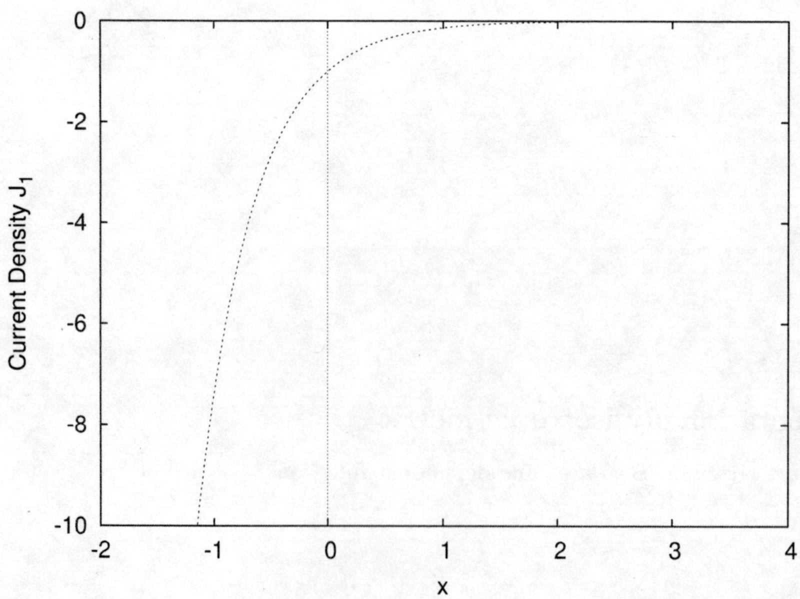

Fig. 4.93: Petrov type N metric, current density J_1 for $\rho = -1$.

Metric

$$g_{\mu\nu} = \begin{pmatrix} -\varepsilon & 0 & 0 & 0 \\ 0 & \varepsilon\, A^2 & 0 & 0 \\ 0 & 0 & e^{2\,x}\, B^2 & 0 \\ 0 & 0 & 0 & e^{2\,x}\, B^2 \end{pmatrix}$$

Contravariant Metric

$$g^{\mu\nu} = \begin{pmatrix} -\frac{1}{\varepsilon} & 0 & 0 & 0 \\ 0 & \frac{1}{\varepsilon\, A^2} & 0 & 0 \\ 0 & 0 & \frac{e^{-2\,x}}{B^2} & 0 \\ 0 & 0 & 0 & \frac{e^{-2\,x}}{B^2} \end{pmatrix}$$

Christoffel Connection

$$\Gamma^0{}_{11} = A\left(\frac{d}{d\,t}\,A\right)$$

$$\Gamma^0{}_{22} = \frac{e^{2\,x}\,B\,\left(\frac{d}{d\,t}\,B\right)}{\varepsilon}$$

$$\Gamma^0{}_{33} = \frac{e^{2\,x}\,B\,\left(\frac{d}{d\,t}\,B\right)}{\varepsilon}$$

$$\Gamma^1{}_{01} = \frac{\frac{d}{d\,t}\,A}{A}$$

$$\Gamma^1{}_{10} = \Gamma^1{}_{01}$$

$$\Gamma^1{}_{22} = -\frac{e^{2\,x}\,B^2}{\varepsilon\,A^2}$$

$$\Gamma^1{}_{33} = -\frac{e^{2\,x}\,B^2}{\varepsilon\,A^2}$$

$$\Gamma^2{}_{02} = \frac{\frac{d}{d\,t}\,B}{B}$$

$$\Gamma^2{}_{12} = 1$$

$$\Gamma^2{}_{20} = \Gamma^2{}_{02}$$

$$\Gamma^2{}_{21} = \Gamma^2{}_{12}$$

$$\Gamma^3{}_{03} = \frac{\frac{d}{d\,t}\,B}{B}$$

$$\Gamma^3{}_{13} = 1$$

$$\Gamma^3{}_{30} = \Gamma^3{}_{03}$$

$$\Gamma^3{}_{31} = \Gamma^3{}_{13}$$

Metric Compatibility

——— o.k.

Riemann Tensor

$$R^0{}_{101} = A\left(\frac{d^2}{d\,t^2}\,A\right)$$

$$R^0{}_{110} = -R^0{}_{101}$$

$$R^0{}_{202} = \frac{e^{2\,x}\,B\,\left(\frac{d^2}{d\,t^2}\,B\right)}{\varepsilon}$$

$$R^0{}_{212} = \frac{e^{2\,x}\,B\,\left(A\,\left(\frac{d}{d\,t}\,B\right) - \frac{d}{d\,t}\,A\,B\right)}{\varepsilon\,A}$$

$$R^0{}_{220} = -R^0{}_{202}$$

$$R^0{}_{221} = -R^0{}_{212}$$

$$R^0{}_{303} = \frac{e^{2\,x}\,B\,\left(\frac{d^2}{d\,t^2}\,B\right)}{\varepsilon}$$

$$R^0{}_{313} = \frac{e^{2\,x}\,B\,\left(A\,\left(\frac{d}{d\,t}\,B\right) - \frac{d}{d\,t}\,A\,B\right)}{\varepsilon\,A}$$

$$R^0{}_{330} = -R^0{}_{303}$$

$$R^0{}_{331} = -R^0{}_{313}$$

$$R^1{}_{001} = \frac{\frac{d^2}{d\,t^2}\,A}{A}$$

$$R^1{}_{010} = -R^1{}_{001}$$

$$R^1{}_{202} = -\frac{e^{2\,x}\,B\,\left(A\,\left(\frac{d}{d\,t}\,B\right) - \frac{d}{d\,t}\,A\,B\right)}{\varepsilon\,A^3}$$

$$R^1{}_{212} = \frac{e^{2\,x}\,B\,\left(A\,\left(\frac{d}{d\,t}\,A\right)\,\left(\frac{d}{d\,t}\,B\right) - B\right)}{\varepsilon\,A^2}$$

$$R^1{}_{220} = -R^1{}_{202}$$

$$R^1{}_{221} = -R^1{}_{212}$$

$$R^1{}_{303} = -\frac{e^{2x}\, B\, \left(A\, \left(\frac{d}{d\,t}\, B\right) - \frac{d}{d\,t}\, A\, B\right)}{\varepsilon\, A^3}$$

$$R^1{}_{313} = \frac{e^{2x}\, B\, \left(A\, \left(\frac{d}{d\,t}\, A\right)\, \left(\frac{d}{d\,t}\, B\right) - B\right)}{\varepsilon\, A^2}$$

$$R^1{}_{330} = -R^1{}_{303}$$

$$R^1{}_{331} = -R^1{}_{313}$$

$$R^2{}_{002} = \frac{\frac{d^2}{d\,t^2}\, B}{B}$$

$$R^2{}_{012} = \frac{A\, \left(\frac{d}{d\,t}\, B\right) - \frac{d}{d\,t}\, A\, B}{A\, B}$$

$$R^2{}_{020} = -R^2{}_{002}$$

$$R^2{}_{021} = -R^2{}_{012}$$

$$R^2{}_{102} = \frac{A\, \left(\frac{d}{d\,t}\, B\right) - \frac{d}{d\,t}\, A\, B}{A\, B}$$

$$R^2{}_{112} = -\frac{A\, \left(\frac{d}{d\,t}\, A\right)\, \left(\frac{d}{d\,t}\, B\right) - B}{B}$$

$$R^2{}_{120} = -R^2{}_{102}$$

$$R^2{}_{121} = -R^2{}_{112}$$

$$R^2{}_{323} = \frac{e^{2x}\, \left(A\, \left(\frac{d}{d\,t}\, B\right) - B\right)\, \left(A\, \left(\frac{d}{d\,t}\, B\right) + B\right)}{\varepsilon\, A^2}$$

$$R^2{}_{332} = -R^2{}_{323}$$

$$R^3{}_{003} = \frac{\frac{d^2}{d\,t^2}\, B}{B}$$

$$R^3{}_{013} = \frac{A\, \left(\frac{d}{d\,t}\, B\right) - \frac{d}{d\,t}\, A\, B}{A\, B}$$

$$R^3{}_{030} = -R^3{}_{003}$$

$$R^3{}_{031} = -R^3{}_{013}$$

$$R^3{}_{103} = \frac{A\, \left(\frac{d}{d\,t}\, B\right) - \frac{d}{d\,t}\, A\, B}{A\, B}$$

$$R^3{}_{113} = -\frac{A\, \left(\frac{d}{d\,t}\, A\right)\, \left(\frac{d}{d\,t}\, B\right) - B}{B}$$

$$R^3{}_{130} = -R^3{}_{103}$$

$$R^3{}_{131} = -R^3{}_{113}$$

$$R^3{}_{223} = -\frac{e^{2x}\, \left(A\, \left(\frac{d}{d\,t}\, B\right) - B\right)\, \left(A\, \left(\frac{d}{d\,t}\, B\right) + B\right)}{\varepsilon\, A^2}$$

$$R^3{}_{232} = -R^3{}_{223}$$

Ricci Tensor

$$\mathrm{Ric}_{00} = -\frac{2\,A\,\left(\frac{d^2}{d\,t^2}\,B\right) + \frac{d^2}{d\,t^2}\,A\,B}{A\,B}$$

$$\mathrm{Ric}_{01} = -\frac{2\,\left(A\,\left(\frac{d}{d\,t}\,B\right) - \frac{d}{d\,t}\,A\,B\right)}{A\,B}$$

$$\mathrm{Ric}_{10} = \mathrm{Ric}_{01}$$

$$\mathrm{Ric}_{11} = \frac{2\,A\,\left(\frac{d}{d\,t}\,A\right)\,\left(\frac{d}{d\,t}\,B\right) + A\,\left(\frac{d^2}{d\,t^2}\,A\right)\,B - 2\,B}{B}$$

$$\mathrm{Ric}_{22} = \frac{e^{2\,x}\,\left(A^2\,B\,\left(\frac{d^2}{d\,t^2}\,B\right) + A^2\,\left(\frac{d}{d\,t}\,B\right)^2 + A\,\left(\frac{d}{d\,t}\,A\right)\,B\,\left(\frac{d}{d\,t}\,B\right) - 2\,B^2\right)}{\varepsilon\,A^2}$$

$$\mathrm{Ric}_{33} = \frac{e^{2\,x}\,\left(A^2\,B\,\left(\frac{d^2}{d\,t^2}\,B\right) + A^2\,\left(\frac{d}{d\,t}\,B\right)^2 + A\,\left(\frac{d}{d\,t}\,A\right)\,B\,\left(\frac{d}{d\,t}\,B\right) - 2\,B^2\right)}{\varepsilon\,A^2}$$

Ricci Scalar

$$R_{sc} = \frac{2\,\left(2\,A^2\,B\,\left(\frac{d^2}{d\,t^2}\,B\right) + A^2\,\left(\frac{d}{d\,t}\,B\right)^2 + 2\,A\,\left(\frac{d}{d\,t}\,A\right)\,B\,\left(\frac{d}{d\,t}\,B\right) + A\,\left(\frac{d^2}{d\,t^2}\,A\right)\,B^2 - 3\,B^2\right)}{\varepsilon\,A^2\,B^2}$$

Bianchi identity (Ricci cyclic equation $R^{\kappa}{}_{[\mu\nu\sigma]} = 0$)

——— o.k.

Einstein Tensor

$$G_{00} = \frac{A^2\,\left(\frac{d}{d\,t}\,B\right)^2 + 2\,A\,\left(\frac{d}{d\,t}\,A\right)\,B\,\left(\frac{d}{d\,t}\,B\right) - 3\,B^2}{A^2\,B^2}$$

$$G_{01} = -\frac{2\,\left(A\,\left(\frac{d}{d\,t}\,B\right) - \frac{d}{d\,t}\,A\,B\right)}{A\,B}$$

$$G_{10} = G_{01}$$

$$G_{11} = -\frac{2\,A^2\,B\,\left(\frac{d^2}{d\,t^2}\,B\right) + A^2\,\left(\frac{d}{d\,t}\,B\right)^2 - B^2}{B^2}$$

$$G_{22} = -\frac{e^{2\,x}\,B\,\left(A^2\,\left(\frac{d^2}{d\,t^2}\,B\right) + A\,\left(\frac{d}{d\,t}\,A\right)\,\left(\frac{d}{d\,t}\,B\right) + A\,\left(\frac{d^2}{d\,t^2}\,A\right)\,B - B\right)}{\varepsilon\,A^2}$$

$$G_{33} = -\frac{e^{2\,x}\,B\,\left(A^2\,\left(\frac{d^2}{d\,t^2}\,B\right) + A\,\left(\frac{d}{d\,t}\,A\right)\,\left(\frac{d}{d\,t}\,B\right) + A\,\left(\frac{d^2}{d\,t^2}\,A\right)\,B - B\right)}{\varepsilon\,A^2}$$

Hodge Dual of Bianchi Identity

——— (see charge and current densities)

Scalar Charge Density ($-R^{0}{}_{i}{}^{i0}$)

$$\rho = -\frac{2\,A\,\left(\frac{d^2}{d\,t^2}\,B\right) + \frac{d^2}{d\,t^2}\,A\,B}{\varepsilon^2\,A\,B}$$

Current Density Class 1 ($-R^{i}{}_{\mu}{}^{\mu j}$)

$$J_1 = -\frac{2\,A\,\left(\frac{d}{d\,t}\,A\right)\,\left(\frac{d}{d\,t}\,B\right) + A\,\left(\frac{d^2}{d\,t^2}\,A\right)\,B - 2\,B}{\varepsilon^2\,A^4\,B}$$

$$J_2 = -\frac{e^{-2\,x}\,\left(A^2\,B\,\left(\frac{d^2}{d\,t^2}\,B\right) + A^2\,\left(\frac{d}{d\,t}\,B\right)^2 + A\,\left(\frac{d}{d\,t}\,A\right)\,B\,\left(\frac{d}{d\,t}\,B\right) - 2\,B^2\right)}{\varepsilon\,A^2\,B^4}$$

$$J_3 = -\frac{e^{-2\,x}\,\left(A^2\,B\,\left(\frac{d^2}{d\,t^2}\,B\right) + A^2\,\left(\frac{d}{d\,t}\,B\right)^2 + A\,\left(\frac{d}{d\,t}\,A\right)\,B\,\left(\frac{d}{d\,t}\,B\right) - 2\,B^2\right)}{\varepsilon\,A^2\,B^4}$$

Current Density Class 2 ($-R^{i}{}_{\mu}{}^{\mu j}$)

$$J_1 = 0$$

$$J_2 = 0$$

$$J_3 = 0$$

Current Density Class 3 ($-R^{i}{}_{\mu}{}^{\mu j}$)

$$J_1 = 0$$

$$J_2 = 0$$

$$J_3 = 0$$

4.4.52 Electrovacuum metric

In the metric of electrovacuum there is only a constant current density J_x.

Coordinates

$$\mathbf{x} = \begin{pmatrix} t \\ x \\ y \\ z \end{pmatrix}$$

Metric

$$g_{\mu\nu} = \begin{pmatrix} -1 & 0 & 0 & 0 \\ 0 & \frac{4}{3} & 0 & 0 \\ 0 & 0 & e^{-2x} & 0 \\ 0 & 0 & 0 & e^{2x} \end{pmatrix}$$

Contravariant Metric

$$g^{\mu\nu} = \begin{pmatrix} -1 & 0 & 0 & 0 \\ 0 & \frac{3}{2^2} & 0 & 0 \\ 0 & 0 & e^{2x} & 0 \\ 0 & 0 & 0 & e^{-2x} \end{pmatrix}$$

Christoffel Connection

$$\Gamma^1{}_{22} = \frac{3\,e^{-2x}}{4}$$

$$\Gamma^1{}_{33} = -\frac{3\,e^{2x}}{4}$$

$$\Gamma^2{}_{12} = -1$$

$$\Gamma^2{}_{21} = \Gamma^2{}_{12}$$

$$\Gamma^3{}_{13} = 1$$

$$\Gamma^3{}_{31} = \Gamma^3{}_{13}$$

Metric Compatibility

——— o.k.

Riemann Tensor

$$R^1{}_{212} = -\frac{3\,e^{-2x}}{4}$$

$$R^1{}_{221} = -R^1{}_{212}$$

$$R^1{}_{313} = -\frac{3\,e^{2x}}{4}$$

$$R^1{}_{331} = -R^1{}_{313}$$

$$R^2{}_{112} = 1$$

$$R^{2}{}_{121} = -R^{2}{}_{112}$$

$$R^{2}{}_{323} = \frac{3\,e^{2\,x}}{4}$$

$$R^{2}{}_{332} = -R^{2}{}_{323}$$

$$R^{3}{}_{113} = 1$$

$$R^{3}{}_{131} = -R^{3}{}_{113}$$

$$R^{3}{}_{223} = -\frac{3\,e^{-2\,x}}{4}$$

$$R^{3}{}_{232} = -R^{3}{}_{223}$$

Ricci Tensor

$$\mathrm{Ric}_{11} = -2$$

Ricci Scalar

$$R_{sc} = -\frac{3}{2}$$

Bianchi identity (Ricci cyclic equation $R^{\kappa}{}_{[\mu\nu\sigma]} = 0$)

——— o.k.

Einstein Tensor

$$G_{00} = -\frac{3}{2^2}$$

$$G_{11} = -1$$

$$G_{22} = \frac{3\,e^{-2\,x}}{4}$$

$$G_{33} = \frac{3\,e^{2\,x}}{4}$$

Hodge Dual of Bianchi Identity

——— (see charge and current densities)

Scalar Charge Density (-$R^{0}{}_{i}{}^{i0}$)

$$\rho = 0$$

Current Density Class 1 ($-R^{i}{}_{\mu}{}^{\mu j}$)

$$J_1 = \frac{3^2}{2^3}$$

$$J_2 = 0$$

$$J_3 = 0$$

Current Density Class 2 ($-R^{i}{}_{\mu}{}^{\mu j}$)

$$J_1 = 0$$

$$J_2 = 0$$

$$J_3 = 0$$

Current Density Class 3 ($-R^{i}{}_{\mu}{}^{\mu j}$)

$$J_1 = 0$$

$$J_2 = 0$$

$$J_3 = 0$$

4.4.53 Spatially homogeneous perfect fluid cosmologies

Metric of spatially homogeneous perfect fluid cosmologies. Coordinate functions are $a(t)$ and $\Sigma(r,k)$ where k is a parameter. The coordinate dependence of charge and current density is implicitly defined by the models of a and Σ. There is no explicit coordinate dependence.

Coordinates

$$\mathbf{x} = \begin{pmatrix} t \\ r \\ \vartheta \\ \varphi \end{pmatrix}$$

Metric

$$g_{\mu\nu} = \begin{pmatrix} -1 & 0 & 0 & 0 \\ 0 & a^2 & 0 & 0 \\ 0 & 0 & a^2\,\Sigma^2 & 0 \\ 0 & 0 & 0 & a^2\,\Sigma^2 \sin^2\vartheta \end{pmatrix}$$

Contravariant Metric

$$g^{\mu\nu} = \begin{pmatrix} -1 & 0 & 0 & 0 \\ 0 & \frac{1}{a^2} & 0 & 0 \\ 0 & 0 & \frac{1}{a^2\,\Sigma^2} & 0 \\ 0 & 0 & 0 & \frac{1}{a^2\,\Sigma^2\,\sin^2\vartheta} \end{pmatrix}$$

Christoffel Connection

$$\Gamma^0{}_{11} = a\left(\frac{d}{d\,t}\,a\right)$$

$$\Gamma^0{}_{22} = a\left(\frac{d}{d\,t}\,a\right)\,\Sigma^2$$

$$\Gamma^0{}_{33} = a\left(\frac{d}{d\,t}\,a\right)\,\Sigma^2\,\sin^2\vartheta$$

$$\Gamma^1{}_{01} = \frac{\frac{d}{d\,t}\,a}{a}$$

$$\Gamma^1{}_{10} = \Gamma^1{}_{01}$$

$$\Gamma^1{}_{22} = -\Sigma\left(\frac{d}{d\,r}\,\Sigma\right)$$

$$\Gamma^1{}_{33} = -\Sigma\left(\frac{d}{d\,r}\,\Sigma\right)\,\sin^2\vartheta$$

$$\Gamma^2{}_{02} = \frac{\frac{d}{d\,t}\,a}{a}$$

$$\Gamma^2{}_{12} = \frac{\frac{d}{d\,r}\,\Sigma}{\Sigma}$$

$$\Gamma^2{}_{20} = \Gamma^2{}_{02}$$

$$\Gamma^2{}_{21} = \Gamma^2{}_{12}$$

$$\Gamma^2{}_{33} = -\cos\vartheta\,\sin\vartheta$$

$$\Gamma^3{}_{03} = \frac{\frac{d}{d\,t}\,a}{a}$$

$$\Gamma^3{}_{13} = \frac{\frac{d}{d\,r}\,\Sigma}{\Sigma}$$

$$\Gamma^3{}_{23} = \frac{\cos\vartheta}{\sin\vartheta}$$

$$\Gamma^3{}_{30} = \Gamma^3{}_{03}$$

$$\Gamma^3{}_{31} = \Gamma^3{}_{13}$$

$$\Gamma^3{}_{32} = \Gamma^3{}_{23}$$

Metric Compatibility

——— o.k.

Riemann Tensor

$$R^0{}_{101} = a \left(\frac{d^2}{d\,t^2}\, a \right)$$

$$R^0{}_{110} = -R^0{}_{101}$$

$$R^0{}_{202} = a \left(\frac{d^2}{d\,t^2}\, a \right) \Sigma^2$$

$$R^0{}_{220} = -R^0{}_{202}$$

$$R^0{}_{303} = a \left(\frac{d^2}{d\,t^2}\, a \right) \Sigma^2 \sin^2 \vartheta$$

$$R^0{}_{330} = -R^0{}_{303}$$

$$R^1{}_{001} = \frac{\frac{d^2}{d\,t^2}\, a}{a}$$

$$R^1{}_{010} = -R^1{}_{001}$$

$$R^1{}_{212} = -\Sigma \left(\frac{d^2}{d\,r^2}\, \Sigma - \left(\frac{d}{d\,t}\, a \right)^2 \Sigma \right)$$

$$R^1{}_{221} = -R^1{}_{212}$$

$$R^1{}_{313} = -\Sigma \left(\frac{d^2}{d\,r^2}\, \Sigma - \left(\frac{d}{d\,t}\, a \right)^2 \Sigma \right) \sin^2 \vartheta$$

$$R^1{}_{331} = -R^1{}_{313}$$

$$R^2{}_{002} = \frac{\frac{d^2}{d\,t^2}\, a}{a}$$

$$R^2{}_{020} = -R^2{}_{002}$$

$$R^2{}_{112} = \frac{\frac{d^2}{d\,r^2}\, \Sigma - \left(\frac{d}{d\,t}\, a \right)^2 \Sigma}{\Sigma}$$

$$R^2{}_{121} = -R^2{}_{112}$$

$$R^2{}_{323} = -\left(\left(\frac{d}{dr}\Sigma\right)^2 - \left(\frac{d}{dt}a\right)^2 \Sigma^2 - 1\right)\sin^2\vartheta$$

$$R^2{}_{332} = -R^2{}_{323}$$

$$R^3{}_{003} = \frac{\frac{d^2}{dt^2}a}{a}$$

$$R^3{}_{030} = -R^3{}_{003}$$

$$R^3{}_{113} = \frac{\frac{d^2}{dr^2}\Sigma - \left(\frac{d}{dt}a\right)^2 \Sigma}{\Sigma}$$

$$R^3{}_{131} = -R^3{}_{113}$$

$$R^3{}_{223} = \left(\frac{d}{dr}\Sigma\right)^2 - \left(\frac{d}{dt}a\right)^2 \Sigma^2 - 1$$

$$R^3{}_{232} = -R^3{}_{223}$$

Ricci Tensor

$$\mathrm{Ric}_{00} = -\frac{3\left(\frac{d^2}{dt^2}a\right)}{a}$$

$$\mathrm{Ric}_{11} = -\frac{2\left(\frac{d^2}{dr^2}\Sigma\right) - a\left(\frac{d^2}{dt^2}a\right)\Sigma - 2\left(\frac{d}{dt}a\right)^2\Sigma}{\Sigma}$$

$$\mathrm{Ric}_{22} = -\left(\Sigma\left(\frac{d^2}{dr^2}\Sigma\right) + \left(\frac{d}{dr}\Sigma\right)^2 - a\left(\frac{d^2}{dt^2}a\right)\Sigma^2 - 2\left(\frac{d}{dt}a\right)^2\Sigma^2 - 1\right)$$

$$\mathrm{Ric}_{33} = -\left(\Sigma\left(\frac{d^2}{dr^2}\Sigma\right) + \left(\frac{d}{dr}\Sigma\right)^2 - a\left(\frac{d^2}{dt^2}a\right)\Sigma^2 - 2\left(\frac{d}{dt}a\right)^2\Sigma^2 - 1\right)\sin^2\vartheta$$

Ricci Scalar

$$R_{sc} = -\frac{2\left(2\Sigma\left(\frac{d^2}{dr^2}\Sigma\right) + \left(\frac{d}{dr}\Sigma\right)^2 - 3a\left(\frac{d^2}{dt^2}a\right)\Sigma^2 - 3\left(\frac{d}{dt}a\right)^2\Sigma^2 - 1\right)}{a^2\Sigma^2}$$

Bianchi identity (Ricci cyclic equation $R^{\kappa}{}_{[\mu\nu\sigma]} = 0$)

——— o.k.

Einstein Tensor

$$G_{00} = -\frac{2\,\Sigma\,\left(\frac{d^2}{d\,r^2}\,\Sigma\right) + \left(\frac{d}{d\,r}\,\Sigma\right)^2 - 3\,\left(\frac{d}{d\,t}\,a\right)^2\,\Sigma^2 - 1}{a^2\,\Sigma^2}$$

$$G_{11} = \frac{\left(\frac{d}{d\,r}\,\Sigma\right)^2 - 2\,a\,\left(\frac{d^2}{d\,t^2}\,a\right)\,\Sigma^2 - \left(\frac{d}{d\,t}\,a\right)^2\,\Sigma^2 - 1}{\Sigma^2}$$

$$G_{22} = \Sigma\,\left(\frac{d^2}{d\,r^2}\,\Sigma - 2\,a\,\left(\frac{d^2}{d\,t^2}\,a\right)\,\Sigma - \left(\frac{d}{d\,t}\,a\right)^2\,\Sigma\right)$$

$$G_{33} = \Sigma\,\left(\frac{d^2}{d\,r^2}\,\Sigma - 2\,a\,\left(\frac{d^2}{d\,t^2}\,a\right)\,\Sigma - \left(\frac{d}{d\,t}\,a\right)^2\,\Sigma\right)\,\sin^2\vartheta$$

Hodge Dual of Bianchi Identity

——— (see charge and current densities)

Scalar Charge Density ($-R^{0}{}_{i}{}^{i0}$)

$$\rho = -\frac{3\,\left(\frac{d^2}{d\,t^2}\,a\right)}{a}$$

Current Density Class 1 ($-R^{i}{}_{\mu}{}^{\mu j}$)

$$J_1 = \frac{2\,\left(\frac{d^2}{d\,r^2}\,\Sigma\right) - a\,\left(\frac{d^2}{d\,t^2}\,a\right)\,\Sigma - 2\,\left(\frac{d}{d\,t}\,a\right)^2\,\Sigma}{a^4\,\Sigma}$$

$$J_2 = \frac{\Sigma\,\left(\frac{d^2}{d\,r^2}\,\Sigma\right) + \left(\frac{d}{d\,r}\,\Sigma\right)^2 - a\,\left(\frac{d^2}{d\,t^2}\,a\right)\,\Sigma^2 - 2\,\left(\frac{d}{d\,t}\,a\right)^2\,\Sigma^2 - 1}{a^4\,\Sigma^4}$$

$$J_3 = \frac{\Sigma\,\left(\frac{d^2}{d\,r^2}\,\Sigma\right) + \left(\frac{d}{d\,r}\,\Sigma\right)^2 - a\,\left(\frac{d^2}{d\,t^2}\,a\right)\,\Sigma^2 - 2\,\left(\frac{d}{d\,t}\,a\right)^2\,\Sigma^2 - 1}{a^4\,\Sigma^4\,\sin^2\vartheta}$$

Current Density Class 2 ($-R^{i}{}_{\mu}{}^{\mu j}$)

$$J_1 = 0$$

$$J_2 = 0$$

$$J_3 = 0$$

Current Density Class 3 ($-R^{i}{}_{\mu}{}^{\mu j}$)

$$J_1 = 0$$

$$J_2 = 0$$

$$J_3 = 0$$

4.4.54 The main cosmological models

Metric of the main cosmological models. Coordinate functions are $A(t)$, $B(t)$, and $\Sigma(y, k)$ where k is a parameter. The coordinate dependence of charge and current density is implicitly defined by the models of A, B, and Σ. There is no explicit coordinate dependence.

Coordinates

$$\mathbf{x} = \begin{pmatrix} t \\ x \\ y \\ z \end{pmatrix}$$

Metric

$$g_{\mu\nu} = \begin{pmatrix} -1 & 0 & 0 & 0 \\ 0 & A^2 & 0 & 0 \\ 0 & 0 & B^2 & 0 \\ 0 & 0 & 0 & \Sigma^2 B^2 \end{pmatrix}$$

Contravariant Metric

$$g^{\mu\nu} = \begin{pmatrix} -1 & 0 & 0 & 0 \\ 0 & \frac{1}{A^2} & 0 & 0 \\ 0 & 0 & \frac{1}{B^2} & 0 \\ 0 & 0 & 0 & \frac{1}{\Sigma^2 B^2} \end{pmatrix}$$

Christoffel Connection

$$\Gamma^{0}{}_{11} = A \left(\frac{d}{dt} A\right)$$

$$\Gamma^{0}{}_{22} = B \left(\frac{d}{dt} B\right)$$

$$\Gamma^{0}{}_{33} = \Sigma^2 B \left(\frac{d}{dt} B\right)$$

$$\Gamma^{1}{}_{01} = \frac{\frac{d}{dt} A}{A}$$

$${\Gamma^{1}}_{10} = {\Gamma^{1}}_{01}$$

$${\Gamma^{2}}_{02} = \frac{\frac{d}{dt} B}{B}$$

$${\Gamma^{2}}_{20} = {\Gamma^{2}}_{02}$$

$${\Gamma^{2}}_{33} = -\Sigma \left(\frac{d}{dy} \Sigma \right)$$

$${\Gamma^{3}}_{03} = \frac{\frac{d}{dt} B}{B}$$

$${\Gamma^{3}}_{23} = \frac{\frac{d}{dy} \Sigma}{\Sigma}$$

$${\Gamma^{3}}_{30} = {\Gamma^{3}}_{03}$$

$${\Gamma^{3}}_{32} = {\Gamma^{3}}_{23}$$

Metric Compatibility

———— o.k.

Riemann Tensor

$${R^{0}}_{101} = A \left(\frac{d^2}{dt^2} A \right)$$

$${R^{0}}_{110} = -{R^{0}}_{101}$$

$${R^{0}}_{202} = B \left(\frac{d^2}{dt^2} B \right)$$

$${R^{0}}_{220} = -{R^{0}}_{202}$$

$${R^{0}}_{303} = \Sigma^2 B \left(\frac{d^2}{dt^2} B \right)$$

$${R^{0}}_{330} = -{R^{0}}_{303}$$

$${R^{1}}_{001} = \frac{\frac{d^2}{dt^2} A}{A}$$

$${R^{1}}_{010} = -{R^{1}}_{001}$$

$$R^1{}_{212} = \frac{\frac{d}{dt} A\, B\, \left(\frac{d}{dt} B\right)}{A}$$

$$R^1{}_{221} = -R^1{}_{212}$$

$$R^1{}_{313} = \frac{\Sigma^2\, \left(\frac{d}{dt} A\right)\, B\, \left(\frac{d}{dt} B\right)}{A}$$

$$R^1{}_{331} = -R^1{}_{313}$$

$$R^2{}_{002} = \frac{\frac{d^2}{dt^2} B}{B}$$

$$R^2{}_{020} = -R^2{}_{002}$$

$$R^2{}_{112} = -\frac{A\, \left(\frac{d}{dt} A\right)\, \left(\frac{d}{dt} B\right)}{B}$$

$$R^2{}_{121} = -R^2{}_{112}$$

$$R^2{}_{323} = \Sigma\, \left(\Sigma\, \left(\frac{d}{dt} B\right)^2 - \frac{d^2}{dy^2} \Sigma\right)$$

$$R^2{}_{332} = -R^2{}_{323}$$

$$R^3{}_{003} = \frac{\frac{d^2}{dt^2} B}{B}$$

$$R^3{}_{030} = -R^3{}_{003}$$

$$R^3{}_{113} = -\frac{A\, \left(\frac{d}{dt} A\right)\, \left(\frac{d}{dt} B\right)}{B}$$

$$R^3{}_{131} = -R^3{}_{113}$$

$$R^3{}_{223} = -\frac{\Sigma\, \left(\frac{d}{dt} B\right)^2 - \frac{d^2}{dy^2} \Sigma}{\Sigma}$$

$$R^3{}_{232} = -R^3{}_{223}$$

Ricci Tensor

$$\mathrm{Ric}_{00} = -\frac{2\, A\, \left(\frac{d^2}{dt^2} B\right) + \frac{d^2}{dt^2} A\, B}{A\, B}$$

$$\mathrm{Ric}_{11} = \frac{A\, \left(2\, \left(\frac{d}{dt} A\right)\, \left(\frac{d}{dt} B\right) + \frac{d^2}{dt^2} A\, B\right)}{B}$$

$$\mathrm{Ric}_{22} = \frac{\Sigma\, A\, B\, \left(\frac{d^2}{dt^2} B\right) + \Sigma\, A\, \left(\frac{d}{dt} B\right)^2 + \Sigma\, \left(\frac{d}{dt} A\right)\, B\, \left(\frac{d}{dt} B\right) - \frac{d^2}{dy^2} \Sigma\, A}{\Sigma\, A}$$

$$\mathrm{Ric}_{33} = \frac{\Sigma\, \left(\Sigma\, A\, B\, \left(\frac{d^2}{dt^2} B\right) + \Sigma\, A\, \left(\frac{d}{dt} B\right)^2 + \Sigma\, \left(\frac{d}{dt} A\right)\, B\, \left(\frac{d}{dt} B\right) - \frac{d^2}{dy^2} \Sigma\, A\right)}{A}$$

Ricci Scalar

$$R_{sc} = \frac{2\left(2\,\Sigma\,A\,B\,\left(\frac{d^2}{d\,t^2}\,B\right) + \Sigma\,A\,\left(\frac{d}{d\,t}\,B\right)^2 + 2\,\Sigma\,\left(\frac{d}{d\,t}\,A\right)\,B\,\left(\frac{d}{d\,t}\,B\right) + \Sigma\,\left(\frac{d^2}{d\,t^2}\,A\right)\,B^2 - \frac{d^2}{d\,y^2}\,\Sigma\,A\right)}{\Sigma\,A\,B^2}$$

Bianchi identity (Ricci cyclic equation $R^{\kappa}{}_{[\mu\nu\sigma]} = 0$)

——— o.k.

Einstein Tensor

$$G_{00} = \frac{\Sigma\,A\,\left(\frac{d}{d\,t}\,B\right)^2 + 2\,\Sigma\,\left(\frac{d}{d\,t}\,A\right)\,B\,\left(\frac{d}{d\,t}\,B\right) - \frac{d^2}{d\,y^2}\,\Sigma\,A}{\Sigma\,A\,B^2}$$

$$G_{11} = -\frac{A^2\,\left(2\,\Sigma\,B\,\left(\frac{d^2}{d\,t^2}\,B\right) + \Sigma\,\left(\frac{d}{d\,t}\,B\right)^2 - \frac{d^2}{d\,y^2}\,\Sigma\right)}{\Sigma\,B^2}$$

$$G_{22} = -\frac{B\,\left(A\,\left(\frac{d^2}{d\,t^2}\,B\right) + \frac{d}{d\,t}\,A\,\left(\frac{d}{d\,t}\,B\right) + \frac{d^2}{d\,t^2}\,A\,B\right)}{A}$$

$$G_{33} = -\frac{\Sigma^2\,B\,\left(A\,\left(\frac{d^2}{d\,t^2}\,B\right) + \frac{d}{d\,t}\,A\,\left(\frac{d}{d\,t}\,B\right) + \frac{d^2}{d\,t^2}\,A\,B\right)}{A}$$

Hodge Dual of Bianchi Identity

——— (see charge and current densities)

Scalar Charge Density (-$R^{0}{}_{i}{}^{i0}$)

$$\rho = -\frac{2\,A\,\left(\frac{d^2}{d\,t^2}\,B\right) + \frac{d^2}{d\,t^2}\,A\,B}{A\,B}$$

Current Density Class 1 (-$R^{i}{}_{\mu}{}^{\mu j}$)

$$J_1 = -\frac{2\,\left(\frac{d}{d\,t}\,A\right)\,\left(\frac{d}{d\,t}\,B\right) + \frac{d^2}{d\,t^2}\,A\,B}{A^3\,B}$$

$$J_2 = -\frac{\Sigma\,A\,B\,\left(\frac{d^2}{d\,t^2}\,B\right) + \Sigma\,A\,\left(\frac{d}{d\,t}\,B\right)^2 + \Sigma\,\left(\frac{d}{d\,t}\,A\right)\,B\,\left(\frac{d}{d\,t}\,B\right) - \frac{d^2}{d\,y^2}\,\Sigma\,A}{\Sigma\,A\,B^4}$$

$$J_3 = -\frac{\Sigma\,A\,B\,\left(\frac{d^2}{d\,t^2}\,B\right) + \Sigma\,A\,\left(\frac{d}{d\,t}\,B\right)^2 + \Sigma\,\left(\frac{d}{d\,t}\,A\right)\,B\,\left(\frac{d}{d\,t}\,B\right) - \frac{d^2}{d\,y^2}\,\Sigma\,A}{\Sigma^3\,A\,B^4}$$

Current Density Class 2 (-$R^{i}{}_{\mu}{}^{\mu j}$)

$$J_1 = 0$$

$$J_2 = 0$$

$$J_3 = 0$$

Current Density Class 3 (-$R^{i}{}_{\mu}{}^{\mu j}$)

$$J_1 = 0$$

$$J_2 = 0$$

$$J_3 = 0$$

4.4.55 Petrov type D fluid

a and n are parameters. The charge and current densities are partly increasing exponentially with x.

Coordinates

$$\mathbf{x} = \begin{pmatrix} t \\ x \\ y \\ z \end{pmatrix}$$

Metric

$$g_{\mu\nu} = \begin{pmatrix} -e^{-2\,a\,x} & 0 & 0 & 0 \\ 0 & 1 & 0 & 0 \\ 0 & 0 & t^{n+1}\,e^{-2\,a\,x} & 0 \\ 0 & 0 & 0 & t^{1-n}\,e^{-2\,a\,x} \end{pmatrix}$$

Contravariant Metric

$$g^{\mu\nu} = \begin{pmatrix} -e^{2\,a\,x} & 0 & 0 & 0 \\ 0 & 1 & 0 & 0 \\ 0 & 0 & t^{-n-1}\,e^{2\,a\,x} & 0 \\ 0 & 0 & 0 & t^{n-1}\,e^{2\,a\,x} \end{pmatrix}$$

Christoffel Connection

$$\Gamma^0{}_{01} = -a$$

$$\Gamma^0{}_{10} = \Gamma^0{}_{01}$$

$$\Gamma^0{}_{22} = \frac{(n+1)\, t^n}{2}$$

$$\Gamma^0{}_{33} = -\frac{n-1}{2\, t^n}$$

$$\Gamma^1{}_{00} = -a\, e^{-2\, a\, x}$$

$$\Gamma^1{}_{22} = a\, t^{n+1}\, e^{-2\, a\, x}$$

$$\Gamma^1{}_{33} = a\, t^{1-n}\, e^{-2\, a\, x}$$

$$\Gamma^2{}_{02} = \frac{n+1}{2\, t}$$

$$\Gamma^2{}_{12} = -a$$

$$\Gamma^2{}_{20} = \Gamma^2{}_{02}$$

$$\Gamma^2{}_{21} = \Gamma^2{}_{12}$$

$$\Gamma^3{}_{03} = -\frac{n-1}{2\, t}$$

$$\Gamma^3{}_{13} = -a$$

$$\Gamma^3{}_{30} = \Gamma^3{}_{03}$$

$$\Gamma^3{}_{31} = \Gamma^3{}_{13}$$

Metric Compatibility

——— o.k.

Riemann Tensor

$$R^{0}{}_{101} = -a^2$$

$$R^{0}{}_{110} = -R^{0}{}_{101}$$

$$R^{0}{}_{202} = \frac{t^{n-1}\,e^{-2\,a\,x}\,\left(n^2\,e^{2\,a\,x} - e^{2\,a\,x} - 4\,a^2\,t^2\right)}{4}$$

$$R^{0}{}_{220} = -R^{0}{}_{202}$$

$$R^{0}{}_{303} = \frac{t^{-n-1}\,e^{-2\,a\,x}\,\left(n^2\,e^{2\,a\,x} - e^{2\,a\,x} - 4\,a^2\,t^2\right)}{4}$$

$$R^{0}{}_{330} = -R^{0}{}_{303}$$

$$R^{1}{}_{001} = -a^2\,e^{-2\,a\,x}$$

$$R^{1}{}_{010} = -R^{1}{}_{001}$$

$$R^{1}{}_{212} = -a^2\,t^{n+1}\,e^{-2\,a\,x}$$

$$R^{1}{}_{221} = -R^{1}{}_{212}$$

$$R^{1}{}_{313} = -a^2\,t^{1-n}\,e^{-2\,a\,x}$$

$$R^{1}{}_{331} = -R^{1}{}_{313}$$

$$R^{2}{}_{002} = \frac{e^{-2\,a\,x}\,\left(n^2\,e^{2\,a\,x} - e^{2\,a\,x} - 4\,a^2\,t^2\right)}{4\,t^2}$$

$$R^{2}{}_{020} = -R^{2}{}_{002}$$

$$R^{2}{}_{112} = a^2$$

$$R^{2}{}_{121} = -R^{2}{}_{112}$$

$$R^{2}{}_{323} = -\frac{t^{-n-1}\,e^{-2\,a\,x}\,\left(n^2\,e^{2\,a\,x} - e^{2\,a\,x} + 4\,a^2\,t^2\right)}{4}$$

$$R^{2}{}_{332} = -R^{2}{}_{323}$$

$$R^{3}{}_{003} = \frac{e^{-2\,a\,x}\,\left(n^2\,e^{2\,a\,x} - e^{2\,a\,x} - 4\,a^2\,t^2\right)}{4\,t^2}$$

$$R^{3}{}_{030} = -R^{3}{}_{003}$$

$$R^{3}{}_{113} = a^2$$

$$R^{3}{}_{131} = -R^{3}{}_{113}$$

$$R^{3}{}_{223} = \frac{t^{n-1}\,e^{-2\,a\,x}\,\left(n^2\,e^{2\,a\,x} - e^{2\,a\,x} + 4\,a^2\,t^2\right)}{4}$$

$$R^{3}{}_{232} = -R^{3}{}_{223}$$

Ricci Tensor

$$\mathrm{Ric}_{00} = -\frac{e^{-2\,a\,x}\left(n^2\,e^{2\,a\,x} - e^{2\,a\,x} - 6\,a^2\,t^2\right)}{2\,t^2}$$

$$\mathrm{Ric}_{11} = -3\,a^2$$

$$\mathrm{Ric}_{22} = -3\,a^2\,t^{n+1}\,e^{-2\,a\,x}$$

$$\mathrm{Ric}_{33} = -3\,a^2\,t^{1-n}\,e^{-2\,a\,x}$$

Ricci Scalar

$$R_{sc} = \frac{n^2\,e^{2\,a\,x} - e^{2\,a\,x} - 24\,a^2\,t^2}{2\,t^2}$$

Bianchi identity (Ricci cyclic equation $R^{\kappa}{}_{[\mu\nu\sigma]} = 0$)

——— o.k.

Einstein Tensor

$$G_{00} = -\frac{e^{-2\,a\,x}\left(n^2\,e^{2\,a\,x} - e^{2\,a\,x} + 12\,a^2\,t^2\right)}{4\,t^2}$$

$$G_{11} = -\frac{n^2\,e^{2\,a\,x} - e^{2\,a\,x} - 12\,a^2\,t^2}{4\,t^2}$$

$$G_{22} = -\frac{t^{n-1}\,e^{-2\,a\,x}\left(n^2\,e^{2\,a\,x} - e^{2\,a\,x} - 12\,a^2\,t^2\right)}{4}$$

$$G_{33} = -\frac{t^{-n-1}\,e^{-2\,a\,x}\left(n^2\,e^{2\,a\,x} - e^{2\,a\,x} - 12\,a^2\,t^2\right)}{4}$$

Hodge Dual of Bianchi Identity

——— (see charge and current densities)

Scalar Charge Density (-$R^{0}{}_{i}{}^{i0}$)

$$\rho = -\frac{e^{2\,a\,x}\left(n^2\,e^{2\,a\,x} - e^{2\,a\,x} - 6\,a^2\,t^2\right)}{2\,t^2}$$

Current Density Class 1 (-$R^{i}{}_{\mu}{}^{\mu j}$)

$$J_1 = 3\,a^2$$

$$J_2 = 3\,a^2\,t^{-n-1}\,e^{2\,a\,x}$$

$$J_3 = 3\,a^2\,t^{n-1}\,e^{2\,a\,x}$$

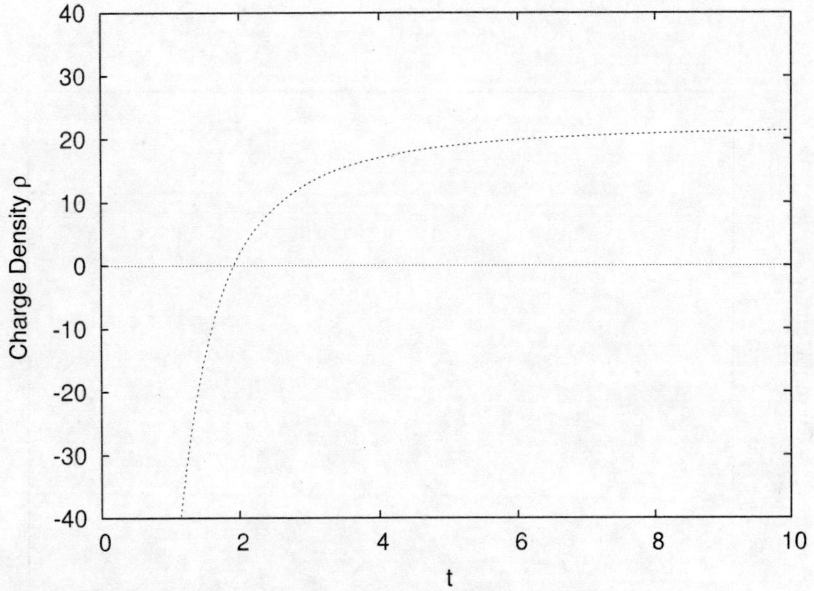

Fig. 4.94: Petrov type D fluid, charge density ρ for $a = 1, n = 2, x = 1$.

Current Density Class 2 (-$R^i{}_\mu{}^{\mu j}$)

$$J_1 = 0$$

$$J_2 = 0$$

$$J_3 = 0$$

Current Density Class 3 (-$R^i{}_\mu{}^{\mu j}$)

$$J_1 = 0$$

$$J_2 = 0$$

$$J_3 = 0$$

4.4.56 Spherically symmetric electromagnetic field with $\Lambda = 0$

m and e are parameters.

Coordinates

$$\mathbf{x} = \begin{pmatrix} t \\ r \\ \vartheta \\ \varphi \end{pmatrix}$$

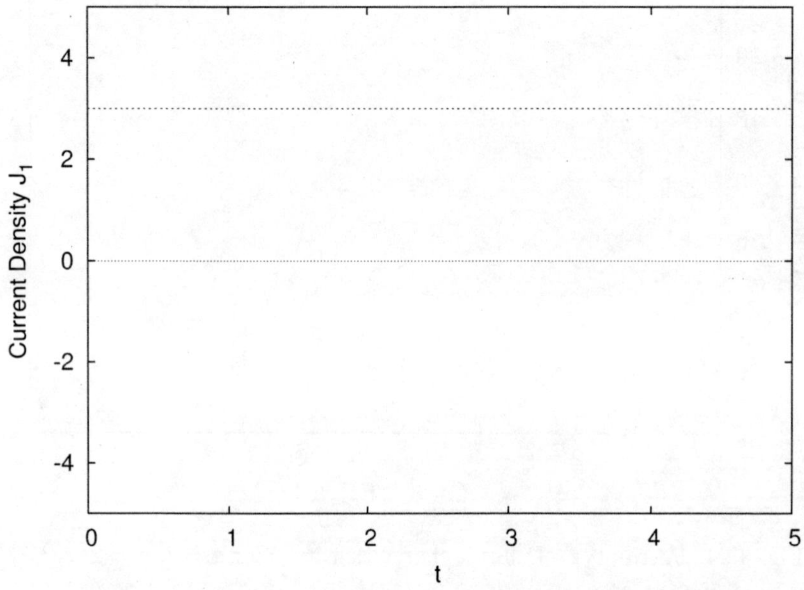

Fig. 4.95: Petrov type D fluid, current density J_1 for $a = 1, n = 2$, x=1.

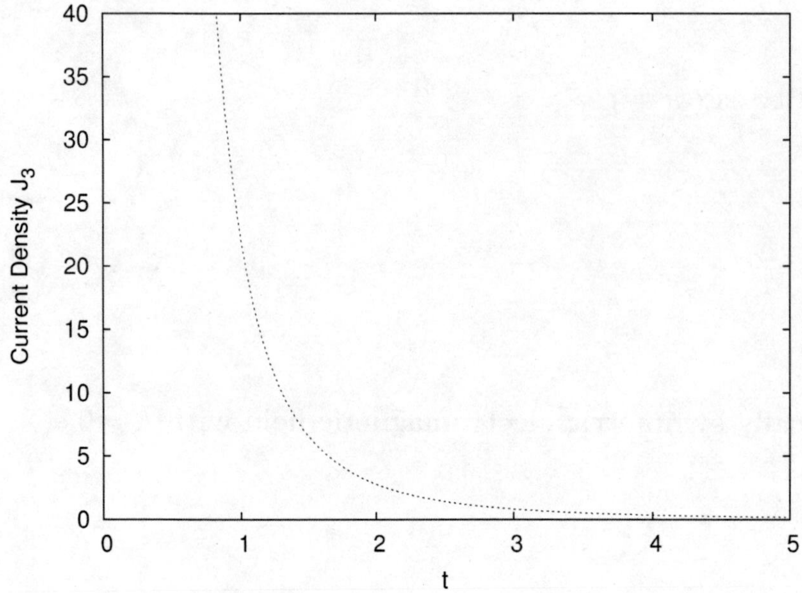

Fig. 4.96: Petrov type D fluid, current density J_2, J_3 for $a = 1, n = 2$, x=1.

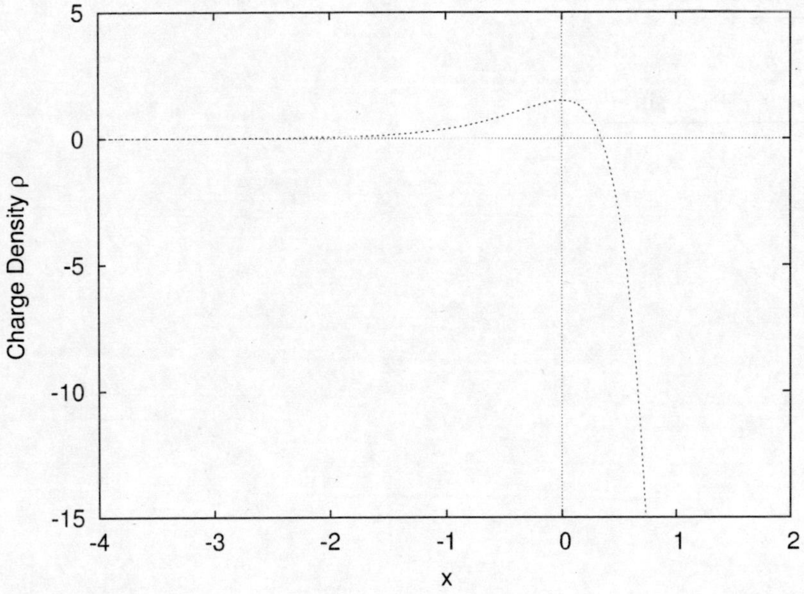

Fig. 4.97: Petrov type D fluid, charge density, x dependence $\rho(x)$ for $a = 1, n = 2, t = 1$.

Metric

$$g_{\mu\nu} = \begin{pmatrix} \frac{2m}{r} - \frac{e^2}{r^2} - 1 & 0 & 0 & 0 \\ 0 & \frac{1}{-\frac{2m}{r}+\frac{e^2}{r^2}+1} & 0 & 0 \\ 0 & 0 & r^2 & 0 \\ 0 & 0 & 0 & r^2 \sin^2\vartheta \end{pmatrix}$$

Contravariant Metric

$$g^{\mu\nu} = \begin{pmatrix} -\frac{r^2}{r^2-2mr+e^2} & 0 & 0 & 0 \\ 0 & \frac{r^2-2mr+e^2}{r^2} & 0 & 0 \\ 0 & 0 & \frac{1}{r^2} & 0 \\ 0 & 0 & 0 & \frac{1}{r^2\sin^2\vartheta} \end{pmatrix}$$

Christoffel Connection

$$\Gamma^0{}_{01} = \frac{m\,r - e^2}{r\,(r^2 - 2\,m\,r + e^2)}$$

$$\Gamma^0{}_{10} = \Gamma^0{}_{01}$$

$$\Gamma^1{}_{00} = \frac{(m\,r - e^2)\,(r^2 - 2\,m\,r + e^2)}{r^5}$$

$$\Gamma^1{}_{11} = -\frac{m\,r - e^2}{r\,(r^2 - 2\,m\,r + e^2)}$$

$$\Gamma^1{}_{22} = -\frac{r^2 - 2\,m\,r + e^2}{r}$$

$$\Gamma^1{}_{33} = -\frac{\left(r^2 - 2\,m\,r + e^2\right)\,\sin^2\vartheta}{r}$$

$$\Gamma^2{}_{12} = \frac{1}{r}$$

$$\Gamma^2{}_{21} = \Gamma^2{}_{12}$$

$$\Gamma^2{}_{33} = -\cos\vartheta\,\sin\vartheta$$

$$\Gamma^3{}_{13} = \frac{1}{r}$$

$$\Gamma^3{}_{23} = \frac{\cos\vartheta}{\sin\vartheta}$$

$$\Gamma^3{}_{31} = \Gamma^3{}_{13}$$

$$\Gamma^3{}_{32} = \Gamma^3{}_{23}$$

Metric Compatibility

——— o.k.

Riemann Tensor

$$R^0{}_{101} = \frac{2\,m\,r - 3\,e^2}{r^2\,\left(r^2 - 2\,m\,r + e^2\right)}$$

$$R^0{}_{110} = -R^0{}_{101}$$

$$R^0{}_{202} = -\frac{m\,r - e^2}{r^2}$$

$$R^0{}_{220} = -R^0{}_{202}$$

$$R^0{}_{303} = -\frac{\left(m\,r - e^2\right)\,\sin^2\vartheta}{r^2}$$

$$R^0{}_{330} = -R^0{}_{303}$$

$$R^1{}_{001} = \frac{\left(2\,m\,r - 3\,e^2\right)\,\left(r^2 - 2\,m\,r + e^2\right)}{r^6}$$

$$R^1{}_{010} = -R^1{}_{001}$$

$$R^1{}_{212} = -\frac{m\,r - e^2}{r^2}$$

$$R^1{}_{221} = -R^1{}_{212}$$

$$R^1{}_{313} = -\frac{\left(m\,r - e^2\right)\,\sin^2\vartheta}{r^2}$$

$$R^1{}_{331} = -R^1{}_{313}$$

$$R^2{}_{002} = -\frac{\left(m\,r - e^2\right)\,\left(r^2 - 2\,m\,r + e^2\right)}{r^6}$$

$$R^2{}_{020} = -R^2{}_{002}$$

$$R^2{}_{112} = \frac{m\,r - e^2}{r^2\,\left(r^2 - 2\,m\,r + e^2\right)}$$

$$R^2{}_{121} = -R^2{}_{112}$$

$$R^2{}_{323} = \frac{\left(2\,m\,r - e^2\right)\,\sin^2\vartheta}{r^2}$$

$$R^2{}_{332} = -R^2{}_{323}$$

$$R^3{}_{003} = -\frac{\left(m\,r - e^2\right)\,\left(r^2 - 2\,m\,r + e^2\right)}{r^6}$$

$$R^3{}_{030} = -R^3{}_{003}$$

$$R^3{}_{113} = \frac{m\,r - e^2}{r^2\,\left(r^2 - 2\,m\,r + e^2\right)}$$

$$R^3{}_{131} = -R^3{}_{113}$$

$$R^3{}_{223} = -\frac{2\,m\,r - e^2}{r^2}$$

$$R^3{}_{232} = -R^3{}_{223}$$

Ricci Tensor

$$\mathrm{Ric}_{00} = \frac{e^2\,\left(r^2 - 2\,m\,r + e^2\right)}{r^6}$$

$$\mathrm{Ric}_{11} = -\frac{e^2}{r^2\,\left(r^2 - 2\,m\,r + e^2\right)}$$

$$\mathrm{Ric}_{22} = \frac{e^2}{r^2}$$

$$\mathrm{Ric}_{33} = \frac{e^2\,\sin^2\vartheta}{r^2}$$

Ricci Scalar

$$R_{sc} = 0$$

Bianchi identity (Ricci cyclic equation $R^{\kappa}{}_{[\mu\nu\sigma]} = 0$)

——— o.k.

Einstein Tensor

$$G_{00} = \frac{e^2\left(r^2 - 2\,m\,r + e^2\right)}{r^6}$$

$$G_{11} = -\frac{e^2}{r^2\left(r^2 - 2\,m\,r + e^2\right)}$$

$$G_{22} = \frac{e^2}{r^2}$$

$$G_{33} = \frac{e^2 \sin^2\vartheta}{r^2}$$

Hodge Dual of Bianchi Identity

——— (see charge and current densities)

Scalar Charge Density (-$R^{0}{}_{i}{}^{i0}$)

$$\rho = \frac{e^2}{r^2\left(r^2 - 2\,m\,r + e^2\right)}$$

Current Density Class 1 (-$R^{i}{}_{\mu}{}^{\mu j}$)

$$J_1 = \frac{e^2\left(r^2 - 2\,m\,r + e^2\right)}{r^6}$$

$$J_2 = -\frac{e^2}{r^6}$$

$$J_3 = -\frac{e^2}{r^6 \sin^2\vartheta}$$

Current Density Class 2 (-$R^{i}{}_{\mu}{}^{\mu j}$)

$$J_1 = 0$$

$$J_2 = 0$$

$$J_3 = 0$$

Current Density Class 3 ($-R^{i}{}_{\mu}{}^{\mu j}$)

$$J_1 = 0$$

$$J_2 = 0$$

$$J_3 = 0$$

4.4.57 Plane-symmetric vacuum metric

Metric of a plane-symmetric vacuum. This is a true vacuum metric.

Coordinates

$$\mathbf{x} = \begin{pmatrix} t \\ x \\ y \\ z \end{pmatrix}$$

Metric

$$g_{\mu\nu} = \begin{pmatrix} -\frac{1}{\sqrt{z}} & 0 & 0 & 0 \\ 0 & z & 0 & 0 \\ 0 & 0 & z & 0 \\ 0 & 0 & 0 & \frac{1}{\sqrt{z}} \end{pmatrix}$$

Contravariant Metric

$$g^{\mu\nu} = \begin{pmatrix} -\sqrt{z} & 0 & 0 & 0 \\ 0 & \frac{1}{z} & 0 & 0 \\ 0 & 0 & \frac{1}{z} & 0 \\ 0 & 0 & 0 & \sqrt{z} \end{pmatrix}$$

Christoffel Connection

$$\Gamma^{0}{}_{03} = -\frac{1}{4\,z}$$

$$\Gamma^{0}{}_{30} = \Gamma^{0}{}_{03}$$

$$\Gamma^{1}{}_{13} = \frac{1}{2\,z}$$

$$\Gamma^{1}{}_{31} = \Gamma^{1}{}_{13}$$

$$\Gamma^{2}{}_{23} = \frac{1}{2\,z}$$

$$\Gamma^{2}{}_{32} = \Gamma^{2}{}_{23}$$

$$\Gamma^3{}_{00} = -\frac{1}{4\,z}$$

$$\Gamma^3{}_{11} = -\frac{\sqrt{z}}{2}$$

$$\Gamma^3{}_{22} = -\frac{\sqrt{z}}{2}$$

$$\Gamma^3{}_{33} = -\frac{1}{4\,z}$$

Metric Compatibility

——— o.k.

Riemann Tensor

$$R^0{}_{101} = \frac{1}{8\,\sqrt{z}}$$

$$R^0{}_{110} = -R^0{}_{101}$$

$$R^0{}_{202} = \frac{1}{8\,\sqrt{z}}$$

$$R^0{}_{220} = -R^0{}_{202}$$

$$R^0{}_{303} = -\frac{1}{4\,z^2}$$

$$R^0{}_{330} = -R^0{}_{303}$$

$$R^1{}_{001} = \frac{1}{8\,z^2}$$

$$R^1{}_{010} = -R^1{}_{001}$$

$$R^1{}_{212} = -\frac{1}{4\,\sqrt{z}}$$

$$R^1{}_{221} = -R^1{}_{212}$$

$$R^1{}_{313} = \frac{1}{8\,z^2}$$

$$R^1{}_{331} = -R^1{}_{313}$$

$$R^2{}_{002} = \frac{1}{8\,z^2}$$

$$R^2{}_{020} = -R^2{}_{002}$$

$$R^2{}_{112} = \frac{1}{4\sqrt{z}}$$

$$R^2{}_{121} = -R^2{}_{112}$$

$$R^2{}_{323} = \frac{1}{8\,z^2}$$

$$R^2{}_{332} = -R^2{}_{323}$$

$$R^3{}_{003} = -\frac{1}{4\,z^2}$$

$$R^3{}_{030} = -R^3{}_{003}$$

$$R^3{}_{113} = -\frac{1}{8\sqrt{z}}$$

$$R^3{}_{131} = -R^3{}_{113}$$

$$R^3{}_{223} = -\frac{1}{8\sqrt{z}}$$

$$R^3{}_{232} = -R^3{}_{223}$$

Ricci Tensor

——— all elements zero

Ricci Scalar

$$R_{sc} = 0$$

Bianchi identity (Ricci cyclic equation $R^\kappa{}_{[\mu\nu\sigma]} = 0$)

——— o.k.

Einstein Tensor

——— all elements zero

Hodge Dual of Bianchi Identity

——— (see charge and current densities)

Scalar Charge Density ($-R^{0\ \ i0}_{\ \ i}$)

$$\rho = 0$$

Current Density Class 1 ($-R^{i\ \ \mu j}_{\ \ \mu}$)

$$J_1 = 0$$

$$J_2 = 0$$

$$J_3 = 0$$

Current Density Class 2 ($-R^{i\ \ \mu j}_{\ \ \mu}$)

$$J_1 = 0$$

$$J_2 = 0$$

$$J_3 = 0$$

Current Density Class 3 ($-R^{i\ \ \mu j}_{\ \ \mu}$)

$$J_1 = 0$$

$$J_2 = 0$$

$$J_3 = 0$$

4.4.58 Sheared dust metric

Metric of a sheared dust. n is a parameter. Functions A and B depend on r.

Coordinates

$$\mathbf{x} = \begin{pmatrix} t \\ r \\ \vartheta \\ \varphi \end{pmatrix}$$

Metric

$$g_{\mu\nu} = \begin{pmatrix} 1-n^2 & 0 & 0 & 0 \\ 0 & \left(\frac{B}{t^n} + t^n\, A\right)^2 & 0 & 0 \\ 0 & 0 & t^2 & 0 \\ 0 & 0 & 0 & t^2\, \sin^2\vartheta \end{pmatrix}$$

Contravariant Metric

$$g^{\mu\nu} = \begin{pmatrix} -\frac{1}{(n-1)\,(n+1)} & 0 & 0 & 0 \\ 0 & \frac{t^{2\,n}}{(B+t^{2\,n}\,A)^2} & 0 & 0 \\ 0 & 0 & \frac{1}{t^2} & 0 \\ 0 & 0 & 0 & \frac{1}{t^2\,\sin^2\vartheta} \end{pmatrix}$$

Christoffel Connection

$$\Gamma^{0}{}_{11} = -\frac{n\,t^{-2\,n-1}\,\left(B - t^{2\,n}\,A\right)\,\left(B + t^{2\,n}\,A\right)}{(n-1)\,(n+1)}$$

$$\Gamma^{0}{}_{22} = \frac{t}{(n-1)\,(n+1)}$$

$$\Gamma^{0}{}_{33} = \frac{t\,\sin^2\vartheta}{(n-1)\,(n+1)}$$

$$\Gamma^{1}{}_{01} = -\frac{n\,\left(B - t^{2\,n}\,A\right)}{t\,\left(B + t^{2\,n}\,A\right)}$$

$$\Gamma^{1}{}_{10} = \Gamma^{1}{}_{01}$$

$$\Gamma^{1}{}_{11} = \frac{\frac{d}{d\,r}\,B + t^{2\,n}\,\left(\frac{d}{d\,r}\,A\right)}{B + t^{2\,n}\,A}$$

$$\Gamma^{2}{}_{02} = \frac{1}{t}$$

$$\Gamma^{2}{}_{20} = \Gamma^{2}{}_{02}$$

$$\Gamma^{2}{}_{33} = -\cos\vartheta\,\sin\vartheta$$

$$\Gamma^{3}{}_{03} = \frac{1}{t}$$

$$\Gamma^{3}{}_{23} = \frac{\cos\vartheta}{\sin\vartheta}$$

$$\Gamma^{3}{}_{30} = \Gamma^{3}{}_{03}$$

$$\Gamma^{3}{}_{32} = \Gamma^{3}{}_{23}$$

Metric Compatibility

——— o.k.

Riemann Tensor

$$R^{0}{}_{101} = \frac{n\,t^{-2\,n-2}\,\left(B+t^{2\,n}\,A\right)\,\left(n\,B+B+n\,t^{2\,n}\,A-t^{2\,n}\,A\right)}{\left(n-1\right)\,\left(n+1\right)}$$

$$R^{0}{}_{110} = -R^{0}{}_{101}$$

$$R^{1}{}_{001} = \frac{n\,\left(n\,B+B+n\,t^{2\,n}\,A-t^{2\,n}\,A\right)}{t^{2}\,\left(B+t^{2\,n}\,A\right)}$$

$$R^{1}{}_{010} = -R^{1}{}_{001}$$

$$R^{1}{}_{212} = -\frac{n\,\left(B-t^{2\,n}\,A\right)}{\left(n-1\right)\,\left(n+1\right)\,\left(B+t^{2\,n}\,A\right)}$$

$$R^{1}{}_{221} = -R^{1}{}_{212}$$

$$R^{1}{}_{313} = -\frac{n\,\sin^{2}\vartheta\,\left(B-t^{2\,n}\,A\right)}{\left(n-1\right)\,\left(n+1\right)\,\left(B+t^{2\,n}\,A\right)}$$

$$R^{1}{}_{331} = -R^{1}{}_{313}$$

$$R^{2}{}_{112} = \frac{n\,t^{-2\,n-2}\,\left(B-t^{2\,n}\,A\right)\,\left(B+t^{2\,n}\,A\right)}{\left(n-1\right)\,\left(n+1\right)}$$

$$R^{2}{}_{121} = -R^{2}{}_{112}$$

$$R^{2}{}_{323} = \frac{n^{2}\,\sin^{2}\vartheta}{\left(n-1\right)\,\left(n+1\right)}$$

$$R^{2}{}_{332} = -R^{2}{}_{323}$$

$$R^{3}{}_{113} = \frac{n\,t^{-2\,n-2}\,\left(B-t^{2\,n}\,A\right)\,\left(B+t^{2\,n}\,A\right)}{\left(n-1\right)\,\left(n+1\right)}$$

$$R^{3}{}_{131} = -R^{3}{}_{113}$$

$$R^{3}{}_{223} = -\frac{n^{2}}{\left(n-1\right)\,\left(n+1\right)}$$

$$R^{3}{}_{232} = -R^{3}{}_{223}$$

Ricci Tensor

$$\mathrm{Ric}_{00} = -\frac{n\,\left(n\,B + B + n\,t^{2\,n}\,A - t^{2\,n}\,A\right)}{t^2\,\left(B + t^{2\,n}\,A\right)}$$

$$\mathrm{Ric}_{11} = \frac{n\,t^{-2\,n-2}\,\left(B + t^{2\,n}\,A\right)\,\left(n\,B - B + n\,t^{2\,n}\,A + t^{2\,n}\,A\right)}{(n-1)\,(n+1)}$$

$$\mathrm{Ric}_{22} = \frac{n\,\left(n\,B - B + n\,t^{2\,n}\,A + t^{2\,n}\,A\right)}{(n-1)\,(n+1)\,\left(B + t^{2\,n}\,A\right)}$$

$$\mathrm{Ric}_{33} = \frac{n\,\sin^2\vartheta\,\left(n\,B - B + n\,t^{2\,n}\,A + t^{2\,n}\,A\right)}{(n-1)\,(n+1)\,\left(B + t^{2\,n}\,A\right)}$$

Ricci Scalar

$$R_{sc} = \frac{2\,n\,\left(2\,n\,B - B + 2\,n\,t^{2\,n}\,A + t^{2\,n}\,A\right)}{(n-1)\,(n+1)\,t^2\,\left(B + t^{2\,n}\,A\right)}$$

Bianchi identity (Ricci cyclic equation $R^{\kappa}{}_{[\mu\nu\sigma]} = 0$)

———— o.k.

Einstein Tensor

$$G_{00} = \frac{n\,\left(n\,B - 2\,B + n\,t^{2\,n}\,A + 2\,t^{2\,n}\,A\right)}{t^2\,\left(B + t^{2\,n}\,A\right)}$$

$$G_{11} = -\frac{n^2\,t^{-2\,n-2}\,\left(B + t^{2\,n}\,A\right)^2}{(n-1)\,(n+1)}$$

$$G_{22} = -\frac{n^2}{(n-1)\,(n+1)}$$

$$G_{33} = -\frac{n^2\,\sin^2\vartheta}{(n-1)\,(n+1)}$$

Hodge Dual of Bianchi Identity

———— (see charge and current densities)

Scalar Charge Density (-$R^{0}{}_{i}{}^{i0}$)

$$\rho = -\frac{n\,\left(n\,B + B + n\,t^{2\,n}\,A - t^{2\,n}\,A\right)}{(n-1)^2\,(n+1)^2\,t^2\,\left(B + t^{2\,n}\,A\right)}$$

Current Density Class 1 ($-R^{i}{}_{\mu}{}^{\mu j}$)

$$J_1 = -\frac{n\,t^{2\,n-2}\,(n\,B - B + n\,t^{2\,n}\,A + t^{2\,n}\,A)}{(n-1)\,(n+1)\,(B + t^{2\,n}\,A)^3}$$

$$J_2 = -\frac{n\,(n\,B - B + n\,t^{2\,n}\,A + t^{2\,n}\,A)}{(n-1)\,(n+1)\,t^4\,(B + t^{2\,n}\,A)}$$

$$J_3 = -\frac{n\,(n\,B - B + n\,t^{2\,n}\,A + t^{2\,n}\,A)}{(n-1)\,(n+1)\,t^4\,\sin^2\vartheta\,(B + t^{2\,n}\,A)}$$

Current Density Class 2 ($-R^{i}{}_{\mu}{}^{\mu j}$)

$$J_1 = 0$$

$$J_2 = 0$$

$$J_3 = 0$$

Current Density Class 3 ($-R^{i}{}_{\mu}{}^{\mu j}$)

$$J_1 = 0$$

$$J_2 = 0$$

$$J_3 = 0$$

4.4.59 Plane-symmetric perfect fluid metric

a and b are parameters. There is a symmetry in tensors for x and y.

Coordinates

$$\mathbf{x} = \begin{pmatrix} t \\ x \\ y \\ z \end{pmatrix}$$

Metric

$$g_{\mu\nu} = \begin{pmatrix} -e^{b\,e^{-2\,(t-z)}+2\,z} & 0 & 0 & 0 \\ 0 & e^{2\,(z+t)} & 0 & 0 \\ 0 & 0 & e^{2\,(z+t)} & 0 \\ 0 & 0 & 0 & a^2\,e^{b\,e^{-2\,(t-z)}+2\,z} \end{pmatrix}$$

Contravariant Metric

$$g^{\mu\nu} = \begin{pmatrix} -e^{-b\,e^{2\,z-2\,t}-2\,z} & 0 & 0 & 0 \\ 0 & e^{-2\,z-2\,t} & 0 & 0 \\ 0 & 0 & e^{-2\,z-2\,t} & 0 \\ 0 & 0 & 0 & \frac{e^{-b\,e^{2\,z-2\,t}-2\,z}}{a^2} \end{pmatrix}$$

Christoffel Connection

$$\Gamma^0{}_{00} = -b\,e^{2\,z-2\,t}$$

$$\Gamma^0{}_{03} = e^{-2\,t}\,\left(b\,e^{2\,z} + e^{2\,t}\right)$$

$$\Gamma^0{}_{11} = e^{2\,t-b\,e^{2\,z-2\,t}}$$

$$\Gamma^0{}_{22} = e^{2\,t-b\,e^{2\,z-2\,t}}$$

$$\Gamma^0{}_{30} = \Gamma^0{}_{03}$$

$$\Gamma^0{}_{33} = -a^2\,b\,e^{2\,z-2\,t}$$

$$\Gamma^1{}_{01} = 1$$

$$\Gamma^1{}_{10} = \Gamma^1{}_{01}$$

$$\Gamma^1{}_{13} = 1$$

$$\Gamma^1{}_{31} = \Gamma^1{}_{13}$$

$$\Gamma^2{}_{02} = 1$$

$$\Gamma^2{}_{20} = \Gamma^2{}_{02}$$

$$\Gamma^2{}_{23} = 1$$

$$\Gamma^2{}_{32} = \Gamma^2{}_{23}$$

$$\Gamma^3{}_{00} = \frac{e^{-2\,t}\,\left(b\,e^{2\,z} + e^{2\,t}\right)}{a^2}$$

$$\Gamma^3{}_{03} = -b\,e^{2\,z-2\,t}$$

$$\Gamma^3{}_{11} = -\frac{e^{2\,t-b\,e^{2\,z-2\,t}}}{a^2}$$

$$\Gamma^3{}_{22} = -\frac{e^{2\,t-b\,e^{2\,z-2\,t}}}{a^2}$$

$$\Gamma^3{}_{30} = \Gamma^3{}_{03}$$

$$\Gamma^3{}_{33} = e^{-2\,t}\,\left(b\,e^{2\,z} + e^{2\,t}\right)$$

Metric Compatibility

——— o.k.

Riemann Tensor

$$R^0{}_{101} = \frac{(a-1)\,(a+1)\,\left(b\,e^{2\,z} + e^{2\,t}\right)\,e^{-b\,e^{2\,z-2\,t}}}{a^2}$$

$$R^0{}_{110} = -R^0{}_{101}$$

$$R^0{}_{202} = \frac{(a-1)\,(a+1)\,\left(b\,e^{2\,z} + e^{2\,t}\right)\,e^{-b\,e^{2\,z-2\,t}}}{a^2}$$

$$R^0{}_{220} = -R^0{}_{202}$$

$$R^0{}_{303} = 2\,(a-1)\,(a+1)\,b\,e^{2\,z-2\,t}$$

$$R^0{}_{330} = -R^0{}_{303}$$

$$R^1{}_{001} = \frac{(a-1)\,(a+1)\,e^{-2\,t}\,\left(b\,e^{2\,z} + e^{2\,t}\right)}{a^2}$$

$$R^1{}_{010} = -R^1{}_{001}$$

$$R^1{}_{212} = \frac{(a-1)\,(a+1)\,e^{2\,t-b\,e^{2\,z-2\,t}}}{a^2}$$

$$R^1{}_{221} = -R^1{}_{212}$$

$$R^1{}_{313} = -\,(a-1)\,(a+1)\,b\,e^{2\,z-2\,t}$$

$$R^1{}_{331} = -R^1{}_{313}$$

$$R^2{}_{002} = \frac{(a-1)\,(a+1)\,e^{-2\,t}\,\left(b\,e^{2\,z} + e^{2\,t}\right)}{a^2}$$

$$R^2{}_{020} = -R^2{}_{002}$$

$$R^2{}_{112} = -\frac{(a-1)\,(a+1)\,e^{2\,t-b\,e^{2\,z-2\,t}}}{a^2}$$

$$R^2{}_{121} = -R^2{}_{112}$$

$$R^2{}_{323} = -\,(a-1)\,(a+1)\,b\,e^{2\,z-2\,t}$$

$$R^2{}_{332} = -R^2{}_{323}$$

$$R^3{}_{003} = \frac{2\,(a-1)\,(a+1)\,b\,e^{2\,z-2\,t}}{a^2}$$

$$R^3{}_{030} = -R^3{}_{003}$$

$$R^3{}_{113} = \frac{(a-1)\,(a+1)\,b\,e^{2\,z-b\,e^{2\,z-2\,t}}}{a^2}$$

$$R^3{}_{131} = -R^3{}_{113}$$

$$R^3{}_{223} = \frac{(a-1)\,(a+1)\,b\,e^{2\,z-b\,e^{2\,z-2\,t}}}{a^2}$$

$$R^3{}_{232} = -R^3{}_{223}$$

Ricci Tensor

$$\mathrm{Ric}_{00} = -\frac{2\,(a-1)\,(a+1)\,e^{-2\,t}\,\left(2\,b\,e^{2\,z}+e^{2\,t}\right)}{a^2}$$

$$\mathrm{Ric}_{11} = \frac{2\,(a-1)\,(a+1)\,e^{2\,t-b\,e^{2\,z-2\,t}}}{a^2}$$

$$\mathrm{Ric}_{22} = \frac{2\,(a-1)\,(a+1)\,e^{2\,t-b\,e^{2\,z-2\,t}}}{a^2}$$

Ricci Scalar

$$R_{sc} = \frac{2\,(a-1)\,(a+1)\,\left(2\,b\,e^{2\,z}+3\,e^{2\,t}\right)\,e^{-b\,e^{2\,z-2\,t}-2\,z-2\,t}}{a^2}$$

Bianchi identity (Ricci cyclic equation $R^{\kappa}{}_{[\mu\nu\sigma]} = 0$)

——— o.k.

Einstein Tensor

$$G_{00} = -\frac{(a-1)\,(a+1)\,e^{-2\,t}\,\left(2\,b\,e^{2\,z}-e^{2\,t}\right)}{a^2}$$

$$G_{11} = -\frac{(a-1)\,(a+1)\,\left(2\,b\,e^{2\,z}+e^{2\,t}\right)\,e^{-b\,e^{2\,z-2\,t}}}{a^2}$$

$$G_{22} = -\frac{(a-1)\,(a+1)\,\left(2\,b\,e^{2\,z}+e^{2\,t}\right)\,e^{-b\,e^{2\,z-2\,t}}}{a^2}$$

$$G_{33} = -(a-1)\,(a+1)\,e^{-2\,t}\,\left(2\,b\,e^{2\,z}+3\,e^{2\,t}\right)$$

Hodge Dual of Bianchi Identity

——— (see charge and current densities)

Scalar Charge Density ($-R^{0}{}_{i}{}^{i0}$)

$$\rho = -\frac{2\,(a-1)\,(a+1)\,\left(2\,b\,e^{2\,z}+e^{2\,t}\right)\,e^{-2\,b\,e^{2\,z-2\,t}-4\,z-2\,t}}{a^2}$$

Current Density Class 1 ($-R^{i}{}_{\mu}{}^{\mu j}$)

$$J_1 = -\frac{2\,(a-1)\,(a+1)\,e^{-b\,e^{2\,z-2\,t}-4\,z-2\,t}}{a^2}$$

$$J_2 = -\frac{2\,(a-1)\,(a+1)\,e^{-b\,e^{2\,z-2\,t}-4\,z-2\,t}}{a^2}$$

$$J_3 = 0$$

Current Density Class 2 ($-R^{i}{}_{\mu}{}^{\mu j}$)

$$J_1 = 0$$

$$J_2 = 0$$

$$J_3 = 0$$

Current Density Class 3 ($-R^{i}{}_{\mu}{}^{\mu j}$)

$$J_1 = 0$$

$$J_2 = 0$$

$$J_3 = 0$$

4.4.60 Spherically symmetric perfect fluid metric (static)

Metric of a static spherically symmetric perfect fluid. δ and ν are functions of r.

Coordinates

$$\mathbf{x} = \begin{pmatrix} t \\ r \\ \vartheta \\ \varphi \end{pmatrix}$$

Metric

$$g_{\mu\nu} = \begin{pmatrix} -e^{2\,\nu} & 0 & 0 & 0 \\ 0 & e^{2\,\lambda} & 0 & 0 \\ 0 & 0 & r^2 & 0 \\ 0 & 0 & 0 & r^2\,\sin^2\vartheta \end{pmatrix}$$

Contravariant Metric

$$g^{\mu\nu} = \begin{pmatrix} -e^{-2\,\nu} & 0 & 0 & 0 \\ 0 & e^{-2\,\lambda} & 0 & 0 \\ 0 & 0 & \frac{1}{r^2} & 0 \\ 0 & 0 & 0 & \frac{1}{r^2\,\sin^2\vartheta} \end{pmatrix}$$

Christoffel Connection

$$\Gamma^0{}_{01} = \frac{d}{d\,r}\,\nu$$

$$\Gamma^0{}_{10} = \Gamma^0{}_{01}$$

$$\Gamma^1{}_{00} = \frac{d}{d\,r}\,\nu\,e^{2\,\nu-2\,\lambda}$$

$$\Gamma^1{}_{11} = \frac{d}{d\,r}\,\lambda$$

$$\Gamma^1{}_{22} = -r\,e^{-2\,\lambda}$$

$$\Gamma^1{}_{33} = -r\,\sin^2\vartheta\,e^{-2\,\lambda}$$

$$\Gamma^2{}_{12} = \frac{1}{r}$$

$$\Gamma^2{}_{21} = \Gamma^2{}_{12}$$

$$\Gamma^2{}_{33} = -\cos\vartheta\,\sin\vartheta$$

$$\Gamma^3{}_{13} = \frac{1}{r}$$

$$\Gamma^3{}_{23} = \frac{\cos\vartheta}{\sin\vartheta}$$

$$\Gamma^3{}_{31} = \Gamma^3{}_{13}$$

$$\Gamma^3{}_{32} = \Gamma^3{}_{23}$$

Metric Compatibility

——— o.k.

Riemann Tensor

$$R^0{}_{101} = \frac{d}{d\,r}\,\nu \left(\frac{d}{d\,r}\,\lambda\right) - \frac{d^2}{d\,r^2}\,\nu - \left(\frac{d}{d\,r}\,\nu\right)^2$$

$$R^0{}_{110} = -R^0{}_{101}$$

$$R^0{}_{202} = -\frac{d}{d\,r}\,\nu\,r\,e^{-2\,\lambda}$$

$$R^0{}_{220} = -R^0{}_{202}$$

$$R^0{}_{303} = -\frac{d}{d\,r}\,\nu\,r\,\sin^2\vartheta\,e^{-2\,\lambda}$$

$$R^0{}_{330} = -R^0{}_{303}$$

$$R^1{}_{001} = e^{2\,\nu-2\,\lambda}\left(\frac{d}{d\,r}\,\nu \left(\frac{d}{d\,r}\,\lambda\right) - \frac{d^2}{d\,r^2}\,\nu - \left(\frac{d}{d\,r}\,\nu\right)^2\right)$$

$$R^1{}_{010} = -R^1{}_{001}$$

$$R^1{}_{212} = r\,e^{-2\,\lambda}\left(\frac{d}{d\,r}\,\lambda\right)$$

$$R^1{}_{221} = -R^1{}_{212}$$

$$R^1{}_{313} = r\,\sin^2\vartheta\,e^{-2\,\lambda}\left(\frac{d}{d\,r}\,\lambda\right)$$

$$R^1{}_{331} = -R^1{}_{313}$$

$$R^2{}_{002} = -\frac{\frac{d}{d\,r}\,\nu\,e^{2\,\nu-2\,\lambda}}{r}$$

$$R^2{}_{020} = -R^2{}_{002}$$

$$R^2{}_{112} = -\frac{\frac{d}{d\,r}\,\lambda}{r}$$

$$R^2{}_{121} = -R^2{}_{112}$$

$$R^2{}_{323} = \sin^2\vartheta\,e^{-2\,\lambda}\,\left(e^{\lambda}-1\right)\,\left(e^{\lambda}+1\right)$$

$$R^2{}_{332} = -R^2{}_{323}$$

$$R^3{}_{003} = -\frac{\frac{d}{d\,r}\,\nu\,e^{2\,\nu-2\,\lambda}}{r}$$

$$R^3{}_{030} = -R^3{}_{003}$$

$$R^3{}_{113} = -\frac{\frac{d}{d\,r}\,\lambda}{r}$$

$$R^3{}_{131} = -R^3{}_{113}$$

$$R^3{}_{223} = -e^{-2\,\lambda}\,\left(e^{\lambda}-1\right)\,\left(e^{\lambda}+1\right)$$

$$R^3{}_{232} = -R^3{}_{223}$$

Ricci Tensor

$$\mathrm{Ric}_{00} = -\frac{e^{2\,\nu-2\,\lambda}\,\left(\frac{d}{d\,r}\,\nu\,r\,\left(\frac{d}{d\,r}\,\lambda\right)-\frac{d^2}{d\,r^2}\,\nu\,r-\left(\frac{d}{d\,r}\,\nu\right)^2\,r-2\,\left(\frac{d}{d\,r}\,\nu\right)\right)}{r}$$

$$\mathrm{Ric}_{11} = \frac{\frac{d}{d\,r}\,\nu\,r\,\left(\frac{d}{d\,r}\,\lambda\right)+2\,\left(\frac{d}{d\,r}\,\lambda\right)-\frac{d^2}{d\,r^2}\,\nu\,r-\left(\frac{d}{d\,r}\,\nu\right)^2\,r}{r}$$

$$\mathrm{Ric}_{22} = e^{-2\,\lambda}\,\left(r\,\left(\frac{d}{d\,r}\,\lambda\right)+e^{2\,\lambda}-\frac{d}{d\,r}\,\nu\,r-1\right)$$

$$\mathrm{Ric}_{33} = \sin^2\vartheta\,e^{-2\,\lambda}\,\left(r\,\left(\frac{d}{d\,r}\,\lambda\right)+e^{2\,\lambda}-\frac{d}{d\,r}\,\nu\,r-1\right)$$

Ricci Scalar

$$R_{sc} = \frac{2\,e^{-2\,\lambda}\,\left(\frac{d}{d\,r}\,\nu\,r^2\,\left(\frac{d}{d\,r}\,\lambda\right)+2\,r\,\left(\frac{d}{d\,r}\,\lambda\right)+e^{2\,\lambda}-\frac{d^2}{d\,r^2}\,\nu\,r^2-\left(\frac{d}{d\,r}\,\nu\right)^2\,r^2-2\,\left(\frac{d}{d\,r}\,\nu\right)\,r-1\right)}{r^2}$$

Bianchi identity (Ricci cyclic equation $R^{\kappa}{}_{[\mu\nu\sigma]} = 0$)

——— o.k.

Einstein Tensor

$$G_{00} = \frac{e^{2\nu-2\lambda}\left(2\,r\left(\frac{d}{d\,r}\lambda\right)+e^{2\lambda}-1\right)}{r^2}$$

$$G_{11} = -\frac{e^{2\lambda}-2\left(\frac{d}{d\,r}\nu\right)r-1}{r^2}$$

$$G_{22} = -r\,e^{-2\lambda}\left(\frac{d}{d\,r}\nu\,r\left(\frac{d}{d\,r}\lambda\right)+\frac{d}{d\,r}\lambda-\frac{d^2}{d\,r^2}\nu\,r-\left(\frac{d}{d\,r}\nu\right)^2 r-\frac{d}{d\,r}\nu\right)$$

$$G_{33} = -r\,\sin^2\vartheta\,e^{-2\lambda}\left(\frac{d}{d\,r}\nu\,r\left(\frac{d}{d\,r}\lambda\right)+\frac{d}{d\,r}\lambda-\frac{d^2}{d\,r^2}\nu\,r-\left(\frac{d}{d\,r}\nu\right)^2 r-\frac{d}{d\,r}\nu\right)$$

Hodge Dual of Bianchi Identity

——— (see charge and current densities)

Scalar Charge Density (-$R^{0}{}_{i}{}^{i0}$)

$$\rho = -\frac{e^{-2\lambda-2\nu}\left(\frac{d}{d\,r}\nu\,r\left(\frac{d}{d\,r}\lambda\right)-\frac{d^2}{d\,r^2}\nu\,r-\left(\frac{d}{d\,r}\nu\right)^2 r-2\left(\frac{d}{d\,r}\nu\right)\right)}{r}$$

Current Density Class 1 (-$R^{i}{}_{\mu}{}^{\mu j}$)

$$J_1 = -\frac{e^{-4\lambda}\left(\frac{d}{d\,r}\nu\,r\left(\frac{d}{d\,r}\lambda\right)+2\left(\frac{d}{d\,r}\lambda\right)-\frac{d^2}{d\,r^2}\nu\,r-\left(\frac{d}{d\,r}\nu\right)^2 r\right)}{r}$$

$$J_2 = -\frac{e^{-2\lambda}\left(r\left(\frac{d}{d\,r}\lambda\right)+e^{2\lambda}-\frac{d}{d\,r}\nu\,r-1\right)}{r^4}$$

$$J_3 = -\frac{e^{-2\lambda}\left(r\left(\frac{d}{d\,r}\lambda\right)+e^{2\lambda}-\frac{d}{d\,r}\nu\,r-1\right)}{r^4\,\sin^2\vartheta}$$

Current Density Class 2 (-$R^{i}{}_{\mu}{}^{\mu j}$)

$$J_1 = 0$$

$$J_2 = 0$$

$$J_3 = 0$$

Current Density Class 3 ($-R^{i}{}_{\mu}{}^{\mu j}$)

$$J_1 = 0$$

$$J_2 = 0$$

$$J_3 = 0$$

4.4.61 Spherically symmetric perfect fluid metric (dynamic)

Metric of a dynamical spherically symmetric perfect fluid. λ, ν, and y are functions of r and t.

Coordinates

$$\mathbf{x} = \begin{pmatrix} t \\ r \\ \vartheta \\ \varphi \end{pmatrix}$$

Metric

$$g_{\mu\nu} = \begin{pmatrix} -e^{2\nu} & 0 & 0 & 0 \\ 0 & e^{2\lambda} & 0 & 0 \\ 0 & 0 & y^2 & 0 \\ 0 & 0 & 0 & \sin^2\vartheta\, y^2 \end{pmatrix}$$

Contravariant Metric

$$g^{\mu\nu} = \begin{pmatrix} -e^{-2\nu} & 0 & 0 & 0 \\ 0 & e^{-2\lambda} & 0 & 0 \\ 0 & 0 & \frac{1}{y^2} & 0 \\ 0 & 0 & 0 & \frac{1}{\sin^2\vartheta\, y^2} \end{pmatrix}$$

Christoffel Connection

$$\Gamma^{0}{}_{00} = \frac{d}{d\,t}\nu$$

$$\Gamma^{0}{}_{01} = \frac{d}{d\,r}\nu$$

$$\Gamma^{0}{}_{10} = \Gamma^{0}{}_{01}$$

$$\Gamma^{0}{}_{11} = e^{2\lambda-2\nu}\left(\frac{d}{d\,t}\lambda\right)$$

$$\Gamma^{0}{}_{22} = e^{-2\nu}\,y\left(\frac{d}{d\,t}y\right)$$

$$\Gamma^{0}{}_{33} = e^{-2\nu}\sin^2\vartheta\,y\left(\frac{d}{d\,t}y\right)$$

$$\Gamma^{1}{}_{00} = \frac{d}{d\,r}\nu\,e^{2\nu-2\lambda}$$

$$\Gamma^{1}{}_{01} = \frac{d}{d\,t}\lambda$$

$$\Gamma^1{}_{10} = \Gamma^1{}_{01}$$

$$\Gamma^1{}_{11} = \frac{d}{d\,r}\,\lambda$$

$$\Gamma^1{}_{22} = -y\,\left(\frac{d}{d\,r}\,y\right)\,e^{-2\,\lambda}$$

$$\Gamma^1{}_{33} = -\sin^2\vartheta\,y\,\left(\frac{d}{d\,r}\,y\right)\,e^{-2\,\lambda}$$

$$\Gamma^2{}_{02} = \frac{\frac{d}{d\,t}\,y}{y}$$

$$\Gamma^2{}_{12} = \frac{\frac{d}{d\,r}\,y}{y}$$

$$\Gamma^2{}_{20} = \Gamma^2{}_{02}$$

$$\Gamma^2{}_{21} = \Gamma^2{}_{12}$$

$$\Gamma^2{}_{33} = -\cos\vartheta\,\sin\vartheta$$

$$\Gamma^3{}_{03} = \frac{\frac{d}{d\,t}\,y}{y}$$

$$\Gamma^3{}_{13} = \frac{\frac{d}{d\,r}\,y}{y}$$

$$\Gamma^3{}_{23} = \frac{\cos\vartheta}{\sin\vartheta}$$

$$\Gamma^3{}_{30} = \Gamma^3{}_{03}$$

$$\Gamma^3{}_{31} = \Gamma^3{}_{13}$$

$$\Gamma^3{}_{32} = \Gamma^3{}_{23}$$

Metric Compatibility

——— o.k.

Riemann Tensor

$$R^0{}_{101} = e^{-2\,\nu}\left(e^{2\,\lambda}\left(\frac{d^2}{d\,t^2}\,\lambda\right) + e^{2\,\lambda}\left(\frac{d}{d\,t}\,\lambda\right)^2 - \frac{d}{d\,t}\,\nu\,e^{2\,\lambda}\left(\frac{d}{d\,t}\,\lambda\right) + e^{2\,\nu}\left(\frac{d}{d\,r}\,\nu\right)\left(\frac{d}{d\,r}\,\lambda\right) - e^{2\,\nu}\left(\frac{d^2}{d\,r^2}\,\nu\right) - e^{2\,\nu}\left(\frac{d}{d\,r}\,\nu\right)^2\right)$$

$$R^0{}_{110} = -R^0{}_{101}$$

$$R^0{}_{202} = y\,e^{-2\,\lambda-2\,\nu}\left(\frac{d^2}{d\,t^2}\,y\,e^{2\,\lambda} - \frac{d}{d\,t}\,\nu\left(\frac{d}{d\,t}\,y\right)e^{2\,\lambda} - e^{2\,\nu}\left(\frac{d}{d\,r}\,\nu\right)\left(\frac{d}{d\,r}\,y\right)\right)$$

$$R^0{}_{212} = -e^{-2\,\nu}\,y\left(\frac{d}{d\,r}\,y\left(\frac{d}{d\,t}\,\lambda\right) + \frac{d}{d\,r}\,\nu\left(\frac{d}{d\,t}\,y\right) - \frac{d^2}{d\,r\,d\,t}\,y\right)$$

$$R^0{}_{220} = -R^0{}_{202}$$

$$R^0{}_{221} = -R^0{}_{212}$$

$$R^0{}_{303} = \sin^2\vartheta\, y\, e^{-2\lambda-2\nu}\left(\frac{d^2}{d\,t^2}\, y\, e^{2\lambda} - \frac{d}{d\,t}\nu\left(\frac{d}{d\,t}y\right)e^{2\lambda} - e^{2\nu}\left(\frac{d}{d\,r}\nu\right)\left(\frac{d}{d\,r}y\right)\right)$$

$$R^0{}_{313} = -e^{-2\nu}\sin^2\vartheta\, y\left(\frac{d}{d\,r}y\left(\frac{d}{d\,t}\lambda\right) + \frac{d}{d\,r}\nu\left(\frac{d}{d\,t}y\right) - \frac{d^2}{d\,r\,d\,t}y\right)$$

$$R^0{}_{330} = -R^0{}_{303}$$

$$R^0{}_{331} = -R^0{}_{313}$$

$$R^1{}_{001} = e^{-2\lambda}\left(e^{2\lambda}\left(\frac{d^2}{d\,t^2}\lambda\right) + e^{2\lambda}\left(\frac{d}{d\,t}\lambda\right)^2 - \frac{d}{d\,t}\nu\, e^{2\lambda}\left(\frac{d}{d\,t}\lambda\right) + e^{2\nu}\left(\frac{d}{d\,r}\nu\right)\left(\frac{d}{d\,r}\lambda\right) - e^{2\nu}\left(\frac{d^2}{d\,r^2}\nu\right) - e^{2\nu}\left(\frac{d}{d\,r}\nu\right)^2\right)$$

$$R^1{}_{010} = -R^1{}_{001}$$

$$R^1{}_{202} = y\, e^{-2\lambda}\left(\frac{d}{d\,r}y\left(\frac{d}{d\,t}\lambda\right) + \frac{d}{d\,r}\nu\left(\frac{d}{d\,t}y\right) - \frac{d^2}{d\,r\,d\,t}y\right)$$

$$R^1{}_{212} = y\, e^{-2\lambda-2\nu}\left(\frac{d}{d\,t}y\, e^{2\lambda}\left(\frac{d}{d\,t}\lambda\right) + e^{2\nu}\left(\frac{d}{d\,r}y\right)\left(\frac{d}{d\,r}\lambda\right) - e^{2\nu}\left(\frac{d^2}{d\,r^2}y\right)\right)$$

$$R^1{}_{220} = -R^1{}_{202}$$

$$R^1{}_{221} = -R^1{}_{212}$$

$$R^1{}_{303} = \sin^2\vartheta\, y\, e^{-2\lambda}\left(\frac{d}{d\,r}y\left(\frac{d}{d\,t}\lambda\right) + \frac{d}{d\,r}\nu\left(\frac{d}{d\,t}y\right) - \frac{d^2}{d\,r\,d\,t}y\right)$$

$$R^1{}_{313} = \sin^2\vartheta\, y\, e^{-2\lambda-2\nu}\left(\frac{d}{d\,t}y\, e^{2\lambda}\left(\frac{d}{d\,t}\lambda\right) + e^{2\nu}\left(\frac{d}{d\,r}y\right)\left(\frac{d}{d\,r}\lambda\right) - e^{2\nu}\left(\frac{d^2}{d\,r^2}y\right)\right)$$

$$R^1{}_{330} = -R^1{}_{303}$$

$$R^1{}_{331} = -R^1{}_{313}$$

$$R^2{}_{002} = \frac{e^{-2\lambda}\left(\frac{d^2}{d\,t^2}\, y\, e^{2\lambda} - \frac{d}{d\,t}\nu\left(\frac{d}{d\,t}y\right)e^{2\lambda} - e^{2\nu}\left(\frac{d}{d\,r}\nu\right)\left(\frac{d}{d\,r}y\right)\right)}{y}$$

$$R^2{}_{012} = -\frac{\frac{d}{d\,r}y\left(\frac{d}{d\,t}\lambda\right) + \frac{d}{d\,r}\nu\left(\frac{d}{d\,t}y\right) - \frac{d^2}{d\,r\,d\,t}y}{y}$$

$$R^2{}_{020} = -R^2{}_{002}$$

$$R^2{}_{021} = -R^2{}_{012}$$

$$R^2{}_{102} = -\frac{\frac{d}{d\,r}y\left(\frac{d}{d\,t}\lambda\right) + \frac{d}{d\,r}\nu\left(\frac{d}{d\,t}y\right) - \frac{d^2}{d\,r\,d\,t}y}{y}$$

$$R^2{}_{112} = -\frac{e^{-2\nu}\left(\frac{d}{dt}\, y\, e^{2\lambda}\, \left(\frac{d}{dt}\,\lambda\right) + e^{2\nu}\, \left(\frac{d}{dr}\, y\right)\, \left(\frac{d}{dr}\,\lambda\right) - e^{2\nu}\, \left(\frac{d^2}{dr^2}\, y\right)\right)}{y}$$

$$R^2{}_{120} = -R^2{}_{102}$$

$$R^2{}_{121} = -R^2{}_{112}$$

$$R^2{}_{323} = \sin^2\vartheta\, e^{-2\lambda-2\nu}\left(e^{2\lambda+2\nu} + \left(\frac{d}{dt}\, y\right)^2 e^{2\lambda} - e^{2\nu}\left(\frac{d}{dr}\, y\right)^2\right)$$

$$R^2{}_{332} = -R^2{}_{323}$$

$$R^3{}_{003} = \frac{e^{-2\lambda}\left(\frac{d^2}{dt^2}\, y\, e^{2\lambda} - \frac{d}{dt}\,\nu\, \left(\frac{d}{dt}\, y\right)\, e^{2\lambda} - e^{2\nu}\, \left(\frac{d}{dr}\,\nu\right)\, \left(\frac{d}{dr}\, y\right)\right)}{y}$$

$$R^3{}_{013} = -\frac{\frac{d}{dr}\, y\, \left(\frac{d}{dt}\,\lambda\right) + \frac{d}{dr}\,\nu\, \left(\frac{d}{dt}\, y\right) - \frac{d^2}{dr\, dt}\, y}{y}$$

$$R^3{}_{030} = -R^3{}_{003}$$

$$R^3{}_{031} = -R^3{}_{013}$$

$$R^3{}_{103} = -\frac{\frac{d}{dr}\, y\, \left(\frac{d}{dt}\,\lambda\right) + \frac{d}{dr}\,\nu\, \left(\frac{d}{dt}\, y\right) - \frac{d^2}{dr\, dt}\, y}{y}$$

$$R^3{}_{113} = -\frac{e^{-2\nu}\left(\frac{d}{dt}\, y\, e^{2\lambda}\, \left(\frac{d}{dt}\,\lambda\right) + e^{2\nu}\, \left(\frac{d}{dr}\, y\right)\, \left(\frac{d}{dr}\,\lambda\right) - e^{2\nu}\, \left(\frac{d^2}{dr^2}\, y\right)\right)}{y}$$

$$R^3{}_{130} = -R^3{}_{103}$$

$$R^3{}_{131} = -R^3{}_{113}$$

$$R^3{}_{223} = -e^{-2\lambda-2\nu}\left(e^{2\lambda+2\nu} + \left(\frac{d}{dt}\, y\right)^2 e^{2\lambda} - e^{2\nu}\left(\frac{d}{dr}\, y\right)^2\right)$$

$$R^3{}_{232} = -R^3{}_{223}$$

Ricci Tensor

$$\mathrm{Ric}_{00} = -\frac{e^{-2\lambda}\left(y\, e^{2\lambda}\left(\frac{d^2}{dt^2}\,\lambda\right) + y\, e^{2\lambda}\, \left(\frac{d}{dt}\,\lambda\right)^2 - \frac{d}{dt}\,\nu\, y\, e^{2\lambda}\, \left(\frac{d}{dt}\,\lambda\right) + e^{2\nu}\, \left(\frac{d}{dr}\,\nu\right)\, y\, \left(\frac{d}{dr}\,\lambda\right) + 2\left(\frac{d^2}{dt^2}\, y\right) e^{2\lambda} - \ldots\right)}{y}$$

$$\mathrm{Ric}_{01} = \frac{2\left(\frac{d}{dr}\, y\, \left(\frac{d}{dt}\,\lambda\right) + \frac{d}{dr}\,\nu\, \left(\frac{d}{dt}\, y\right) - \frac{d^2}{dr\, dt}\, y\right)}{y}$$

$$\mathrm{Ric}_{10} = \mathrm{Ric}_{01}$$

$$\mathrm{Ric}_{11} = \frac{e^{-2\nu}\left(y\, e^{2\lambda}\left(\frac{d^2}{dt^2}\,\lambda\right) + y\, e^{2\lambda}\, \left(\frac{d}{dt}\,\lambda\right)^2 + 2\, \left(\frac{d}{dt}\, y\right)\, e^{2\lambda}\, \left(\frac{d}{dt}\,\lambda\right) - \frac{d}{dt}\,\nu\, y\, e^{2\lambda}\, \left(\frac{d}{dt}\,\lambda\right) + 2\, e^{2\nu}\, \left(\frac{d}{dr}\, y\right)\, \left(\frac{d}{dr}\,\lambda\right) + \ldots\right)}{y}$$

$$\mathrm{Ric}_{22} = e^{-2\lambda-2\nu}\left(y\left(\frac{d}{dt}\, y\right) e^{2\lambda}\left(\frac{d}{dt}\,\lambda\right) + e^{2\nu}\, y\left(\frac{d}{dr}\, y\right)\left(\frac{d}{dr}\,\lambda\right) + e^{2\lambda+2\nu} + y\left(\frac{d^2}{dt^2}\, y\right) e^{2\lambda} + \left(\frac{d}{dt}\, y\right)^2 e^{2\lambda} - \ldots\right)$$

$$\mathrm{Ric}_{33} = \sin^2\vartheta\, e^{-2\lambda-2\nu}\left(y\left(\frac{d}{dt}\, y\right) e^{2\lambda}\left(\frac{d}{dt}\,\lambda\right) + e^{2\nu}\, y\left(\frac{d}{dr}\, y\right)\left(\frac{d}{dr}\,\lambda\right) + e^{2\lambda+2\nu} + y\left(\frac{d^2}{dt^2}\, y\right) e^{2\lambda} + \left(\frac{d}{dt}\, y\right)^2 e^{2\lambda} - \ldots\right)$$

Ricci Scalar

$$R_{sc} = \frac{2\,e^{-2\,\lambda-2\,\nu}\left(y^2\,e^{2\,\lambda}\left(\frac{d^2}{d\,t^2}\,\lambda\right)+y^2\,e^{2\,\lambda}\left(\frac{d}{d\,t}\,\lambda\right)^2+2\,y\left(\frac{d}{d\,t}\,y\right)e^{2\,\lambda}\left(\frac{d}{d\,t}\,\lambda\right)-\frac{d}{d\,t}\,\nu\,y^2\,e^{2\,\lambda}\left(\frac{d}{d\,t}\,\lambda\right)+2\,e^{2\,\nu}\,y\left(\frac{d}{d\,r}\,y\right)\left(\frac{d}{d\,r}\,\lambda\right)+...\right)}{y^2}$$

Bianchi identity (Ricci cyclic equation $R^{\kappa}{}_{[\mu\nu\sigma]} = 0$)

——— o.k.

Einstein Tensor

$$G_{00} = \frac{e^{-2\,\lambda}\left(2\,y\left(\frac{d}{d\,t}\,y\right)e^{2\,\lambda}\left(\frac{d}{d\,t}\,\lambda\right)+2\,e^{2\,\nu}\,y\left(\frac{d}{d\,r}\,y\right)\left(\frac{d}{d\,r}\,\lambda\right)+e^{2\,\lambda+2\,\nu}+\left(\frac{d}{d\,t}\,y\right)^2e^{2\,\lambda}-2\,e^{2\,\nu}\,y\left(\frac{d^2}{d\,r^2}\,y\right)-e^{2\,\nu}\left(\frac{d}{d\,r}\,y\right)^2\right)}{y^2}$$

$$G_{01} = \frac{2\left(\frac{d}{d\,r}\,y\left(\frac{d}{d\,t}\,\lambda\right)+\frac{d}{d\,r}\,\nu\left(\frac{d}{d\,t}\,y\right)-\frac{d^2}{d\,r\,d\,t}\,y\right)}{y}$$

$$G_{10} = G_{01}$$

$$G_{11} = -\frac{e^{-2\,\nu}\left(e^{2\,\lambda+2\,\nu}+2\,y\left(\frac{d^2}{d\,t^2}\,y\right)e^{2\,\lambda}+\left(\frac{d}{d\,t}\,y\right)^2e^{2\,\lambda}-2\left(\frac{d}{d\,t}\,\nu\right)y\left(\frac{d}{d\,t}\,y\right)e^{2\,\lambda}-e^{2\,\nu}\left(\frac{d}{d\,r}\,y\right)^2-2\,e^{2\,\nu}\left(\frac{d}{d\,r}\,\nu\right)y\left(\frac{d}{d\,r}\,y\right)\right)}{y^2}$$

$$G_{22} = -y\,e^{-2\,\lambda-2\,\nu}\left(y\,e^{2\,\lambda}\left(\frac{d^2}{d\,t^2}\,\lambda\right)+y\,e^{2\,\lambda}\left(\frac{d}{d\,t}\,\lambda\right)^2+\frac{d}{d\,t}\,y\,e^{2\,\lambda}\left(\frac{d}{d\,t}\,\lambda\right)-\frac{d}{d\,t}\,\nu\,y\,e^{2\,\lambda}\left(\frac{d}{d\,t}\,\lambda\right)+e^{2\,\nu}\left(\frac{d}{d\,r}\,y\right)\left(\frac{d}{d\,r}\,\lambda\right)+...\right)$$

$$G_{33} = -\sin^2\vartheta\,y\,e^{-2\,\lambda-2\,\nu}\left(y\,e^{2\,\lambda}\left(\frac{d^2}{d\,t^2}\,\lambda\right)+y\,e^{2\,\lambda}\left(\frac{d}{d\,t}\,\lambda\right)^2+\frac{d}{d\,t}\,y\,e^{2\,\lambda}\left(\frac{d}{d\,t}\,\lambda\right)-\frac{d}{d\,t}\,\nu\,y\,e^{2\,\lambda}\left(\frac{d}{d\,t}\,\lambda\right)+e^{2\,\nu}\left(\frac{d}{d\,r}\,y\right)\left(\frac{d}{d\,r}\,\lambda\right)+...\right)$$

Hodge Dual of Bianchi Identity

——— (see charge and current densities)

Scalar Charge Density (-$R^{0}{}_{i}{}^{i0}$)

$$\rho = -\frac{e^{-2\,\lambda-4\,\nu}\left(y\,e^{2\,\lambda}\left(\frac{d^2}{d\,t^2}\,\lambda\right)+y\,e^{2\,\lambda}\left(\frac{d}{d\,t}\,\lambda\right)^2-\frac{d}{d\,t}\,\nu\,y\,e^{2\,\lambda}\left(\frac{d}{d\,t}\,\lambda\right)+e^{2\,\nu}\left(\frac{d}{d\,r}\,\nu\right)y\left(\frac{d}{d\,r}\,\lambda\right)+2\left(\frac{d^2}{d\,t^2}\,y\right)e^{2\,\lambda}-...\right)}{y}$$

Current Density Class 1 (-$R^{i}{}_{\mu}{}^{\mu j}$)

$$J_1 = -\frac{e^{-4\,\lambda-2\,\nu}\left(y\,e^{2\,\lambda}\left(\frac{d^2}{d\,t^2}\,\lambda\right)+y\,e^{2\,\lambda}\left(\frac{d}{d\,t}\,\lambda\right)^2+2\left(\frac{d}{d\,t}\,y\right)e^{2\,\lambda}\left(\frac{d}{d\,t}\,\lambda\right)-\frac{d}{d\,t}\,\nu\,y\,e^{2\,\lambda}\left(\frac{d}{d\,t}\,\lambda\right)+2\,e^{2\,\nu}\left(\frac{d}{d\,r}\,y\right)\left(\frac{d}{d\,r}\,\lambda\right)+...\right)}{y}$$

$$J_2 = -\frac{e^{-2\,\lambda-2\,\nu}\left(y\left(\frac{d}{d\,t}\,y\right)e^{2\,\lambda}\left(\frac{d}{d\,t}\,\lambda\right)+e^{2\,\nu}\,y\left(\frac{d}{d\,r}\,y\right)\left(\frac{d}{d\,r}\,\lambda\right)+e^{2\,\lambda+2\,\nu}+y\left(\frac{d^2}{d\,t^2}\,y\right)e^{2\,\lambda}+\left(\frac{d}{d\,t}\,y\right)^2e^{2\,\lambda}-\frac{d}{d\,t}\,\nu\,y\left(\frac{d}{d\,t}\,y\right)e^{2\,\lambda}-...\right)}{y^4}$$

$$J_3 = -\frac{e^{-2\,\lambda-2\,\nu}\left(y\left(\frac{d}{d\,t}\,y\right)e^{2\,\lambda}\left(\frac{d}{d\,t}\,\lambda\right)+e^{2\,\nu}\,y\left(\frac{d}{d\,r}\,y\right)\left(\frac{d}{d\,r}\,\lambda\right)+e^{2\,\lambda+2\,\nu}+y\left(\frac{d^2}{d\,t^2}\,y\right)e^{2\,\lambda}+\left(\frac{d}{d\,t}\,y\right)^2e^{2\,\lambda}-\frac{d}{d\,t}\,\nu\,y\left(\frac{d}{d\,t}\,y\right)e^{2\,\lambda}-...\right)}{\sin^2\vartheta\,y^4}$$

Current Density Class 2 ($-R^{i}{}_{\mu}{}^{\mu j}$)

$$J_1 = 0$$

$$J_2 = 0$$

$$J_3 = 0$$

Current Density Class 3 ($-R^{i}{}_{\mu}{}^{\mu j}$)

$$J_1 = 0$$

$$J_2 = 0$$

$$J_3 = 0$$

4.4.62 Collision of plane waves

This metric does not contain terms du^2, dv^2. Nevertheless the matrix of the metric elements has rank 4 due to the off-diagonal elements.

Coordinates

$$\mathbf{x} = \begin{pmatrix} u \\ v \\ x \\ y \end{pmatrix}$$

Metric

$$g_{\mu\nu} = \begin{pmatrix} 0 & -2 & 0 & 0 \\ -2 & 0 & 0 & 0 \\ 0 & 0 & \cos^2(b\,v - a\,u) & 0 \\ 0 & 0 & 0 & \cos^2(b\,v + a\,u) \end{pmatrix}$$

Contravariant Metric

$$g^{\mu\nu} = \begin{pmatrix} 0 & -\frac{1}{2} & 0 & 0 \\ -\frac{1}{2} & 0 & 0 & 0 \\ 0 & 0 & \frac{1}{\cos^2(b\,v - a\,u)} & 0 \\ 0 & 0 & 0 & \frac{1}{\cos^2(b\,v + a\,u)} \end{pmatrix}$$

Christoffel Connection

$$\Gamma^0{}_{22} = -\frac{b\,\cos(b\,v - a\,u)\,\sin(b\,v - a\,u)}{2}$$

$$\Gamma^0{}_{33} = -\frac{b\,\cos(b\,v + a\,u)\,\sin(b\,v + a\,u)}{2}$$

$$\Gamma^1{}_{22} = \frac{a\,\cos(b\,v - a\,u)\,\sin(b\,v - a\,u)}{2}$$

$$\Gamma^1{}_{33} = -\frac{a\,\cos(b\,v + a\,u)\,\sin(b\,v + a\,u)}{2}$$

$$\Gamma^2{}_{02} = \frac{a\,\sin(b\,v - a\,u)}{\cos(b\,v - a\,u)}$$

$$\Gamma^2{}_{12} = -\frac{b\,\sin(b\,v - a\,u)}{\cos(b\,v - a\,u)}$$

$$\Gamma^2{}_{20} = \Gamma^2{}_{02}$$

$$\Gamma^2{}_{21} = \Gamma^2{}_{12}$$

$$\Gamma^3{}_{03} = -\frac{a\,\sin(b\,v + a\,u)}{\cos(b\,v + a\,u)}$$

$$\Gamma^3{}_{13} = -\frac{b\,\sin(b\,v + a\,u)}{\cos(b\,v + a\,u)}$$

$$\Gamma^3{}_{30} = \Gamma^3{}_{03}$$

$$\Gamma^3{}_{31} = \Gamma^3{}_{13}$$

Metric Compatibility

——— o.k.

Riemann Tensor

$$R^0{}_{202} = \frac{a\,b\,\cos^2(b\,v - a\,u)}{2}$$

$$R^0{}_{212} = -\frac{b^2\,\cos^2(b\,v - a\,u)}{2}$$

$$R^0{}_{220} = -R^0{}_{202}$$

$$R^0{}_{221} = -R^0{}_{212}$$

$$R^0{}_{303} = -\frac{a\,b\,\cos^2(b\,v + a\,u)}{2}$$

$$R^0{}_{313} = -\frac{b^2\,\cos^2(b\,v + a\,u)}{2}$$

$$R^0{}_{330} = -R^0{}_{303}$$

$$R^0{}_{331} = -R^0{}_{313}$$

$$R^1{}_{202} = -\frac{a^2\,\cos^2(b\,v - a\,u)}{2}$$

$$R^1{}_{212} = \frac{a\,b\,\cos^2(b\,v - a\,u)}{2}$$

$$R^1{}_{220} = -R^1{}_{202}$$

$$R^1{}_{221} = -R^1{}_{212}$$

$$R^1{}_{303} = -\frac{a^2\,\cos^2(b\,v + a\,u)}{2}$$

$$R^1{}_{313} = -\frac{a\,b\,\cos^2(b\,v + a\,u)}{2}$$

$$R^1{}_{330} = -R^1{}_{303}$$

$$R^1{}_{331} = -R^1{}_{313}$$

$$R^2{}_{002} = -a^2$$

$$R^2{}_{012} = a\,b$$

$$R^2{}_{020} = -R^2{}_{002}$$

$$R^2{}_{021} = -R^2{}_{012}$$

$$R^2{}_{102} = a\,b$$

$$R^2{}_{112} = -b^2$$

$$R^2{}_{120} = -R^2{}_{102}$$

$$R^2{}_{121} = -R^2{}_{112}$$

$$R^3{}_{003} = -a^2$$

$$R^3{}_{013} = -a\,b$$

$$R^3{}_{030} = -R^3{}_{003}$$

$$R^3{}_{031} = -R^3{}_{013}$$

$$R^3{}_{103} = -a\,b$$

$$R^3{}_{113} = -b^2$$

$$R^3{}_{130} = -R^3{}_{103}$$

$$R^3{}_{131} = -R^3{}_{113}$$

Ricci Tensor

$$\mathrm{Ric}_{00} = 2\,a^2$$

$$\mathrm{Ric}_{11} = 2\,b^2$$

$$\mathrm{Ric}_{22} = a\,b\,\cos^2(b\,v - a\,u)$$

$$\mathrm{Ric}_{33} = -a\,b\,\cos^2(b\,v + a\,u)$$

Ricci Scalar

$$R_{sc} = 0$$

Bianchi identity (Ricci cyclic equation $R^{\kappa}{}_{[\mu\nu\sigma]} = 0$)

——— o.k.

Einstein Tensor

$$G_{00} = 2\,a^2$$

$$G_{11} = 2\,b^2$$

$$G_{22} = a\,b\,\cos^2(b\,v - a\,u)$$

$$G_{33} = -a\,b\,\cos^2(b\,v + a\,u)$$

Hodge Dual of Bianchi Identity

——— (see charge and current densities)

Scalar Charge Density ($-R^{0}{}_{i}{}^{i0}$)

$$\rho = \frac{b^2}{2}$$

Current Density Class 1 ($-R^{i}{}_{\mu}{}^{\mu j}$)

$$J_1 = -\frac{a^2}{2}$$

$$J_2 = -\frac{a\,b}{\cos^2(b\,v - a\,u)}$$

$$J_3 = \frac{a\,b}{\cos^2(b\,v + a\,u)}$$

Current Density Class 2 ($-R^{i}{}_{\mu}{}^{\mu j}$)

$$J_1 = 0$$

$$J_2 = 0$$

$$J_3 = 0$$

Current Density Class 3 ($-R^{i}{}_{\mu}{}^{\mu j}$)

$$J_1 = 0$$

$$J_2 = 0$$

$$J_3 = 0$$

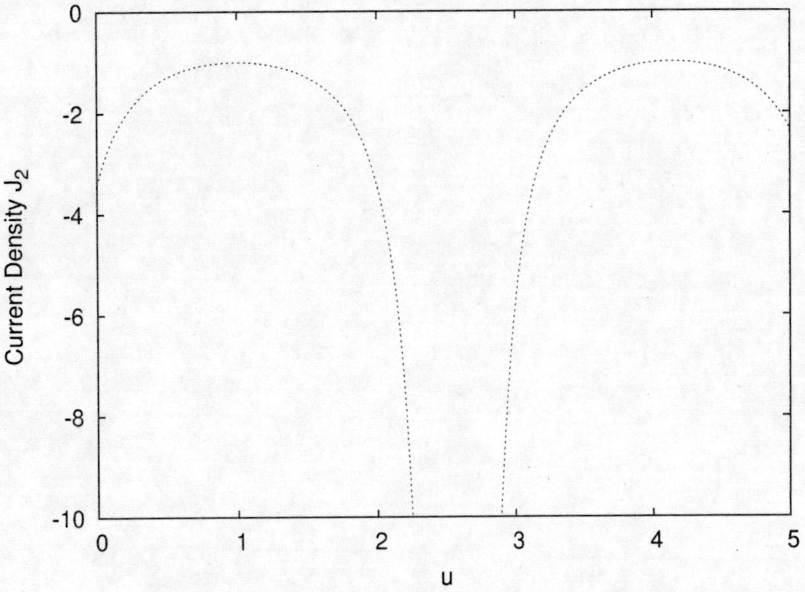

Fig. 4.98: Collision of plane waves, current density $J_x(u)$ for $v = 1, a = b = 1$.

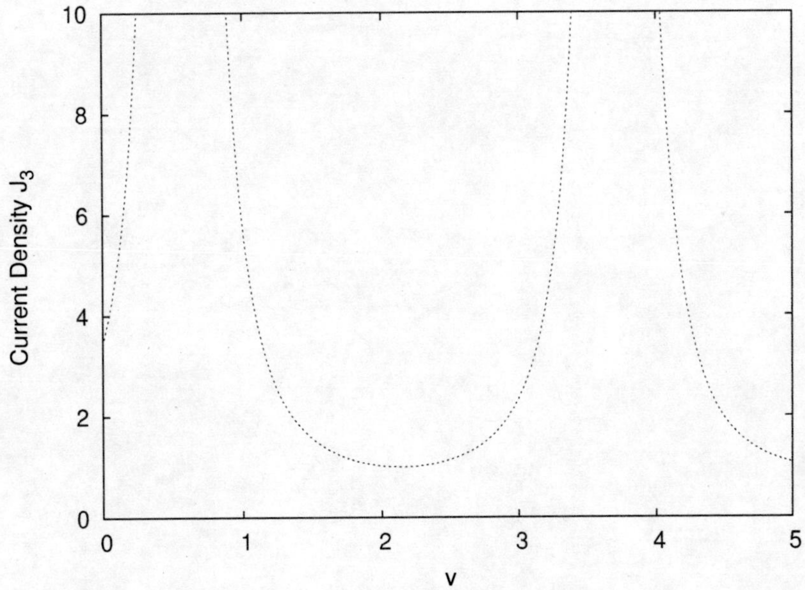

Fig. 4.99: Collision of plane waves, current density $J_y(v)$ for $u = 1, a = b = 1$.

Bibliography

[1] M. W. Evans, "Generally Covariant Unified Field Theory" (Abramis Academic, 2005 onwards), in six volumes to date, seventh in prep.

[2] M. W. Evans et alii, the ECE websites (www.aias.us, www.atomicprecision.com, www.unifiedfieldtheoy.info).

[3] L. Felker, "The Evans Equations of Unified Field Theory" (Abramis 2007).

[4] K. Pendergast, "Crystal Spheres" (Abramis in prep and www.aias.us).

[5] K. Pendergast, "The Life of Myron Evans" (Abramis in prep. and www.aias.us).

[6] F. Fucilla (Director), "The Universe of Myron Evans" (52 minute film an youtube trailer, in distribution stage, made 2008).

[7] M. W. Evans (ed.), "Modern Non-Linear Optics" (Wiley Intersceince, 2001), second edition.

[8] ibid., M. W. Evans and S. Kielich (eds.), first edition (1992, 1993 and 1997).

[9] M. W. Evans and L. B. Crowell, "Classical and Quantum Electrodynamics and the B(3) Field" (World Scientific, 2001).

[10] M. W. Evans and J. P. Vigier, "The Enigmatic Photon" (Kluwer, Doredrecht, 1994 to 2002 hardback and softback) in five volumes.

[11] M. W. Evans and A. A. Hasanein, "The Photomagneton in Quantum Field Theory" (World Scientific 1994).

[12] M. W. Evans, "The Photon's Magnetic Field, Optical NMR Spectrscopy" (World Scientific 1992).

[13] S. P. Carroll, "Spacetime and Geometry: an Introductrion to General Relativity" (Addison Wesley, New York, 2004, free downloadable notes of 1997).

[14] H. Stephani, D. Kramer, M. McCallum, C. Hoenselaers and E. Hertl, "Exact Solutions of Einstein's Field Equation" (Cambridge University Press, 2nd Edition, 2003).

Chapter 5

Einstein's Great Contributions to Physics, New Cosmologies and the Alternating Theory of the Universe, as a Replacement for the Flawed Big Bang Theory

by

Kerry Pendergast[1],

Alpha Institute for Advance Study (AIAS)
(www.aias.us, www.atomicprecision.com)

5.1 Introduction

Einstein is famous for his five great papers of 1905, which changed physics forever. He is also famous for special and general relativity, new cosmologies and his quest to unify gravity with quantum theory. This chapter gives a brief summary of Einstein's early work which transformed physics and describes how this work, has been taken forward by workers at AIAS. It then goes on to describe how Einstein was able to see that space was curved and how he formulated the Einstein-Hilbert equation, to mathematically describe gravity as the curvature of space. Next the methods used to test Einstein's theory of general relativity through astronomical observations are described. The evidence for the existence of black holes is then given, before casting doubt on the existence of black holes as singularities. Some speculations are then given on the nature of new cosmologies and the Alternating Theory of the Universe is outlined, as a replacement for the flawed Big Bang theory. The chapter ends with a devastating criticism of dark matter and concludes that dark matter is a figment of mathematician's imagination!

5.2 Einstein's early work and how it has been extended by workers at AIAS

Einstein's early work at the turn of the twentieth century includes his five miracle year papers in 1905, which changed physics. These five papers led:

[1]e-mail: kp.phys@btinternet.com

1. To our understanding of the Photoelectric effect and quantum theory,
2. Proved to physicists that atoms existed,
3. Contributed to our understanding of special relativity and
4. Gave us the famous equation $E = mc^2$.

5.2.1 Einstein's miracle year and subsequent work

Einstein attended Zurich Polytechnic from October 1896 to July 1900. Zurich polytechnic was a teachers' and technical college, which sported new state of the art science laboratories paid for by Werner Siemens. By 1911, the polytechnic was so well thought of that its status was elevated and was soon renamed ETH (Eidgenössische Technische Hochschule Zurich).

After graduating Einstein eventually found a well paid job, at the Bern Patent Office (where he worked until 1909), which was well suited to both his demeanor and talents. Einstein's father had owned his own electric lighting and generating company, which had involved Albert in patent work and the practical design of electromagnetic devices. Albert found he was able to do his patent work in a fraction of his working day, leaving him free to surreptitiously work on his theories while still at his work desk. Michele Besso, Einstein's life long friend also worked at the patent office, giving Albert a sounding board for his theories.

During this time the great advances in theoretical physics were being made by applying statistical mechanics and kinetic theory to problems of radiation and thermodynamics. Professor Ludwig Boltzmann had produced influential work, showing how maths and statistics could be applied to problems involving molecules in physics and chemistry. In 1901, Einstein had his first paper published which described capillary action in terms of the attraction between large numbers of molecules. Einstein then spent time considering how statistical mechanics could be applied to molecules undergoing diffusion and other processes. This paved the way for Einstein's miracle year in 1905, in which he presented to the world his five great papers that changed physics.

1. Einstein's miracle year 1905, started with, 'On a Heuristic Point of View Concerning the production and transformation of light', which explained the photoelectric effect and was to lead to quantum theory and Einstein's only Nobel Prize.

2. Einstein's second paper; 'A New Determination of Molecular Dimensions' used the processes of viscosity and diffusion to formulate two simultaneous equations for the unknown particle sizes and numbers of particles. Feeding in the data for the case of sugar dissolved in water and solving the simultaneous equations then produced Avogadro's number and the size of the molecules involved.

3. Einstein's third paper explained the phenomena of Brownian motion and effectively proved atoms and molecules existed. The British Civil List Scientist, Robert Brown had discovered in 1828 that pollen grains in water could be seen in a microscope altering their speeds and directions, as if they were receiving random kicks from different directions. Einstein used statistical mechanics to show that invisible particles many orders of magnitude smaller at the size of molecules could randomly act together to produce the random kicks seen. Furthermore, Einstein showed that the distance that the pollen grains would move away from each starting position, was proportional to the square root of the time between observations, which could be proved by simply observing the grains through a microscope.

4. Einstein's fourth paper changed our understanding of space and time and was entitled, 'On the Electrodynamics of moving bodies'. This was Einstein's theory of special relativity.

5. Einstein's fifth paper developed from his fourth paper and was entitled, Does the Inertia of a Body Depend on its Energy Content? This introduced the world to Einstein's famous equation: $E = mc^2$.

Einstein went on to extend his 1905 theory of special relativity to general relativity, which included the effects of acceleration and gravity. Einstein finally developed his famous Einstein-Hilbert field equation in 1915, but the strain of learning the advanced mathematics of curved surfaces inherent in the theory, badly affected his health and came close to killing him. Luckily his cousin and future second wife Elsa Einstein was there to nurse him back to health.

In the second half of his life, Einstein attempted to unify his theory of quantum theory with his theory of general relativity to produce a unified field theory, which is often referred to as his theory of everything. After being taken ill at his Princeton home, Einstein spent the last two days of his life in the Princeton Hospital where even hours before dying, he continued to work on the mathematics required to complete his theory of everything. This work was subsequently said to be impossible, until Myron Evans proved the critics wrong by producing ECE theory in 2003.

In formulating the great Einstein-Hilbert field equation of general relativity, both Hilbert and Einstein had been forced to make a simplification to the mathematics of space-time. The simplification involved making an assumption of symmetry, because the Riemann mathematics of curved surfaces they were building upon, could not incorporate the required asymmetry. The tensor mathematics Hilbert and Einstein required was however, being pioneered by the great French mathematician Elie Cartan. Cartan's tensor mathematics could incorporate the effect of spin into space-time and could therefore; hold out the possibility of combining the electromagnetism of light with the curvature caused gravity. Einstein knew the 1915 Einstein-Hilbert field equation of relativity had shortcomings and it was only a step towards his goal to describe physics through geometry. He therefore collaborated unsuccessfully in the twenties with Elie Cartan to try to combine the curvature of gravity with the spin of light and space-time. However, he could not arrive at a solution in his lifetime and it would take till the twenty-first century, for this ultimate goal of chemical physics to be achieved, through the advent of Einstein-Cartan-Evans Grand Unified Field Theory.

5.2.2 The photoelectric effect, quantum theory and the photon

1905 was Einstein's 'Miracle Year' in which he presented to the world his five great papers that changed physics.

- The first paper of Einstein's miracle year 1905 was, 'On a Heuristic Point of View Concerning the production and transformation of light'. In this first seminal paper of 1905, Einstein gave his explanation of the photoelectric effect which led to quantum theory and Einstein's only Nobel Prize.

Einstein studied for his degree in physics at ETH University in Zurich, which at that time was called ETH Polytechnic. There, Einstein was so obsessed with the nature of light that at times he failed to turn up for lectures, because he was following his own path to 'reveal the enigma' of light! Einstein's lecturers were aware of his undoubted talents in physics, but were put off by his arrogance and failure to abide by the etiquette dictated by his position as a student. As a result, Einstein could not get a position in a university after he graduated, but did find an excellent position to suit his temperament and practical knowledge of electrical devices, by working as a patent clerk in the Bern Patent Office. This 'hands on' practical knowledge of electrical generation came from his father's business enterprises. His father had been given contracts to light the streets of Munich in Albert's school days. However, the business was based on DC current and faltered when the follow on contracts were awarded to Einstein's father's rival, Siemens. Siemens used the profits generated by his thriving business, to turn ETH Polytechnic's physics department into a state of the art facility for training physicists and physics teachers. In 1905, while working as a patent clerk, Einstein published five papers which revolutionized physics. His paper on the photoelectric effect, asserted that rays of light are not continuous, but travel as discrete quantities or packets of energy, which we now call quanta or photons. This paper was to earn Einstein belatedly the 1921 Nobel Prize for physics and enabled him to return to Zurich as a Physics Professor at Zurich University, the 'sister' university to ETH. In 1905 Einstein's explanation of Brownian motion proved to physicists that atoms and molecules existed. Einstein had previously studied the work of leading physics philosophers, such as Mach who did not believe in atoms, and Einstein's work with Brownian motion brought home to the maturing Albert, that physics was unnecessarily abstract and Boltzmann was right to believe in the deterministic nature of matter at the atomic level! Boltzmann's equations for describing the particulate nature of matter by collisions between countless numbers of spheres did indeed have its root in real collisions between atoms and molecules.

Evans had since his early years in Aberystwyth, also become obsessed by the nature of light in its chemical manifestation in spectroscopy. Far infrared spectroscopy had become the main thrust of his work, as soon as he had started working for his Ph D. This work required Evans to consider how light in the far infrared low energy range, would affect the movement of molecules. In the infrared, light causes the atoms in molecules to vibrate about their bonds. The enigma in the far infrared was how light caused molecules to translate, rotate and oscillate. Evans' move into computer simulation was essentially an extension of Einstein's work on Brownian motion. Thus, for years Evans had been following a similar path to Einstein's, but this connection was not obvious. When Evans moved to Zurich, he would not simply cross paths with Einstein geographically, but also through the mutual desire of the two physicists to understand more fully the nature of light. Poetically then, Myron made the discovery at Zurich, which would further our understanding of light and the photon and in time would facilitate the completion of Einstein's life's work! Einstein's understanding of light built on the work of the Civil List Scientist Michael Faraday. Now, Evans would build on the work of both men with his work on the Inverse Faraday Effect, which fittingly was carried out in Zurich with Wagniere research group.

5.2.3 The existence and motion of atoms

- Einstein's second miracle year paper was, 'A New Determination of Molecular Dimensions'. In this paper Einstein used the processes of viscosity and diffusion to calculate the size of molecules and the value of Avogadro's number.

- Einstein soon followed up this paper with his third miracle year paper, which used Brownian motion to prove atoms did actually exist.

Robert Brown (1773-1858) was the Scottish botanist on the Matthew Flinders Expedition (1801-1803) to New Holland (now called Western Australia) aboard the Investigator, where he discovered several thousand new plant species. Interestingly, Brown also discovered and named the nucleus found in cells. In 1828, Brown noticed that tiny particles, such as pollen grains in water or dust in air lit up in a beam of light, are in random motion even though the air or the water appears still. This random motion is called Brownian motion. Robert Brown's contributions to botany and cell biology, together with his discovery of Brownian motion, led to his appointment to the Civil List.

Brown was not able to explain Brownian motion, but in 1905 Einstein was. Einstein stated that the random motion of tiny pollen grains, is due to collisions with even smaller, fast moving water particles. Einstein had already derived Avogadro's number from a diffusion experiment in his second great paper of 1905, enabling him to calculate the approximate sizes of atoms and molecules and thereby providing important proof for their existence and deriving Avogadro's number. Einstein's third great paper in 1905 allowed him through his explanation of Brownian motion, to prove to physicists that atoms did actually exist. The great thing about Albert's paper on Brownian motion, was the proof did not simply rely on mathematics, but could be proven experimentally simply by watching pollen grains down a microscope tube and timing their motion. This was a great feat of Baconian science that is sadly lacking from much of modern theoretical physics, which is too reliant on abstract mathematics at the expense of experimentation.

Evans took Einstein's work on Brownian motion further, while at Aberystwyth and Oxford and this led to two separate awards from the Royal Society of Chemistry within the space of a year. These were the Harrison Memorial Prize and Meldola Medal and were awarded for an extension of the Brownian motion theory called Mori theory (See the early papers of Evans' Omnia Opera and the monograph "Molecular Dynamics" which is available on www.aias.us). Evans has worked on Brownian motion with Gareth Evans, Bill Coffey and Paolo Grigolini. The overview description in Evan's Omnia Opera also gives an account of this work, and one or two of Evans' "top ten papers" is on the subject. Myron and his colleagues worked on the Smoluchowski, Fokker Planck and Kramer's equations, and also on the Euler-Langevin equation. Mori theory extends the friction coefficient of the Langevin equation into a memory function and continued fraction. Evans' and his team tested this with far infrared data obtained at Aberystwyth. Later this work developed into the Pisa Algorithm (see Omnia Opera). Myron used a combination of simulation, theory and data. The "collection of positive opinion" on www.aias.us shows this work had great impact, for example the letter from Max Maglashan of University College London, of which Myron was Ramsay Memorial Fellow (1974 to 1976) based at Aberystwyth.

In 1905 in his second great paper; 'A New Determination of Molecular Dimensions' Einstein had used the processes of viscosity and diffusion to formulate two simultaneous equations for the unknown particle sizes and numbers of particles. By feeding in the data for the case of sugar dissolved in water and solving the simultaneous equations, Einstein produced Avogadro's number and the size of the molecules involved. In his third great paper in 1905 Einstein had gone on to use observations of the Brownian motion, to prove that pollen grains are moving due to interaction with much smaller particles called molecules. This is similar to a Langevin equation approach.

Jean Perrin (1870-1942) had shown in 1895 that cathode rays had a negative charge, preparing the way for the discovery of the electron and calculated Avogadro's number by several methods. Perrin also did the experimental work to test Einstein's Brownian motion prediction on the movement of atoms and in so doing proved to physicists, that John Dalton's atomic theory was correct. He also proposed in 1909, that the constant should be named in honor of Avogadro, who had proposed it in 1811. Perrin went on to receive the 1926 Nobel Prize for Physics, largely for his work on Avogadro's constant. Perrin's and Einstein's work vindicated Boltzmann's belief in the existence of atoms and showed clearly that even at the atomic level, nature is deterministic and follows the laws of physics, so well described by Newton in his book 'The Principia".

Paul Langevin (1872-1946) worked closely with the Curies and like Pierre and Marie studied magnetism in detail, which led to Paul giving its explanation in terms of electrons in atoms. He is also famous for the 'Langevin Equation' which is a stochastic differential equation used to describe Brownian motion. Einstein frequently visited Paul Langevin and Marie Curie in Paris and also met them at the Solvay conferences, from 1911. Langevin found that magnetism is due to the motion of electrons in atoms. In chemistry the spin of electrons in orbitals give rise to magnetism. As orbitals are progressively filled with electrons the magnetism builds up. However, once orbitals are half full the electrons pair up, with the additional electrons spinning in the opposite direction to the electrons already in the orbital, causing the observed magnetism to be cancelled out. Iron is a strongly magnetic material (ferromagnetic), because each atom sports five unpaired electrons with their spin lined up to produce the extra strong magnetism

In later years, diffusion equations were developed by Smoluchowski, Fokker, Planck and Kramers. In the seventies there was strong interest in this at the Dublin Institute for Advanced Studies, clearly following the research guidance of people like Schrödinger and Synge at DIAS. Evans' tested out these very complicated equations in the far infrared, using the Elliot 4130 and CDC 7600. This can all be traced back to those first years of the twentieth century, which were so important to physics.

Evans developed the Langevin equation for the far infrared using memory function methods in his Ph. D. Work (see early Omnia Opera papers on www.aias.us). This work was later extensively developed with Bill Coffey at Trinity College Dublin and Paolo Grigolini in Pisa. The Debye bell shaped dielectric loss is produced from the rotational Langevin equation, but gives the Debye plateau in the far infrared. This was first recognized and named at the EDCL by Myron's Ph. D. supervisor, Professor Mansel Davies. The friction coefficient of the Langevin equation was developed into a continued fraction of memory functions. Gareth Evans, Colin Reid and Myron used this method extensively in the far infrared, giving a first explanation of the far infrared spectra of materials. Myron added computer simulation to this technique in 1975, using one of the very first simulation methods developed by the Konrad Singer group at Royal Holloway College, and the Oxford Group of Professor Sir John Rowlinson where Evans worked from 1974 to 1976. With the greatly improved power of modern computers, it should be possible to simulate millions of molecules now and allow the far infrared to be described with ever greater refinement, using the same basic methods worked out in the seventies by Myron's EDCL "Hall of Fame" group (BBC Mid Wales link on www.aias.us).

5.2.4 Special relativity

- Einstein's fourth miracle year paper changed our understanding of space and time and was entitled, 'On the Electrodynamics of moving bodies'. This was Einstein's theory of special relativity. Special relativity came about as a result of the speed of light being measured and tested experimentally.

In 1675, the Dutch astronomer Ole Roemer noticed that the times of the eclipses of Jupiter's moons, were affected by the position of the Earth in its orbit around the Sun in relation to Jupiter. When Jupiter was at opposition, at its closest approach to the Earth, the position of the satellites was ahead of the predicted times. This showed that the time taken for light to cross the Earth's orbit could be determined by the variations in the eclipses of Jupiter's moons, due to the Earth's position. The timings work best with the moon Io, because it is one of the smaller Galilean moons and being the nearest to Jupiter is the fastest to emerge from eclipse.

The time and distance could then be used to calculate the speed of light. Using the accepted diameter of the Earth's orbit of the time, the speed of light was calculated as 200, 000 km/s.

In 1849 the French physicist Hippolyte Louis Fizeau used a rapidly rotating toothed wheel to reduce the need for vast distances to be used for speed of light experiments. Light was shone through the teeth of a first wheel, through a second wheel five miles away and reflected back along its path through the same gap. With the toothed wheels rotating at hundreds of times a second, time intervals of fractions of a second could be measured. This method gave a better value for the speed of light as 313, 300 km/s. In 1926 Leon Foucault's improved method using rotating mirrors, achieved a figure of 299, 796 km/s for the speed of light.

In the second half of the nineteenth century, the full importance of the speed of light to physics, astronomy and cosmology was starting to be appreciated. Maxwell's equations asserted that the speed of light could not be exceeded and light would always move away from an observer at the speed of light, no matter how fast the observer was moving. Oliver Heaviside's experiments with electricity were also throwing up important results.

In 1887 Albert Abraham Michelson (1852-1931) and Edward Morley (1838-1923), carried out the famous Michelson-Morley experiment. An observer on Earth travels at 30 km per second in the direction of the Earth's orbit and Michelson and Morley designed an experiment to measure the effect on the speed of light. They expected to find light traveled faster when assisted by the Earth and slower in the opposite direction, but were perplexed to find the speed of the observer and apparatus had no effect.

In 1892 George Fitzgerald (1851-1901) working at Trinity College, Dublin explained the results of the Michelson-Morley experiment by suggesting that as objects approached the speed of light, their length in the direction of motion would become progressively shorter and this could account for the constancy of the speed of light seen by a moving observer. Heaviside's new equations and experiments helped Fitzgerald come to this conclusion. Fitzgerald's qualitative hypothesis was adopted almost immediately by Hendrik Lorentz (1853-1928) who set about making it mathematically precise.

In 1904, Hendrik Lorentz produced the Lorentz transform which could quantify the contraction of length at high speeds. Jules Poincaré (1854-1912), the great French mathematician and physicist was able to see the importance of Lorentz transformations to the synchronization of time and relativity. The scene was set for Einstein to formulate his theory of special relativity, which he did in 1905.

Michael Faraday, the father of classical electrodynamics, believed all forms of light were composed of electromagnetic waves moving at the speed of light. The Faraday Effect provided the first experimental evidence that light and magnetism were related and inspired Faraday to champion his views to the public, which he did in 1846, with his lecture, 'Thoughts on Ray Vibrations'. However, his views on the nature of light were not widely accepted. James Clerk Maxwell (1831-79) the great physicist and mathematician did believe Faraday's ideas however and set about proving them.

The German Wilhelm Weber was also on the case. In 1858 Weber had measured the ratio of magnetic to electric forces and when Maxwell fed Weber's ratio result into his own equations, a velocity equal to the speed of light appeared.

In 1868, Maxwell managed to reduce his mathematical ideas concerning light, into the four Maxwell equations, which unified electricity and magnetism as a wave traveling in the ether at the speed of light. Maxwell's equations of electromagnetism mathematically confirmed what Faraday had been saying about the nature of light for the many years and forever changed our views on the nature of light.

In addition in 1864, Maxwell was able to state that visible light was only one form of light within an electromagnetic spectrum which included invisible forms of light of longer and shorter wavelengths. In the next thirty years radio, X and gamma rays would be discovered and added to the electromagnetic spectrum. Herschel had already discovered infrared light in 1800, when he saw that a thermometer placed beyond the red end of light from a prism, indicated it was being heated by a previously unknown form of light

Infrared telescopes such as UKIRT (United Kingdom Infra Red Telescope) observe objects that are relatively cold, such as dust clouds (nebulae) containing stars at the start of their lives. Infrared light can penetrate dust clouds, because it has wavelengths longer than visible light. This is similar to red light being able to penetrate fog better than white light. The biggest and best telescope in the world operating in the far infrared is in Hawaii and is named in honor of James Clerk Maxwell or JCMT for short.

Maxwell's great mathematical ability was appreciated from his earliest days in Cambridge and in 1852; he was invited to become an apostle. The Apostles were a secret society (founded in 1820), composed of the twelve students who were considered to have the greatest intellect of those currently at Cambridge. On leaving, the apostles became Angels and met in secret every few years, at a Cambridge College. Many angels such as Bertrand Russell went on to work in the media, government and church. Maynard Keynes and the Bloomsbury group were well known before the First World War. The apostles came to the attention of the public again in 1951, with the exposure of the Cambridge spy ring. Anthony Blunt (MI5) and Guy Burgess (MI6) were both spies passing information to the Soviets.

Maxwell's equations predicted that light would always move away from an observer at the speed of light, no matter how fast the observer was moving. When Einstein was a student he was intrigued by Maxwell's equations and their implications. This stimulated him to formulate his great theory of special relativity.

The theory proposed by Maxwell was developed using William Hamilton's quanternions. Quanternions are a non-commutative extension of complex numbers, which were subsequently converted into vector form by Oliver Heaviside, to produce the Maxwell-Heaviside (MH) theory. Sir William Rowan Hamilton (1805-1865) was a great Irish mathematician and physicist. His work later became relevant to quantum physics, where the 'Hamiltonian' bears his name.

Oliver Heaviside (1850 to1925) from Camden town, London, made major contributions to physics and mathematics. In 1885 he presented Maxwell's equations in their present form, which is why they can be referred to as the Maxwell-Heaviside equations of field theory. He also developed operational calculus, the theory of vectors, the Heaviside step function, the Heaviside equation of telegraphy, and (in 1889) the basic equation for the motion of charge in a magnetic field. He developed m.k.s. units, the use of complex numbers for circuit theory, and the modern theory of Laplace transforms through his powerful operational calculus method. He deduced the denominator root (1 - (v/c) squared) fifteen years before Einstein in his studies of the speed of light in circuits.

In 1896 Oliver Heaviside was awarded a Civil List pension on the recommendation of Rayleigh, Kelvin, Fitzgerald and others. He is one of the greatest physicists and mathematicians in history and deserves to be better known.

Heaviside also proposed that the vector potential A in classical electrodynamics be an abstract entity, only the fields E and B are physical in that view. The latter is the opposite of the view of Faraday and Maxwell, who considered the potential (electrotonic state of Faraday) to be physical. The Heaviside point of view evolved into the gauge principle in the twentieth century, mainly through the ideas of Weyl. This principle states that the action is invariant under the gauge transformation of any field. This appeared to be effective until Evans proposed the ECE spin field B(3) in Physica B, 182, 227 and 237 (1992). The B(3) field implies an O(3) gauge invariance for electrodynamics, as in the work Evans did from about 1992 to 2003 (see collected papers on www.aias.us).

In paper 71 an invariance principle was introduced with the intention of replacing the gauge principle. In papers 71 and 72 some applications were developed of the invariance principle. In paper 73 some details of its advantages over the gauge principle were developed. Evans proposed the invariance principle in paper 71, because it is based directly on an invariant of ECE theory and Cartan geometry - the tetrad postulate. The latter is frame invariant (the covariant derivative of the tetrad always vanishes in all frames of reference). It was shown from this property that the tetrad field (i.e. all fields) must include a phase alpha, which cannot depend on distance and time and is therefore "global", in the rather vague but received terminology, which Evans follows for convenience - "local" and "global". The local and global connection can be used to explain effects of quantum entanglement, showing that the Heisenberg uncertainty principle is wrong and that Einstein was correct to assert that nature is deterministic and 'God does not play dice'.

In 2005, Myron W. Evans followed in the footsteps of William Herschel, Robert Brown, John Couch Adams, Michael Faraday, Alfred Wallace and Oliver Heaviside by being awarded a civil list pension. At the present time, Myron Evans is the only scientist in Britain or the Commonwealth to hold this high honor.

5.2.5 $E = mc^2$

- Einstein's fifth miracle year paper developed from his fourth paper and was entitled, Does the Inertia of a Body Depend on its Energy Content? This introduced the world to Einstein's famous equation: - $E = mc^2$.

Einstein is perhaps best known for his equation $E = mc^2$, which shows that a small amount of mass can be converted into a vast amount of energy. E stands for energy, m is the mass and c stands for the speed of light. The speed of light is the fastest speed achievable and the speed of light squared is a vast number. Therefore the equation shows that when a small amount of mass or matter is converted into energy, a vast amount of energy is created as is seen in the atomic bomb. This conversion of matter into vast amounts of energy, through nuclear reactions turning one element into another, is the power of the stars and is the source of the Sun's energy that allows life to thrive on Earth.

In the nineteenth century two major breakthroughs were made, which took chemistry and biology forward at a great rate. In chemistry, Dalton, Davy and Faraday did much to show atoms existed, allowing chemistry to develop and expand at an ever increasing rate, which did much to fuel the industrial revolution. In biology, it was the work of Darwin, who was born two hundred years ago in Shrewsbury, along with Alfred Wallace from Usk in Monmouthshire, whose theories on 'survival of the fittest' led to an understanding of the theory of evolution, which had already been recognized by scientists for hundreds of years, including Darwin's grandfather. Obviously this had implications for the power of the church and religious individuals were obviously going to oppose evolution and strive to maintain creationism, as the only explanation for the ascent of man. Lord Kelvin opposed evolution on the grounds that evolution would require billions of years for the changes to animal and plant life to be incorporated, but Kelvin believed that the Sun could not shine for more than a few million years by chemical combustion. This is where the work of the Curies, along with Frederick Soddy and Ernest Rutherford would come into play. Einstein's famous equation $E = mc^2$, would then be used to show that the Sun would easily be able to shine, for the billions of years needed for the effects of survival of the fittest to lead to the evolution, as witnessed in the fossil record.

Madam and Pierre Curie in Paris at the turn of the century did the important work of separating uranium and radium from pitchblende, from mines in Czechoslovakia. The Curies then supplied Soddy and Rutherford with radioactive materials to work on in McGill University in Canada. Here the chemist Soddy and the physicist Rutherford were able to observe the transmutation of elements for the first time and to discover that radioactive isotopes have a characteristic half life, the time which is taken for half the atoms of a particular isotope to undergo radioactive decay, into other elements. In 1904, Rutherford came to London to give a lecture at the Royal Institution, where he showed that the age of rocks could be determined by measuring how much radioactive decay had occurred since the rocks had been formed or crystallized. Uranium has a half life of around a billion years and over this time; half the atoms would have decayed to form lead. Rutherford had simply measured the proportion of lead to uranium and related this to the known half life of uranium, to give the age of the rock samples. This gave a figure in the billion plus range for the oldest rock samples and showed the Earth had indeed been around long enough for evolution to take place. Furthermore, Einstein's 1905 equation $E = mc^2$, was able subsequently to explain that through nuclear reactions and transmutations of the elements, such a vast amount of energy was created, that the Sun was easily able to shine steadily for the time required to support life on Earth, for this evolution to take place.

Soddy returned to Britain in 1903, to work with Lord Ramsay at University College, London, where he was able to show alpha particles are the nuclei of helium emitted during nuclear fission, the splitting in two of unstable nuclear isotopes. Soddy also showed subsequently that elements could exist in alternative forms which he called isotopes. Soddy also went on to show, that alpha particle emission causes an element to shift two spaces to the left in the periodic table and the emission of a beta particle, causes the element to shift one space to the right. Thus, Soddy could be called the world's first successful alchemist. In 1904, Soddy left London to continue his great work at the University of Glasgow. Soddy's work is of great importance in chemical physics and society, as described by Aberystwyth's Professor Mansel Davies in his 1992 article in the 'Annals of Science', entitled 'Frederick Soddy: The Scientist as Prophet'. His work greatly influenced H. G. Wells, particularly concerning the use of radioactivity to make atomic weapons, which is considered in Well's books. H. G. Well's in turn greatly liked the writings of Soddy. Interestingly, Mansel Davies wrote a book on the history of science at the request of Wells. Soddy studied chemistry initially in Aberystwyth, before continuing his studies at Oxford. This connection was celebrated in Aberystwyth, when the radiation laboratory set up by Dr. Cecil Monk at the Edward Davies Chemical Laboratories, was named 'The Soddy Laboratory'. Sadly, Dr. Monk who had lived at Borth, outside Aberystwyth for many years, passed away early in 2009. Soddy's work with Ramsay is again celebrated in Aberystwyth, through Myron's award of a University College London, Ramsay Fellowship, which

Myron used to finance his researches in Aberystwyth. H. G. Wells was also well regarded in Aberystwyth and its former head of chemistry, Professor John Meurig Thomas ended his book on Michael Faraday's life at the Royal Institution by quoting Well's work.

Well's work used science to predict what the future could be like and how the folly of man could lead to anarchy. This is the theme of the great film, 'Things to Come' in which after the collapse of civilizations, petrol needed to be refined from coal to allow planes to fly again: putting coal mines into the front of new power struggles. The pit shown in the 1936 film is the Griffin Colliery in Blaina, which was closed due to problems with too much gas in its coal seams. Later the colliery became entirely covered with the earth extracted from the nearby Rose Heyworth Colliery, in order to get to the coal underground. Now nearly every pit in South Wales has gone, but the colliery shown in the 1936 film, is still there hidden beneath its tomb of earth: A time capsule for future generations to uncover.

When Soddy the alchemist left London for Glasgow, he was replaced as Lord Ramsay's assistant at University College, London in 1904 by Otto Hahn, the German chemist. Hahn in 1905 went on to work with Rutherford at McGill University in Canada. Following in Soddy's footsteps, Hahn would become a major figure in nuclear chemical physics, in the first half of the twentieth century and at the outbreak of World War 2 produced a paper describing how the uranium atom could be split by bombardment with neutrons. This was a key experiment in producing both nuclear energy and creating an atomic bomb, through the process of fission and gave Hahn the 1944 Nobel Prize for Chemistry. At the outbreak of War, Niels Bohr was in New York and on reading Hahn's Berlin paper, recognized that splitting the uranium nucleus could lead to a chain reaction which could release the power described by the equation $E = mc^2$, as a war winning weapon. Bohr warned the American government about it, before returning to Copenhagen, where he would eventually meet up with his old friend Heisenberg, at a time when Denmark was under occupation.

Hahn had written a book on radiochemistry during his time at Cornell University, before returning to Germany. In a twist of fait, Hahn managed to keep out of the Nazis atomic bomb project, but his knowledge of radiochemistry described in his book, was used by the Americans to achieve the separation of the isotopes of uranium during the Manhattan Project.

After the war, the world's knowledge of the nature of the atom had come on in leaps and bounds and major laboratories were founded, such as Aldermaston in Britain and Los Alamos in the USA, in addition to their equivalents in France and Russia. It was soon recognized that fusing hydrogen atoms as occurs in the Sun and stars, would release a thousand times more energy in a nuclear bomb, than the fission of uranium and plutonium does. In the space of a decade, both Russia and America would master the process of fusion which is the secret behind the Sun's longevity and is the source of energy for life on Earth.

Niels Bohr developed good links with the allies during World War 2. At the outbreak of war, he had alerted the Americans to the possibility of a uranium chain reaction leading to a Nazi atomic weapon. By the time Heisenberg visited Bohr in occupied Copenhagen in 1941, their famous special relationship was faltering. In October 1943, Bohr escaped from occupied Denmark and was flown from Sweden to Britain in a De Havilland Mosquito bomber, to assist with the British and American bomb projects.

The Mosquito was an all wood bomber designed by De Havilland at the outbreak of war and put into services in only twenty-one months. It was one of the iconic planes of World War 2, being twenty miles an hour faster than the spitfire and the fastest plane in the world when first produced. Its all wood design meant it was easy to build and did not need the metal, which was in short supply during this time. Later versions of the Mosquito were given a nose cannon or powerful guns and were a menace to enemy submarines and to trains crossing occupied territory. One Mosquito attacked the train carrying Jean-Pierre Vigier for interrogation, thereby saving his life.

Jean-Pierre Vigier was born in the Sorbonne in Paris and educated in Geneva, where he took doctoral degrees in physics and mathematics. He was a member of the general staff of the French Resistance in the Savoy Mountains. Vigier having been betrayed to the Vichy Gestapo was being taken to Lyons for interrogation by Klaus Barbie, the war criminal not known for his gentility when the train was bombed by the R.A.F. allowing Vigier to escape from captivity to rejoin the Resistance. He joined the French Army in 1944 and was wounded at the Remagen bridgehead across the Rhine, but soon repatriated by a US mechanized infantry division. He was awarded the Medaille de Resistance and the Legion d'Honneur.

Vigier joined Joliot-Curie's staff after the war, but did not want to be involved in this atom bomb work and resigned in protest at the French nuclear bomb. He was invited to work with Einstein at the Princeton Institute of Advanced Study, but was refused a visa because he was a member of the French Communist Party and the McCarthy witch hunts were on at the time. Vigier however was able to work with Prince Louis de Broglie at the Institut Henri Poincaré in Paris and together they were able to continue Einstein's work in deterministic physics into the latter part of the twentieth century.

Both Vigier and de Broglie were champions of Einstein's deterministic physics in the latter part of the twentieth century. Vigier linked up with Myron to write the series of books 'The Enigmatic Photon'. This link up gave Myron vital links back to Einstein and de Broglie allowing him to receive insights into the nature of their work, which would stimulate advances in theory, which would ultimately lead to the Einstein-Cartan-Evans unified field theory in 2003.

One valuable insight was the revelation that Einstein was an ardent believer in photon mass. In the standard model, photon mass has been discounted from the time of the 1927 Solvay Conference, in which theoretical physicists moved away from the deterministic physics of Einstein, de Broglie and Newton into an obtuse mathematical representation of the atomic world, which progressively moved further and further away from Baconian Science based on matching theories with experiment. Richard Feynman eloquently summed up the standard model of physics in the latter part of the twentieth century, by stating, 'that you should not be worried if you do not understand quantum electrodynamics (QED), because he did not understand it and nor did anyone else!' Theoretical physicists see this statement as testament to Feynman's greatness whereas AIAS scientists see this as a clear admission that the standard model has spun itself a web of mathematical deceit, which has captured the Copenhagenists in a Tholian Web, from which they cannot extract themselves, no matter how elaborate they make their mathematics. Supporters of the standard model tell us that the theory of quantum electrodynamics is the most accurate theory ever thought up; and yet it can be shown that Planck's constant, which is used in their calculations, is not known to anything like the accuracy required for producing their astonishingly accurate results.

Heisenberg transformed Schrödinger's ground breaking equation into his mathematical matrix mechanics, which lost its valuable connection to deterministic and real science. Mathematicians were thrilled, because Heisenberg matrix mechanics could be used as a job creation scheme to keep them employed, without the need to understand deterministic and real physics. The Heisenberg uncertainty principle was certainly not a principle of physics, but rather a con to confuse the world of science into funding, physics which was not soundly based. The uncertainty principle was clearly wrong, because Compton had already shown that X-rays could be observed colliding with electrons and their paths could be observed before and after collision. Compton received the Nobel Prize for physics for this work, which clearly showed momentum is conserved in collisions between photons and electrons and that light as Einstein had claimed, had mass. The conservation of momentum also showed that there was no uncertainty in the system at the atomic level as claimed by Heisenberg. The Copenhagenists and advocates of the standard model have conveniently forgotten Compton's seminal experimental work and have conveniently been given substantial funding to pursue their quaint and idiosyncratic theories, at the expense of the tax payer and real physicists!

Clearly then it can be seen that photons do indeed have mass. Furthermore mass can be converted into light to give truly vast amounts of energy, which can be calculated by Einstein's famous equation $E = mc^2$. Likewise, light can be converted to mass.

The mass of an electron changing into energy according to the equation $E = mc^2$, produces light of a wavelength and frequency corresponding to the X-ray region of the electromagnetic spectrum.

In the standard model of physics there exists the well known 'Measurement Problem'. Theoretical physicists of the Copenhagen school believe wrongly, that photons and electrons exist in a limbo state until observed and then turn into either a wave state or a particle state, depending on the way the observation is conducted. This is the interpretation of wave particle duality that dates back to Bohr and Einstein. Einstein would try to convince Bohr how ludicrous this interpretation was by asking, 'Does that mean the Moon is not there when I am not looking at it!?

In chemistry and ECE theory, wave particle duality means that electrons have simultaneously a particle nature and a wave nature, as first recognized by Prince Louis de Broglie. In high resolution electron microscopy, the wave nature of an electron is customized by controlling the speed of the electron, by accelerating it through the desired potential after it has been emitted from the electron gun at the top of the microscope. As the electrons moves faster, their wavelength shortens in a controllable way and after hitting the sample the electrons are diffracted and their exact position can be seen as a pattern on a fluorescent screen inside the microscope. The diffraction pattern is 'played with' by tilting the sample in two directions in the beam, until the diffraction pattern suddenly becomes fully illuminated. When this occurs the atoms in the crystals are lined up in columns and the sample is face on. Now the electron lenses can be switched on and the lattice can be imaged directly and the position of the atoms in the unit cell of the crystal visualized directly! The required resolution for lattice imaging to be made possible is of the order of just over an angstrom and the wavelength of the electron beam is simply set, by setting the voltage for the acceleration of electrons to the order of one hundred kilovolts. There is no uncertainty here of the type fantasized about by Heisenberg! The wavelength of the electrons is defined and the position of the collision in the sample is defined right down to the atomic level!

The electron microscope shows that electrons develop increasing wave characteristics as they accelerate through the electric field, as they head towards the sample after leaving the electron gun at the top of the microscope. Prince Louis de Broglie had predicted this in the twenties, when he suggested for his Ph. D thesis in Paris that electrons could exhibit wave particle duality in the same way that a photon does. de Broglie believed that an electron in orbit around the nucleus of an atom was guided by a pilot wave. The electron microscope shows that electrons do indeed become wavy when they are accelerated to high speeds. It is therefore quite obvious that an electron being captured by an ion or becoming trapped by the electric potential well of an atomic nucleus will accelerate as it goes into a dive towards the positively charged nucleus (in a dive that is analogous to a Sun dive of a comet from the Oort cloud, changing direction towards the Sun) and the pilot wave will develop and adapt and guide the electron into a suitable allowed orbit or orbital around the nucleus. When an allowed orbit is reached, the electron then emits a photon of the corresponding wavelength to allow a suitable quantity or quanta of energy to be dissipated, to trap the electron into that stable permitted orbit. The electron is then guided around the nucleus by the pilot wave that has emerged as a result of the encounter of the electron with the atom or ion. The electron now has a particle aspect being guided around the nucleus with its associated wave aspect. The wave and particle are intimately interlinked to display a simultaneous wave particle duality.

This wave particle duality is an inherent property of an electron in orbit around an atom and does not need to be observed in order for it to exist, just as the Moon exists and maintains its orbit whether we look at it or not! This is why Einstein came up with his EPR thought experiment to show 'quantum entanglement' was impossible. The standard model is crippled by its support for the measurement problem, which maintains that electrons in orbit around an atom are in a limbo state (as are photons supposed to be) until observed and the measurement causes the electron to take on either a particle or wave nature. This is obviously absurd and Einstein's $E = mc^2$, can be rolled out to prove it. If an electron existing as a particle was to change to a pure wave motion when observed, the resulting energy of the wave corresponding to the mass of an electron would be in the X-ray region of the electromagnetic spectrum! This would mean that anyone observing an atom or indeed any material would risk being bathed in X-rays. This simply does not occur, showing that the standard model is both wrong and redundant. Prince Louis de Broglie and Einstein were right after all. At the Solvay conference Wolfgang Pauli had told de Broglie to 'Shut Up!' History then recorded that de Broglie then went away and did shut up!

However, on the train back to Paris after the conference, Einstein told de Broglie that it was up to him to show deterministic physics reigned supreme. de Broglie did shut up for many years, but in the fifties Vigier encouraged Louis to once more take up the challenge. The work of Einstein was conveyed to Vigier by de Broglie and Vigier eventually wrote books with Myron in the nineties to continue the cause of deterministic physics. This work was then taken to its conclusion by the Alpha Institute for advanced Study. Notably, through the advent of the electron microscope, Prince Louis de Broglie's unique contribution to physics can be remembered and it can be seen that his insight into the wave particle duality of the electron was spot on. Furthermore, Einstein's equation $E = mc^2$, can be used to demonstrate that the Copenhagenists were utterly wrong and that the standard model is no more! Q.E.D.

5.3 Einstein and general relativity

- Special relativity did not incorporate the effects of acceleration and gravity on space-time, so in 1906 Einstein turned his mind to a theory of general relativity to overcome this shortfall. Through one of his famous thought experiments he was able to see that acceleration and gravity are equivalent and space is curved. However he did not see the importance of torsion to curvature until much later, when he tried unsuccessfully with the French mathematician Elie Cartan to extend his theory to incorporate light into his theory to create his fabled unified field theory.

The story of general relativity starts with Euclid and his Euclidean geometry dealing with flat surfaces, because general relativity allows us to understand the force of gravity in terms of the geometry. Euclid gave us the means to draw lines and angles and relate them together by theorems, which explained how they interacted and depended on one another. Euclid gave us the means to mathematically construct a box and Einstein was able to make his breakthrough in general relativity, by considering how an observer inside a box would perceive the actions of acceleration and gravity. This thought experiment led Einstein to formulate his equivalence principle, in which he made the important step of realizing that the acceleration due to gravity has something to do with geometry - the equivalence principle. The development of general relativity from the equivalence principle to the famous Einstein-Hilbert Equation field equation of general relativity required Einstein to become acquainted with developments in mathematics describing how objects move in time and space.

Vectors came into use at the turn of the eighteenth century and the term is derived from the Latin verb to carry. The vector points in the specified direction, with its length giving the magnitude of the force required to 'carry' in that direction. It was first used by astronomers to describe how the 'radius vector', a line drawn from a planet to the focus of an ellipse, 'carries' the planet around the centre. Vector usually appeared in the phrase radius vector. The French term was rayon vecteur as seen in Laplace's 'Celestial Mechanics, which was translated by the British Civil List Scientist, astronomer and mathematician Mary Fairfax-Somerville 1780-1872.

The modern meanings of the terms 'vector' and 'scalar' were introduced by William Rowan Hamilton (1805-1865) of Trinity College Dublin, in his paper to the Royal Irish Academy in 1844 entitled 'On Quaternions'. Quaternions are a non-commutative extension of complex numbers which still find use in three dimensional rotations, but have largely been replaced by vectors. Hamilton also introduced the term 'tensor' in 1846.

In 1906, when Einstein started thinking about general relativity, he turned to his old classmate Marcel Grossmann from his days in Zurich's ETH University for advice on how to proceed. Grossmann was a mathematical genius and was able to acquaint Einstein with the work of Riemann, Christoffel, Ricci and Levi-Civita on a then new kind of geometry, generally known as Riemann geometry, in which space and time were merged together in space-time, and in which the framework or frame of reference could be dynamic and curve. Bianchi's work was also to be of seminal importance, in Einstein's quest to extend special relativity, to include the effects of acceleration and gravity.

Professor Luigi Bianchi (1856-1928) was a great Italian mathematician, who worked in Pisa with Gregorio Ricci-Curbastro (1853-1925) who invented tensor calculus and Tullio Levi-Civita (1873-1941) who was born in and worked from Padua. All three mathematicians developed ground breaking mathematical treatments which were needed for the geometrically based Einstein-Hilbert Equation of general relativity.

In 1900, 'Ricci' and Levi-Civita published their theory of tensors, which Einstein studied to help him understand the spherical geometry needed for general relativity. In 1915 Levi-Civita corresponded with Einstein to correct some errors in his calculus and also contributed work to Paul Dirac's equations in 1933. Levi-Civita became a professor in Rome in 1918, where he worked successfully until he was sacked by the Fascist government.

Einstein's breakthrough to general relativity and curved space started in 1907, when he realized that the effect of gravity and acceleration are the same. This is his equivalence principle. This can be shown by considering a box that could be isolated in space or a lift suspended by a cable in Earth's gravity. A person in the box, who could feel the effect of gravity, would not be able to tell if he was stationary in a lift or being accelerated by a rocket in space. Similarly if the person was weightless, he would not know if he was isolated in space or if the lift was in free fall. This is Einstein's equivalence principle.

In 1907, Einstein's old mathematics tutor Minkowski from ETH University, put Einstein's theory of special relativity into a new mathematical framework, which put space and time together as space-time. It was now seen that space and time were not independent of each other, but together formed the fabric of space. However, Einstein soon came to see that space and time were curved or warped by massive objects. Special relativity showed how objects behaved when approaching the speed of light, but gravity and acceleration was not included in this treatment. Einstein wanted to incorporate gravity and acceleration into special relativity, to formulate general relativity. To do this he came up with an amazing thought experiment, which showed gravity and acceleration are equivalent. This is called his equivalence principle and led him to realize space was curved. If a light beam entered sideways into a spacecraft that was accelerating upwards, an observer in the spacecraft would see the beam bending downwards as it crossed the cabin. By the equivalence principle, gravity and acceleration are equivalent, so massive objects would cause space to bend by virtue of their gravitational field.

Einstein considered what a light beam would look like to an observer in the box, if the box was moving or stationary. In the case of the box being stationary, the beam would travel horizontally across the box to the far wall. However, if the spacecraft or box was accelerating rapidly upwards, then the box would move upwards as the beam of light traveled towards the far wall. The light beam would be seen by the observer to curve downwards. Einstein saw that gravity and acceleration were equivalent, meaning that if the spacecraft or box was not accelerating, but rather was in a powerful gravitational field, the beam of light would once again be seen to bend. This thought experiment led Einstein to the conclusion, that space bends in a gravitational field and this bending of space-time produced the gravitational field.

Before Galileo's time it was assumed that heavy objects fell to Earth faster than light objects. The Feather and Guinea experiment shows that this is not the case. The acceleration due to gravity is the same for light objects and heavy objects. Consider astronauts arriving and entering a space station orbiting the Earth. The density of the space station increases when the astronauts are inside the space station, but this does not affect the orbital speed or diameter of the orbit of the space station.

The bending of space as described by Einstein gives an alternative explanation to the elliptical path of the planets, given by Newton after Kepler deduced his laws of planetary motion from observational data. Both Newton's and Einstein's explanations of planetary motion can closely predict the orbits of the planets, even though their theories are radically different.

General relativity tells us that massive bodies cause space and time to warp and this is what gives rise to Newton's force of gravity. Objects moving close to this warped space follow the best straight line they can, which becomes more curved when close to a massive body or when the body is more massive. The resulting curved path is what gives rise to orbital motion. The mass, volume or density of the moving object does not affect the diameter of the object's orbit around the massive body. It is only the objects speed through the curved space that determines the objects orbit.

David Hilbert was a late starter in the 'general relativity' stakes. In June 1915 in Göttingen, Einstein gave lectures on how he was going to get to the equations of relativity. Hilbert was present at those lectures and Einstein gave him the run down of how he was going to finally solve the problem. Soon after, Einstein realised he had wasted several years in his quest, by not following up work with Grossman on Riemann tensors.

Hilbert meanwhile set out to beat Einstein to the post by finding the equations first. Einstein was horrified when he found out that Hilbert had joined the race. Einstein became very worried that Hilbert would beat him to the punch and worked furiously to complete his 'Einstein equation' even to the point of risking his health. In November 1915 both physicists completed the tasks within days of each other. Einstein was however burnt out with the exhaustion of running the race and pictures taken of Einstein in early 1915 and early 1916 show the youthful looking Einstein had changed to the older looking Einstein that we are all familiar with. Einstein was so exhausted that he had to spend the early months of 1916 being nursed back to health by his partner Elsa Einstein. This great contest in mathematical physics has parallels to Ali's famous 'Rumble in the Jungle fight', which exhausted Ali and Foreman so much that it could have cost either boxer their lives. Hilbert however thrived on the competition and took the view that General Relativity was inherently Einstein's theory and that he (Hilbert) had only worked on the final mathematical steps to the finish.

Einstein took about ten years to find the Einstein field equation, from 1906 to 1915. Einstein made several, entirely understandable, false turns as is well known, because he had nothing to guide him. He finally realized that the second Bianchi identity is proportional to the Noether Theorem. This was all expressed in terms of Riemann geometry and curvature, but we now know (paper 88) that it can all be expressed in terms of torsion. Hilbert derived the same equation using a lagrangian.

David Hilbert (1862-1943) independently inferred the EH field equation in 1915 using the Lagrangian method. A Lagrangian is a function that summarizes the dynamics of the system and is named after Joseph-Louis Lagrange (1736-1813) who was born in Turin and worked in France, Italy and Prussia. Lagrange was one of the greatest mathematicians of the eighteenth century and made contributions to number theory, as well as celestial mechanics. Lagrange followed Leonhard Paul Euler (1707-1783) as director of mathematics at the Prussian Academy of Sciences in Berlin, on the recommendation of Jean le Rond d'Alembert (1717-1783). His students included Joseph Fourier and Simeon Poisson.

Einstein made two major discoveries en route to the famous field equation: 1. That the second Bianchi equation then available had to be used, and 2. That the covariant derivative (not the flat space-time ordinary derivative) had to be used in the Noether Theorem. Einstein's theory was tested by light seen to be bending around the Sun during the 1919 total solar eclipse. When the measurements where declared correct by the Royal Society, Einstein became the world's first scientific superstar and was nominated for the 1921 Nobel Prize. However after the accuracy of the solar eclipse measurements were questioned, Einstein was given the 1921 Nobel Prize, not for relativity, but for his explanation of the photoelectric effect, which had led to the realization that light was composed of photons and had led to quantum theory. Niels Bohr was given the 1922 Nobel Prize, for his application of Einstein's quantum theory to explain the emission spectrum of the hydrogen atom. This description showed that electrons occupy concentric shells in atoms, with the innermost shells being more tightly held than shells further out. Prince Louis de Broglie then realized that electrons are guided around the atom with a pilot wave, so that electrons simultaneously exhibited a wave and particle nature known as wave particle duality. The icing on the cake came in 1925 when Erwin Schrödinger produced his famous wave equation, which mathematically described the motion of the electron around the atom. All should have been now straight forward in physics, but at the 1927 Solvay Conference mathematicians muddied the water and the advance of theoretical physics was stopped in its tracks!

5.4 Testing relativity, by observing light bending around the sun

Ernest Rutherford's daughter Eileen married Ralph Fowler (1889-1944) in 1921. Fowler lectured mathematics from 1920 at Cambridge University and wrote an important book on stellar spectra, temperatures and pressures. Fowler worked with Dirac at Cambridge, introducing him to quantum theory in 1923 and collaborating with him on the statistical mechanics of white dwarf stars. Fowler also worked with the great English astronomer Arthur Eddington.

Arthur Stanley Eddington (1882-1944) replaced Darwin's son as the Plumian Professor of Astronomy in 1913 and became the Director of the Cambridge Observatory in 1914. Eddington showed that the matter in stars is ionized due to the high temperatures involved and that the pull of gravity on the matter is balanced by the gas and radiation pressure. Eddington showed that the gas pressure required to balance the star's gravity indicates that the core temperatures of stars must be millions of degrees. He went on to support the idea that a star's energy is produced by the nuclear fusion of hydrogen into helium.

The equation $E = mc^2$, demonstrates that light has mass and as such should be deflected by a gravitational field. In 1911, Einstein calculated the deflection of light from a star, caused by proximity to the Sun. He took into account that time is slowed down by a strong gravitational field and this effect would increase the deflection.

In 1914 a German astronomical expedition set off to observe the total eclipse of the Sun in the Crimea. However, the World War started and the team was arrested by the Russians before they could observe it. Luckily for the team however, they were soon released on an exchange of prisoners. If the team had been successful they would not have proved Einstein right, because he had not yet calculated the deflection expected for the light beam correctly. Later he realized that gravity also bends space as well as slowing down time, making the path the light takes longer and doubling the angle of deflection in his initial calculations. Why Einstein decided that the angle should be exactly doubled is not clear and so must be put down to his great scientific insight or otherwise to an educated guess to allow his calculations to continue and to provide a figure that could be tested experimentally. Either way, Einstein was correct in his assumption as has been shown by the NASA Cassini probe, which is now able to show that the deflection is indeed twice the Newtonian value experimentally to a very high precision.

General relativity came from the equivalence principle in which acceleration and gravity were considered to be equivalent by considering an observer in a box. The observer could be in a lift or in a spaceship. If the observer felt the pull of gravity in the box then this could simply be the pull of gravity on the stationary lift. However, if he was in a spaceship he could be feeling the ship accelerating at a rate corresponding to the Earth's gravity 'g'. Similarly, if the spaceship was stationary the observer would become weightless, but if the box was actually a lift with the cable snapped he would also feel weightless and would not be aware the box was falling with an acceleration of 'g', because he to would be accelerating at the same rate as the box.

Einstein conceived the bending of light and space by gravity from this thought experiment. If light enters through a window in the box inside the rocket as it accelerates upwards, then the box would move up as the light crossed to the other side, causing the observer to see the light bending as it crossed the box. The bending of the light would naturally be greater if the acceleration was greater. Einstein's great insight was then to realize that the effect would be equivalent to an observer held stationary in a gravitational field inside the lift. The greater the gravitational field the greater the bending. Einstein had conceived general relativity, extending special relativity of objects moving at high speeds, close to the speed of light, to include the effects of gravity and acceleration also. Einstein only needed now to put his ideas into mathematical form to produce his famous Einstein equation. However, this involved cutting edge mathematics and Einstein had been a 'lazy dog' in this respect as a student in ETH University in Zurich, opting to take the easier mathematic options, despite being capable of taking the much harder mathematical courses. Luckily, the mathematical genius Marcel Grossmann was in Einstein's class at ETH University, so Einstein was able to pick his brains on the nature of the mathematics required to describe curved space. The mathematics Einstein was advised to study was that of Riemann and Noether. Albert Einstein then went on to derive his Einstein field equation from the second Bianchi identity without torsion (Riemann geometry) and the Noether Theorem.

In 1915, when Einstein introduced the world to General Relativity, his insights were so fantastic, that he needed to gain support through experiment, if this new theory of gravity was to be believed. The bending of light by the Sun's gravity, as photons passed close by, was the test. In May 1919, Arthur Eddington the British astronomer led an expedition to the Island of Principe off the coast of Africa to observe the total eclipse of the sun, while a second British expedition set off to Sobral in Brazil. Photographs were taken during totality to see if bright stars close to the sun had shifted their position. At the same total eclipse, measurements were also recorded by the expedition in Sobral in Brazil. At the time of totality the Sun was in the constellation of Taurus the bull. The Bull's head in Taurus is marked by a 'V' shaped cluster, known as the Hyades, consisting of reasonably bright stars, with the much brighter and nearer star Aldebran marking the bull's eye.

As it turns out some of the results for the bending of light recorded on the Brazil photographic plates were in better agreement with Newtonian theory, while Eddington's measurements were supportive of Einstein. After consulting with the Astronomer Royal and J. J. Thomson (President of the Royal Society), it was decided to disregard the results from Brazil and so in November 1919, Eddington was able to state that Einstein's theory of relativity had passed this test of validity. Einstein then became a world wide celebrity. The experiment for calculating the bending of light close by the Sun does not now need a total eclipse. In October each year two quasar radio sources pass close to the Sun and the apparent changes in the angle between them, can be followed by radio telescope in broad daylight.

Einstein-Cartan-Evans (ECE) theory takes the prediction of the deflection of light a step further, by predicting that the light will also twist. The polarization of the light is predicted to change as it passes close to massive objects such as white dwarfs with strong gravitational fields. This can be seen in M. W. Evans, "Generally Covariant Unified Field Theory" (volume three, Abramis, Oct. 2006), where the dielectric theory is developed, and "ECE Theory of Gravity Induced Polarization Changes" on the www.aias.us homepage.

In ECE theory, the essential new thing is the homogeneous current j, which adds a term on the right hand side of the Faraday Law of induction. This results in various optical effects such as changes of polarization, observed in a white dwarf as in the paper.

ECE theory reduces to Einstein-Hilbert theory when the Cartan torsion vanishes, so producing all that Eddington saw. Calculations were helped by Freeman Dyson, who later moved to the Princeton Institute of Advanced Studies. The accuracy of the light bending now is 1: 100,000 (NASA Cassini, 2003 onwards), supporting EH. The extra effects of ECE come from the Cartan torsion, which is entirely missing in EH. The homogeneous current j gives the way in which torsion affects EH.

In 2004, Kerry Pendergast was awarded a Royal Society Partnership Grant in relation to an astronomical project entitled 'Daylight Astronomy'. As part of the award, Kerry was invited to a special evening viewing of the Royal Society's Summer Exhibition, which highlights groundbreaking projects in education and industry. Impressively representatives from Jodrell Bank were there, describing their discovery of a double pulsar in a binary orbit, in which both pulsar beams would over time eclipse each other in their orbits, as seen from earth. This allowed an unparalleled opportunity to study the bending of space and the slowing of time in giant gravitational fields and was dubbed the test bed of relativity theory. Later, Kerry reflected that this double pulsar was just the ticket for taking the Eddington Experiment and general relativity forward into the 21st century, since it could be used to test and study the prediction made by ECE theory, that a polarized light beam would not only bend around a high gravitational field, but would also tilt. In 2007, Kerry was invited for the day to Jodrell Bank and took the opportunity to discuss the problem with two professors there. They described the experiment as 'cute'.

It is only a matter of time before workers at Jodrell Bank or elsewhere get around to carrying out this experiment, which has the capability of proving ECE theory beyond doubt and could well lead to a Nobel Prize for the first team to accomplish the task. It is very possible that the data required has already been collected, but without its significance being realised. This experiment is an intriguing update to Eddington's great total solar eclipse work that brought Einstein world fame!

The double pulsar was discovered by Jodrell Bank in 2003. The rotation times are 2.8 seconds for one and in the millisecond range for the other and their orbit get smaller by 7mm a day as predicted. The 2.8 second pulsar has a mass of 1.25 and the 23 millisecond pulsar has a mass of 1.34 solar masses. They orbit each other every 2.4 hours with an orbit slightly smaller than the diameter of our Sun and are traveling at 0.01 the speed of light. The 7 mm decrease per revolution is due in ECE to T/R not being quite constant. In fact in ECE theory, T/R replaces the universal constant G/c squared for a given M. The "ECE Paradox" is that the EH equation does not obey the Bianchi identity, and as with all paradoxes this is leading to wholly new information - notably that gravitation is not quite universal in the sense that G/c squared is not quite constant. This shows up in binary pulsars and also in the Pioneer anomalies now perplexing NASA. In paper 106 the orbit of a binary pulsar is described without using gravitational radiation, which has never been directly observed.

ECE theory replaces Riemann geometry with the much more powerful Cartan geometry, which allows not only the bending of space-time to be taken into account, but also the effect of the twisting or torsion of space-time, allowing all the forces of physics to be described in terms of geometry. When light grazes the Sun the slowing down of time causes the light to bend. Additionally the bending of space by the Sun's gravity causes the light beam to bend also. Nevertheless, even in ECE theory there is no explanation as to why the slowing time of light in a gravitational field and the bending of light should both produce the same degree of bending, so that the angle of deflection is double the angle calculated by Einstein for Newtonian gravitational theory. However, Myron has proposed that $R = \omega T$ as another form of the null geodesic condition and believes this will explain why the light deflection is twice the Newtonian value whenever light (electromagnetic radiation) is deflected by gravity, i.e. by light grazing an object of any mass M.

"There is no doubt from NASA Cassini that the deflection of light in the Eddington type experiment is twice the prediction of the Newtonian theory to 0.001% or thereabouts. It is now known however (papers 93 to 105 of ECE theory on www.aias.us) that the use of the Christoffel symbol is incompatible with the Bianchi identity of Cartan. The version of the Bianchi identity used by Einstein omits the torsion, which is unfortunately an irretrievable flaw. This showed up in the Hodge dual of the Bianchi identity, and it turns out that we cannot just set $T = 0$ in the Bianchi identity. Therefore in paper 103 onwards an attempt has been initiated to make the great Einsteinian theory fully compatible with the Cartan torsion. The conditions for the observed deflection have been defined and described in paper 105. This work can account for the Pioneer anomaly while the Einstein-Hilbert theory cannot, because it cannot adjust M. The problem is that the so called "Schwarzschild metric" used routinely in astronomy is not the one devised in 1916 by Schwarzschild himself in two exact solutions. Stephen Crothers has shown this definitively (see www.aias.us). In paper 105 a first attempt was made to explain the true origin of the so called Schwarzschild metric. Also, ECE theory predicts in a simple way that the polarization of light is changed by gravitation, as are all the optical and electrodynamical properties, whereas the Einstein-Hilbert theory is a pure kinematic theory based on the mass of the photon being attracted by an object of mass M.

5.5 Black holes, singularities and large masses

Observationally large masses are detectable at the centres of galaxies, which are usually referred to as black holes. However, the existence of black holes is contested in this book, because it is shown, that contrary to popular belief, they do not emerge from the flawed Einstein field equation. The connection used by Einstein was incorrect (see papers 122 onwards on www.aias.us). To keep on using this incorrect correction symmetry is misguided and here Einstein's original calculations have been wholly replaced by ECE theory. Since "black holes" are not therefore predicted mathematically, it may be better to replace this term with the term "large mass".

Einstein completed his attempt to formulate his field equation of general relativity at the end of 1915, in a photo finish with David Hilbert. However, this Einstein-Hilbert field equation was not really the definitive equation of curved space-time that he had set out to formulize ten years earlier. It only marked the furthest point that Einstein was able to reach, in his quest for his great equation. In order not to let Hilbert take all the credit, the race to formulate the equation stopped at the end of 1915, with both Einstein the physicist and Hilbert the mathematician sharing the credit. The equation appeared to be good enough to satisfy the needs of a relativistic theory of gravity, with regards to describing the corrections needed to Newton's work, in order to describe the orbits of satellites and trajectories of space probes, but led to problems of interpretation in some areas of cosmology. However, it failed to describe the orbital motions of stars in spiral galaxies, but ECE theory has been able to account for this motion in terms of torsion.

The shortcomings of the Einstein-Hilbert field equation, resulted from false assumptions and simplifications being made right at the start of the work, regarding how Riemann mathematics could be applied to the curvature of space. A symmetric assumption was made and the role of torsion was not incorporated. Whilst gravity could be described as the curving of space by mass, the possible twisting forces due to electromagnetic forces was omitted and as a result, the famous E-H field equation is not as comprehensive in its descriptions, in the way that was intended. The omissions and incorrect assumptions also prevented Einstein from completing his later work, of combing light and gravity to produce his fabled theory of everything. In the field of cosmology, Einstein's field equation predicted the existence of black holes, but short comings in the equation meant that the nature of these black holes was not defined, since as black holes were predicted to collapse, infinities would occur in the calculations, which meant that the meaningfulness of the calculations would break down and the equation, would no longer describe physical reality in the vicinity of the black hole.

Einstein accepted the limits inherent in his field equation in relation to black holes and singularities and even wrote a paper in 1939, to assert that black holes and singularities could not exist in nature. It is can now be seen that Einstein's papers on general relativity are incorrect, because of the use of an incorrect connection. So the era 1915-2009 in cosmology, can now be moved forward by using theory such as ECE, that by virtue of uses the correct geometry, can overcome the shortcomings inherent in the Einstein-Hilbert equation. Many physicists and mathematicians over the last hundred years have used Einstein's equation, to predict black holes will collapse to zero volume to form singularities, with infinite gravity and energies. This is not a physical reality however and recently physicists and theoreticians have woken up to the situation and are now coming to believe, that while there is good observational evidence for the existence of large masses, singularities do not necessarily exist and new work needs to be done, to produce a more realistic mathematical and theoretical description of the situation.

Astronomy is predominantly an observational science and with the advent of astronomical telescopes, our view on the universe has been expanded and with the vast distances involved in space, we have even been able to look back in time. With the naked eye we are able to just about glimpse the Andromeda galaxy, from which it takes the light two million years to reach our eyes. With modern telescopes and photographic images, it is now possible to see billions of years into the past and to the edge of the known universe, when viewing distant galaxies. Augmenting the worlds biggest and best land based telescopes are NASA's great space telescopes and on the ground and in space different telescopes have been built, that can give us views across the cosmos not only in the visible, but across all the different regions of the electromagnetic spectrum. Notable amongst these telescopes, are the two hundred inch Palomar Telescope, Gemini, Keck and the Very Large Telescope (VLT) in Chile, which operate predominantly in the visible. These are augmented by the Jodrell Bank radio observatory, UKIRT and the JCMT in the infrared and in space by Spitzer (infrared), Hubble (visible, UV and infrared),

Chandra (X-ray) and Fermi at gamma ray frequencies.

It is believed that black holes are formed at the end of the lives of massive stars. However, in this book a new description, such as 'large mass density' is preferred to black hole. For stars of average mass, the stalling of energy producing nuclear fusion reactions, results in gravitational forces overwhelming the outward radiation pressure and an implosion occurs, which causes the nuclear core of the star to collapse into a white dwarf star, while simultaneously, the outer remnants of the star are ejected at high velocity into inter stellar space to form a planetary nebula. These nebulae have all kinds of shapes, just as snowflakes do, with the characteristic white dwarf star seen at their centres and are well seen in the visible region of the electromagnetic spectrum. The collapse of the star's nuclear core to form a white dwarf, results in the mass of the star's core being contained in a volume similar to that of a city and Newton's inverse square law of gravitation can be used to show that gravitational force has increased exponentially as the radius of the core has shrunk. However, stars around ten times or more massive than the Sun end their lives even more spectacularly in a supernova and the collapse of the core goes further!

It is thought that stars going supernova, produce a neutron star or pulsar. These supernova remnants have been greatly studied at radio frequencies by observatories such as Jodrell Bank in Cheshire, England. The most famous of these pulsars, is probably the one at the centre of the Crab Nebula, which was the result of the supernova first seen in 1054. Jodrell Bank has a radio dish dedicated to observing the crab pulsar, whenever it is above the horizon. Over time, it has been seen that the rate of rotation of this pulsar is very gradually decreasing.

NASA's Fermi Gamma-ray Space Telescope (formerly known as the Gamma-ray Large Area Space Telescope or GLAST for short), was launched into space in June, 2008. It has several instruments on board and has already discovered a number of pulsars by detecting gamma ray emissions. A pulsar's gamma ray emissions account for around ten percent of a pulsars total energy emissions, which is over a million times more than its radio emissions.

The Vela and Crab Nebula pulsars are the brightest sources of persistent gamma ray emission in the sky. The Vela pulsar spins eleven times every second, but despite its brightness only delivers one gamma ray photon to the Fermi Space Telescope every two minutes and the faintest pulsars only deliver two gamma ray photons to the Telescope in an entire day!

It is believed that when the most massive stars undergo supernova events, they collapse beyond the point of forming white dwarfs or pulsar neutron stars, to form black holes. Black holes are believed to have such a vast mass density, due to the collapse, that the gravitational forces become so large, that not even light can escape. It is therefore speculated that extremely massive stars undergoing supernova, collapse to produce black holes. Whether black holes exist is questionable however, because the collapse forms an object with a large, but finite mass density. The gravitational forces become extremely large, but cannot produce singularities and this process, certainly cannot be the physical result of a mathematically incorrect equation, the Einstein field equation.

If black holes are truly black, no light should be able to escape from them and it could be assumed that they cannot be seen or detected. However, one method of detecting black holes or large mass objects is by analyzing images that exhibit gravitational lensing, such as Einstein's Cross. Black holes unseen in the foreground of an image, can distort the light from objects and galaxies much further back in space, to act like a lens, to give a magnification to the more distant objects and can even provide alternative paths for the light to take. Gravitational lensing is thought to provide evidence for the existence of black hole or large mass supernova remnants and for supermassive black holes or super dense objects at the centres of galaxies.

Black holes are thought to release high energy radiation, as matter falls into them and this high energy radiation can be detected in the X-ray and gamma regions of the spectrum, to provide further evidence for the existence of black holes. The Chandra X-ray space observatory provides valuable data here. Chandra can examine the X-ray emitted from 'feeding' black holes, be they in our own our own neighborhood of space or in nearby galaxies. The Chandra X-ray images of the Andromeda galaxy can be superimposed on optical images, to show the location of the black holes studded throughout the galaxy.

Galaxies are now thought to have super massive black holes at their centre, but these are more often than not shrouded in dust and gas, making it rather difficult to observe them. Our own spiral galaxy is believed to have a supermassive black hole at its centre, but this is obscured from view, by vast amounts of dust, because we have to view the centre of our galaxy, through the galactic plane, in which the stars, gas and dust of the galaxy is concentrated. In the last decade or two, the obscured centre of our galaxy, has been rendered visible by observing it in the infrared, with telescopes, such as the twin 10 metre Keck telescopes (near infrared), the 3.8 metre, United Kingdom Infra red Telescope (UKIRT) and in the submillimetre (between the far infrared and microwave), with the 15 metre, James Clerk Maxwell Telescope (JCMT) on Manau Kea, in Hawaii. The JCMT is able to look further into the infrared, by cooling its detecting instrument close to absolute zero, by using a tank of liquid helium. Importantly, these instruments have allowed the thirty or so stars, closest to the galactic centre to be observed for sufficient time to have their orbits plotted. These stars are seen to be rapidly orbiting an invisible, supermassive object and their orbits can each be used to calculate the mass of the object, which they are orbiting!

Four hundred years ago in 1609, Galileo made himself a refracting telescope and pointed it to the sky to observe the planets, Moon and Sun in greater detail than had been previously achieved. He quickly produced one of the most important books in scientific history, called 'The Starry Messenger' and established himself as the world's greatest astronomer. One of his most important revelations was that Jupiter had four giant moons, which over the course of a few days could be seen orbiting Jupiter. By noting the position of Jupiter's four great Galilean moons (Io, Europa, Ganymede and Callisto), in terms of Jupiter diameters from Jupiter, it is a simple process to find their periods (time to complete one orbit) and this can be easily used to calculate the mass of Jupiter. The four moons, give four independent results for the mass of Jupiter, which are in good agreement. The average of the four masses provided by observing the four moons, gives confidence in the data obtained and a more accurate value for the estimate of Jupiter's mass. This method has been adapted in recent years, to find the mass of the supermassive black hole at the centre of our galaxy. Observations of the orbits of the thirty or so stars nearest to the galactic centre provide an estimate for the mass of the supermassive object, as several million solar masses! This object is heavy, but cannot be a singularity. This book shows in different ways that the metrics of the Einstein field equation are incorrect, so nothing can be deduced from them.

Thus, it appears that there is good observational evidence for the existence of black holes or large masses and indeed for supermassive objects. However, it is true to say that no one has ever seen such an object directly and it is not known how compacted a the object can become, before repulsive and relativistic effects stop the contraction going further. Black holes are believed to develop spin as the contraction occurs and this would help to balance contraction forces, as would the vast temperatures that would be generated. What actually goes on inside a black hole is anybody's guess, since the traditional equations of physics break down under these conditions. General relativity and Einstein's field equation breaks down, because as contraction occurs, infinities would occur as the radius of the black hole contracts. For 1/r, where the radius r tends to zero, then gravitational fields would go to infinity and time would stop and space would curve in on itself. This has been acceptable to mathematicians for the last hundred years or so, but is not real physics, since infinities do not agree with the realities of the physical world we observe.

Attempts have been made to describe the inner workings of black holes in terms of a combination of quantum theory, which describe the contributions from the very small, with the Einstein field equation, describing gravitational forces as the curving of space on the large scale, in a combined theory called quantum relativity. However, the two theories do not fit well together in any meaningful way and there is no reason to believe that matter would contract to form singularities, which only have meaning to mathematicians. We can conclude that although there is good evidence for the existence of these dense objects, no one has ever seen one and no one can actually say for sure, what is going on inside a black hole and what exactly a black hole is and if the collapse actually stops, before the dense object can become a black hole. Solving the black hole problem is now being tackled by physicists, astronomers and mathematicians working together to try to come up with a solution. What is required here is a grand unified field theory of physics, which can combine general relativity with quantum theory and electromagnetism. The leading grand unified field theory to date is ECE theory, which combines the torsion of spinning space-time with the curving of space by gravitational forces.

It is interesting to consider the roles that supermassive objects play in the creation and evolution of galaxies. M87 is a nearby galaxy in the Virgo cluster, in which gas spinning at its centre can be used to indicate that a supermassive object resides within the galactic core. It is now believed that supermassive objects can be found at the centre of the vast majority or even all galaxies. Furthermore, a relationship has been found between the mass of galaxies and the mass of the supermassive objects residing at their centres. The mass of a galaxy is found to be one thousand times greater, than the mass of the supermassive object at its centre. This indicates that the supermassive objects at the centres of galaxies are not the destructive entities that they may thought to be, but on the contrary are an integral part of the make up of the galaxy. What role these so called 'black holes' play in the creation and evolution of galaxies is open to debate. The most active supermassive black holes at galactic centres are believed to be responsible for the extremely active galaxies, known as quasars. What quasars actually are has puzzled astronomers for many years.

In the autumn of 2009, the active blazer galaxy 3C 454.3 some seven billion light years away in Pegasus, brightened to overtake the Vela pulsar as the brightest gamma ray source in the sky. Effectively, we are looking down the barrel of this object's particle jet, which is powered by the galaxy's central supermassive black hole. Traditionally, it is believed that such jets are the result of oppositely directed jets of particles, traveling close to the speed of light, being produced as a result of matter falling into the supermassive black hole at the galactic centre. However, it is perhaps worthwhile considering other possibilities for these events, which emit energy and matter from these supermassive black holes.

Rather than the supermassive entity at the galactic centre, feeding on nearby stellar material, the interpretation could be turned on its head, by considering the possibility that the mass concentration is on the contrary a source of high energy discharge, in the form of a spewing white hole. The white hole could have an energy source we have not yet recognized and being unable to contain this burgeoning energy, is expelling it as high energy gamma rays beams, radiating from the source, in opposite directions, as is seen in the Pegasus blazer. The beam of gamma rays reaching the Earth may only be a fraction of the gamma ray energy leaving the galactic centre. As the high energy gamma rays leave the immense electric, magnetic and gravitational fields that exist in the poorly understood black hole region, the vast majority of the gamma rays could perhaps undergo conversion into matter. Whereas X-rays would produce electrons by conversion to particle form, gamma rays two thousand times more energetic would be the right mass to be converted to the well known nuclear particles of neutrons and protons. Neutrons are unstable outside the atomic nucleus and over a timescale of minutes would degrade steadily to form proton and electron pairs. Such pairs are more commonly known as hydrogen atoms. On taking their particle form, the speed of the gamma rays would drop from light speed and be captured by the gravitational field of the galaxy and provide the source of hydrogen for the next generation of stars. Perhaps this is what gives rise to the familiar bar structures that are seen in the central bulge regions of galaxies and is the source of the young stars evolving to form the spiral arms of blue spiral galaxies!

This could also be the source of cosmic rays, which happen to be ninety percent made up of protons and ten percent helium and alpha particles. After traveling across space for millions or billions of years, after being emitted from the cores of blazers and active galaxies, the high energy gamma rays could be stimulated to convert to particle form as protons, and alpha particles (four times the energy) as they interact with the solar wind, the Earth's magnetic field and the atmosphere to be detected as the particles we call cosmic rays.

Gamma rays have energies which correspond to the masses of neutrons and protons and therefore, the conversion of gamma rays to their particle form, could account for the abundance of hydrogen atoms in the universe. Perhaps the centre of new active galaxies, such as quasars possess the magnetic or electric field that can cause a kind of resonance to occur, where a gamma ray reservoir or primordial voltage produces or interacts with a super massive white hole, to provide the source of hydrogen, which is the most abundant element found in nature. Similarly, gamma rays for times more energetic could produce helium, the second most abundant element in nature and would explain it being present in higher abundances than is explained by its formation in nuclear fusion in stars.

5.6 New cosmologies

Fred Hoyle (1915-2001) was an English astronomer from Yorkshire, who founded the Institute of Astronomy at Cambridge University and was also an honorary professor at Cardiff University from 1975, where the 'panspermia theory' that life on Earth is in part due to the transport of viable cells from space, has been developed by Professor Chandra Wickramasinghe. Micro-organisms have now been detected as high as forty-one kilometres up in the stratosphere. Hoyle developed the theory of the structure of stars and made key advances in the understanding of how chemical elements in stars are formed by nuclear reactions.

In the 1950's, Hoyle proposed the Steady-State theory as an alternative to the notion that the Universe began with a 'Big Bang' some 13.8 billion years ago. Infact, the term 'Big Bang' was coined by Hoyle as a derogatory description for the expanding universe theory. The battle between the two theories took place in the fifties and sixties, before the discovery of the cosmological background radiation in 1964, and delivered Hoyle's theory a supposed knockout punch. However, Hoyle's theory has not been dead but only dormant and new ideas are being developed, which could offer an alternative explanation for the cosmic background radiation. One idea is that light of galactic origin, is absorbed by iron 'whiskers' produced by supernovae and then reemitted as microwaves.

The steady state theory hinges on the creation of matter between galaxies over time. As galaxies move apart new galaxies evolve to fill the space in between them. Just as a river flows, but the river remains the same, the universe could be considered to be expanding, but unchanging. A problem with this theory today is that the Hubble Telescope's deep field image shows that the most distant and oldest galaxies are different to galaxies in our neighborhood.

Hoyle believed that solutions to major unsolved problems were best solved by exploring radical hypotheses, whilst at the same time showing due reference to well-attested scientific tools and methods. This was because if scientific breakthroughs were always orthodox in nature, they would already have been discovered.

In 1929 Edwin Hubble started work at the newly completed telescope at Mount Wilson, in California. The telescope he was about to use was the world's most modern and biggest. Many people have pondered the question, 'Are there any other universes, beside our own?' At the time the known universe was the Milky Way galaxy, which contained some nebulae which had a characteristic spiral shape and which were thought to be gas clouds within the Milky Way. In 1923, Edwin made his first great discovery. While looking at the spiral Andromeda nebula, known as M31 with the world's biggest telescope, he was able to see that the gas nebula was not made of gas, but was actually made of stars. This had already been confirmed in the previous century, by the Earl of Ross's observatory in Ireland, with his 'Leviathan Telescope'. Hubble went further, when he identified a special type of variable star, known as a Cepheid variable, in the Andromeda galaxy, which was able to give him its distance. Thus, the spiral nebulae that were already discovered at that time could infact, be regarded as other universes at a great distance from the Milky Way. Consequently, these other universes became classified as galaxies and it became known that our Milky Way was only one of a great number of other galaxies, some of which like our own had a characteristic spiral shape.

The distances of stars to a distance of around seventy light years are determined by the method of solar parallax. Beyond this distance, special variable stars can be used as a yardstick. At still father distances, the Doppler Effect is brought into play to determine distances by the red shift method.

Stephan's Quintet was discovered by Frenchman Jean Marie Eduard Stephan, with his 40-cm refractor at Marseilles in 1877, with five galaxies contained in a mere 5' field and was something of a historical curiosity. Four had measured redshifts in the range 5750-6750 km/s, while one, NGC 7320, measured a mere 790 km/s. Conventional cosmology would, by associating redshift with distance, place this one galaxy much closer than its "companions", and argue it to be unassociated with them; their closeness on the sky would be thought purely coincidental. However, in the 1960s Halton Arp and others proposed that NGC 7320 could be seen to be interacting with its companions, and therefore must be associated with them. Arp argued that the redshifts of objects might not measure their distances so reliably as many thought. Halton Arp worked at one time with Sir Fred Hoyle. In chapter five of volume three of GCUFT (paper 49), the red shift is explained with the dielectric version of ECE, without the need for the Big Bang.

The cosmic microwave background radiation hypothesis came into play, after the horn radio antenna, constantly recorded a signal no matter what direction it was facing. After all possible causes for the radio interference were discounted, the sky and deep space became the only source left and the rest is history. However, Professor Pierre-Marie Robitaille of Ohio State University, who is an expert in radiological studies, now believes the cosmological 2.8 K background is due to radiation from the oceans, which could remove a great supporting plank from the need for the Big Bang theory as reported in paper 93. The PLANCK satellite is to be sent to the second Langrage point and could well confirm Professor Robitaille's new interpretation of the cause of the radio interference. Arp and Roscoe have also collected a great deal of data against Big Bang; see Roscoe in vol. 119(3) of ACP (2001). The age of galaxies is another vast set of data against Big Bang, as has been pointed out by Norman Page.

What is needed now is a new way to look at the problem afresh, with the emphasis on what can be clearly seen across the universe. It is now widely assumed that at the centre of galaxies there are black holes, which feed on stars which fall into them. However, it is important to keep an open mind to allow observed data to be interpreted properly. Often history shows scientists are led by good reasoning to take the opposite view to what the information is actually telling them. In ECE theory, cosmological equation (52) brings spin-coupled resonance into play. Spin-coupled resonance offers the possibility of interpreting old data in new ways. In the evolution of galaxies, spin-coupled resonance could give a mechanism by which matter could actually be created at the centre of galaxies, giving an opposite view compared to the standard view accepted today. In the middle of a galaxy there may not be an all devouring black hole, but a matter producing white hole (in other words: a region of massive spin coupled resonance).

The advent of powerful robotic telescope has opened up the field of direct imaging of galaxies over huge distances to analysis by computer or humans. Millions of galactic images are now being taken robotically and many images are also provided with spectrographic data, which gives distances via the red shift method. Foremost amongst the robotic telescopes for revealing much needed data for cosmology is the 'Sloan Digital Sky Survey'. The Sloan Survey has provided millions of pictures of an area covering all of the northern sky and provides a means to allow deductions to be made from simple observations. It is now time to make sense of this myriad of images to explain how galaxies form and evolve. It may soon be seen as obvious, that small galaxies evolve from much smaller clusters of stars and gas which clump into bigger conurbations.

Analysis emerging from the Sloan Survey is leaning towards galaxies starting out, as spinning spiral galaxies emerging inside a cocoon of dust and matter. The direction of spin of the galaxy is revealed by examining which way the spiral arms are wound, but the direction of spin clockwise or anticlockwise would of course be the opposite if viewed from the opposite side of the disc. Young spiral galaxies would tend to be smaller and sport a blue colour due to large fast lived giant blue stars. Blue giants live life in the fast lane, but only last a few million years and so are found in young galaxies, where star formation is happening apace. Short-lived blue stars produce second generation stars, many of which will be smaller and redder in colour and which contain elements formed, during the supernova destruction of the blue stars. Galaxies are not that far apart compared with their sizes and so collisions between them are not unusual, but go with the territory. Spiral galaxies can frequently be seen in various stages of collisions and collisions can cause them to change shape, but if they are young enough and forming new stars, rotation can again predominate to restore the spiral footprint. It is seen that in galactic clusters, spirals tend to be found with other spirals and large elliptical galaxies are found with other large elliptical galaxies. The elliptical galaxies are thought to be large, through feeding on other galaxies and reddish in colour because their star making days are coming to an end and the fast lived blue stars have already burnt out.

A major question that needs answering is, why there appear be two distinctly different types of galaxies; namely the giant elliptical galaxies and the spiral galaxies. Elliptical galaxies are large and star formation has stopped, so appear to be much older than the blue spiral galaxies, which are also seen in large numbers. Strangely, elliptical galaxies tend to be seen in large clusters, whereas spirals tend to be seen by themselves.

The Sloan survey is on track to reveal new insights into the nature and evolution of galaxies and may confirm that blue spirals galaxies are young galaxies in the beginning of aggregation and red elliptical galaxies, have reached the end of their active lives. It is now being recognized that red elliptical galaxies at the end of their lives have lost the cocoon of gas that once fuelled star creation and this could be, because large older galaxies possess a 'galactic wind' along the lines of the solar wind, but acting on a galactic scale to blow dust and light elements, away from old galaxies and into the vacuum of intergalactic space. Thus, older galaxies could be in a state of 'evaporation', seeding space with new material for new young galaxies to form and providing 'fuel' for the steady state production and birth of new stars and galaxies.

Horst Eckardt has described galactic evolution with ECE theory by using the equations:

$$\boldsymbol{\nabla} \cdot \mathbf{g} = 4\pi\, G\, \rho \tag{5.1}$$

$$\mathbf{g} = -\boldsymbol{\nabla}\phi + \boldsymbol{\omega}\phi \tag{5.2}$$

Using (5.2) in (5.1) produces an Euler-Bernoulli equation. Under well defined conditions this can have a resonant initial condition, which would be the start of an evolution of some type. This could be a galactic evolution as described above and the equation could be used to produce analytical models which it is possible to animate. The resonant initial condition would be the birth of a galaxy, which then evolves out to the condition:

$$\boldsymbol{\nabla} \cdot \mathbf{g} = 0 \tag{5.3}$$

as observed in the flat part of the velocity curve where the dynamics are totally non-Newtonian, but easily described by ECE.

It is puzzling why blue spiral galaxies are segregated from red elliptical galaxies. What has been recently realized is, that as blue spirals approach red elliptical galaxies, they change into red spirals, showing that their star making ability has been lost. There is evidence that the cocoon of star forming gas in blue spirals is blown away as they approach the red elliptical galaxies, giving evidence of a kind of galactic wind, which could be a feature of galaxies approaching the later stages of galactic evolution.

In the late thirteenth and early fourteenth century, Marco Polo and his father and uncle made contact and traded with Kublai Khan and the Mongol empire as far as China. Marco Polo's book of his travels, gave the inspiration to Christopher Columbus, to find a route to the Eastern riches by means of the sea, traveling west. This of course resulted in Columbus's discovery of America in 1492! With time the need to train navigators for English voyages of discovery, led to the foundation of Gresham College and subsequently the Royal Society, institutions in which the great experimental scientist Robert Hooke made his name.

Robert Hooke is one of the world's greatest ever scientists and was very different in his approach to his work compared to Newton. Newton was the great mathematical physicist, whereas Hooke relied upon his great insight into nature and was like an English Leonardo De Vinci, with the ability to apply himself to all sorts of problems. Hooke came up with the famous law, which describes how elastic materials such as rubber and springs, stretch when a force is applied. Hooke's law produces a straight line graph, when extension is plotted against load. Once the graph has been plotted, it can be used to interpolate, to find accurately other values for extension or loads for which the measurements were not actually measured. The important thing is that the value of the mathematical function lies between known values!

In reaching out, to discover the nature of the universe across the vastness of space, it is not always possible to interpolate in the manner just described for a Hooke's Law graph and so the process of extrapolation must be used, where the line is extended past the points at the extremes of the graph. In cosmology, this extrapolation can give the cosmologist a false sense of security with the validity of his data and lead to wrong or invalid conclusions, which are not well founded on observational data. This could well be the case for red shift measurements, taken from light which has traveled extreme distances. Over smaller cosmic distances, the red shift and blue shift can be used to show that objects are moving towards or away from the Earth. This is verified by observing double stars, with the motion of the stellar partners alternatively giving red and blue shifts to their spectrum, corresponding to their position in their orbit with respect to the Earth. It is usually assumed, that even at truly vast distances, the red shift is simply due to the speed of the object (in this case a distant galaxy) moving away from the Earth. However, over extreme distances, this relationship may not hold.

If we take the case of Hooke's Law for a simple spring, increasing the loads initially corresponds to the extensions found by extrapolation, but beyond the limit of proportionality, the spring goes past the elastic limit where Hooke's Law is no longer obeyed. Thus, for extreme red shifts, it could be argued that other factors come into play. It is known that when light travels in a vacuum that it travels at the speed of light, but when it travels through a medium such as glass slows down. Over the truly vast distances of distant galaxies, the fact that space is not a true vacuum, but contains dust gas and subatomic particles could come into play. As a result the big picture could be that on the vast scale, light travels slower or loses energy through interactions with the interstellar medium and as a result, the light shifts to the red, for reasons not related to the speed of the light source away from the Earth. ECE paper 49 was built around the interaction of light with inter-stellar gravitation via the homogeneous current of ECE theory. This is the cause of Beer-Lambert absorption and refraction (change of frequency as in a prism, the red shift).

In Hooke's Law extrapolating past the limit of proportionality is clearly not valid. Experimentally it can be seen adding too many weights leads to unexpected results, in the form of a big bang as the material snaps and the weights crash to the floor. In cosmology, the validity of the big bang is by no means certain and the evidence should be periodically reexamined, to see if there are alternative explanations to account for the red shift and the cosmic microwave background. Surprisingly, this reexamination of the evidence rarely takes place in cosmological circles and the evidence is considered to be beyond question. However, it is good scientific practice to present alternatives to the predominant theory where appropriate!

Radio absorption of the inter-galactic medium is known experimentally. Vigier et al. have shown that this type of absorption casts doubt on the Big Bang theory, by virtue of five critical tests:

1. Angular size;
2. Redshift;
3. Hubble diagram test;
4. Galaxy number count;
5. Differential log N log S test.

If we accept the background Penzias-Wilson radiation (a big if!), the 2.7 K microwave background radiation is the mean temperature of the Universe, indicating a dark night sky. In ECE cosmology, this background (if it exists) would be due to the Beer-Lambert law as in paper 49. Interaction of gravitation and light, with the homogeneous current j, causes absorption and heating. From the Stefan Boltzmann law, the mean black body temperature emitted form the heated universe is 2.7 K, which resolves Olbers' paradox. In general red shifts are spectra in ECE theory. The origin and distribution of elements, is as described by the Pinter Hypothesis, some parts of the universe may be locally expanding or contracting, but overall the universe is boundless and without beginning or end.

Kepler's analysis of the data for the movement of the planets through the zodiac, led to Kepler's laws and the realization the planets orbited the Sun in ellipses. Kepler's laws also helped Newton formulate his theory of universal gravitation and the inverse square law. The inverse law tells us that gravitation does not simply decrease linearly with distance, but as the distance doubles, the force of gravity reduces to only a quarter of its initial strength (1/2x1/2 =1/4), so that gravitational force reduces rapidly with distance, similar to that experienced, when two opposite poles of a pair of magnets are progressively moved further apart and released.

The planets were formed from a cloud that began rotating as it condensed, to form the massive Sun at its centre, with proto-planets called planetisimals around it. The gravitational pull between planetisimals, caused collisions and eventually resulted in the stable orbits of the remaining planets. When the Civil List Scientist Herschel discovered the planet Uranus, the size of the solar system was doubled at a stroke. It was naturally assumed that the flat disc occupied by the planets of the solar system as far as Uranus, could be extrapolated beyond the orbit of Uranus, into the orbit of as yet unknown planets. Anomalies in the orbit of Uranus were used by the Civil List Scientist John Couch Adams to mathematically pinpoint the position of the planet Neptune, showing that extrapolation of the plane of the planets orbits held to that distance. However, further extrapolation to the orbital distance of Pluto, which was discovered in 1930, was not valid.

Pluto was known to have a much more elliptical orbit than the other planets and its orbit, was at much more of a tilt to the orbital plane, than that of the planets. Instead, beyond Neptune, it is now known the Kuiper belt exists, where thousands of Pluto like, giant 'dirty snowballs' are presumed to exist and where the plane of the planets, opens out becoming 'thicker and thicker', with the plane opening out into an increasingly wide belt, eventually transforming into the Ort Cloud, where millions of comets are held in a loose orbit in a sphere around the Sun. Inside the Ort cloud is the region of space, which is closer to the Sun and which was swept clean by gravity, four and a half billion years ago, to produce the Solar System. When the Sun ignited, its solar wind swept away remains of the cloud from which it was formed, with its local domination of space, being seen to lessen significantly firstly at the Kuiper belt and almost completely as the region of the Ort cloud is reached. This emphasizes the lessons learned from Hooke's law graphs, which are plotted beyond the limit of proportionality. The law may break down for the system, but for reasons which can be readily understood, by looking at the bigger picture and applying Baconian rules to determine the reasons why.

The great question about gravity is 'how far can we extrapolate Kepler's laws and the force of gravity away from the Sun, before the limit of proportionality is reached and other factors come into play?' The Pioneer and Voyager probes have been heading out of the solar system for many years now and already there are signs that the limits of proportionality have been passed. This is shown by data on the motion of the probes that are described to as the Pioneer and Voyager anomalies. The anomalies may as yet be small, but are believed to be real. In ECE theory, it is presumed that at this distance, the gravitational force caused by the curvature of space, is decreasing to the point, where the torsion of the spinning space-time of electromagnetism is starting to become significant in comparison.

Could our theories for the start and evolution of the universe need changing or updating? Is our universe really 13.8 billion years old as predicted by the Big Bang theory and did the universe start from a singularity?

The latest Hubble deep field image can see galaxies back in time to only 700 million years after the supposed Big Bang. The first stars and galaxies are said to have been forming around 500 million years after the Big Bang, but in the 700 million year old image provided by the Hubble space telescope, the stars are already similar or identical to those seen in our local galaxies today, being composed of second generation stars. Galaxies being composed of second generation stars, only two hundred million years after the first stars were formed, could be seen as evidence, that the oldest galaxies imaged by the Hubble telescope are older than allowed for by the Big Bang theory.

The Hubble Space Telescope deep field images show galaxies, which are billions of light years away, at the edge of the observable universe. It is said that these image, show the galaxies formed far in our past, do not look identical to nearby galaxies, which have been formed in more recent times. This suggests that there has been an evolution of galaxies, since the time of the big bang and is strong evidence against Hoyle's steady state theory. However, there are good reasons why the image of galaxies from billions of years ago is different from those of today:

Firstly, the image of galaxies from the edge of the observable universe may suffer from a small degree of gravitational lensing, by material along its optical path to the telescope. Secondly, dust in the intervening space may cause a degree of distortion through refraction. Thirdly, objects at extreme distances, undergo a high degree of red shifting of their light, before it reaches an observer. Therefore, for the most distant galaxies, this red shifting will cause the original visible light to shift into the infrared, to be replaced by ultraviolet light that has moved out of the ultraviolet and into the visible part of the spectrum, before it reaches the observer. The Hubble deep field image of the most distant galaxies would therefore not be identical to nearby galaxies, because the visible image of the distant galaxies, would effectively be comprised of light of shorter wavelengths and would therefore be a comprised more of light from the blue end of the visible spectrum and ultraviolet light. This 'ultraviolet image' of distant galaxies, would therefore be expected to be different, from the truly visible image of galaxies from our own neighborhood.

The cosmic microwave background could be no more, than radiation showing the temperature of gas and dust that has cooled to the ambient temperature of space, so maps of this background simply show the mass densities of space in various directions.

Redshift data only suggests the universe is expanding and in no way tells us, that the universe expanded from a singularity at the start of time! The age of the universe is indicated to be 13.8 billion years, but this is simply the time estimate, for when the now expanding universe, would be in its smallest state of contraction. It does not follow that before this time, the universe was all contained in a singularity and this was where time started! The question is not 'What happened before the mythical Big Bang', but rather, 'what happened before the expansion!'

This question is easily answered. Before the big expansion there was the great contraction! What we call the big bang, is merely the cross over point of maximum matter density, marking the minimum contraction point of the known universe at that time. To go back further in time, you just go backwards in time to the previous state of the universe, which then was undergoing a contraction.

Now, the universe is no longer seen as 13.8 billion years old at all. That time in the past is merely the switch over event, from the general contraction of old, to the general expansion that we see today. We can consider a time in the past when the universe was double its supposed age now. At a time 27.6 billion years in the past, stars, galaxies and matter in general would be undergoing a contraction, towards the centre of gravity of the region of space, which we call our universe. Interactions between galactic clusters and super clusters will be taking place in complex ways, so just as the motion of icebergs and weather patterns follow unpredictable patterns, by the time the universe reaches its point of greatest matter density, some 13.8 billion years ago, the galaxies will be spread out in such a way that it will not meet at a point, to form a singularity, causing time to stop and the universe to meet an unphysical end! Rather, the bulk of the galaxies, stars and matter will simply pass by one another, due to the vastness of space, even in its contracted form and pass into the state of expansion that we see today. Some material may go into a melting pot at the centre of this contraction and enter a super contracted state, heat up to extremely high temperatures, resulting in emission of high energy electromagnetic radiation, which in time, could be observed as the cosmic background echo of this supposed event. In another, 13.8 billion years time, this 'change over universe' will be identical to the one we see today, without the need to invoke a big bang, emanating from a singularity'. We have here a universe, alternating between periods of expansion and contraction.

This 'Alternating Universe' explains everything that can be explained by the 'Big Bang' theory and more" and does not encounter the problem of singularities and infinities. It neatly side steps and answers the problem, 'what happened before the Big Bang?' It gives a mechanism for the creation of matter from energy 13.8 billion years ago and can explain why two types of galaxies exist. The new spiral galaxies may have emerged after the contraction was complete 13.8 billion years ago and the old elliptical galaxies, may have been formed long ago, when the universe was collapsing and have simply passed by one another as the universe reached its greatest mass density and switched into the state of expansion that we see today! Spiral galaxies, already in existence would merge and form further elliptical galaxies around this time!

Why globular clusters are associated with galaxies is a problem that needs explaining. Globular clusters orbit the central bulge of spiral galaxies in highly elliptical paths, which are outside the plane of the galaxy. Furthermore, estimates of the age of globular clusters indicate that they may be older than their 'parent' galaxies and may predate the time of the supposed big bang. With the age of globular clusters being estimated between fourteen and eighteen billion years old, they may be a product of a time when the known universe, was approaching its most contracted state and thus globular clusters, may be a by-product of special star forming processes, which were possible or even prevalent at this 'change over' time of the universe. The composition of the stars in globular clusters, indicate that they are population II stars, which are thought to be composed of matter, which was forming stars at this time, because they lack the heavier elements found in stars that are found in galactic spiral arms for instance, that have been formed from matter from previous generations of stars, that have gone supernova.

Our Milky Way galaxy contains about 150 globular clusters and with ninety percent of them lying in directions close to the centre of our galaxy, led earlier astronomers to realize that the centre of our galaxy lay in the constellation of Sagittarius. The origin of globular clusters is open to debate and their supposed age, suggests that they could have been already been around before their parent galaxy was formed and so they could simply have been captured by the gravitational fields of newly formed galaxies. This could certainly provide a reason for globular clusters having idiosyncratic, elliptical orbits, outside the plane of spiral galaxies.

Globular clusters are usually composed of between one hundred thousand and a millions equally spaced stars of similar size, age and composition. The arrangement and nature of the stars, makes them quite unlike other stars and star clusters in galaxies. They are particularly prevalent in the giant elliptical galaxy, M87. A galaxy noted for its strong X-rays emissions.

The centre of our Milky Way galaxy is in the constellation of Sagittarius and was found at the dawn of radio astronomy in 1932, to be a strong source of radio waves. This central, radio region, known as Sagittarius A, is about fifty light years across and is associated with magnetic field lines. The physical centre of our galaxy is found within Sagittarius A and is known as Sagittarius A*. This massive central core is over ten billion miles across (120 astronomical units) and is big enough to engulf the orbit of Pluto (forty AU from our Sun). It does not emit much radiation, but contains around three million solar masses of radiation and matter. What happens inside Sagittarius A*, the central core of our galaxy is anybody's guess. The Einstein-Hilbert equation of general relativity has no validity here, but what is evident is that, since this region, containing three million solar masses, has not shrunk to a point; it cannot be described as a singularity! The lack of strong emissions from Sagittarius A*, means that our Milky Way is described as being a non-active galaxy. We do however; have an active galaxy as a neighbor, in the form of the giant, elliptical galaxy M87!

It is believed by many that red shift measurements show that all galaxies are moving away from each other. However, this is not strictly true. Galaxies are grouped together in clusters and clusters are grouped together in superclusters. Nearby galaxies can be moving towards or away from their neighbors. Clusters of galaxies interact with each other and even superclusters exert measurable gravitational effects on the superclusters of galaxies that surround them. It is only on the giant scale, that galaxies are indicated to be moving away from one another.

Chapter 6

Index

Popular science has been too focused on the Big Bang theory and as a result, advances in our knowledge of the way the galaxies are structured, on the intermediate scale of the universe, has not been given the attention it deserves. This new knowledge has been brought about in the last few decades, by bigger telescopes, improved electronics and the amazing advances in computer power.

The cosmological red shift and expansion of galaxies, is often explained by drawing galaxies on a balloon. As more air is blown into the balloon, the galaxies move apart from each other and none of them is actually at the centre of the universe. The balloon illustrates how in theory the galaxies emerged from a common centre which is the location of the Big Bang. The balloon gives a simple, but clever representation to illustrate the theory. However, with new data on the motion and location of galaxies becoming available, it is no longer the revealing model it once was. A new model is needed to describe the eloquent structure that is now seen, percolating through the firmament.

The arrangement of galaxies is no longer best described by a single balloon, but is now given a three-dimensional structure by describing it in terms of soap bubbles. It is obvious that the bulk of the vastness of space is empty and this structure is represented, by the empty space within the soap bubbles, but the real action takes place where the bubbles interact along their surfaces. It is now known that galaxies pan out in this fashion. Computer simulations of this structure, produced by image and distance data, gives a structure that looks like a three dimensional representation of nerve cells, with the galactic voids represented by the body of the cells and the location of galaxies being strung out along the nerve fibres. Another way of describing the arrangement of galaxies would be as having a kind of honeycomb structure. Now if we compare the balloon model with the honeycomb model, we can see that if some time in the past the universe was more compressed (like a compressed sponge), the voids in the cells of the honeycomb would get smaller and the galaxies and matter in the universe would become more concentrated, but there would not necessarily be a centre to the vastness of space and the universe.

The new alternating model of the universe, expounded here, could be visualized by compressing and releasing a three-dimensional, rubber-based version of bubble wrap or honeycombed structured material. Here, there is no centre to the compression and expansions, but in its compressed state, there are stored energy implications, which could affect the production and evolution of galaxies. This can be compared with the oscillations of a pendulum, where the total energy of the system stays the same, but gravitational potential energy and kinetic energy, exchange in a periodic fashion. Globular clusters and quasars are thought to have been formed, in the time frame of the Big Bang or time of maximum compression of our universe. Some estimates of the age of globular clusters are from fourteen to eighteen billion years, which is older than the supposed age of our universe and the big bang of 13.8 billion years. Quasars are very distant galaxies that are so far away that they only look like stars in telescopes. Quasars are thought to be active galaxies, outpouring vast amounts of energy and by virtue of their measured redshifts, are believed to be ten billion or more years old. Quasars may be an early product of the universe, after the compression stage of the universe had changed to the expansion stage, we see today.

Our Milky Way and the Andromeda galaxy are the largest members of our closest group of forty or so galaxies. Our nearby galaxies are referred to as our local group. Looking outwards from our galaxy in various directions, we see groups of galaxies, concentrated in certain directions, which reveal the locations of our nearest galactic superclusters. Our nearest supercluster and the one of which we are a member, is called the Virgo cluster. Bang in the middle of the Virgo supercluster, is our nearest galaxy with an active nucleus. This galaxy is the giant elliptical galaxy M87 and it may be asked why such a remarkable galaxy, just happens to be at the centre of our galactic supercluster. Perhaps it is just chance, but there may be a reason that is as yet not clear to us.

M87 is 55 million light years away, is the largest giant elliptical galaxy to Earth and is one hundred and twenty light years across. At its centre is a supermassive, compact structure, which at six billion solar masses, is two thousand times the mass of Sagittarius A*, at the centre of our galaxy. It is a strong source of both radio waves and X-rays. It is an active galaxy, emitting matter and radiation from its central region, in a jet seen sprawling at least 5,000 light years into space at speeds close to the speed of light. This monumental elliptical galaxy is also notable for its plethora of ancient globular clusters. There are 12,000 globular clusters in M87, nearly a thousand times more, than are found in our Milky Way galaxy. There are also narrow X-ray emitting filaments to be found in the galaxy, stretching across 100,000 light years of space. The X-rays could be a source of electrons, helping to explain the strong radio emissions from the galaxy. The volume of the supermassive structure at the centre of M87 is thought to be similar to the size of Sagittarius A, of the order of the volume of the Pluto's orbit around the Sun. Long exposure photographs of M87 give it a size in the sky, as seen from Earth, as slightly bigger than our Moon!

An intriguing twist to Fred Hoyle's Steady State Theory would be that matter is created inside active galactic nuclei, by space-time resonance with the primordial voltage. X-ray and gamma rays, spewing from these structures could be converted to their particle forms of electrons, protons and alpha particles to seed the universe, with the matter needed for the creation of new galaxies.

5.7 Dark matter in focus

Recently spacially discerning spectroscopy has been applied to globular clusters, to determine the Doppler shift of individual stars. The results show that globular clusters do not obey Newton's inverse gravitational law and do not fit the predictions of the dark matter hypothesis. However, ECE theory can easily explain this anomaly and also the reason that there are so may spiral galaxies to be seen. There is an excellent simulation by Dr. Horst Eckardt of the torsion based theory of spiral galaxies, which can be viewed on www.aias.us . This is preferred by Ockham's razor of natural philosophy to the standard model, because it is much simpler and more powerful and is developed from the first principles of general relativity. "Dark matter" on the other hand is a phenomenological, unscientific idea that is used to try to compensate for the apparent "missing mass" of a spiral galaxy. In fact, there is no "missing mass"; the galaxy is a spiral, because of the torsion of space-time. This torsional theory adheres to the basic principles of general relativity, that physics is geometry. "Dark matter" does not exist in nature. Maps of "dark matter" are in fact maps of the torsion of space-time throughout the universe, thus proving Einstein-Cartan-Evans (ECE) theory.

In paper 76, Horst Eckardt shows that the structure and velocity curve of a spiral galaxy can be described in terms of a constant Cartan torsion (spinning space-time) using ECE theory. It is shown, that in the central bulge region of a spiral galaxy, gravitational attraction predominates as described by Riemann curvature. However, it can be seen that in the spiral arms of the galaxy, the Cartan torsion becomes predominant, so that the structure of the arms becomes a hyperbolic spiral due to the underlying constant spinning of space-time. In this generally covariant unified field theory, dark matter does not exist and is replaced by the Cartan torsion missing from the Einstein-Hilbert field theory of gravitation and its weak field limit, Newtonian dynamics.

The Einstein-Hilbert equation is one of the greatest achievements of human thought in any area, because nature is shown to be geometry. Paper 88 makes this very clear, through the use of the simpler torsion tensor. The spinning of a spiral galaxy is a beautiful, visible example of the torsion at work. Dark matter does not exist at all and the theory of dark matter is completely wrong. It belongs in the same category as phlogiston or epicycles. Cosmologists, who naively believe in dark matter, are now seeking to compound their ill-conceived ideas, by linking them to particle physics and string theory. String theory initially showed some promise, but physicists failed to ditch it when it became obvious, that its use produced no new insights in nature and did not produce any testable hypotheses. Instead more and more complex mathematics was introduced to hide its failings and extra dimensions added to explain away anomalous results, as the web of their own mathematical strings tightened in on them. ECE can explain the data and has none of the short comings of string theory.

The main advantage is that ECE is generally covariant in all sectors, and only needs to be developed in four dimensions. ECE theory is already being used for engineering and to facilitate the start up of new companies, while string theory is unable to predict anything, that is not already better and more simply understood by ECE theory. ECE theory can be applied to phenomena of all kinds, such as non-linear optics, in which string theory cannot be applied. String theory grew out of particle and field theory and is still largely confined to that area of physics, after about forty years of trying to predict anything that is really new. In five years, ECE theory has been widely applied to physics, chemistry and electrical engineering. String theory is not used at all in chemistry and engineering and is not used in physics, except in specialized circumstances. String theory carries with it all the flaws of standard general relativity, so is no advance on 1915 and neglects the Cartan torsion, which has been so eloquently used in ECE theory and which allows new theoretical insights to be used, to develop new technologies and to found new companies!

Proponents of dark matter have invented dark matter to account for stars orbiting a galactic centre too quickly. Dark matter has been eloquently contrived to explain the anomaly, but has no basis in the experimental science of the last five hundred years. Dark matterist's simply play think of a number and play round with it until it fits the data they are trying to simulate. This is merely a mathematical exercise, with no foundation on Baconian principles. In other words it is pseudoscience or gobbledygook! It is like a wheel being balanced in a garage. The forces around the wheel centre can be seen to be out of balance, simply by placing the wheel horizontally on a spirit level. The spirit level then indicates where and how dense weights need to be placed, in order to balance the wheel. In this case the 'missing' mass is determined and calculated experimentally so the forces balance. However, it is ludicrous to think this process can be done to balance the forces in the spiral arms of spinning galaxies. The mathematical balancing of a galaxy, by placing a spherical shell of 'lead' of a certain mass around it, is simply a mathematical exercise with no foundation in the natural world. It could be used however, to indicate the degree to which our understanding of the movement of stars in spiral galaxies is outside the limits of proportionality of Newton's Law of gravitation and to give an indication of the size of the unknown forces which are acting. It is ludicrous however, to go against five hundred years of Baconian science and bring in the concept of 'the dark matter halo' that cannot be detected, because it does not produce visible light or any other form of electromagnetic radiation. This smacks of introducing the workings of a deity, who arranges the universe by putting invisible structures in place, which human's cannot see, in order to drive the universe by a kind of celestial clockwork. Effectively, all that dark matter proponents have done, is reinvent the crystal spheres used in ancient times, to explain the movements of the planets, Moon and Sun in the Heavens driven by invisible angels, with the movements being accompanied around the cosmos by the 'Music of the Spheres'. All very comforting and inspiring, but hardly twenty-first century rocket science!

Chapter 6

Index